Alphaherpesviruses

Molecular Biology, Host Interactions and Control

Editors:

Ekaterina E. Heldwein
Tufts University School of Medicine
Boston, MA
USA

Gregory A. Smith
Feinberg School of Medicine
Northwestern University
Chicago, IL
USA

Caister Academic Press

www.caister.com

Caister Academic Press, UK
www.caister.com

ISBN: 978-1-913652-55-5 (paperback)
ISBN: 978-1-913652-56-2 (ebook)

DOI: https://doi.org/10.21775/9781913652555

Cover Illustration

SplitsTree diagrams of the three human alphaherpesviruses: HSV-1, HSV-2 and VZV (the illustration is from Chapter 1, Figure 1).

Contents

Preface

A little over half a century ago, Richard Johnson and Cedric Mims commented that "it must be explained, for example, why a virus causing no more than a cold sore in one person can produce fatal encephalitis in another". Although their thoughts were directed at herpes simplex virus type 1, this inquiry can be extended to all neuroinvasive alphaherpesviruses. How do these viruses gain access to the nervous system? How do they establish and reactivate from the latent state? And why do these typically benign infections sometimes have severe or lethal outcomes?

Herpesviruses are among the most complex viruses that infect humans and, with genomes in excess of 100 kilobases, their biology remains dauntingly complex. Nonetheless, since Johnson and Mims proffered their insightful question in 1968, a wealth of knowledge on alphaherpesviruses has been gained. In particular, since the last edition of this book in 2011, our understanding of the alphaherpesvirus infectious cycle has been accelerating. With this expanded knowledge comes the recognition that a comprehensive overview of all research, including the clinical and veterinary pathogens belonging to the six recognized alphaherpesvirus genera, is beyond the scope of this text. Instead, in this edition of *Alphaherpesviruses: Molecular Biology, Host Interactions and Control*, researchers with expertise in different aspects of the host-pathogen interface offer their perspectives on the current state of the field using herpes simplex virus as the focus.

The book begins with the topic of genomics (Chapter 1), which was in its infancy ten years ago and has since yielded new insights into virus diversity and evolution. It then moves on to the early stages of the infectious cycle, a period prior to *de novo* viral gene expression, with chapters dedicated to viral entry into the cell (Chapter 2), genome delivery from the cell periphery to the nucleus (Chapter 3), and the fate of the genome in the nucleus (Chapter 4). With this knowledge in hand, the book then turns to how infection progresses in neurons (Chapter 5) and how latency is established there (Chapter 6). Because latency is tied to host response, the book next covers cellular responses to infection and corresponding viral immune evasion strategies (Chapter 7). Discussion of the infectious cycle concludes with the late stages of infection, including capsid egress from the nucleus (Chapter 8) and subsequent virion assembly in the cytoplasm (Chapter 9). The book ends with the two translational topics, vaccines (Chapter 10) and oncolytic vectors (Chapter 11).

We thank Caister Academic Press for supporting the field of Alphaherpesviruses and especially thank all the contributing authors that dedicated their time and effort to compile this comprehensive overview of the alphaherpesviruses.

Ekaterina Heldwein
Greg Smith

Current books of interest

- Legionellosis Diagnosis and Control in the Genomic Era 2020
- Bacterial Viruses: Exploitation for Biocontrol and Therapeutics 2020
- Microbial Biofilms: Current Research and Practical Implications 2020
- Astrobiology: Current, Evolving and Emerging Perspectives 2020
- *Chlamydia* Biology: From Genome to Disease 2020
- Bats and Viruses: Current Research and Future Trends 2020
- SUMOylation and Ubiquitination: Current and Emerging Concepts 2019
- Avian Virology: Current Research and Future Trends 2019
- Microbial Exopolysaccharides: Current Research and Developments 2019
- Polymerase Chain Reaction: Theory and Technology 2019
- Pathogenic Streptococci: From Genomics to Systems Biology and Control 2019
- Insect Molecular Virology: Advances and Emerging Trends 2019
- Methylotrophs and Methylotroph Communities 2019
- Prions: Current Progress in Advanced Research (Second Edition) 2019
- Microbiota: Current Research and Emerging Trends 2019
- Microbial Ecology 2019
- Porcine Viruses: From Pathogenesis to Strategies for Control 2019
- *Lactobacillus* Genomics and Metabolic Engineering 2019
- Cyanobacteria: Signaling and Regulation Systems 2018
- Viruses of Microorganisms 2018
- Protozoan Parasitism: From Omics to Prevention and Control 2018
- Genes, Genetics and Transgenics for Virus Resistance in Plants 2018
- Plant-Microbe Interactions in the Rhizosphere 2018
- DNA Tumour Viruses: Virology, Pathogenesis and Vaccines 2018
- Pathogenic *Escherichia coli*: Evolution, Omics, Detection and Control 2018
- Postgraduate Handbook 2018
- Enteroviruses: Omics, Molecular Biology, and Control 2018
- Molecular Biology of Kinetoplastid Parasites 2018
- Bacterial Evasion of the Host Immune System 2017
- Illustrated Dictionary of Parasitology in the Post-Genomic Era 2017
- Next-generation Sequencing and Bioinformatics for Plant Science 2017
- Brewing Microbiology: Current Research, Omics and Microbial Ecology 2017
- Metagenomics: Current Advances and Emerging Concepts 2017
- The CRISPR/Cas System: Emerging Technology and Application 2017
- *Bacillus*: Cellular and Molecular Biology (Third edition) 2017
- Cyanobacteria: Omics and Manipulation 2017
- Foot-and-Mouth Disease Virus: Current Research and Emerging Trends 2017

www.caister.com

Chapter 1

Alphaherpesvirus Genomics: Past, Present and Future

Chad V. Kuny and Moriah L. Szpara*

Departments of Biology, and Biochemistry and Molecular Biology, Center for Infectious Disease Dynamics, and the Huck Institutes of the Life Sciences, Pennsylvania State University, University Park, Pennsylvania, 16802, USA

*moriah@psu.edu, cvk108@psu.edu

DOI: https://doi.org/10.21775/9781913652555.01

Abstract

Alphaherpesviruses, as large double-stranded DNA viruses, were long considered to be genetically stable and to exist in a homogeneous state. Recently, the proliferation of high-throughput sequencing (HTS) and bioinformatics analysis has expanded our understanding of herpesvirus genomes and the variations found therein. Recent data indicate that herpesviruses exist as diverse populations, both in culture and *in vivo*, in a manner reminiscent of RNA viruses. In this chapter, we discuss the past, present, and potential future of alphaherpesvirus genomics, including the technical challenges that face the field. We also review how recent data has enabled genome-wide comparisons of sequence diversity, recombination, allele frequency, and selective pressures, including those introduced by cell culture. While we focus on the human alphaherpesviruses, we draw key insights from related veterinary species and from the beta- and gamma-subfamilies of herpesviruses. Promising technologies and potential future directions for herpesvirus genomics are highlighted as well, including the potential to link viral genetic differences to phenotypic and disease outcomes.

Introduction

Herpesviruses are ubiquitous worldwide, with nine species that infect humans and dozens more infecting other hosts (Davison, 2010). The lifelong nature of herpesvirus infection creates a burden for individual health as well as for public health policy (Gantt et al., 2016; Jansen et al., 2016; Looker et al., 2015a, 2015b). They also have a global impact through their effects on veterinary species involved in food production and those kept as companion animals (Davison, 2010; Loncoman et al., 2017). Within the *Herpesviridae* family, the *alphaherpesvirinae* subfamily is characterized by common genomic features, and epithelial and mucosal sites of active replication (Pellett and Roizman, 2013; Roizman et al., 2013). Most alphaherpesviruses establish lifelong latency in neurons, with a few non-neuronal exceptions such as the *Mardivirus* genus (e.g. Marek's disease virus (MDV) or *gallid alphaherpesvirus 2*). The human alphaherpesviruses consist of herpes simplex virus 1 and 2 (HSV-1,2 or *human herpesvirus 1,2*, HHV-1,2) and varicella zoster virus (VZV or HHV-3) (Davison, 2010). These viruses, as with all herpesviruses, have large dsDNA genomes that have presented challenges for genomic characterization, particularly in comparison to other viral families (Koonin et al., 2015).

Herpesviruses have co-existed with humanity for as long as we have written records. Ancient Egyptian hieroglyphics record the presence of herpetic lesions, implying that these viruses have been a notable health issue for millennia (Pellett and Roizman, 2013). However, detailed scientific studies of these viruses, especially those investigating the entirety of the viral genome, have occurred in quite recent times (Davison, 2010; Depledge et al., 2018a; Loncoman et al., 2017; Renner and Szpara, 2018). The progression of herpesvirus genomics studies has been greatly expanded by the advancement of DNA sequencing technologies. Sanger sequencing enabled the study of individual viral genes, or with painstaking effort, a composite full-length viral genome (Klupp et al., 2004; McGeoch et al., 1985, 1988). With the advent of high throughput sequencing (HTS), entire viral genomes can now be assembled from one sample and its sequencing "run", and in a matter of days rather than years (Greninger et al., 2018; Parsons et al., 2015). HTS approaches also yield a depth of sequence information that provides insight into the genetic diversity within a viral sample. This relative wealth of data has revealed a substantial amount of genetic diversity between strains of any given alphaherpesvirus, as well as genetic diversity indicative of a population of viruses within a single sample. The effects that this level of sequence diversity may have on virus biology or clinical outcomes remain an area of active research (Houldcroft et al., 2017; Loncoman et al., 2017; Renner and Szpara, 2018).

Despite the recent expansion of alphaherpesvirus genomics, substantial challenges remain in the field. Alphaherpesvirus genomes have multiple features that are difficult to analyze with current short-read HTS technology, including highly repetitive areas, regions of high G+C content, and even the presence of genomic isomers. Handling these challenges is a key focus of the future of herpesvirus genomics. In this chapter, we will describe the progression of alphaherpesvirus genomics, including historical context, current areas of research, and potential future directions for the field.

The inherent (in)stability of herpesvirus genomes

A prevailing opinion among virologists is that DNA viruses are inherently stable and RNA viruses are inherently variable (Sanjuan et al., 2010). Indeed, RNA viruses are often described as existing as a quasispecies swarm of varying genomes, rather than as a defined genetic species (Andino and Domingo, 2015). Mechanistically, this view stems from the lack of error correction in most RNA-dependent RNA polymerases (Sanjuán and Domingo-Calap, 2016). By contrast, most DNA viruses have high fidelity polymerases with error correction, and have lower reported mutational rates (Hall and Almy, 1982; Drake and Hwang, 2005). Early studies of the mutation rate of HSV-1 examined single genes, and detected mutation rates on the order of 1×10^{-7} or 1×10^{-8} mutations per base per infectious cycle (Hall and Almy, 1982; Drake and Hwang, 2005). The rates identified in such studies are those that are then subsequently quoted in comparisons between RNA and DNA viruses (Drake, 1991; Sanjuan et al., 2010; Sanjuán and Domingo-Calap, 2016). While those mutation rates fit well with older restriction-fragment length polymorphism (RFLP) comparisons of herpesvirus genomes, these mutation rates fail to explain the ease with which herpesvirus variants are selected or revealed under strong selective pressures. For example, analysis of multiple HSV-1 and HSV-2 strains revealed the presence of drug-resistance mutations in 1 out of 10^3 to 10^4 PFU in the overall virus population (Sarisky et al., 2000). This result suggests that alphaherpesviruses, even in culture, exist as populations of diverse genomes - a phenomenon otherwise known as standing variation (Firth et al., 2010; Renzette et al., 2014). An alternative explanation is that the rate of evolution observed under these laboratory conditions is substantially different than natural conditions, although the frequency with which drug-resistant viruses are detected *in vivo* suggests otherwise (Burrel et al., 2010; Sauerbrei et al., 2010).

Multiple lines of evidence support the idea that alphaherpesviruses exist in populations with standing variation. Genome-wide HTS comparisons of HSV-1 have revealed approximately 3-4% nucleotide variation genome-wide, with differences of 1-2% observed between plaque-picked subclones of a given strain (Parsons et al., 2015; Szpara et al., 2014; Bowen et al., 2016). Additionally, a study that inferred mutation rates from phylogenetic history found a higher than expected frequency of observed mutations in herpesvirus populations (Firth et al., 2010). There is also experimental evidence to support a hypothesis that genetic diversity is inherently beneficial to herpesviruses. In an investigation of Muller's ratchet - the hypothesis that small asexual populations will accumulate deleterious mutations - Jaramillo *et al.* subjected ten individual sub-clones of HSV-1 to sequential plaque-to-plaque transfers (Jaramillo et al., 2013). As a result of the extreme genetic bottlenecks induced by this experimental approach, two clonal lineages were completely lost, with the remaining clones exhibiting an attenuation of host mortality after intracerebral inoculation into mice (Jaramillo et al., 2013). Genetic analysis of these clones revealed a mutation frequency of 3.6×10^{-4} substitutions per base per plaque transfer, a far higher rate than those described in single gene studies (Hall and Almy, 1982; Drake and Hwang, 2005; Drake, 1991; Jaramillo et al., 2013). Additionally, while not the focus of this chapter, studies of the beta-herpesvirus human cytomegalovirus (HCMV) have described the existence and expansion of standing variation within immunocompromised hosts (Gorzer et al., 2010; Sijmons et al., 2015; Houldcroft et al., 2016; Hage et al., 2017). On the whole, while studies of the mutation rate of the herpesvirus polymerases have been useful, the breadth of genetic diversity within herpesviruses is substantial and has relevance to both laboratory and clinical settings (Houldcroft et al., 2017; Loncoman et al., 2017; Renner and Szpara, 2018). With the advent of advanced sequencing technologies, researchers are beginning to appreciate the effects that genetic diversity can have on disease outcomes, drug resistance, and other viral phenotypes of interest.

Comparative genomics reveals the true diversity of herpesviruses

Two seminal examples of VZV genome analysis by Peters *et al.* 2006 and Tyler *et al.* 2007 predate the era of HTS, but nonetheless set the tone for many later analyses (Peters et al., 2006; Tyler et al., 2007). We applied many of the same comparative genomics analyses laid out in these papers in our first analyses of HSV-1 and pseudorabies virus (PRV) genomes (Szpara et al., 2010, 2011). These included phylogenetic and recombination analyses, as well as comparisons of coding diversity, tandem repeats, and specific genetic loci (Peters et al., 2006; Szpara et al., 2014; Tyler et al., 2007). Many subsequent studies have followed the same approach, albeit for different alphaherpesvirus species or for interesting subsets of viruses within a species (see for example (Depledge et al., 2014a; Newman et al., 2015; Bryant et al., 2018; Lewin et al., 2018)). Studies using these methods have illuminated the evolutionary history of several alphaherpesvirus species (Zell et al., 2012; Norberg et al., 2015; Kolb et al., 2015; Browning et al., 2016; Vaz et al., 2016a; Burrel et al., 2017; Koelle et al., 2017; Trimpert et al., 2017), demonstrated that recombination can occur between and within a single species (Loncoman et al., 2017; Lewin et al., 2018; Zell et al., 2012; Norberg et al., 2015; Kolb et al., 2015; Burrel et al., 2017; Koelle et al., 2017; Vaz et al., 2016b; Kolb et al., 2017) (as detailed below in "Mechanisms that drive genetic diversity"), and shown that vaccine strains can regenerate virulent viruses if given the opportunity (Lee et al., 2012; Ye et al., 2016). What remains under-represented in many HTS studies is a connection of the comparative genomics data to the measurement of biological phenotypes. These phenotypes could include classic virological measures, such as plaque morphology or replication fitness (Bowen et al., 2018; Parsons et al., 2015), or *in vivo* measures such as neuroinvasion and mortality rate (Bryant et al., 2018;

Pandey et al., 2017). The alphaherpesvirus literature reflects decades of research on the *in vitro* and *in vivo* phenotypic effects of individual gene deletions, mutations, or over-expression (Pellett and Roizman, 2013; Roizman et al., 2013). The next frontier for comparative genomics is to integrate these prior data with the insights from comparative genomics, as detailed below (see "Future Frontiers").

Mechanisms that drive genetic diversity

As we gain insight into the level of genetic diversity within the *alphaherpesvirinae* subfamily, one outstanding question is how such diversity is generated. One straightforward source of genetic diversity is DNA replication error by the viral polymerase (Hall and Almy, 1982; Drake and Hwang, 2005). However, it is unclear how much of a role DNA polymerase error plays in the generation of alphaherpesvirus genetic diversity. Studies of the error rate of herpesvirus polymerases have indicated a very low mutation rate, on the order of 1×10^{-7} or 1×10^{-8} mutations per base per infectious cycle, which would imply that many rounds of genomic replication are necessary if polymerase error were the sole or primary source of genetic diversity (Hall and Almy, 1982; Drake and Hwang, 2005). However, those studies were also done in the context of a single coding region, which may well underestimate polymerase error rates (Brown, 2004). Polymerases are known to have higher error rates in contexts of high G +C content or repetitive regions, and these sequence contexts are over-represented in alphaherpesvirus genomes with high inter-strain diversity, such as HSV-1 (Figure 1). In contrast the smaller VZV genome has an overall lower rate of inter-strain diversity. The VZV genome has reduced intergenic regions and ~50% G+C content as compared to ~68% G+C content for HSV-1 (peaking at ~80% average G+C in highly repetitive regions) (Figure 1) (Peters et al., 2006; Tyler et al., 2007; Szpara et al., 2011; Sijmons et al., 2015).

Beyond polymerase error, it is now clear that mutation and evolution in herpesviruses result not only from base substitutions, but also from recombination between strains of a given species, and to a less frequent extent, recombination between species (Loncoman et al., 2017; Norberg, 2010; Renner and Szpara, 2018). As with segmented RNA virus reassortment, recombination

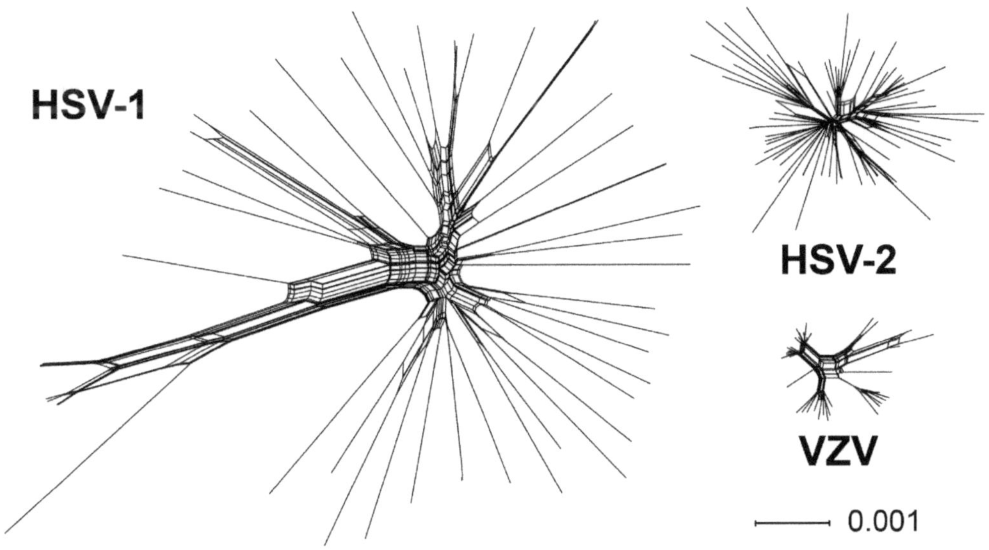

Figure 1. Comparison of the known genetic diversity within the human alphaherpesviruses. We generated SplitsTree diagrams (Huson, 1998) with recently published analyses of known genetic diversity in the three human alphaherpesviruses (Akhtar et al., 2019; Shipley et al., 2019; Zell et al., 2012). Each virus' tree is presented at the same scale, so that the size of the tree is proportional to the observed genetic diversity for each virus. The HSV-1 diagram includes 51 strains (Shipley et al., 2019), HSV-2 includes 68 strains (Akhtar et al., 2019), and VZV includes 42 strains (Zell et al., 2012).

in herpesviruses can provide a strong force for evolutionary shifts. Studies of lab-generated recombinants of HSV-1 have revealed a bias towards recombination breakpoints occurring in repetitive regions, areas of locally high G+C content, and intergenic regions (Lee et al., 2015). However, most studies of recombination within the alphaherpesviruses have focused on naturally circulating variants, and have inferred recombination and phylogenetic relationships for historical sites of recombination from a comparison of extant strains (Burrel et al., 2017; Koelle et al., 2017; Loncoman et al., 2017; Norberg et al., 2015). For example, as more strains of VZV have been sequenced, the number of phylogenetic clades have expanded and there is now evidence for ancient, inter-clade recombination in addition to modern recombination between strains (Norberg et al., 2015; Zell et al., 2012). Within circulating populations of

HSV-1 and HSV-2, there is evidence of rampant recombination among strains within each species, although it is unknown if the observed recombination is historical or recent (Szpara et al., 2014; Kolb et al., 2015; Norberg, 2010; Norberg et al., 2007). Recently, two separate groups have found evidence for inter-species recombination between HSV-1 and HSV-2 (Burrel et al., 2015, 2017; Casto et al., 2019; Koelle et al., 2017). These findings are based on the presence of several regions in the HSV-2 genome that have a high degree of sequence similarity to extant HSV-1 genomes, whereas the remainder of the HSV-2 genome is more closely related to a chimpanzee herpesvirus (ChHV) than to HSV-1 (Severini et al., 2013; Wertheim et al., 2014). For the non-human alphaherpesviruses, multiple examples of recombination both *in vivo* and *in vitro* have been reported, with intriguing results. Devlin and colleagues found that a virulent strain of Gallid herpesvirus 1 (infectious laryngotracheitis virus or ILTV) found in an agricultural setting was a spontaneous recombinant from two live-attenuated vaccine strains (Lee et al., 2012). As reviewed recently by Loncoman *et al.* (Loncoman et al., 2017), recombination of equine herpesviruses has been noted within-species (intrastrain, e.g. between EHV-1 isolates (Bryant et al., 2018; Kolb et al., 2017; Vaz et al., 2016b)) and between species (inter-strain, e.g. between EHV-1 and EHV-4 (Bryant et al., 2018)), and also within-species for feline herpesvirus (Kolb et al., 2017; Lewin et al., 2018), PRV (Kolb et al., 2017; Szpara et al., 2011; Ye et al., 2016), and MDV (Loncoman et al., 2017; Trimpert et al., 2017). The next challenge for understanding alphaherpesvirus recombination will be to determine where and how often this occurs *in vivo*. Since recombination requires the co-occurrence of two distinct viral lineages in one infected cell in the host, this may be rare and difficult to detect in clinical or field settings.

Other contributions to functional diversity

While single-nucleotide changes, recombination, and even horizontal gene transfer are generally accepted mechanisms for genetic variation within herpesviruses, data indicate that other mechanisms may also contribute to genetic diversity. These mechanisms include ribosome slippage and novel coding and non-coding RNAs revealed by transcriptome analysis or ribosome profiling. These genomic outputs cannot be predicted solely by analysis of genomic DNA, and instead require approaches that integrate transcriptional and/or translational data in combination with a matched DNA genome comparison.

Several studies have applied sensitive RNA-sequencing-based approaches to demonstrate an expanded range of transcriptional impacts of HSV-1 infection on host cells. A recent series of papers used directional and nascent RNA-sequencing, in combination with ribosome profiling, to reveal previously undetected effects on the host transcriptome during HSV-1 replication (Rutkowski et al., 2015; Wyler et al., 2017; Hennig et al., 2018). These included the disruption of transcription termination (DoTT) or transcriptional read-through, which leads to the production of many run-on RNA transcripts (Rutkowski et al., 2015; Hennig et al., 2018). It also included a large number of antisense transcripts from the host genome (Wyler et al., 2017). More recently, these analyses have been applied to several alphaherpesvirus genomes, revealing many non-canonical transcripts and novel open reading frames (ORFs) expressed by these viruses, along with previously undetected transcript extensions, truncations, and splicing events (Tombácz et al., 2017a; Prazsák et al., 2018; Tombácz et al., 2019; Depledge et al., 2019; Whisnant et al., 2020). Future studies will need to explore the expression and potential function(s) of these transcripts in different cellular and *in vivo* models of infection, and their conservation across other herpesviruses (Arias et al., 2014; Tirosh et al., 2015).

Ribosome frameshifting is a necessary aspect of translation for retroviruses like HIV, where production of the nucleocapsid and polymerase proteins are directed by the same RNA transcript and ribosome frameshifting is required to obtain both products. While it occurs at a lower frequency, ribosome frameshifting has been demonstrated on thymidine kinase (TK) transcripts in HSV-1 (Griffiths, 2011; Griffiths et al., 2003; Pan and Coen, 2012). This can be detected in genomes with a homopolymer-based frameshift in the TK gene, which often arise in response to treatment with the antiviral drug acyclovir (Burrel et al., 2010; Griffiths, 2011; Sauerbrei et al., 2010). While a direct translation of the RNA encoded by such altered genes would lead to non-functional proteins, ribosomal frameshifting enables a low level of functional protein production in these drug-resistant viruses (Griffiths, 2011; Griffiths et al., 2003; Pan and Coen, 2012).

RNA editing is a mechanism by which viral transcripts may be recoded at the single-nucleotide level to differ from the existing viral genome. While host-defensive or therapeutic CRISPR-based RNA editing is a destructive process that may be used against herpesviruses (Nakaya et al., 2016; Oh et al., 2019; Suspene et al., 2011), there is also the potential for other outcomes of RNA editing. Potential pro-viral effects of RNA editing of selected transcripts have been explored for the gammaherpesviruses Epstein-Barr Virus, EBV (Lin et al., 2013) and Kaposi's Sarcoma-Associated Herpesvirus, KSHV (Arias et al., 2014). The only example of RNA editing thus far for alphaherpesviruses involves editing of a long non-coding RNA of MDV, which the authors propose could contribute to the viral lytic cycle via a role in regulating host innate immune defenses (Figueroa et al., 2016). Further analyses of alphaherpesvirus transcriptomes may reveal more such examples and enable further testing of their potential functional effects (Tombácz et al., 2017a; Prazsák et al., 2018; Tombácz et al., 2019; Depledge et al., 2019; Whisnant et al., 2020).

Clear nomenclature to document viral strains and genome sources

As increasing numbers of alphaherpesvirus isolates of a given species are studied at the genomic level, it has become increasingly important to document the different sources of material used to obtain viral genomes. While descriptive terms such as "clinical isolate" and "laboratory strain" are often used to denote either low or high numbers of cell culture passages for a given viral isolate, these terms are interpreted variably by different research groups (Kuhn et al., 2013; Wilkinson et al., 2015). There are no standards for how many times a viral sample can be passaged in culture before it is no longer considered a "clinical isolate", nor are there standards as to whether a viral population should be plaque purified before calling it a strain. The introduction of sensitive DNA isolation and amplification techniques has also enabled whole viral genomes to be collected and sequenced directly from hosts, avoiding cell culture entirely (Greninger et al., 2018; Depledge et al., 2011; Watson et al., 2013; Johnston et al., 2017a; Shipley et al., 2018, 2020). It is important that researchers

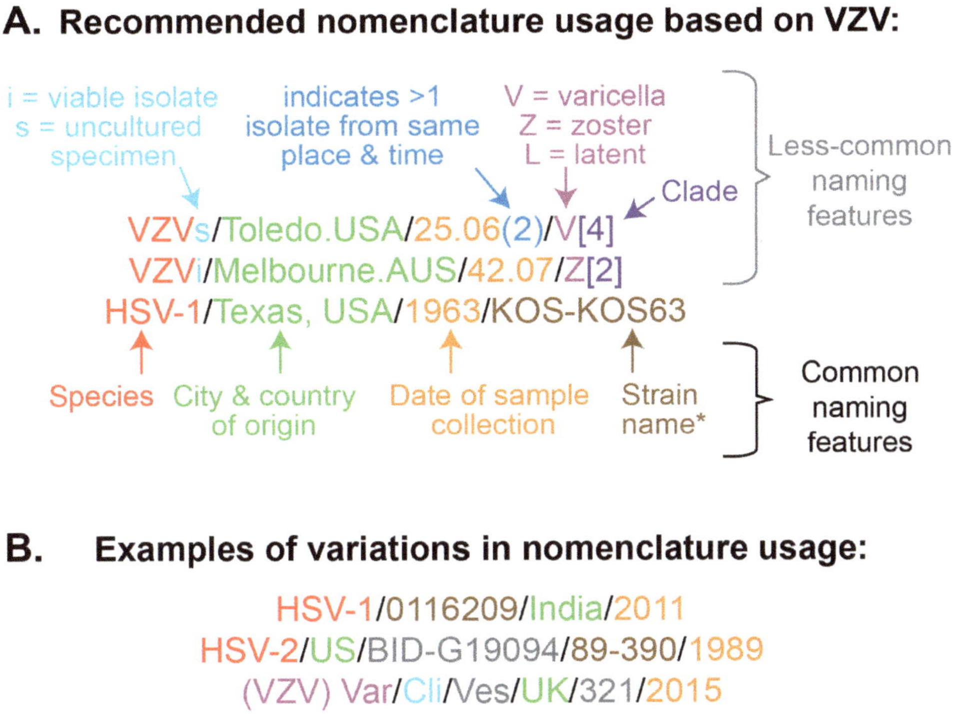

Figure 2. Viral nomenclature presents an opportunity to record crucial details about strain origin and history. (A) Common aspects of viral strain identification are shown here, based on the 2008 recommendations of VZV Nomenclature Meeting (Breuer et al., 2010). The recommended nomenclature includes the following data for each strain (moving from left to right in the diagram): viral status as a cultured isolate or an uncultured sequence, the geographic city and country of origin, collection date, whether or not any related sequences were collected, the disease presentation (e.g. varicella, zoster or latent), and the strain clade based on phylogenetic clustering or single-nucleotide polymorphism (SNP) profile (Breuer et al., 2010). The nomenclature components are color-coded consistently throughout the diagram. In a review article recommending this approach (Renner and Szpara, 2018), we suggested the addition of strain name to this nomenclature, in keeping with recommended practices for RNA virus nomenclature (Kuhn et al., 2013). **(B)** However, the current usage of this nomenclature is limited, with variable applications whose content is not always straightforward to interpret. A selection of recent variations in nomenclature usage are shown from the following GenBank records and publications: KJ847330 (Bondre et al., 2016), KR135321 (Newman et al., 2015), and MG764307 (Depledge et al., 2018b). A more consistent application of the recommended viral strain nomenclature would provide clear benefits for sample data retrieval and the accurate linking of sequence data with published results. Figure adapted from Breuer *et al.* 2008 (Breuer et al., 2010).

generating new genomic data, and those using sequences from the GenBank databases for purely computational studies, appreciate these distinctions. The VZV community has recommended a viral naming system that is similar to those used for RNA viruses or bacteria to help record and clarify these details for each new genome (Figure 2A) (Breuer et al., 2010). This standard emphasizes defining not only a name for each strain, but also its geographic origin, collection date, disease type (e.g. varicella, zoster or latent), and its status as a cultured isolate or an uncultured sequence (Breuer et al., 2010). Whenever possible the strain clade, based on phylogenetic clustering or single-nucleotide polymorphism (SNP) profile, is included as well. While not yet widely or consistently adopted (Figure 2B), this nomenclature approach would clearly be beneficial if applied to other alphaherpesviruses as well. This would improve data-accessibility across the primary literature and sequence databases, as well as enabling easier comparison of data across research groups.

Genomes: consensus vs. population

Most prior studies that catalog herpesvirus diversity have produced a single, consensus genome for each new sample. The consensus genome is representative of the most common allele or nucleotide at each position in the genome (Figure 3). In the simplest scenario, the consensus genome from a sample is built from the most common nucleotide at each position throughout the genome (Figure 3A). However, this scenario is complicated by evidence that herpesvirus isolates generally exist as populations that contain a degree of sequence diversity. Because more than one nucleotide may exist at a given location across the viral population in a sample, the creation of a single consensus genome necessarily removes the details of underlying components of the viral genetic population. The consensus genome for a given sample may also be an amalgamation of frequent but non-contiguous alleles (Figure 3B). Genetic variants or alleles that exist at a sub-consensus level are often referred to as minor variants. Depending on the intended purpose of a study, an analysis of the presence and identity of minor variants may be crucial to a proper interpretation of the data, or that information may be viewed as a luxury. For example, studies of transmission or evolution of HSV-1 (Greninger et al., 2018; Pandey et al., 2017; Shipley et al., 2018; Lassalle et al., 2020), HSV-2 (Akhtar

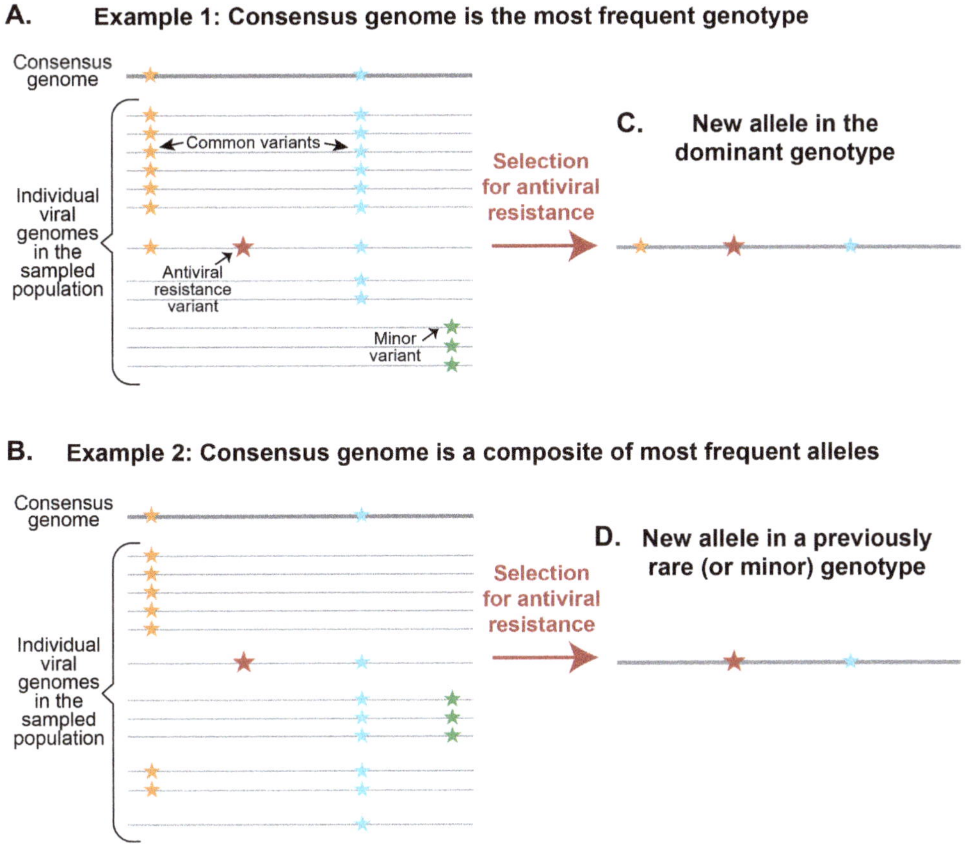

Figure 3. A sample of virus contains a population of viral genomes, which includes both common and rare or low-frequency (minor) genetic variants. A viral consensus genome represents the most common allele or nucleotide every position in the genome. In this diagram, lines represent individual genomes, with alleles or loci that vary between genomes denoted as stars. The thicker line represents the consensus genome that results from deep sequencing and assembly of the most common (or most frequently observed) alleles. (**A**) In example 1, the consensus genome is the most common or frequent genotype present in the viral sample. (**B**) In example 2, the consensus genome contains all of the most common alleles, but this amalgamation of alleles does not exist at high frequency in the viral sample. (**C-D**) In either case, selection pressures such as the application of an antiviral drug can lead to a genome that carries a rare drug-resistance variant (denoted by the red star) being positively selected. The drug-resistant genotype may then become the most commonly observed allele. Depending on the genomic context of the drug-resistance allele, this allele may appear in the same genetic context as the previously dominant genotype (**C**), or it may appear alongside other variants that were previously rare (**D**). Figure adapted from Renner and Szpara 2018 (Renner and Szpara, 2018).

et al., 2019; Minaya et al., 2017a), and VZV (Depledge et al., 2016a, 2016b, 2018b) have compared minor variants detected in samples of related virus populations to draw their conclusions. From a clinical perspective, minor variants may include drug-resistant variants in the viral population that could flourish following treatment (Figure 3CD) (Burrel et al., 2010; Houldcroft et al., 2017; Sauerbrei et al., 2010). Phylogenetic studies, by contrast, would have little use for analyzing the minor variants in their samples (Zell et al., 2012; Kolb et al., 2017; Johnston et al., 2017a; Pfaff et al., 2016). So, when designing a genetic study of herpesviruses, careful consideration must be given to the analysis of minor variants and the depth of sequence coverage needed to do so.

Another consideration that must be made when analyzing consensus sequences from deep sequencing data relates to the large size of alphaherpesvirus genomes. Since the sequence read length of commonly used HTS or deep sequencing approaches are substantially shorter than alphaherpesvirus genomes (e.g. ~300 bases vs hundreds of kilobases), it is difficult to determine which genetic variants coexist within one intact genome. If each variant is the most common at its location, it will be included as an amalgamation of several genotypes as the consensus genome (see Figure 3). For smaller viruses, researchers have the ability to clone and sequence individual genomes, which has enabled the measurement of viral genotypes in a given population and the development of software to infer likely haplotypes from the HTS data (Lou et al., 2013; Töpfer et al., 2014; Jayasundara et al., 2015). Barcoded methods offer the potential to improve the linking of variant alleles into haplotypes (Lauring and Andino, 2011; Keys et al., 2015), but these have not yet been widely applied to herpesviruses. Trimpert *et al* recently applied a combination of HTS-based assessment, with bacterial artificial chromosome (BAC)-based genome capture, to validate MDV population diversity in a series of polymerase mutants (Trimpert et al., 2019). This combination enabled them to confirm the co-location of minor variants observed by HTS, in single BAC-cloned viral genomes (Trimpert et al., 2019). Other technologies such as nanopore-based (MinION©, Oxford Nanopore) or single-molecule real-time sequencing (SMRT©), Pacific Biosystems (PacBio)) offer the promise of longer read lengths and the resulting ability to link minor alleles, although they have not yet been widely applied to analyze minor variants in herpesvirus populations (Tombácz et al., 2017a; Prazsák et al., 2018; Tombácz et al., 2019; Depledge et al., 2019; Whisnant et al., 2020). However, both long-read technologies currently exhibit a degree of sequencing error that is sufficient to limit their useful application at this point in their development. Thus short-read HTS platforms such as Illumina© (e.g. MiSeq or HiSeq) remain the most common choice for herpesvirus genomics, where the technology is sufficient to identify the location and prevalence of minor variants, but not to evaluate their co-occurrence on single genomes (Hage et al., 2017; Shipley et al., 2018; Depledge et al., 2016b; Pandey et al., 2016). Newer hybrid sequencing approaches may help to bridge this gap, by employing both the sensitivity of Illumina short-read HTS, and the longer scaffold accuracy of long-read Nanopore or PacBio HTS.

The path to obtaining a viral genome

Most herpesvirus sequencing projects, whether from laboratory experiments or clinical samples, begin with infection of a cell line to generate a sufficient level of DNA for the preparation of sequencing libraries. This approach enables the generation of large quantities of DNA, including the amplification of DNA from a potentially scarce clinical sample. However, growing viruses in cell culture is inherently different from how the virus would replicate *in vivo*, with the potential to introduce unexpected genetic bottlenecks or selective pressures (Parsons et al., 2015; Wilkinson et al., 2015; Kuny et al., 2020). The potential for cell culture-induced bias in sequencing results has led to the development of oligonucleotide-based enrichment methods, with the goal of sequencing uncultured samples to the same quality as cultured stocks (Greninger et al., 2018; Depledge et al., 2011; Johnston et al., 2017a; Shipley et al., 2018, 2020). This "oligo-enrichment" approach uses synthetic RNA or DNA probes, also known as baits, that are designed to hybridize with sparse amounts of the targeted viral genome(s) using sequence complementarity. These are mixed into solution with the sample of interest. Once they have hybridized with viral genetic material, the baits can be isolated from the mixed sample by virtue of an attached tag, such as biotin. The sample that results from this isolation is greatly enriched for the targeted DNA and can be amplified and sequenced using standard HTS approaches. Oligo-enrichment has enabled the sequencing of alphaherpesviruses directly from samples such as skin swabs, saliva, blood, vesicle fluid, and tissue samples, among others (Greninger et al., 2018; Depledge et al., 2011; Johnston et al., 2017a; Shipley et al., 2018). The sensitivity of viral genome capture with this method has been used to gain new insights into viral population diversity in infected humans. This has included tracking changes at the consensus genome level in multiple reactivation episodes (Shipley et al., 2018; Greninger et al., 2018; Casto et al., 2019; Depledge et al., 2014b), or during transmission between human hosts (Depledge et al., 2016b; Shipley et al., 2019), and examining how specific allelic variants (i.e. minor variants) change in frequency over time within individual hosts (Minaya et al., 2017b; Shipley et al., 2018).

Once viral genomic DNA is prepared for sequencing, the next step for a researcher is to decide which platform they will choose to generate sequence data. The most common approach is to use short-read HTS platforms, such as Illumina©. A detailed examination of how HTS data is processed into full-length consensus genomes and/or minor variants has been described in depth elsewhere (Houldcroft et al., 2017; Posada-Cespedes et al., 2016). The basic approach is for the short sequencing reads to be stitched, via bits of overlapping sequence, into longer and longer contiguous pieces that eventually form a new consensus genome. In a second quality-control step, individual sequencing reads are then aligned (or "mapped") back to their matching locations on the newly constructed consensus genome, usually with a great number of sequencing reads present per nucleotide (i.e. the sequencing or coverage depth). When sequencing coverage is deep (e.g. >300-fold coverage), genomes can be created with greater confidence than with lesser coverage. Deep sequence coverage also allows low-frequency minor variants to be identified with greater confidence. Publications generally only report minor variants above a certain minimum frequency threshold, such as 2% (Greninger et al., 2018; Hage et al., 2017; Shipley et al., 2018; Depledge et al., 2014b). It is important to appreciate that with 100X coverage, a 2% threshold for minor variant detection would mean that only 2 sequencing reads were sufficient to identify (or "call") the minor variant. For this reason, we recommend a sequencing depth far exceeding 100-fold for the confident and accurate detection of minor variants. Greater coverage depth means that more sequence reads must support the minor variant for it to be detected or called, thus minimizing the chance of an incorrectly identified minor variant. The cost of high coverage depth is that fewer viral strain genomes can be assembled from a given sequencing "run" (e.g. one iteration of operating an Illumina MiSeq). Again, this is where the priority of the researcher factors into decisions made in the sequencing process. If the priority is to sequence as many genomes as possible in one sequencing run, the limitations of existing technology mean that such genomes will have lower coverage than a sequencing run that involves only a limited number of unique samples. However research that is focused solely on species identification or serotype determination may benefit from increased efficiency and cost-savings by foregoing coverage depth and the detection of minor variants (Greninger et al., 2018; Zell et al., 2012; Vaz et al., 2016a; Pfaff et al., 2016).

Since the analysis of sequenced genomes often depends on how accurately individual sequencing reads can be overlapped to create the scaffold of a full-length genome, repetitive regions present a distinct problem (Figure 4) (Treangen and Salzberg, 2011). If a given sequencing read consists entirely of a perfectly repeating sequence, it may be unclear how the read overlaps with its neighbors, i.e. how a long a given repeating array of sequences truly is. Many alphaherpesvirus genomes have multiple extremely long repetitive arrays, such as the classic "reiterations" recognized in early studies of HSV-1 and VZV (Davison and Wilkie, 1981; Moss et al., 1981). Length variations in these tandem repeats have been recognized since the era of restriction fragment length polymorphism (RFLP) analyses, and these fluctuations further confound the true length determination of these repeats (Moss et al., 1981; Simon et al., 1989; Sakaoka et al., 1994). In HTS genome assemblies, the issue of uncertain overlaps in repetitive sequence arrays can cause the termination of *de novo* assembled fragments, creating gaps in the resulting consensus genome. It may also confound the ability to map sequence reads back to repeated regions in alignment-based methods of genome analysis (Parsons et al., 2015; Treangen and Salzberg, 2011). In addition to having large structural repeats at internal

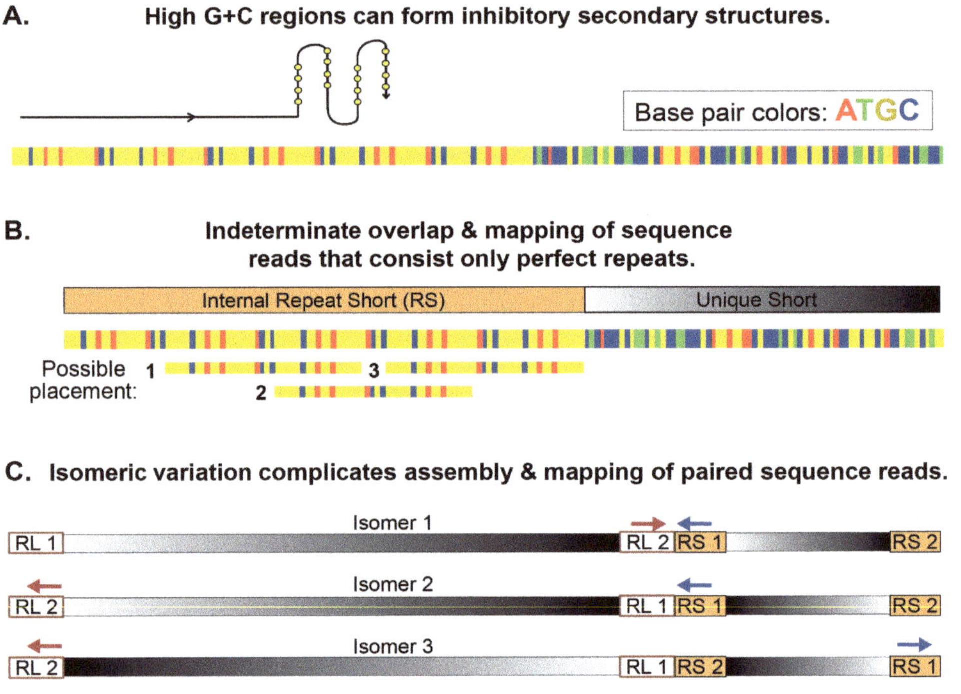

Figure 4. Repetitive and/or G+C rich regions complicate the process of genome sequencing and assembly. (**A**) Challenges in the physical sequencing of G+C-rich regions and/or areas containing DNA secondary structures such as the stem-loop at the origin of replication (Ori) can lead to a lower quantity of raw sequence-read data being produced in during high-throughput sequencing (HTS). (**B**) During *de novo* assembly or alignment of viral genome sequence data, challenges arise from ambiguities in the overlap and/or mapping of sequence reads that contain tandem repeats, homopolymers, or adjacent mixtures of these motifs. **(C)** Alphaherpesvirus genomes generally have two unique regions (termed unique long and unique short; shown as gradients of gray), each of which are flanked by repeats (known as repeat long (RL) or repeat (RS); shown as white and orange boxes). RL and RS exist in two copies, one terminal and one internal and inverted, on each genome. Isomers of these genomes - which appear to have either the left or right unique and repeated regions in the reverse orientation - exist in viral populations (Mahiet et al., 2012; Roizman, 1979). Sequence read mapping issues can result from ambiguities in mapping or placement of paired-end sequences that span the isomerization boundary of the long and short regions of alphaherpesvirus genomes (gradients of gray denote orientation of the unique regions). The combination of decreased sequence quantity (**A**) and ambiguities in sequence read mapping (**B**) combine to create regions of low-coverage and gaps in the final consensus genome produced by HTS (**C**). This phenomenon is frequently observed in the terminal and internal repeats of HSV-1 and HSV-2 consensus genomes (Greninger et al., 2018; Parsons et al., 2015; Szpara et al., 2011; Newman et al., 2015; Norberg et al., 2011).

and terminal regions, and the classically-defined large reiterations, many alphaherpesviruses are rich in short tandem repeated sequences (also known as short sequence repeats or SSRs, and variable number tandem repeats, or VNTRs) (Szpara et al., 2011). The G+C content tends to be even higher within these repeats (Szpara et al., 2011, 2014). This leads to secondary structures such as stem-loops and G-quadruplexes that decrease the overall yield of sequence data in these areas and further complicates the assembly of these high G+C repetitive regions (Figure 4) (Szpara et al., 2011; Parsons et al., 2015; Artusi et al., 2015; Biswas et al., 2016).

Alphaherpesviruses often have latency- or neurovirulence-associated genes encoded in the large structural repeated regions of the genome (e.g. LAT and ICP34.5 in HSV-1), in addition to the key transcriptional regulator ICP4 (also known as IE180 or IE62, depending on the virus). Therefore, researchers must balance the desire to study the biological functions associated with these regions of the genome, with the technical difficulties of sequencing them. Based on the priorities of each study, some groups have decided that omission of the repeats was the most useful (Greninger et al., 2018; Newman et al., 2015; Norberg et al., 2011). Other studies include an analysis of the repetitive regions, with certain caveats given to the analysis of the data (Szpara et al., 2014, 2011; Kolb et al., 2015; Pandey et al., 2016). Other groups seek to address the technical difficulties of using HTS approaches in repetitive areas by using Sanger sequencing to resolve troublesome regions of the genome (Minaya et al., 2017a). The challenges surrounding the repetitive regions of these genomes, and how to deal with them, remain one of the most challenging aspects of alphaherpesvirus genomics.

High quality genomes result from dedicated bioinformatics and quality control

While the decisions that go into choosing a sequencing method are not trivial, they mean little without proper bioinformatic efforts to assemble viral genomes and analyze them. The ability to rapidly sequence and assemble large DNA virus genomes like the alphaherpesviruses has been facilitated by advances in software and computational workflows, although the diversity of approaches to the task has led to a wide variety in the quality of finished genomes. One of the first choices in the bioinformatic analysis of viral HTS data is whether to construct a genome by alignment of the sequencing reads to a prior reference genome, or by *de novo* assembly (Posada-Cespedes et al., 2016). Each approach has its own strengths and weaknesses that should be considered. Alignment-based approaches use a previously determined reference genome to achieve faster results, but the end result is biased by the reference genome. As a result, these approaches are prone to mis-calling of minority and structural variants, and may have repeat-lengths that are missing (gapped) or biased by the reference genome chosen for alignment. *De novo* approaches are unbiased by prior data, but they are significantly more computationally intensive, and can require more curation following the initial assembly to obtain optimal results. However, the effort required *by de* novo assembly is often rewarded by a greater ability to detect structural changes in genomes, such as insertions or deletions, or minor variants in the viral population. We have found that a combination of iterative *de novo* assembly and reference-guided ordering of the resulting sequence blocks, as first demonstrated for herpesviruses by Cunningham *et al.* (Cunningham et al., 2010), has been the most useful and robust approach thus far - whether for cultured viral genomes, sparse field samples, or newer oligo-enrichment derived genomes (Parsons et al., 2015; Szpara et al., 2014; Shipley et al., 2018; Pandey et al., 2016)

The tools for performing the steps of genome assembly and analysis are often open-source and include web-based platforms as well as others using command-line (Unix) interfaces. The vast majority of viral *de novo* assembly programs have been developed and tested only for RNA viruses. We have developed both web-based (VirAmp; (Wan et al., 2015)) and command-line based (VirGA; (Parsons et al., 2015)) workflows for *de novo* assembly of herpesvirus genomes. VirAmp is unique in using the Galaxy framework of web-based bioinformatics tools (Blankenberg et al., 2010; Giardine et al., 2005). Most options for viral genome assembly, alignment, annotation and comparison utilize a command-line interface, which has a more difficult learning curve. These Unix-based software options, such as our *de novo* genome assembly suite VirGA (Parsons et al., 2015), the alignment packages Bowtie and BWA

(Langmead and Salzberg, 2012; Posada-Cespedes et al., 2016), and similar script-based workflows (Greninger et al., 2018; Mangul et al., 2014), are usually open source and are generally available through repositories like BitBucket or GitHub. If these programs prove too difficult to use, researchers can also choose from commercial packages (ex. Geneious© Biomatters and CLC Bio© Qiagen) for alignment or *de novo* assembly of genomes.

The proliferation of genomics software has been a boon for the production and publication of genomics data. However, this increased accessibility comes with a potential loss of quality, if these "easier-to-use" programs are used improperly. The rationale for the generation of high-quality genomes has been established by the human genome project and by many microbial genome projects (Fraser et al., 2002). For those that are new to genomics, the question of what constitutes an optimal, "high quality" genome may be a very difficult one to answer. Without delving into a true point-by-point definition of what may be considered a high-quality genome assembly, there are a few guidelines that are helpful to follow, particularly for alphaherpesvirus genomics. Genomes should be fully assembled, ideally without gaps or unfinished regions in the final sequence. This includes the large structural repeat regions, and the genes contained within these. These genomes should also be annotated to the full ability that current research allows and deposited into an international repository such as GenBank or the European Nucleotide Archive (ENA). The genome annotation should include genes and protein coding regions, along with details on the assembly and annotation methods, in addition to strain source and handling (see Figure 2). These details enable future comparisons with other genomes and gene-sequence data. Raw sequence read data should be deposited in an international Short Read Archive (SRA), for others to validate and/or re-analyze in the future. As discussed above ("Clear nomenclature..."), high quality genomes with detailed annotation and nomenclature (Figure 2) enable future scientists to accurately trace the origins of the data and utilize these data in further studies.

Herpesviruses can occasionally present unique challenges to these guidelines, as completion of the sequencing of complex tandem repeats or G+C-rich sequences can be difficult. However, the absence of such sequences hinders future comparative studies, because partial genomes represent an incomplete source of data. This is particularly worrisome when these regions are a source of genetic diversity and rapid evolution, as with the herpesviruses. In a recent comparison of neonatal HSV-2 genomes, for example, we found almost sixty adult-derived HSV-2 genomes in the GenBank sequence database (Akhtar et al., 2018). However less than ten of these GenBank-deposited HSV-2 genomes included the key genes found in the internal and terminal repeat regions: the key transcriptional regulator ICP4 (encoded by the RS1 gene), the neurovirulence protein ICP34.5 (RL1), and the immediate-early protein ICP0 (RL2) (Johnston et al., 2017a, 2017b; Koelle et al., 2017; Kolb et al., 2015; Newman et al., 2015). This missing data prevents a full genetic comparison of coding diversity in these genes, all of which have been demonstrated to be crucial for *in vivo* pathogenesis and spread (Roizman and Campadelli-Fiume, 2007). Since a sequenced viral genome forms the basis for all hybrid sequencing applications like RNAseq, chromatin-immunoprecipitation (ChIP) sequencing, and chromatin conformational capture (CCC or 3C) sequencing, deficiencies in the starting reference genome can cause many unintentional downstream problems. The tremendous insights that hybrid genomic approaches can provide therefore rely upon a strong foundation in the initial deciphering of viral genome populations.

Future Frontiers

Throughout this chapter we have discussed the progression of alphaherpesvirus genomics, from the first assembled genomes to the current state of the art, and some of the insights that these studies have revealed. The technological advancements that have occurred in recent years have driven a substantially greater understanding of the genetic diversity and evolution of herpesviruses. We now have an appreciation of the variations that can be seen within herpesvirus genomes, and that these viruses often exist as populations which contain heterogeneous genomes. This heterogeneity has been documented via the *in vivo* detection of minor variants during genital infection by HSV-1 (Shipley et al., 2018) or HSV-2 (Minaya et al., 2017a), and the selection of drug-resistant variants during antiviral therapy (Burrel et al., 2010; Depledge et al., 2016b;

Houldcroft et al., 2017; Sauerbrei et al., 2010). Studies using newer oligo-enrichment approaches have begun to reveal the genetic diversity present *in vivo*, in natural human herpesvirus infections (Depledge et al., 2016b; Greninger et al., 2018; Johnston et al., 2017a; Shipley et al., 2018), without any potential bias from expanding the virus population in cell culture. The plasticity of alphaherpesvirus populations has also been observed in cultured virus stocks (Greninger et al., 2018; Parsons et al., 2015), and examined over time, as seen in our recent application of *in vitro* evolution to HSV-1 (Kuny et al., 2020). With a renewed appreciation of the connection between herpesvirus genetics and biology, it is only natural to look to the future for the next revelations.

First, the potential for improvement within third-generation sequencing (long-read) technology is tantalizing, and its application to herpesvirus genomics could be revolutionary. Early applications of MinION and SMRT long-read sequencing in herpesvirology have already been useful, providing many new insights into novel transcriptional networks (Depledge et al., 2018a, 2018c; Tombácz et al., 2017b, 2017a). The combination of this approach with approaches discussed above (see "Other contributions to functional diversity" above), such as ribosome profiling or the analysis of novel transcripts, will further illuminate the functional genetic outputs of these viral genomes (Depledge et al., 2018a). Oldfield *et al.* recently demonstrated a completely synthetic approach to the generation of an alphaherpesvirus genome, which provides an opportunity for vast improvements in the efficiency of generating multiple precise mutations in a single viral genome, albeit at a currently higher cost (Oldfield et al., 2017). Maturation and improvement in the accuracy of MinION and SMRT technologies will allow for improved detection of recombinant genomes and structural variants, as well as the ability to define genetic haplotypes in mixed populations. These technologies may also allow for sequencing through the entirety of the large internal and terminal repeat regions of alphaherpesviruses, allowing improved assembly and a higher-throughput analysis of those areas of the genome. These methods may also be able to detect methylated bases or secondary structures in the viral DNA, or even read substrates besides DNA and RNA (Depledge et al., 2018a).

Another intriguing future area of alphaherpesvirus genomics research will be to link viral genetics to observable phenotypes, particularly virulence (Dutilh et al., 2013; Power et al., 2016). As current-generation HTS technologies have matured, genome wide association studies (GWAS) and quantitative trait locus (QTL) studies have been applied to this question for herpesviruses. Brandt and colleagues recently used QTL analyses to examine the contribution of specific viral genotypes to ocular phenotypes of HSV-1 infection in mice (Kolb et al., 2016; Lee et al., 2016). This study used the recombinant progeny of two strains of HSV-1, where the individual recombinants were completely sequenced with HTS and analyzed with comparative genomics (Lee et al., 2015). Examining the phenotypes of randomly generated or naturally occurring variants is also known as forward genetics, and these approaches can complement previous decades of reverse genetic approaches (i.e. disrupting specific genes and studying the results). Reverse genetics was crucial to discover the function(s) of many herpesvirus genes, but the types of mutations that are introduced in most reverse genetic approaches (e.g. entire gene deletion or complete disruption of enzymatic function) are rarely seen in field or clinical samples (Bryant et al., 2018; Depledge et al., 2018b; Pandey et al., 2016; Johnston et al., 2017b). The more subtle changes observed in natural settings may have substantial impacts in human or animal clinical outcomes (Loncoman et al., 2017; Bryant et al., 2018; Mechelli et al., 2015; Arav-Boger, 2015). Extending GWAS analyses to naturally isolated viral variants, particularly in comparison to matched human samples, will shed light on the intersection of human and viral genetics that leads to a spectrum of observed disease outcomes (Thompson et al., 2014; Ramchandani et al., 2019).

Acknowledgements

For the purposes of this chapter on alphaherpesvirus genomics, we have focused primarily on references using high-throughput and deep sequencing-based approaches. However, these data would not be possible without the foundational research of many scientists whose work is not cited here for space reasons. We thank members of the Szpara lab and the Genomics Methods discussion groups at the International Herpesvirus Workshops for insightful conversations and feedback. This work was supported by the Eberly College of

Science and the Huck Institutes for the Life Sciences at the Pennsylvania State University (MLS), and by NIH grant support from R01 AI132692.

References

Akhtar, L.N., Bowen, C.D., Renner, D.W., Pandey, U., Della Fera, A.N., Kimberlin, D.W., Prichard, M.N., Whitley, R.J., Weitzman, M.D., and Szpara, M.L. (2018). Genotypic and phenotypic diversity within the neonatal HSV-2 population. bioRxiv *http://biorxiv.org/lookup/doi/10.1101/262055.*

Akhtar, L.N., Bowen, C.D., Renner, D.W., Pandey, U., Della Fera, A.N., Kimberlin, D.W., Prichard, M.N., Whitley, R.J., Weitzman, M.D., and Szpara, M.L. (2019). Genotypic and Phenotypic Diversity of Herpes Simplex Virus 2 within the Infected Neonatal Population. MSphere *4*, e00590-18.

Andino, R., and Domingo, E. (2015). Viral quasispecies. Virology *479-480*, 46-51.

Arav-Boger, R. (2015). Strain Variation and Disease Severity in Congenital Cytomegalovirus Infection. Infect. Dis. Clin. North Am. *29*, 401-414.

Arias, C., Weisburd, B., Stern-Ginossar, N., Mercier, A., Madrid, A.S., Bellare, P., Holdorf, M., Weissman, J.S., and Ganem, D. (2014). KSHV 2.0: a comprehensive annotation of the Kaposi's sarcoma-associated herpesvirus genome using next-generation sequencing reveals novel genomic and functional features. PLoS Pathog *10*, e1003847.

Artusi, S., Nadai, M., Perrone, R., Biasolo, M.A., Palù, G., Flamand, L., Calistri, A., and Richter, S.N. (2015). The Herpes Simplex Virus-1 genome contains multiple clusters of repeated G-quadruplex: Implications for the antiviral activity of a G-quadruplex ligand. Antiviral Res. *118*, 123-131.

Biswas, B., Kandpal, M., Jauhari, U.K., and Vivekanandan, P. (2016). Genome-wide analysis of G-quadruplexes in herpesvirus genomes. BMC Genomics *17*.

Blankenberg, D., Von Kuster, G., Coraor, N., Ananda, G., Lazarus, R., Mangan, M., Nekrutenko, A., and Taylor, J. (2010). Galaxy: a web-based genome analysis tool for experimentalists. Curr. Protoc. Mol. Biol. *Chapter 19*, Unit 19.10.1-21.

Bondre, V.P., Sankararaman, V., Andhare, V., Tupekar, M., and Sapkal, G.N. (2016). Genetic characterization of human herpesvirus type 1: Full-length genome sequence of strain obtained from an encephalitis case from India. Indian J. Med. Res. *144*, 750-760.

Bowen, C.D., Renner, D.W., Shreve, J.T., Tafuri, Y., Payne, K.M., Dix, R.D., Kinchington, P.R., Gatherer, D., and Szpara, M.L. (2016). Viral forensic genomics reveals the relatedness of classic herpes simplex virus strains KOS, KOS63, and KOS79. Virology *492*, 179-186.

Bowen, C.D., Paavilainen, H., Renner, D.W., Palomäki, J., Lehtinen, J., Vuorinen, T., Norberg, P., Hukkanen, V., and Szpara, M. (2018). HSV-1 strains circulating in Finland demonstrate uncoupling of geographic and phenotypic variation.

Breuer, J., Grose, C., Norberg, P., Tipples, G., and Schmid, D.S. (2010). A proposal for a common nomenclature for viral clades that form the species varicella-zoster virus: summary of VZV Nomenclature Meeting 2008, Barts and the London School of Medicine and Dentistry, 24-25 July 2008. J. Gen. Virol. *91*, 821-828.

Brown, J. (2004). Effect of gene location on the evolutionary rate of amino acid substitutions in herpes simplex virus proteins. Virology *330*, 209-220.

Browning, G.F., Vaz, P.K., Devlin, J.M., Hartley, C.A., Studdert, M.J., Gilkerson, J.R., Horsington, J., and Ficorilli, N.P. (2016). Evidence of widespread natural recombination among field isolates of equine herpesvirus 4 but not among field isolates of equine herpesvirus 1. J. Gen. Virol. *97*, 747-755.

Bryant, N.A., Wilkie, G.S., Russell, C.A., Compston, L., Grafham, D., Clissold, L., McLay, K., Medcalf, L., Newton, R., Davison, A.J., et al. (2018). Genetic diversity of equine herpesvirus 1 isolated from neurological, abortigenic and respiratory disease outbreaks. Transbound. Emerg. Dis. *65*, 817-832.

Burrel, S., Deback, C., Agut, H., and Boutolleau, D. (2010). Genotypic Characterization of UL23 Thymidine Kinase and UL30 DNA Polymerase of Clinical Isolates of Herpes Simplex Virus: Natural Polymorphism and Mutations Associated with Resistance to Antivirals. Antimicrob. Agents Chemother. *54*, 4833-4842.

Burrel, S., Désiré, N., Marlet, J., Dacheux, L., Seang, S., Caumes, E., Bourhy, H., Agut, H., and Boutolleau, D. (2015). Genetic Diversity within Alphaherpesviruses: Characterization of a Novel Variant of Herpes Simplex Virus 2. J. Virol. *89*, 12273-12283.

Burrel, S., Boutolleau, D., Ryu, D., Agut, H., Merkel, K., Leendertz, F.H., and Calvignac-Spencer, S. (2017). Ancient Recombination Events between Human Herpes Simplex Viruses. Mol. Biol. Evol. *34*, 1713-1721.

Casto, A.M., Roychoudhury, P., Xie, H., Selke, S., Perchetti, G.A., Wofford, H., Huang, M.-L., Verjans, G.M.G.M., Gottlieb, G.S., Wald, A., et al. (2019). Large, stable, contemporary interspecies recombination events in circulating human herpes simplex viruses. J. Infect. Dis. jiz199.

Cunningham, C., Gatherer, D., Hilfrich, B., Baluchova, K., Derrick, J., Thomson, M., Griffiths, P.D., Wilkinson, G.W.G., Schulz, T.F., Dargan, D.J., et al. (2010). Sequences of complete human cytomegalovirus genomes from infected cell cultures and clinical specimens. J. Gen. Virol. *91*, 605-615.

Davison, A.J. (2010). Herpesvirus systematics. Vet. Microbiol. *143*, 52-69.

Davison, A.J., and Wilkie, N.M. (1981). Nucleotide sequences of the joint between the L and S segments of herpes simplex virus types 1 and 2. J. Gen. Virol. *55*, 315-331.

Depledge, D.P., Palser, A.L., Watson, S.J., Lai, I.Y.-C., Gray, E.R., Grant, P., Kanda, R.K., Leproust, E., Kellam, P., and Breuer, J. (2011). Specific Capture and Whole-Genome Sequencing of Viruses from Clinical Samples. PLoS ONE *6*, e27805.

Depledge, D.P., Gray, E.R., Kundu, S., Cooray, S., Poulsen, A., Aaby, P., and Breuer, J. (2014a). Evolution of Cocirculating Varicella-Zoster Virus Genotypes during a Chickenpox Outbreak in Guinea-Bissau. J. Virol. *88*, 13936-13946.

Depledge, D.P., Kundu, S., Jensen, N.J., Gray, E.R., Jones, M., Steinberg, S., Gershon, A., Kinchington, P.R., Schmid, D.S., Balloux, F., et al. (2014b). Deep Sequencing of Viral Genomes Provides Insight into the Evolution and Pathogenesis of Varicella Zoster Virus and Its Vaccine in Humans. Mol. Biol. Evol. *31*, 397-409.

Depledge, D.P., Yamanishi, K., Gomi, Y., Gershon, A.A., and Breuer, J. (2016a). Deep Sequencing of Distinct Preparations of the Live Attenuated Varicella-Zoster Virus Vaccine Reveals a Conserved Core of Attenuating Single-Nucleotide Polymorphisms. J. Virol. *90*, 8698-8704.

Depledge, D.P., Brown, J., Macanovic, J., Underhill, G., and Breuer, J. (2016b). Viral Genome Sequencing Proves Nosocomial Transmission of Fatal Varicella. J. Infect. Dis. *214*, 1399-1402.

Depledge, D.P., Mohr, I., and Wilson, A.C. (2018a). Going the distance: optimizing RNA-Seq strategies for transcriptomic analysis of complex viral genomes. J. Virol.

Depledge, D.P., Cudini, J., Kundu, S., Atkinson, C., Brown, J.R., Haque, T., Houldcroft, C.J., Koay, E.S., McGill, F., Milne, R., et al. (2018b). High Viral Diversity and Mixed Infections in Cerebral Spinal Fluid From Cases of Varicella Zoster Virus Encephalitis. J. Infect. Dis.

Depledge, D.P., Kalanghad Puthankalam, S., Sadaoka, T., Beady, D., Mori, Y., Placantonakis, D., Mohr, I., and Wilson, A. (2018c). Native RNA sequencing on nanopore arrays redefines the transcriptional complexity of a viral pathogen.

Depledge, D.P., Srinivas, K.P., Sadaoka, T., Bready, D., Mori, Y., Placantonakis, D.G., Mohr, I., and Wilson, A.C. (2019). Direct RNA sequencing on nanopore arrays redefines the transcriptional complexity of a viral pathogen. Nat. Commun. *10*, 1-13.

Drake, J.W. (1991). A constant rate of spontaneous mutation in DNA-based microbes. Proc. Natl. Acad. Sci. *88*, 7160-7164.

Drake, J.W., and Hwang, C.B.C. (2005). On the mutation rate of herpes simplex virus type 1. Genetics *170*, 969-970.

Dutilh, B.E., Backus, L., Edwards, R.A., Wels, M., Bayjanov, J.R., and van Hijum, S.A.F.T. (2013). Explaining microbial phenotypes on a genomic scale: GWAS for microbes. Brief. Funct. Genomics *12*, 366-380.

Figueroa, T., Boumart, I., Coupeau, D., and Rasschaert, D. (2016). Hyperediting by ADAR1 of a new herpesvirus lncRNA during the lytic phase of the oncogenic Marek's disease virus. J. Gen. Virol. *97*, 2973-2988.

Firth, C., Kitchen, A., Shapiro, B., Suchard, M.A., Holmes, E.C., and Rambaut, A. (2010). Using time-structured data to estimate evolutionary rates of double-stranded DNA viruses. Mol. Biol. Evol. *27*, 2038-2051.

Fraser, C.M., Eisen, J.A., Nelson, K.E., Paulsen, I.T., and Salzberg, S.L. (2002). The Value of Complete Microbial Genome Sequencing (You Get What You Pay For). J. Bacteriol. *184*, 6403-6405.

Gantt, S., Orem, J., Krantz, E.M., Morrow, R.A., Selke, S., Huang, M.-L., Schiffer, J.T., Jerome, K.R., Nakaganda, A., Wald, A., et al. (2016). Prospective Characterization of the Risk Factors for Transmission and

Symptoms of Primary Human Herpesvirus Infections Among Ugandan Infants. J. Infect. Dis. jiw076.

Giardine, B., Riemer, C., Hardison, R.C., Burhans, R., Elnitski, L., Shah, P., Zhang, Y., Blankenberg, D., Albert, I., Taylor, J., et al. (2005). Galaxy: a platform for interactive large-scale genome analysis. Genome Res. *15*, 1451-1455.

Gorzer, I., Guelly, C., Trajanoski, S., and Puchhammer-Stockl, E. (2010). Deep sequencing reveals highly complex dynamics of human cytomegalovirus genotypes in transplant patients over time. J. Virol. *84*, 7195-7203.

Greninger, A.L., Roychoudhury, P., Xie, H., Casto, A., Cent, A., Pepper, G., Koelle, D.M., Huang, M.-L., Wald, A., Johnston, C., et al. (2018). Ultrasensitive Capture of Human Herpes Simplex Virus Genomes Directly from Clinical Samples Reveals Extraordinarily Limited Evolution in Cell Culture. MSphere *3*, e00283-18.

Griffiths, A. (2011). Slipping and sliding: Frameshift mutations in herpes simplex virus thymidine kinase and drug-resistance. Drug Resist. Updat. *14*, 251-259.

Griffiths, A., Chen, S., Horsburgh, B.C., and Coen, D.M. (2003). Translational Compensation of a Frameshift Mutation Affecting Herpes Simplex Virus Thymidine Kinase Is Sufficient To Permit Reactivation from Latency. J. Virol. *77*, 4703-4709.

Hage, E., Wilkie, G.S., Linnenweber-Held, S., Dhingra, A., Suárez, N.M., Schmidt, J.J., Kay-Fedorov, P.C., Mischak-Weissinger, E., Heim, A., Schwarz, A., et al. (2017). Characterization of Human Cytomegalovirus Genome Diversity in Immunocompromised Hosts by Whole-Genome Sequencing Directly From Clinical Specimens. J. Infect. Dis. *215*, 1673-1683.

Hall, J.D., and Almy, R.E. (1982). Evidence for control of herpes simplex virus mutagenesis by the viral DNA polymerase. Virology *116*, 535-543.

Hennig, T., Michalski, M., Rutkowski, A.J., Djakovic, L., Whisnant, A.W., Friedl, M.-S., Jha, B.A., Baptista, M.A.P., L'Hernault, A., Erhard, F., et al. (2018). HSV-1-induced disruption of transcription termination resembles a cellular stress response but selectively increases chromatin accessibility downstream of genes. PLOS Pathog. *14*, e1006954.

Houldcroft, C.J., Bryant, J.M., Depledge, D.P., Margetts, B.K., Simmonds, J., Nicolaou, S., Tutill, H.J., Williams, R., Worth, A.J.J., Marks, S.D., et al. (2016).

Detection of Low Frequency Multi-Drug Resistance and Novel Putative Maribavir Resistance in Immunocompromised Pediatric Patients with Cytomegalovirus. Front. Microbiol. *7*.

Houldcroft, C.J., Beale, M.A., and Breuer, J. (2017). Clinical and biological insights from viral genome sequencing. Nat. Rev. Microbiol. *15*, 183-192.

Huson, D.H. (1998). SplitsTree: analyzing and visualizing evolutionary data. Bioinformatics *14*, 68-73.

Jansen, M.A.E., van den Heuvel, D., Bouthoorn, S.H., Jaddoe, V.W.V., Hooijkaas, H., Raat, H., Fraaij, P.L.A., van Zelm, M.C., and Moll, H.A. (2016). Determinants of Ethnic Differences in Cytomegalovirus, Epstein-Barr Virus, and Herpes Simplex Virus Type 1 Seroprevalence in Childhood. J. Pediatr. *170*, 126-134.e6.

Jaramillo, N., Domingo, E., Munoz-Egea, M.C., Tabares, E., and Gadea, I. (2013). Evidence of Muller's ratchet in herpes simplex virus type 1. J. Gen. Virol. *94*, 366-375.

Jayasundara, D., Saeed, I., Maheswararajah, S., Chang, B.C., Tang, S.-L., and Halgamuge, S.K. (2015). ViQuaS: an improved reconstruction pipeline for viral quasispecies spectra generated by next-generation sequencing. Bioinformatics *31*, 886-896.

Johnston, C., Magaret, A., Roychoudhury, P., Greninger, A.L., Reeves, D., Schiffer, J., Jerome, K.R., Sather, C., Diem, K., Lingappa, J.R., et al. (2017a). Dual-strain genital herpes simplex virus type 2 (HSV-2) infection in the US, Peru, and 8 countries in sub-Saharan Africa: A nested cross-sectional viral genotyping study. PLOS Med. *14*, e1002475.

Johnston, C., Magaret, A., Roychoudhury, P., Greninger, A.L., Cheng, A., Diem, K., Fitzgibbon, M.P., Huang, M., Selke, S., Lingappa, J.R., et al. (2017b). Highly conserved intragenic HSV-2 sequences: Results from next-generation sequencing of HSV-2 U L and U S regions from genital swabs collected from 3 continents. Virology *510*, 90-98.

Keys, J.R., Zhou, S., Anderson, J.A., Eron, J.J., Rackoff, L.A., Jabara, C., and Swanstrom, R. (2015). Primer ID Informs Next-Generation Sequencing Platforms and Reveals Preexisting Drug Resistance Mutations in the HIV-1 Reverse Transcriptase Coding Domain. AIDS Res. Hum. Retroviruses *31*, 658-668.

Klupp, B.G., Hengartner, C.J., Mettenleiter, T.C., and Enquist, L.W. (2004). Complete, Annotated Sequence of the Pseudorabies Virus Genome. J. Virol. *78*, 424-440.

Koelle, D.M., Norberg, P., Fitzgibbon, M.P., Russell, R.M., Greninger, A.L., Huang, M.-L., Stensland, L., Jing, L., Magaret, A.S., Diem, K., et al. (2017). Worldwide circulation of HSV-2 × HSV-1 recombinant strains. Sci. Rep. *7*, 44084.

Kolb, A.W., Larsen, I.V., Cuellar, J.A., and Brandt, C.R. (2015). Genomic, Phylogenetic, and Recombinational Characterization of Herpes Simplex Virus 2 Strains. J. Virol. *89*, 6427-6434.

Kolb, A.W., Lee, K., Larsen, I., Craven, M., and Brandt, C.R. (2016). Quantitative Trait Locus Based Virulence Determinant Mapping of the HSV-1 Genome in Murine Ocular Infection: Genes Involved in Viral Regulatory and Innate Immune Networks Contribute to Virulence. PLOS Pathog. *12*, e1005499.

Kolb, A.W., Lewin, A.C., Moeller Trane, R., McLellan, G.J., and Brandt, C.R. (2017). Phylogenetic and recombination analysis of the herpesvirus genus varicellovirus. BMC Genomics *18*.

Koonin, E.V., Krupovic, M., and Yutin, N. (2015). Evolution of double-stranded DNA viruses of eukaryotes: from bacteriophages to transposons to giant viruses: Evolution of dsDNA viruses of eukaryotes. Ann. N. Y. Acad. Sci. *1341*, 10-24.

Kuhn, J.H., Bao, Y., Bavari, S., Becker, S., Bradfute, S., Brister, J.R., Bukreyev, A.A., Chandran, K., Davey, R.A., Dolnik, O., et al. (2013). Virus nomenclature below the species level: a standardized nomenclature for natural variants of viruses assigned to the family Filoviridae. Arch. Virol. *158*, 301-311.

Kuny, C.V., Bowen, C.D., Renner, D.W., Johnston, C.M., and Szpara, M.L. (2020). In vitro evolution of herpes simplex virus 1 (HSV-1) reveals selection for syncytia and other minor variants in cell culture. Virus Evol. *6*.

Langmead, B., and Salzberg, S.L. (2012). Fast gapped-read alignment with Bowtie 2. Nat. Methods *9*, 357-359.

Lassalle, F., Beale, M.A., Bharucha, T., Williams, C.A., Williams, R.J., Cudini, J., Goldstein, R., Haque, T., Depledge, D.P., and Breuer, J. (2020). Whole genome sequencing of Herpes Simplex Virus 1 directly from human cerebrospinal fluid reveals selective constraints in neurotropic viruses. Virus Evol. *6*.

Lauring, A.S., and Andino, R. (2011). Exploring the Fitness Landscape of an RNA Virus by Using a Universal Barcode Microarray. J. Virol. *85*, 3780-3791.

Lee, K., Kolb, A.W., Sverchkov, Y., Cuellar, J.A., Craven, M., and Brandt, C.R. (2015). Recombination Analysis of Herpes Simplex Virus 1 Reveals a Bias toward GC Content and the Inverted Repeat Regions. J. Virol. *89*, 7214-7223.

Lee, K., Kolb, A.W., Larsen, I., Craven, M., and Brandt, C.R. (2016). Mapping Murine Corneal Neovascularization and Weight Loss Virulence Determinants in the Herpes Simplex Virus 1 Genome and the Detection of an Epistatic Interaction between the UL and IRS/US Regions. J. Virol. *90*, 8115-8131.

Lee, S.-W., Markham, P.F., Coppo, M.J.C., Legione, A.R., Markham, J.F., Noormohammadi, A.H., Browning, G.F., Ficorilli, N., Hartley, C.A., and Devlin, J.M. (2012). Attenuated Vaccines Can Recombine to Form Virulent Field Viruses. Science *337*, 188-188.

Lewin, A.C., Kolb, A.W., McLellan, G.J., Bentley, E., Bernard, K.A., Newbury, S.P., and Brandt, C.R. (2018). Genomic, Recombinational and Phylogenetic Characterization of Global Feline Herpesvirus 1 Isolates. Virology *518*, 385-397.

Lin, Z., Wang, X., Strong, M.J., Concha, M., Baddoo, M., Xu, G., Baribault, C., Fewell, C., Hulme, W., Hedges, D., et al. (2013). Whole-Genome Sequencing of the Akata and Mutu Epstein-Barr Virus Strains. J. Virol. *87*, 1172-1182.

Loncoman, C.A., Vaz, P.K., Coppo, M.J., Hartley, C.A., Morera, F.J., Browning, G.F., and Devlin, J.M. (2017). Natural recombination in alphaherpesviruses: Insights into viral evolution through full genome sequencing and sequence analysis. Infect. Genet. Evol. J. Mol. Epidemiol. Evol. Genet. Infect. Dis. *49*, 174-185.

Looker, K.J., Magaret, A.S., May, M.T., Turner, K.M.E., Vickerman, P., Gottlieb, S.L., and Newman, L.M. (2015a). Global and regional estimates of prevalent and incident herpes simplex virus type 1 infections in 2012. PLoS ONE *10*, e0140765.

Looker, K.J., Magaret, A.S., Turner, K.M.E., Vickerman, P., Gottlieb, S.L., and Newman, L.M. (2015b). Global estimates of prevalent and incident herpes simplex virus type 2 infections in 2012. PloS One *10*, e114989.

Lou, D.I., Hussmann, J.A., McBee, R.M., Acevedo, A., Andino, R., Press, W.H., and Sawyer, S.L. (2013). High-throughput DNA sequencing errors are reduced by orders of magnitude using circle sequencing. Proc. Natl. Acad. Sci. *110*, 19872-19877.

Mahiet, C., Ergani, A., Huot, N., Alende, N., Azough, A., Salvaire, F., Bensimon, A., Conseiller, E., Wain-Hobson, S., Labetoulle, M., et al. (2012). Structural Variability of the Herpes Simplex Virus 1 Genome In Vitro and In Vivo. J. Virol. *86*, 8592-8601.

Mangul, S., Wu, N.C., Mancuso, N., Zelikovsky, A., Sun, R., and Eskin, E. (2014). Accurate viral population assembly from ultra-deep sequencing data. Bioinformatics *30*, i329-i337.

McGeoch, D.J., Dolan, A., Donald, S., and Rixon, F.J. (1985). Sequence determination and genetic content of the short unique region in the genome of herpes simplex virus type 1. J. Mol. Biol. *181*, 1-13.

McGeoch, D.J., Dalrymple, M.A., Davison, A.J., Dolan, A., Frame, M.C., McNab, D., Perry, L.J., Scott, J.E., and Taylor, P. (1988). The complete DNA sequence of the long unique region in the genome of herpes simplex virus type 1. J. Gen. Virol. *69*, 1531-1574.

Mechelli, R., Manzari, C., Policano, C., Annese, A., Picardi, E., Umeton, R., Fornasiero, A., D'erchia, A.M., Buscarinu, M.C., Agliardi, C., et al. (2015). Epstein-Barr virus genetic variants are associated with multiple sclerosis. Neurology *84*, 1362-1368.

Minaya, M.A., Jensen, T., Goll, J., Korom, M., Datla, S.H., Belshe, R.B., and Morrison, L.A. (2017a). Molecular evolution of herpes simplex virus 2 complete genomes: Comparison between primary and recurrent infections. J. Virol. JVI.00942-17.

Minaya, M.A., Jensen, T.L., Goll, J.B., Korom, M., Datla, S.H., Belshe, R.B., and Morrison, L.A. (2017b). Molecular evolution of herpes simplex virus 2 complete genomes: comparison between primary and recurrent infections. J. Virol. *91*, e00942-17.

Moss, B., Winters, E., and Cooper, N. (1981). Instability and reiteration of DNA sequences within the vaccinia virus genome. Proc. Natl. Acad. Sci. U. S. A. *78*, 1614-1618.

Nakaya, Y., Stavrou, S., Blouch, K., Tattersall, P., and Ross, S.R. (2016). In Vivo Examination of Mouse APOBEC3- and Human APOBEC3A- and APOBEC3G-Mediated Restriction of Parvovirus and Herpesvirus Infection in Mouse Models. J. Virol. *90*, 8005-8012.

Newman, R.M., Lamers, S.L., Weiner, B., Ray, S.C., Colgrove, R.C., Diaz, F., Jing, L., Wang, K., Saif, S., Young, S., et al. (2015). Genome Sequencing and Analysis of Geographically Diverse Clinical Isolates of Herpes Simplex Virus 2. J. Virol. JVI.01303-15.

Norberg, P. (2010). Divergence and genotyping of human alpha-herpesviruses: an overview. Infect. Genet. Evol. J. Mol. Epidemiol. Evol. Genet. Infect. Dis. *10*, 14-25.

Norberg, P., Kasubi, M.J., Haarr, L., Bergstrom, T., and Liljeqvist, J.-A. (2007). Divergence and recombination of clinical herpes simplex virus type 2 isolates. J. Virol. *81*, 13158-13167.

Norberg, P., Tyler, S., Severini, A., Whitley, R., Liljeqvist, J.-A., and Bergstrom, T. (2011). A genome-wide comparative evolutionary analysis of herpes simplex virus type 1 and varicella zoster virus. PloS One *6*, 1-8.

Norberg, P., Depledge, D.P., Kundu, S., Atkinson, C., Brown, J., Haque, T., Hussaini, Y., MacMahon, E., Molyneaux, P., Papaevangelou, V., et al. (2015). Recombination of Globally Circulating Varicella-Zoster Virus. J. Virol. *89*, 7133-7146.

Oh, H.S., Neuhausser, W.M., Eggan, P., Angelova, M., Kirchner, R., Eggan, K.C., and Knipe, D.M. (2019). Herpesviral lytic gene functions render the viral genome susceptible to novel editing by CRISPR/Cas9. ELife *8*.

Oldfield, L.M., Grzesik, P., Voorhies, A.A., Alperovich, N., MacMath, D., Najera, C.D., Chandra, D.S., Prasad, S., Noskov, V.N., Montague, M.G., et al. (2017). Genome-wide engineering of an infectious clone of herpes simplex virus type 1 using synthetic genomics assembly methods. Proc. Natl. Acad. Sci. 201700534.

Pan, D., and Coen, D.M. (2012). Net -1 frameshifting on a noncanonical sequence in a herpes simplex virus drug-resistant mutant is stimulated by nonstop mRNA. Proc. Natl. Acad. Sci. *109*, 14852-14857.

Pandey, U., Bell, A.S., Renner, D.W., Kennedy, D.A., Shreve, J.T., Cairns, C.L., Jones, M.J., Dunn, P.A., Read, A.F., and Szpara, M.L. (2016). DNA from dust:

Comparative genomics of large DNA viruses in field surveillance samples. MSphere *1*, e00132-16.

Pandey, U., Renner, D.W., Thompson, R.L., Szpara, M.L., and Sawtell, N.M. (2017). Inferred father-to-son transmission of herpes simplex virus results in near-perfect preservation of viral genome identity and in vivo phenotypes. Sci. Rep. *7*, 13666.

Parsons, L.R., Tafuri, Y.R., Shreve, J.T., Bowen, C.D., Shipley, M.M., Enquist, L.W., and Szpara, M.L. (2015). Rapid genome assembly and comparison decode intrastrain variation in human alphaherpesviruses. MBio *6*, e02213-14.

Pellett, P.E., and Roizman, B. (2013). Herpesviridae. In Fields Virology, D.M. Knipe, and P.M. Howley, eds. (Philadelphia, PA: Lippincott Williams and Wilkins), pp. 1802-1822.

Peters, G. a, Tyler, S.D., Grose, C., Severini, A., Gray, M.J., Upton, C., and Tipples, G. a (2006). A full-genome phylogenetic analysis of varicella-zoster virus reveals a novel origin of replication-based genotyping scheme and evidence of recombination between major circulating clades. J. Virol. *80*, 9850-9860.

Pfaff, F., Groth, M., Sauerbrei, A., and Zell, R. (2016). Genotyping of herpes simplex virus type 1 by whole-genome sequencing. J. Gen. Virol. *97*, 2732-2741.

Posada-Cespedes, S., Seifert, D., and Beerenwinkel, N. (2016). Recent advances in inferring viral diversity from high-throughput sequencing data. Virus Res. pii: S0168-1702(16)30413-0.

Power, R.A., Parkhill, J., and de Oliveira, T. (2016). Microbial genome-wide association studies: lessons from human GWAS. Nat. Rev. Genet. *18*, 41-50.

Prazsák, I., Moldován, N., Balázs, Z., Tombácz, D., Megyeri, K., Szűcs, A., Csabai, Z., and Boldogkői, Z. (2018). Long-read sequencing uncovers a complex transcriptome topology in varicella zoster virus. BMC Genomics *19*.

Ramchandani, M.S., Jing, L., Russell, R.M., Tran, T., Laing, K.J., Magaret, A.S., Selke, S., Cheng, A., Huang, M.-L., Xie, H., et al. (2019). Viral Genetics Modulate Orolabial Herpes Simplex Virus Type 1 Shedding in Humans. J. Infect. Dis. *219*, 1058-1066.

Renner, D.W., and Szpara, M.L. (2018). The impacts of genome-wide analyses on our understanding of human herpesvirus diversity and evolution. J. Virol. *92*, e00908-17.

Renzette, N., Gibson, L., Jensen, J.D., and Kowalik, T.F. (2014). Human cytomegalovirus intrahost evolution—a new avenue for understanding and controlling herpesvirus infections. Curr. Opin. Virol. *8*, 109-115.

Roizman, B. (1979). The structure and isomerization of herpes simplex virus genomes. Cell *16*, 481-494.

Roizman, B., and Campadelli-Fiume, G. (2007). Alphaherpes viral genes and their functions. In Human Herpesviruses: Biology, Therapy, and Immunoprophylaxis, (Cambridge), pp. 70-83.

Roizman, B., Knipe, D.M., and Whitley, R. (2013). Herpes Simplex Viruses. In Fields Virology, D.M. Knipe, and P.M. Howley, eds. (Philadelphia, PA: Lippincott Williams and Wilkins), pp. 1823-1897.

Rutkowski, A.J., Erhard, F., L'Hernault, A., Bonfert, T., Schilhabel, M., Crump, C., Rosenstiel, P., Efstathiou, S., Zimmer, R., Friedel, C.C., et al. (2015). Widespread disruption of host transcription termination in HSV-1 infection. Nat. Commun. *6*, 7126.

Sakaoka, H., Kurita, K., Iida, Y., Takada, S., Umene, K., Kim, Y.T., Ren, C.S., and Nahmias, A.J. (1994). Quantitative analysis of genomic polymorphism of herpes simplex virus type 1 strains from six countries: studies of molecular evolution and molecular epidemiology of the virus. J. Gen. Virol. *75*, 513-527.

Sanjuán, R., and Domingo-Calap, P. (2016). Mechanisms of viral mutation. Cell. Mol. Life Sci. *73*, 4433-4448.

Sanjuan, R., Nebot, M.R., Chirico, N., Mansky, L.M., and Belshaw, R. (2010). Viral Mutation Rates. J. Virol. *84*, 9733-9748.

Sarisky, R.T., Nguyen, T.T., Duffy, K.E., Wittrock, R.J., and Leary, J.J. (2000). Difference in incidence of spontaneous mutations between Herpes simplex virus types 1 and 2. Antimicrob. Agents Chemother. *44*, 1524-1529.

Sauerbrei, A., Deinhardt, S., Zell, R., and Wutzler, P. (2010). Phenotypic and genotypic characterization of acyclovir-resistant clinical isolates of herpes simplex virus. Antiviral Res. *86*, 246-252.

Severini, A., Tyler, S.D., Peters, G.A., Black, D., and Eberle, R. (2013). Genome sequence of a chimpanzee herpesvirus and its relation to other primate alphaherpesviruses. Arch. Virol. *158*, 1825-1828.

Shipley, M.M., Renner, D.W., Ott, M., Bloom, D.C., Koelle, D.M., Johnston, C., and Szpara, M.L. (2018). Genome-wide surveillance of genital herpes simplex virus type 1 from multiple anatomic sites over time. J. Infect. Dis. *218*, 595-605.

Shipley, M.M., Renner, D.W., Pandey, U., Ford, B., Bloom, D.C., Grose, C., and Szpara, M.L. (2019). Personalized viral genomic investigation of herpes simplex virus 1 perinatal viremic transmission with dual fatality. Mol. Case Stud. *5*, a004382.

Shipley, M.M., Rathbun, M.M., and Szpara, M.L. (2020). Oligonucleotide Enrichment of HSV-1 Genomic DNA from Clinical Specimens for Use in High-Throughput Sequencing. In Herpes Simplex Virus, R.J. Diefenbach, and C. Fraefel, eds. (New York, NY: Springer New York), pp. 199-217.

Sijmons, S., Thys, K., Mbong Ngwese, M., Van Damme, E., Dvorak, J., Van Loock, M., Li, G., Tachezy, R., Busson, L., Aerssens, J., et al. (2015). High-Throughput Analysis of Human Cytomegalovirus Genome Diversity Highlights the Widespread Occurrence of Gene-Disrupting Mutations and Pervasive Recombination. J. Virol. *89*, 7673-7695.

Simon, A., Mettenleiter, T.C., and Rziha, H.J. (1989). Pseudorabies virus displays variable numbers of a repeat unit adjacent to the 3' end of the glycoprotein gII gene. J. Gen. Virol. *70*, 1239-1246.

Suspene, R., Aynaud, M.-M., Koch, S., Pasdeloup, D., Labetoulle, M., Gaertner, B., Vartanian, J.-P., Meyerhans, A., and Wain-Hobson, S. (2011). Genetic editing of herpes simplex virus 1 and Epstein-Barr herpesvirus genomes by human APOBEC3 cytidine deaminases in culture and in vivo. J. Virol. *85*, 7594-7602.

Szpara, M.L., Parsons, L., and Enquist, L.W. (2010). Sequence variability in clinical and laboratory isolates of herpes simplex virus 1 reveals new mutations. J. Virol. *84*, 5303-5313.

Szpara, M.L., Tafuri, Y.R., Parsons, L., Shamim, S.R., Verstrepen, K.J., Legendre, M., and Enquist, L.W. (2011). A wide extent of inter-strain diversity in virulent and vaccine strains of alphaherpesviruses. PLoS Pathog. *7*, 1-23.

Szpara, M.L., Gatherer, D., Ochoa, A., Greenbaum, B., Dolan, A., Bowden, R.J., Enquist, L.W., Legendre, M., and Davison, A.J. (2014). Evolution and diversity in human herpes simplex virus genomes. J. Virol. *88*, 1209-1227.

Thompson, R.L., Williams, R.W., Kotb, M., and Sawtell, N.M. (2014). A Forward Phenotypically Driven Unbiased Genetic Analysis of Host Genes That Moderate Herpes Simplex Virus Virulence and Stromal Keratitis in Mice. PLoS ONE *9*, 1-10.

Tirosh, O., Cohen, Y., Shitrit, A., Shani, O., Le-Trilling, V.T.K., Trilling, M., Friedlander, G., Tanenbaum, M., and Stern-Ginossar, N. (2015). The Transcription and Translation Landscapes during Human Cytomegalovirus Infection Reveal Novel Host-Pathogen Interactions. PLoS Pathog. *11*, e1005288.

Tombácz, D., Csabai, Z., Szűcs, A., Balázs, Z., Moldován, N., Sharon, D., Snyder, M., and Boldogkői, Z. (2017a). Long-Read Isoform Sequencing Reveals a Hidden Complexity of the Transcriptional Landscape of Herpes Simplex Virus Type 1. Front. Microbiol. *8*.

Tombácz, D., Balázs, Z., Csabai, Z., Moldován, N., Szűcs, A., Sharon, D., Snyder, M., and Boldogkői, Z. (2017b). Characterization of the Dynamic Transcriptome of a Herpesvirus with Long-read Single Molecule Real-Time Sequencing. Sci. Rep. *7*, 43751.

Tombácz, D., Moldován, N., Balázs, Z., Gulyás, G., Csabai, Z., Boldogkői, M., Snyder, M., and Boldogkői, Z. (2019). Multiple Long-Read Sequencing Survey of Herpes Simplex Virus Dynamic Transcriptome. Front. Genet. *10*.

Töpfer, A., Marschall, T., Bull, R.A., Luciani, F., Schönhuth, A., and Beerenwinkel, N. (2014). Viral Quasispecies Assembly via Maximal Clique Enumeration. PLoS Comput. Biol. *10*, e1003515.

Treangen, T.J., and Salzberg, S.L. (2011). Repetitive DNA and next-generation sequencing: computational challenges and solutions. Nat. Rev. Genet.

Trimpert, J., Groenke, N., Jenckel, M., He, S., Kunec, D., Szpara, M.L., Spatz, S.J., Osterrieder, N., and McMahon, D.P. (2017). A phylogenomic analysis of Marek's disease virus reveals independent paths to virulence in Eurasia and North America. Evol. Appl. *10*, 1091-1101.

Trimpert, J., Groenke, N., Kunec, D., Eschke, K., He, S., McMahon, D.P., and Osterrieder, N. (2019). A proofreading-impaired herpesvirus generates populations with quasispecies-like structure. Nat. Microbiol. 1-9.

Tyler, S.D., Peters, G.A., Grose, C., Severini, A., Gray, M.J., Upton, C., and Tipples, G.A. (2007). Genomic cartography of varicella-zoster virus: a complete genome-based analysis of strain variability with implications for attenuation and phenotypic differences. Virology *359*, 447-458.

Vaz, P.K., Job, N., Horsington, J., Ficorilli, N., Studdert, M.J., Hartley, C.A., Gilkerson, J.R., Browning, G.F., and Devlin, J.M. (2016a). Low genetic diversity among historical and contemporary clinical isolates of felid herpesvirus 1. BMC Genomics *17*.

Vaz, P.K., Horsington, J., Hartley, C.A., Browning, G.F., Ficorilli, N.P., Studdert, M.J., Gilkerson, J.R., and Devlin, J.M. (2016b). Evidence of widespread natural recombination among field isolates of equine herpesvirus 4 but not among field isolates of equine herpesvirus 1. J. Gen. Virol. *97*, 747-755.

Wan, Y., Renner, D.W., Albert, I., and Szpara, M.L. (2015). VirAmp: a galaxy-based viral genome assembly pipeline. GigaScience *4*.

Watson, S.J., Welkers, M.R.A., Depledge, D.P., Coulter, E., Breuer, J.M., de Jong, M.D., and Kellam, P. (2013). Viral population analysis and minority-variant detection using short read next-generation sequencing. Philos. Trans. R. Soc. B Biol. Sci. *368*, 1-9.

Wertheim, J.O., Smith, M.D., Smith, D.M., Scheffler, K., and Pond, S.L.K. (2014). Evolutionary Origins of Human Herpes Simplex Viruses 1 and 2. Mol. Biol. Evol. 1-26.

Whisnant, A.W., Jürges, C.S., Hennig, T., Wyler, E., Prusty, B., Rutkowski, A.J., L'hernault, A., Djakovic, L., Göbel, M., Döring, K., et al. (2020). Integrative functional genomics decodes herpes simplex virus 1. Nat. Commun. *11*.

Wilkinson, G.W.G., Davison, A.J., Tomasec, P., Fielding, C.A., Aicheler, R., Murrell, I., Seirafian, S., Wang, E.C.Y., Weekes, M., Lehner, P.J., et al. (2015). Human cytomegalovirus: taking the strain. Med. Microbiol. Immunol. (Berl.) *204*, 273-284.

Wyler, E., Menegatti, J., Franke, V., Kocks, C., Boltengagen, A., Hennig, T., Theil, K., Rutkowski, A., Ferrai, C., Baer, L., et al. (2017). Widespread activation of antisense transcription of the host genome during herpes simplex virus 1 infection. Genome Biol. *18*.

Ye, C., Guo, J.-C., Gao, J.-C., Wang, T.-Y., Zhao, K., Chang, X.-B., Wang, Q., Peng, J.-M., Tian, Z.-J., Cai, X.-H., et al. (2016). Genomic analyses reveal that partial sequence of an earlier pseudorabies virus in China is originated from a Bartha-vaccine-like strain. Virology *491*, 56-63.

Zell, R., Taudien, S., Pfaff, F., Wutzler, P., Platzer, M., and Sauerbrei, A. (2012). Sequencing of 21 Varicella-Zoster Virus Genomes Reveals Two Novel Genotypes and Evidence of Recombination. J. Virol. *86*, 1608-1622.

Chapter 2

Entry of Alphaherpesviruses

Tina M. Cairns[1] and Sarah A. Connolly[2]*

[1]Department of Microbiology, School of Dental Medicine, University of Pennsylvania, Philadelphia, Pennsylvania, USA
[2]Departments of Health Sciences and Biological Sciences, College of Science and Health, DePaul University, Chicago, Illinois, USA

*sarah.connolly@depaul.edu

DOI: https://doi.org/10.21775/9781913652555.02

Abstract

Alphaherpesviruses are enveloped viruses that enter cells by fusing the viral membrane with a host cell membrane, either within an endocytic vesicle or at the plasma membrane. This entry event is mediated by a set of essential entry glycoproteins, including glycoprotein D (gD), gHgL, and gB. gHgL and gB are conserved among herpesviruses, but gD is unique to the alphaherpesviruses and is not encoded by all alphaherpesviruses. gD is a receptor-binding protein, the heterodimer gHgL serves as a fusion regulator, and gB is a class III viral fusion protein. Sequential interactions among these glycoproteins are thought to trigger the virus to fuse at the right place and time. Structural studies of these glycoproteins from multiple alphaherpesviruses has enabled the design and interpretation of functional studies. The structures of gD in a receptor-bound and in an unliganded form reveal a conformational change in the C terminus of the gD ectodomain upon receptor binding that may serve as a signal for fusion. By mapping neutralizing antibodies to the gHgL structures and constructing interspecies chimeric forms of gHgL, interaction sites for both gD and gB on gHgL have been proposed. A comparison of the postfusion structure of gB and an alternative conformation of gB visualized using cryo-electron tomography suggests that gB undergoes substantial refolding to execute membrane fusion. Although these structures have provided excellent insights into the entry mechanism, many questions remain about how these viruses coordinate the interactions and conformational changes required for entry.

Introduction

As the first step of infection, alphaherpesvirus entry requires the coordinated action of several glycoproteins on the viral surface. Unlike many other viruses, herpesviruses encode receptor-binding and membrane fusion functions on separate proteins. These proteins must cooperate to mediate fusion of the viral and cellular membranes at the right time and place, either at the plasma membrane or within an endosome.

Entry proceeds through three basic steps: initial attachment, entry receptor binding, and membrane fusion (Figure 1A). Initial attachment to a cell is mediated typically by glycoprotein C (gC) and/or gB binding to cell surface proteoglycans (Herold et al., 1994; Herold et al., 1991; Laquerre et al., 1998; Rux et al., 2002). This attachment is reversible and does not trigger membrane fusion or virus entry. Although the attachment enhances infection, it is not required for entry because cells lacking heparan sulfate can be infected, albeit with reduced efficiency (Gruenheid et al., 1993).

After attachment, membrane fusion and virus entry require a set of core entry glycoproteins (Figure 1B). All herpesviruses encode gH, gL, and gB and most alphaherpesviruses also encode gD, with the notable exception of varicella zoster virus (VZV). These four glycoproteins are essential for infection and they are sufficient to mediate cell-cell fusion of receptor-bearing cells expressing these glycoproteins (Browne et al., 2001; Cole and Grose, 2003; Muggeridge, 2000; Pertel et al., 2001; Turner et al., 1998). In the current model of alphaherpesviral entry into cells (Figure 1B), gD binding to an entry receptor triggers a conformational change that transmits a signal to the gHgL

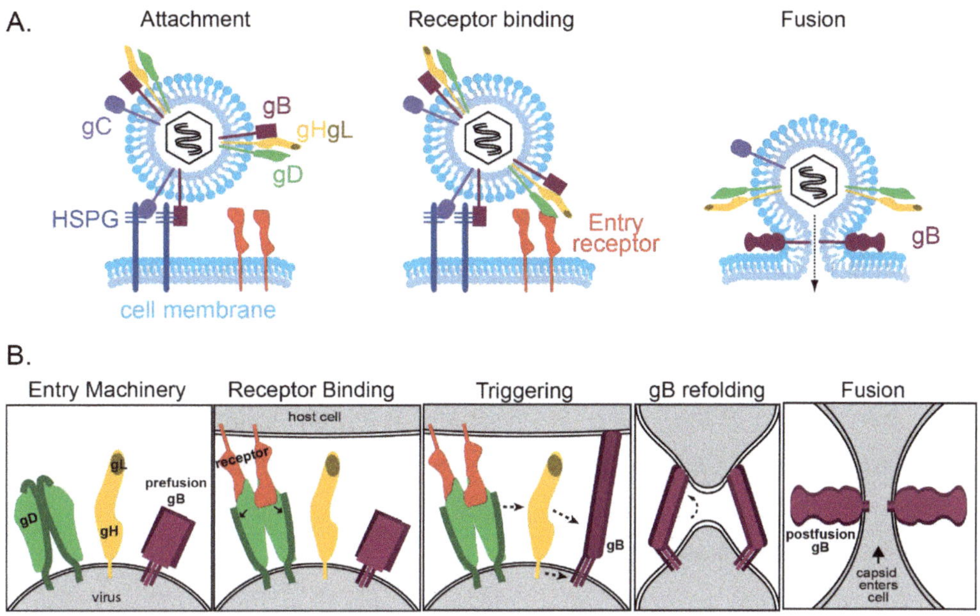

Figure 1. Mechanism of alphaherpesvirus entry. (A) Using HSV as a model for virus entry, heparan sulfate proteoglycans (HSPG), entry receptors, and viral glycoproteins gB, gD, gHgL, and gB are shown. Virus attaches to cells by binding to HSPG via gC and/or gB. Although this attachment step enhances entry, it is not required for fusion. Attachment is followed by the engagement of entry receptors. With the exception of VZV, this receptor-binding is mediated by gD. Functional gD receptors include members of the nectin family, HVEM, and 3-O-sulfonated heparan sulfate. In the current model of entry, gD-receptor binding signals gHgL to trigger gB to mediate fusion. gB inserts into the host cell membrane and refolds to fuse the viral and cellular membranes, allowing the viral capsid and genome to enter the host cell. (B) The core entry glycoproteins that are necessary and sufficient for fusion include the receptor-binding protein gD (green), the gHgL heterodimer (yellow), and the trimeric fusion protein gB (purple). Although VZV does not encode a gD homolog, it does encode gHgL and gB. Prior to receptor binding, the C-terminus of the gD ectodomain occludes the receptor-binding site. gD binding to either the HVEM or nectin-1 receptor (red) displaces the gD C-terminus and transmits a signal to activate gHgL. This signal may come directly from the gD C-terminal region or as a result the C-terminal repositioning to expose another interaction site on gD. gHgL then transmits this signal to trigger the fusion protein gB. The interaction with gB may be mediated by the gHgL ectodomain and/or the gH CT. Upon triggering, gB refolds from a prefusion conformation to insert its fusion loops into the host cell membrane. The orientation of the fusion loops with respect to the viral membrane in the prefusion form of gB is unresolved. gB then collapses back on itself, bringing the fusion loops embedded the cell membrane into proximity with the gB TM, thereby fusing the viral and cell membranes. Most likely, more than one gB trimer must be triggered to create a fusion pore through which the viral capsid can enter the cell, hence two trimers are shown in the later panels.

heterodimer. gHgL serves as a regulator of fusion, interacting with both the gD-receptor complex and the viral fusion protein, gB. Through an undefined mechanism, gHgL triggers the metastable gB to undergo an extensive refolding event, inserting its hydrophobic fusion loops into the host cell membrane and then folding back on itself to bring the viral and cellular membranes together. Fusion of these membranes creates a lipid pore that permits the viral genome to enter the cell.

Structures of each of the core entry glycoproteins from multiple alphaherpesviruses have been determined. These structures have advanced our understanding of the viral entry mechanism greatly. Using these structures as a framework, this chapter will review our current knowledge of alphaherpesvirus entry, relating functional studies to the structures and identifying current areas of interest.

Glycoprotein D: Receptor binding

For most alphaherpesviruses, gD is the receptor-binding protein and one of the core viral glycoproteins necessary to carry out membrane fusion and cell-cell spread. Two exceptions are VZV, which lacks gD and may instead rely on other glycoproteins for cell surface attachment (see below) (Duus et al., 1995; Maresova et al., 2001; Oliver et al., 2016), and pseudorabies virus (PRV), for which gD is essential for entry but not cell-cell spread (Ch'ng et al., 2007; Klupp and Mettenleiter, 1999; Klupp et al., 2000). Interestingly, with the right selection pressure, replication-competent gD deletion mutants can be generated *in vitro* for PRV and bovine herpes virus type 1 (BHV-1) (Schmidt et al., 2001; Schroder and Keil, 1999). These mutant lab strains possess compensatory mutations in gB and gH that circumvent the need for gD; however, in wild-type alphaherpesviruses other than VZV, the binding of gD to specific cellular receptors is required for viral entry. Viral infection can be inhibited by blocking virus-receptor interaction with soluble versions of gD or the gD receptor ectodomains (Johnson et al., 1990; Montgomery et al., 1996; Nicola et al., 1997; Warner et al., 1998; Whitbeck et al., 1997). In addition, antibodies generated against gD or the gD receptors are potent virus neutralizers and block membrane fusion (Atanasiu et al., 2018; Cairns et al., 2017; Cocchi et al., 1998; Fuller and Spear, 1987; Highlander et al., 1987; Krummenacher et al., 2000; Lazear et al., 2012; Montgomery et al., 1996; Saw et al., 2015; Whitbeck et al., 2001). Many neutralizing monoclonal antibodies (MAbs) against gD have been generated that block the binding of one or both receptors (Krummenacher et al., 1998; Lee et al., 2013; Nicola et al., 1998; Whitbeck et al., 1999). Additional neutralizing MAbs that do not block receptor binding have been hypothesized to interfere with the ability of gD to activate gHgL (Atanasiu et al., 2016; Atanasiu et al., 2018; Gallagher et al., 2013; Lazear et al., 2012).

gD receptors

Three classes of gD entry receptors have been identified, and gD has been co-crystallized with two of these receptors (Figure 2A, B). The first receptor class consists of the nectin and nectin-like proteins, which are immunoglobulin (Ig) superfamily members, each having one variable-like (V-like) and two constant-like (C-like) Ig domains (Nakanishi and Takai, 2004). Nectin proteins function as homophilic adhesion molecules and are widely distributed on the surface of most human tissues, including epithelial and neuronal cells, recognized targets for alphaherpesviral infection (Cocchi et al., 1998; Galen et al., 2006; Mata et al., 2001; Mizoguchi et al., 2002; Richart et al., 2003; Takahashi et al., 1999; Warner et al., 1998). The use of nectin-1 as a cell surface receptor is common among alphaherpesviruses. HSV-1, HSV-2, PRV, BHV-1, and herpes B virus all can use nectin-1 for entry (Fan and Longnecker, 2012; Milne et al., 2001; Ono et al., 2004; Warner et al., 1998). HSV and PRV gD bind to nectin-1 with high affinity while the relatively low affinity of BHV-1 for nectin-1 suggests either that low affinity is sufficient for entry or that BHV uses another receptor in its natural host (Connolly et al., 2001; Milne et al., 2001). Nectin-2 has a more limited scope as an alphaherpesvirus receptor and can mediate the entry of PRV, HSV-2, and certain HSV-1 mutants (Lopez et al., 2000; Spear et al., 2006; Warner et al., 1998). Poliovirus receptor (nectin-like molecule 5) permits entry of PRV, BHV-1, and herpes B virus, but not HSV (Fan and Longnecker, 2012; Warner et al., 1998). Under selective pressure, HSV can acquire gain-of-

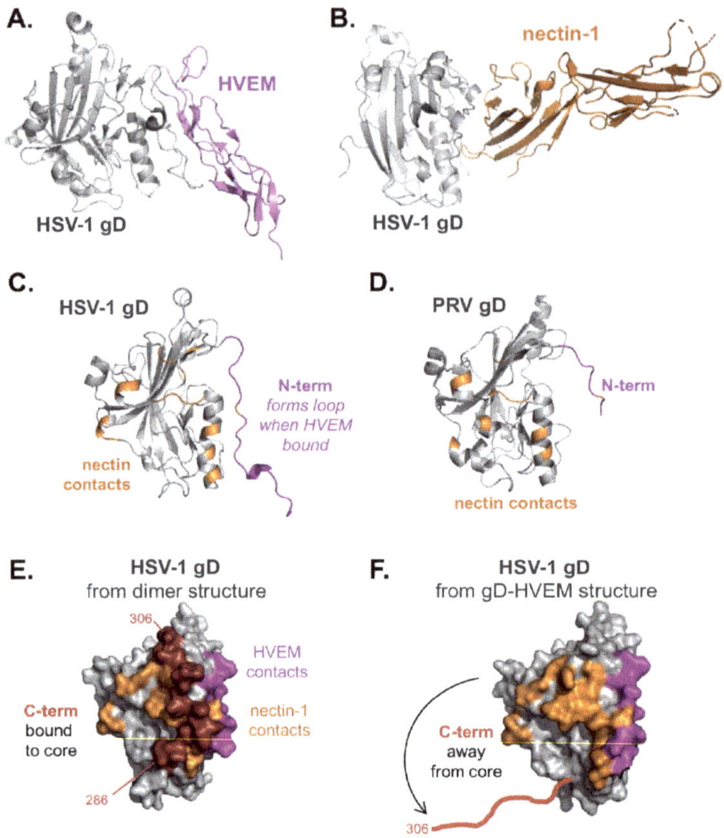

Figure 2. gD structures. (A) The crystal structure of gD (gray) bound to HVEM receptor (purple) is shown (PDB 1JMA) (Carfi et al., 2001). All of the HVEM contacts are contained within the N-terminal loop of gD. (B) The crystal structure of gD bound to nectin-1 receptor (orange) is shown (PDB 3SKU) (Di Giovine et al., 2011). HVEM and nectin-1 bind to the same face of gD. Nectin-1 binding prevents formation of the gD N-terminal loop, thus providing a structural basis for why HVEM and nectin-1 cannot bind to gD simultaneously. (C) The crystal structure of gD alone, not bound to receptor (PDB 1L2G) (Carfi et al., 2001). The nectin-1 contact residues (orange) and HVEM binding site (purple) are shown. The C-terminal region is not present in this structure. (D) The crystal structure of PRV gD alone (PDB 5X5V) (Li et al., 2017b), with nectin-1 contacts colored as in part C. The N-terminus of PRV gD is shorter than that of HSV-1 gD, providing a structural explanation for the failure of HVEM to serve as a PRV entry receptor. (E). Surface rendering of HSV-1 gD alone from a mutant with a stabilized C-terminus (PDB 2C36) (Krummenacher et al., 2005). The C-terminus of the gD ectodomain was stabilized by the addition of a cysteine at residue 307 that resulted in a disulfide-linked gD dimer. The C-terminus of the gD ectodomain (residues 285-306, red) is shown, as well as the contact residues for HVEM (purple) and nectin-1 (orange). The position of the C-terminus occludes both the HVEM and nectin-1 binding sites, providing a structural basis for the increase in affinity for both HVEM and nectin-1 when this portion of the C-terminus is deleted from soluble forms of the gD ectodomain. (F). Surface rendering of HSV-1 gD from the gD-HVEM co-crystal (PDB 1JMA) (Carfi et al., 2001). The contact residues for HVEM (purple) and nectin-1 (orange) are shown. Binding of either receptor would require the C-terminus to move, as depicted by the red tail extending from gD. Movement of the C-terminus may serve as a signal for triggering the downstream fusion events.

function mutations (not only in gD, but also in gB) that allow it to use other nectins for entry (nectins 2 through 4) (Cocchi et al., 2004b; Uchida et al., 2010).

The second class of gD receptors consists of the herpesvirus entry mediator (HVEM), a member of the tumor necrosis factor receptor superfamily, containing four characteristic cysteine-rich domains (CRD) (Carfi et al., 2001; Locksley et al., 2001). HVEM is used as a receptor by HSV-1 and HSV-2, but not by the other alphaherpesviruses including PRV, BHV-1, and herpes B virus. HVEM is expressed primarily on lymphocytes, a cell type that is not a primary target for HSV infection (Montgomery et al., 1996; Raftery et al., 1999; Whitbeck et al., 1997). The affinity of HSV gD for HVEM is the same as it is for nectin-1 (Krummenacher et al., 1998; Willis et al., 1998). Interestingly, despite the fact that a single mutation in the HVEM binding site on gD can abrogate HVEM usage without preventing nectin-1 usage (Yoon et al., 2003), all of the 49 clinical isolates tested were able to use both HVEM and nectin-1 as a receptor (Krummenacher et al., 2004). This finding suggests a selective pressure to retain HVEM usage.

The third class of gD receptor, 3-O-sulfonated derivatives of heparan sulfate (3-OST HS), is generated by D-glucosaminyl-3-O-sulfotransferase-3 modification of heparan sulfate and can mediate the entry of HSV-1 (Shukla et al., 1999; Tiwari et al., 2007). 3-OST HS is present in neurons of the mouse trigeminal ganglion, but its contribution to HSV infection is yet unclear (Lawrence et al., 2007).

The relative importance of the receptors to HSV infection has been studied using mice knocked-out for one or two of the receptors. Intercranial infection of knock-out mice suggests that nectin-1 is required for infection and progression to encephalitis (Kopp et al., 2009). Intercranial infection in the absence of HVEM progressed as in wild-type mice, whereas mice lacking nectin-1 did not show disease. Similarly, when knock-out mice were infected intravaginally in the absence of nectin-1, the virus did not spread to the sensory ganglia, but infection in the absence of HVEM was normal (Taylor et al., 2007). Mice knocked-out for both HVEM and nectin-1 were resistant to infection of the vaginal epithelium, indicating that either receptor alone is sufficient for infection of epithelial cells. The role of HVEM in ocular infection is more complex. Corneal infection of the knockout mice revealed that the loss of either nectin-1 or HVEM attenuated HSV-1 infection of the cornea (Karaba et al., 2011), although infection with HSV-2 was not dependent on HVEM (Karaba et al., 2012). Further studies suggest that the contribution of HVEM to ocular disease is due to its immune modulatory role, rather than a direct role in virus entry (Edwards et al., 2015; Edwards and Longnecker, 2017).

gD and receptor structures

The crystal structures of gD from HSV-1 (Carfi et al., 2001; Krummenacher et al., 2005), HSV-2 (Lee et al., 2013), and PRV (Li et al., 2017a) reveal a highly conserved domain organization (Figure 2C, D). The core of gD consists of a V-like Ig fold flanked by long N- and C-terminal extensions. Structures of gD co-crystallized with the receptors nectin-1 and HVEM have also been resolved, and gD binds to the membrane-distal N-terminal regions of both receptors (Figure 2A, B) (Carfi et al., 2001; Di Giovine et al., 2011; Li et al., 2017a; Lu et al., 2014). The structure of HVEM is similar to that of other members of the tumor necrosis factor receptor family and is composed of four CRDs of approximately 40 residues each (Carfi et al., 2001). Although amino acids in both CRD1 and CRD2 of HVEM contact HSV gD, residue Y23 in CRD1 represents a "hot-spot" for gD binding while CRD2 is required mainly due to its effect on the presentation of the binding site on CRD1 (Carfi et al., 2001; Connolly et al., 2002; Whitbeck et al., 2001).

Nectin-1 and HVEM bind to HSV gD at distinct sites on the same face of gD (Carfi et al., 2001; Di Giovine et al., 2011; Krummenacher et al., 1998; Lu et al., 2014; Whitbeck et al., 1999). When HVEM is bound to gD, the gD N-terminus forms a hairpin loop and this loop comprises the entire HVEM-binding site (Figure 2A). Antibodies against a peptide corresponding to this N-terminal gD loop neutralize viral infectivity (Cohen et al., 1984). The N-terminal sequences of PRV and BHV-1 gD diverge from that of HSV-1 and HSV-2 gD, most likely because HSV gD uses this stretch for HVEM binding, whereas the others do not. In the absence of HVEM, the N-terminal gD residues do not form a hairpin and are disordered (Figure 2C) (Carfi et al., 2001). Therefore, the formation of the hairpin loop represents a conformational change in the gD N-terminus that must occur for HVEM to bind gD (Lazear et al., 2014). PRV gD contains a shortened N-terminal extension, as compared to HSV gD, that is incapable of forming this hairpin (Figure 2D), coinciding with the inability of PRV gD to bind HVEM (Connolly et al., 2001). HVEM and nectin-1 cannot bind to gD simultaneously because the N-terminal hairpin loop formed via HVEM binding partially occludes the nectin-1 binding site (Figure 2A. B) (Di Giovine et al., 2011; Krummenacher et al., 2005). Comparison of the nectin-1 binding sites on HSV and PRV gD reveal a similar footprint on the core domain of gD (Figure 2C, D). The conservation of the nectin-1 binding site between HSV and PRV suggests that this binding site is conserved between most gD homologues.

3-OST HS has not been co-crystallized with gD, however two sulfate ions found in the gD structure may represent the site of 3-OST HS binding, including one ion in a basic cavity on gD that is close to the gD/HVEM interface (Carfi et al., 2001). Further evidence that the 3-OST HS and HVEM binding sites on gD may overlap is that mutations in the N-terminus of gD impact the usage of HVEM and 3-OST HS similarly (Yoon et al., 2003).

The C-terminal region of the gD ectodomain plays an important role in receptor binding and gD function. The structure of a dimeric form of HSV-1 gD, created by the addition of a disulfide bond at the end of the ectodomain, revealed that the C-terminus of the gD ectodomain lies against the gD core, overlapping the nectin-1 binding site and occupying the same space as the gD N-terminal loop residues that form the HVEM binding site (Figure 2E) (Di Giovine et al., 2011; Krummenacher et al., 2005; Lu et al., 2014). The binding of either HVEM or nectin-1 would require the displacement of this C-terminus away from the gD core (Figure 2F). To validate the position of the C-terminus in this structure, an additional gD mutant with a disulfide bond engineered to lock the C-terminus against the gD core was crystallized (Krummenacher et al., 2005). As would be predicted, although this mutant was antigenically intact, it was unable to bind either receptor. Conversely, mutations in HSV and PRV gD that either delete the C-terminal tail or prevent its association with the gD core enhance binding to both nectin-1 and HVEM by 10-50 fold, due to an increase in the rate of complex formation (k_{on}) (Krummenacher et al., 1999; Krummenacher et al., 2005; Rux et al., 1998; Willis et al., 1998) (Li et al., 2017a). Without the gD C-terminal tail bound to the gD core, the nectin-1 binding site is available for receptor binding and the N-terminal hairpin loop can more readily form to permit HVEM binding. So, although the C-terminal tail of gD does not contain any receptor-contacting residues, this region influences gD-receptor binding.

Transmitting a signal for fusion

gD binding to receptor plays a greater role in virus entry than simply tethering the virus to the cell. gD homologs are not interchangeable, even when they bind the same nectin-1 receptor (Bohm et al., 2016; Fan et al., 2014). The C-terminus of gD appears to play a dual role in fusion by both inhibiting fusion when covering the receptor binding site and promoting fusion after receptor binding (Figure 1B). By creating a panel of gD mutants with engineered disulfide bonds that lock the C-terminus to the gD core at different points, a mutant was created that retained the ability to bind the receptors but failed to promote fusion (Lazear et al., 2008). This result suggests that the displacement of the C-terminus does more than just permit receptor binding. Moreover, an engineered gD mutant designed to force the displacement the C-terminal tail is able to mediate fusion at a low level in the absence of gD receptors (Gallagher et al., 2013), suggesting that displacement of this C-terminal tail is sufficient to trigger the downstream events required for fusion.

A region within the C-terminal tail (residues 261-285) of HSV gD has been termed the "profusion domain" due to its requirement for virus entry (Cocchi et al., 2004a; Zago et al., 2004). Since removal of the profusion domain inhibits virus entry without preventing gD-receptor binding, this region was proposed to serve as a binding site for gHgL and/or gB (Cocchi et al., 2004a; Gianni et al.,

2009). Indeed, antibodies that bind at the profusion domain inhibit cell-cell fusion and virus spread (Du et al., 2017; Lazear et al., 2012; Saw et al., 2015). However, the epitopes of several other anti-gD antibodies that neutralize virus without blocking gD/receptor binding map to regions outside the profusion domain (residues 54 and 67) (Atanasiu et al., 2018; Lazear et al., 2012). Both these neutralizing antibodies and those that bind to the profusion domain were shown to block the interaction between soluble gD and gHgL using surface plasmon resonance (SPR) (Cairns et al., 2019). The epitopes of these antibodies therefore outline a potential gHgL binding region on gD that spans a large region on the face adjacent to, but distinct from, the receptor binding site. In fact, both receptor (either HVEM or nectin-1) and gHgL were able to bind gD simultaneously (Cairns et al., 2019). It is currently unknown if, in the context of the virus, gD must disengage from receptor before its interaction with gHgL in order for the fusion cascade to proceed.

Vaccine potential

gD is a highly immunogenic protein and a prime target for vaccine development (Awasthi et al., 2014; Awasthi et al., 2015; Hook et al., 2018). Humans naturally infected with HSV-1, HSV-2, or both serotypes generate strong antibody responses against gD, that can be type-common or type-specific (Cairns et al., 2015; Cairns et al., 2014). In these individuals, the dominant immune response that results in virus-neutralizing antibodies is due to gD or a combination of gD and gB. Anti-gD antibodies generated from natural infection correspond to three distinct gD epitopes, two involved in receptor binding and one involved in gHgL binding, and correlate with total antibody neutralization activity (Cairns et al., 2014).

Retargeting HSV for oncolytic therapy

Defining the receptor-binding sites on gD has stimulated attempts to detarget HSV from its natural receptors and retarget the virus to novel receptors. By inserting the appropriate natural ligands into the N-terminus of gD, HSV has been retargeted to use the interleukin-13 receptor or urokinase plasminogen activator receptor for entry, instead of nectin-1 or HVEM (Zhou and Roizman,

2006, 2007; Zhou et al., 2002). HSV also has been retargeted to use human epidermal growth factor receptor 2 (HER2) and epidermal growth factor receptor (EGFR) for entry by mutating the native HVEM or nectin-1 binding sites in gD and inserting receptor-specific single-chain antibodies into gD (Menotti et al., 2008; Tuzmen et al., 2020; Uchida et al., 2010). Supplementing these gD mutations with additional manipulations in other glycoproteins that impact entry, including gHgL, gB, and gK, may further strengthen the oncolytic potential for HSV (Campadelli-Fiume et al., 2016; Okubo et al., 2016; Petrovic et al., 2018). How the insertion of novel ligands into gD successfully recapitulates the triggering that normally occurs when wild-type gD binds a natural receptor is unclear.

Glycoprotein H and glycoprotein L: Fusion regulators

gH and gL form a heterodimer that is essential for both virus-cell and cell-cell fusion (Babic et al., 1996; Desai et al., 1988). This heterodimer is a target of antibodies that neutralize virus or inhibit cell-cell spread (Birlea et al., 2013; Buckmaster et al., 1984; Cairns et al., 2006; Gompels et al., 1991; Montalvo and Grose, 1986; Peng et al., 1998b; Rodriguez et al., 1993; Showalter et al., 1981; Xing et al., 2015). Although its requirement for herpesvirus entry has long been known, its precise role remains a mystery.

gH is a type I transmembrane protein, whereas gL is not membrane-anchored and, in alphaherpesviruses, associates non-covalently with the gH ectodomain. On mature virions and on the surface of HSV-infected cells, gH and gL are found together in a stable 1:1 complex (Peng et al., 1998a). gL expressed by itself is released into the media as a soluble protein (Dubin and Jiang, 1995; Hutchinson et al., 1992; Peng et al., 1998a). In contrast, for most alphaherpesviruses, gH expressed by itself is not transported to the cell surface and remains unfolded and trapped in the ER (Foa-Tomasi et al., 1991; Hutchinson et al., 1992; Klyachkin et al., 2006; Roberts et al., 1991). gH is able to be transported to the cell surface and incorporated into the virion only when co-expressed with gL (Dubin and Jiang, 1995; Hutchinson et al., 1992; Peng et al., 1998a; Roop et al., 1993; Wu et al., 2001). The exception to this rule is

PRV gH, which can be transported and incorporated into virions without gL (Klupp et al., 1997).

Although gL shares little amino acid identity across herpesviruses, gH is more conserved in sequence, and both are required for virus-cell and cell-cell fusion (Duus and Grose, 1996; Duus et al., 1995). Initially, gL was thought to be merely a chaperone or a scaffold protein for gH. In fact, VZV gL contains a putative ER targeting signal at its N terminus instead of the typical signal sequence, a trait shared with ER chaperone proteins. Furthermore, when VZV gH was co-expressed with gE or gI, gH was transported out of the ER and to the cell surface in the absence of gL (Duus et al., 1995). However, this gH was not processed completely and was most likely non-functional. Likewise, a mutant form of HSV-2 gH (with the N-terminal 29 residues of the mature protein deleted) was transported to the cell surface in the absence of gL but was non-functional (Cairns et al., 2007). Co-expression with gL restored function to this gH mutant and these findings suggest that gL plays a role beyond the transport of gH. In support of this concept, several HSV antibodies specific to gL have been isolated that inhibit cell-cell fusion and virus spread (Cairns et al., 2006; Novotny et al., 1996).

gHgL is not a viral fusion protein

Early on, gHgL was speculated to be a viral fusogen, i.e., an active participant in the mixing of virus and cellular membranes. Several labs identified potential fusion peptides within HSV-1 gH based on sequence analysis (Galdiero et al., 2005; Galdiero et al., 2007; Galdiero et al., 2006; Gianni et al., 2005a; Gianni et al., 2005b; Lopper and Compton, 2004). HSV-1 virions inactivated by neutralizing anti-gH antibodies or through the deletion of gH were reported to attach to cells and form a fusion bridge (Fuller and Lee, 1992). Although these viruses were unable to expand the bridge to allow entry, partial infectivity could be restored to gH-null virions if a fusogenic agent was added to the cell medium (Babic et al., 1996; Forrester et al., 1992; Roop et al., 1993). In VZV, gHgL alone, produced from transfected cells, mediated low levels of cell-cell fusion (Duus et al., 1995).

The proposed viral fusogen role for gHgL was refuted when the first structure of gHgL was determined (Chowdary et al., 2010). This structure of a soluble form of the HSV-2 gH ectodomain in complex with gL did not resemble any known fusogen. Although many helices within the core of gH formed helical bundles, the trimeric hairpin bundle characteristic of fusion proteins was absent from the gHgL structure. In fact, all of the previously hypothesized fusion peptides and heptad repeats were buried in the core of the protein and were unlikely to promote any significant conformational changes (Chowdary et al., 2010; Xing et al., 2015).

gHgL structures

The structure of HSV-2 gHgL revealed an extensive interaction between the two proteins that explained their interdependence for folding, transport, and function. gHgL has the overall shape of a "boot" (Figure 3A), which fits with earlier cryo-electron tomography studies of HSV-1 virions (Grunewald et al., 2003). gH itself is split into three domains: H1, which encompasses the N-terminus and the gL-binding domain; H2, the helical central domain; and H3, the C-terminal β-sandwich "toe" of the boot and the domain closest to the transmembrane anchor (TM). The sequences of gH domain H1 and of gL vary substantially among herpesviruses, probably due to their co-evolution together as a tight-knit binding pair (Chowdary et al., 2010). The N-terminal domain of gH lacks a folded core and requires gL for proper folding and stabilization.

The crystal structure of a soluble form of VZV gHgL was also solved and bears a striking resemblance to that of HSV-2 (Figure 3B) (Xing et al., 2015). In addition, the structure of a core fragment of PRV gH (lacking gL and H1) was solved in complex with a FAb (Figure 3C) (Backovic et al., 2010; Vallbracht et al., 2019). Both the VZV and PRV structures adopt very similar folds to that of HSV-2, especially in domains H2 and H3, which are the most highly conserved in their amino acid sequence. However, whereas HSV-2 and VZV gHgL are boot-shaped, the PRV gH C-terminal fragment adopts a more cylindrical shape, in which the H2 and H3 domains are aligned in the same plane. The

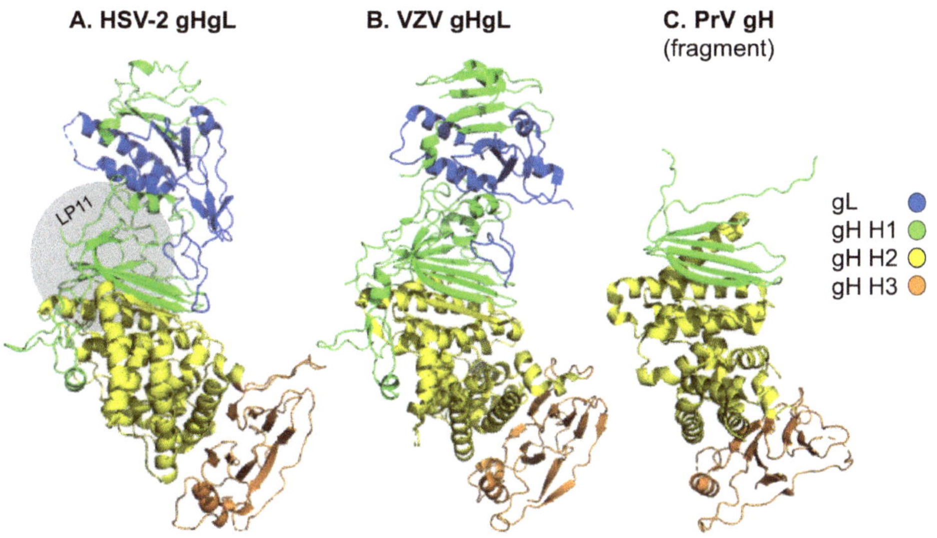

Figure 3. gHgL structures. (A) The crystal structure of the HSV-2 gHgL heterodimer (PDB 3M1C) (Chowdary et al., 2010). The structure is colored to define gL (blue) and three domains of gH, including the gL-interacting domain H1 (green), the central helical domain H2 (yellow) and the C-terminal β-sandwich domain H3 (orange). The gH domains are designated as originally defined in the HSV-2 gHgL structure. The tight association of gL with the N-terminal domain of gH provide a structural explanation for the requirement of gL for proper gH folding. The epitope for the neutralizing MAb LP11 (shaded gray) maps to domains H1 and H2, on the back of the heterodimer, and highlights a possible gB binding site. (B) The crystal structure of the VZV gHgL heterodimer (PDB 4XHJ) (Xing et al., 2015). The domains are colored as in part A. Both the HSV-2 and VZV structures adopt a boot shape. (C) The crystal structure of a fragment of PRV gH (PDB 2XQY) (Backovic et al., 2010). The domains are colored as in part A. The N-terminus of gH was absent from the protein crystallized and the core fragment of gH was co-crystallized with a Fab.

cylindrical shape of PRV gH more closely resembles that of the gammaherpesvirus EBV gHgL heterodimer (Matsuura et al., 2010). This difference in shape among alphaherpesvirus gHgL (boot vs. cylinder) may be due to the PRV structure lacking gL or being bound to a FAb or may it indicate that flexibility exists between the domains.

The PRV gH structure highlighted the existence of an extended "flap" in domain H3, near where the ectodomain would transition to the TM of the protein and encounter the viral membrane (Backovic et al., 2010). This flap covers a conserved hydrophobic patch and the movement of this flap through isomerization of conserved disulfide bonds was hypothesized to expose the hydrophobic patch and allow it to interact with the lipid membrane (Backovic et al., 2010). In contrast, mutational analysis of this region in both PRV and VZV suggests that this "flap" needs to be rigid, not flexible (Fuchs et al., 2012; Vleck et al., 2011). Nevertheless, mutations in gH that abrogate or impair fusion typically map to domain H3 (Cairns et al., 2005; Galdiero et al., 1997; Jackson et al., 2010), highlighting its importance in membrane fusion and entry. This domain is also the most highly conserved, which implies a functional importance.

The gH TM and short cytoplasmic tail (CT) also contribute to gHgL function. Although a soluble form of the HSV gHgL ectodomain is able to trigger fusion of cells expressing gB, gD, and a gD receptor, the efficiency of fusion is lower, suggesting a role for the gH TM and/or CT in this process (Atanasiu et al., 2010a). Indeed, replacement or deletion of certain residues within the gH CT of HSV inhibit fusion (Browne et al., 1996; Jackson et al., 2010; Rogalin and Heldwein, 2015; Silverman and Heldwein, 2013; Wilson et al., 1994). gHgL anchored using a heterologous TM or a GPI-anchor is nonfunctional (Harman et al., 2002; Jones and Geraghty, 2004). In PRV, removal of the gH CT results in decreased membrane fusion activity, while deletion of the TM completely abolishes function (Vallbracht et al., 2018). Conversely, mutations in the VZV gH CT have been shown to enhance fusion (Yang et al., 2014).

gHgL regulates viral fusion

The crystal structure and the functional data indicate that gHgL is a fusion regulator. The core entry glycoproteins from HSV-1 and HSV-2 can be functionally swapped in the context of cell-cell fusion of receptor-bearing cells transfected with these four glycoproteins (Atanasiu et al., 2016; Muggeridge, 2000). These experiments demonstrate that the rate of cell-cell fusion is higher for HSV-2 than HSV-1 and that the gHgL serotype is a rate-limiting factor (Atanasiu et al., 2016). Similarly, several HSV-1 gH insertion mutants display a slow-fusing phenotype (Jackson et al., 2010). The current model proposes that, upon receiving a signal from gD after receptor binding, gHgL regulates the transition of the gB fusion protein from a prefusion to a fusion-activated state (Chowdary et al., 2010). HSV gHgL and gB interact only when gD and a gD receptor are also present (Atanasiu et al., 2007) (see below), suggesting that gB and gHgL are not normally associated and only come into contact when fusion is triggered. In addition, the CT of HSV gH has been implicated in regulating fusion through an interaction with the gB cytodomain (Cooper et al., 2018; Rogalin and Heldwein, 2015). All current models of how gHgL regulates the fusogenic activity of gB involve a direct interaction between the three proteins, but capturing this multi-protein complex has proven challenging.

gHgL as a receptor-binding protein

In beta- and gammaherpesviruses, gHgL, either alone or in complex with other viral proteins, binds to cellular receptors (Chen et al., 2018; Chen et al., 2019; Chesnokova and Hutt-Fletcher, 2011; Chesnokova et al., 2009; Hahn et al., 2012; Nishimura and Mori, 2019; Santoro et al., 2003; Wang and Shenk, 2005). Although most alphaherpesviruses use gD as a receptor binding protein, gHgL may bind its own cellular ligand(s) as well during entry. In VZV, which lacks gD, cell-cell fusion mediated by gB and gHgL is reduced when av integrin expression is knocked down with siRNAs or when these integrins are bound by antibodies, suggesting an interaction between the viral glycoproteins and integrins (Yang et al., 2016). Interestingly, the canonical RGD motif and the disintegrin-like domain, both shown to bind integrins, are absent from VZV gB and gHgL, suggesting that any potential interaction between integrins and the VZV core entry glycoproteins would occur through an as yet unidentified motif

Three different types of integrins have been shown to bind to HSV gH: αvβ3, αvβ6, and αvβ8 integrins (Gianni et al., 2013; Parry et al., 2005). αvβ3 integrin binds HSV gHgL with low affinity and is proposed to serve as a "routing factor" because its silencing altered the route of virus infection but did not inhibit infection (Gianni and Campadelli-Fiume, 2012; Parry et al., 2005). Both αvβ6 and αvβ8 integrins bind gHgL with high affinity and contribute to the attachment of virus to the cell surface although they are less important for viral attachment than heparin sulfate (Gianni et al., 2013). Furthermore, the integrins could not substitute for the gD receptor nectin-1. αvβ6 and αvβ8 integrins have been proposed to act as additional HSV receptors that influence the pathway of HSV entry. The integrins may act as additional "triggers" for gHgL, to determine when and where to signal gB to mediate fusion.

Glycoprotein B: The viral fusogen

Viral fusogens are proteins that execute the final membrane fusion step of viral entry by inserting into a target cell membrane and refolding to bring the viral and cell membranes together (White et al., 2008). Fusogens are initially present in a metastable prefusion conformation that, upon triggering, rearranges into an extended form that inserts hydrophobic residues into the target cell membrane. This extended conformation then refolds into a stable hairpin-like postfusion conformation to bring the membrane-inserted region into proximity with the membrane anchor of the protein, thereby facilitating fusion of the two membranes.

Crystal structures of the gB ectodomain

gB is conserved in all herpesviruses and required for viral entry (Cooper and Heldwein, 2015). Crystal structures of the ectodomains of several gB homologs have been determined, including those from the alphaherpesviruses HSV-1 (Cooper et al., 2018; Heldwein et al., 2006) and PRV (Li et al., 2017b; Vallbracht et al., 2017), as well as from the betaherpesvirus cytomegalovirus (CMV) (Burke and Heldwein, 2015; Chandramouli et al., 2015) and the gammaherpesvirus Epstein-Barr virus (Backovic et al., 2009). These structures provide the strongest evidence that gB is a viral fusogen. gB shares structural similarity with the fusogens from rhabdoviruses (Baquero et al., 2015; Belot et al., 2020; Roche et al., 2006; Yang et al., 2020), baculovirus (Kadlec et al., 2008), and Thogotovirus (Peng et al., 2017) (Figure 4). This structural similarity is remarkable given a lack of sequence similarity among these viral fusogens.

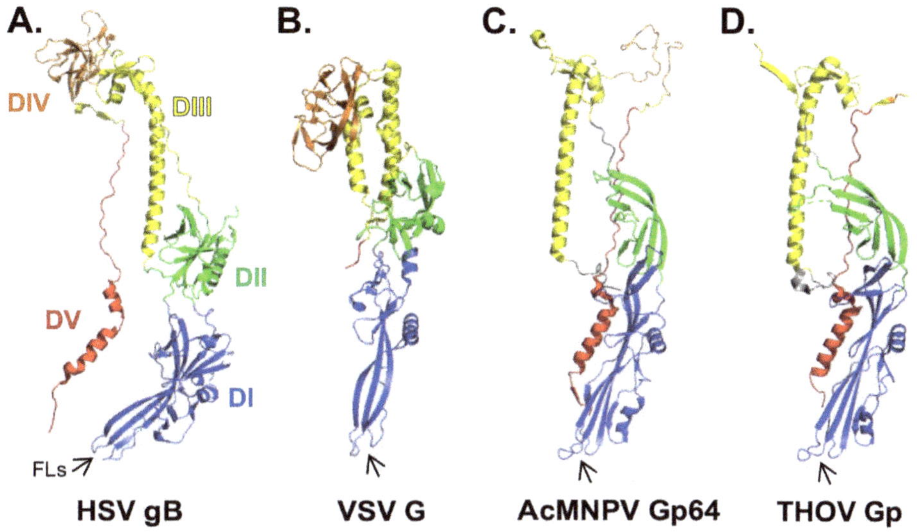

Figure 4. Class III fusogens. (A) The crystal structure of the HSV-1 gB monomeric ectodomain is shown (PDB 2GUM) with each structural domain (DI-DV) colored (Heldwein et al., 2006). The location of the fusion loops (FLs) is indicated with an arrow. The crystal structures of rhabdovirus VSV G (B) (PDB 512M) (Roche et al., 2006), baculovirus Autographa californica nucleopolyhedrovirus Gp64 (C) (PDB 3DUZ) (Kadlec et al., 2008), and Thogotovirus Gp (D) (PDB 5XEA) (Peng et al., 2017) monomeric ectodomains are shown in ribbon diagram, with domains colored as in part A. The VSV G structure is missing domain V, while domain IV was unresolved in Thogotovirus Gp. All four fusion proteins are positioned with their fusion loops pointing down (arrows).

Together, these proteins represent class III of fusogens (Backovic and Jardetzky, 2009). Structures of both the prefusion and postfusion forms of the rhabdovirus fusion protein G from vesicular stomatitis virus (VSV) and rabies virus have been determined, an accomplishment facilitated by the ability of G to undergo reversible, pH-dependent conformational changes (Roche et al., 2007; Yang et al., 2020). Structural similarity to G indicates that the gB crystal structures represent the postfusion conformation.

gB is a trimeric type I transmembrane protein, with a large ectodomain, a single transmembrane domain (TM), and a cytoplasmic tail domain (CT). The gB ectodomain is organized into five domains (Figure 5A) (Heldwein et al.,

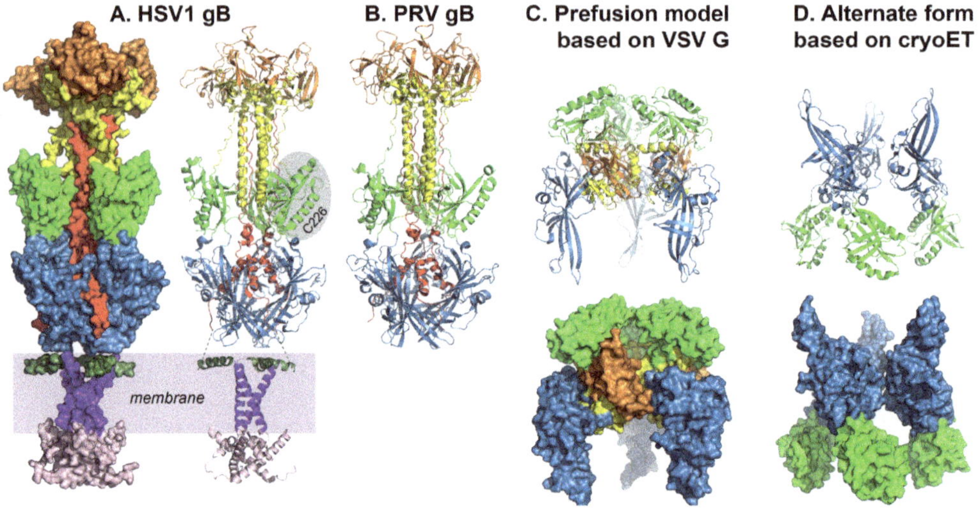

Figure 5. gB structures. (A) The crystal structure of full-length HSV-1 gB is shown, with surface rendering on the left and a ribbon diagram on the right (PDB 5V2S) (Cooper et al., 2018). This structure represents the postfusion conformation and the presumed position of the membrane is indicated by a purple box. Five domains of the ectodomain are colored, including domains I (blue), II (light green), III (yellow), IV (orange), and V (red). C-terminal to the ectodomain, the membrane-proximal region (MPR, dark green), a single-pass transmembrane domain (TM, purple) and cytoplasmic domain (CT, pink) are shown. The fusion loops are located at the tip of domain I, in contact with the membrane. Domain V extends the length of the molecule and packs against the central coiled-coil of domain III before descending into the MPR. The MPR lies parallel to the membrane and the single-pass TM forms an inverted teepee beneath the structure. The CT has trimeric contacts that may influence the activation of gB. The epitope for the neutralizing MAb C226 (shaded gray) maps to domain II (Atanasiu et al., 2010b) and highlights a possible binding site for gHgL. (B) The crystal structure of PRV gB ectodomain is shown in ribbon diagram, with domains colored as in part A (PDB 6ESC) (Vallbracht et al., 2017). The structural similarity between HSV-1 and PRV gB is apparent. (C) A working model of prefusion HSV-1 gB is shown, as a ribbon diagram at the top and surface rendering at the bottom (Gallagher et al., 2014). This model was generated by fitting the domains from the postfusion gB structure to the prefusion crystal structure of VSV G (PBD 2J6J) (Roche et al., 2007). The model presumes that the gB prefusion-to-postfusion conformational change is comprised primarily of tertiary structure rearrangements. The model predicts a break in the extended coil of domain III and domain V is absent because it is not present in the VSV G structure. (D) A model of an alternative conformation of gB, based on cryoET, is shown (Zeev-Ben-Mordehai et al., 2016). Fitting domains I and II from the postfusion gB structure into this form orients the fusion loops away from the membrane. Domains III-V from the postfusion structure were not placed into this form and the structure would not accommodate the intact extended coil of domain III.

2006). Domain I, the "fusion domain", is located at the base of the ectodomain, near the membrane. This domain contains a pleckstrin-homology domain along with two adjacent beta-hairpin extensions that present hydrophobic residues at their tips that can insert into membranes ("fusion loops") (Falanga et al., 2012). Mutational studies demonstrate that the hydrophobic residues in the fusion loops are essential for the gB fusogenic function (Backovic et al., 2007; Hannah et al., 2009; Oliver et al., 2009; Vallbracht et al., 2017). Structural domain II contains a second pleckstrin-homology domain.

The core of gB consists of a long central helix (domain III). In the gB trimer, three helices twist around one another to form an extended coiled-coil (Figure 5A). An extended polypeptide in domain V spans the length of the molecule and packs against the coiled-coil. Domain V from one protomer packs against domain III of two other protomers in an anti-parallel orientation. The packing of domain V against the central coil is analogous to a six-helix bundle, a structural feature that is present in the postfusion form of class I fusogens. Formation of this very stable, low-energy structure has been proposed to contribute energy for membrane fusion (Melikyan et al., 2000). Mutations that are predicted to disrupt the interaction between domains III and V reduce both the rate of viral entry and the extent of cell-cell fusion (Connolly and Longnecker, 2012; Fan et al., 2017). In addition, peptides that include sequences from domain III or V have been shown to inhibit viral entry (Akkarawongsa et al., 2009), supporting the notion that these regions are exposed at some point during conformational rearrangements and come together as gB folds into its final postfusion conformation.

The C terminus of domain V threads through domain I (Figure 5A). In the full-length protein, domain V would then lead into the membrane proximal region (MPR), TM and CT. The close juxtaposition of the C terminus of domain V and the fusion loops of domain I in the postfusion form of gB explains how folding to the postfusion form could bring the two membrane-anchored portions of the protein into proximity, similarly to the postfusion forms of fusogens from other classes (Harrison, 2015).

Transmembrane and cytoplasmic domains of gB

For gB crystallization, removal of the MPR, TM, and CT to create a soluble ectodomain presumably resulted in spontaneous folding into a postfusion form. The TM and/or CT appear to be critical for maintaining the prefusion form of gB, as has been observed for some class I fusogens (Yin et al., 2005; Yin et al., 2006). This postfusion gB structure is remarkably stable. Mutants designed to capture an alternate conformation of gB, by substituting the TM with a trimeric domain or by adding mutations known to prevent fusion, still adopt the postfusion form (Silverman et al., 2010; Vitu et al., 2013).

The crystal structure of the full-length gB, including the MPR, TM, and CT, was determined recently (Figure 5A, B) (Cooper et al., 2018). The ectodomain of this full-length molecule adopted a postfusion conformation, likely due to detergent solubilization. The MPR lies beneath the ectodomain, parallel to the membrane, the TM forms an inverted teepee, and the CT forms a trimeric pedestal. The gB CT is significantly longer than the CT of other class III fusogens, and mutational studies indicate that it plays a crucial role in fusion. Some mutations in the CT inhibit fusion (Cai et al., 1988; Wanas et al., 1999) whereas others can enhance fusion and/or impart a syncytial phenotype (Baghian et al., 1993; Diakidi-Kosta et al., 2003; Engel et al., 1993; Fan et al., 2002; Foster et al., 2001; Gage et al., 1993; Heineman and Hall, 2002; Muggeridge et al., 2004; Oliver et al., 2013; Silverman et al., 2012). Mapping known gB mutations to this gB CT structure suggests that this structure is present in the prefusion form of the protein. The structural and mutagenic data together suggest that the CT stabilizes the prefusion form of gB, interacting with the membrane and acting as a clamp to control the refolding of the protein. The CTs of other fusion proteins impart a similar regulation of fusion function (Waning et al., 2004; Wyss et al., 2005).

Alternate conformations of the gB ectodomain

For gB, only the structure of the postfusion form is known. Computational homology models of the prefusion gB have been generated using the prefusion structure of VSV G (Figure 5C) (Backovic et al., 2009; Gallagher et al., 2014). These models predict that the extended helix of the domain III core is partially unfolded in the prefusion form while domain V could not be modeled because it is absent from the VSV G structure.

Recently, cryo-electron tomography (cryoET) reconstructions of membrane-anchored HSV-1 gB by two research groups have revealed alternative compact forms of gB with trimeric symmetry (Fontana et al., 2017; Zeev-Ben-Mordehai et al., 2016). In the first study, gB was visualized on the surface of extracellular vesicles produced in cells overexpressing HSV-1 gB (Zeev-Ben-Mordehai et al., 2016). Two distinct conformations of gB were detected, one that matched the known postfusion form and the other, a more compact, novel trimeric form, which could represent prefusion gB or an intermediate form between pre- and postfusion. Fitting domains I and II from the postfusion gB structure into the compact form positioned domain I distal to the membrane, with the fusion loops facing away from the membrane (Figure 5D). Interestingly, this fitting resulted in the distances between the fusion loops of each gB protomer being similar to the distances between fusion loops in the prefusion VSV G. Domains III-V did not fit into the model, potentially due to structural rearrangements in those domains.

The second cryoET study used landmarks to orient gB during imaging of full-length gB on vesicles, including large insertions of fluorescence proteins at known sites in gB and antibodies with known binding sites on gB (Fontana et al., 2017). Proper folding of gB on these vesicles was confirmed using a panel of antibodies specific for conformational epitopes, suggesting that this gB is in the prefusion conformation. Using these landmarks, the computational model of HSV gB based on prefusion VSV G (Gallagher et al., 2014) was fit into the structure with the fusion loops oriented towards the membrane (Figure 5C). A recent cryoET study of CMV gB also oriented the fusion loops towards the membrane (Si et al., 2018). Although these cryoET studies propose different orientations of gB with respect to the membrane, the studies all reveal alternate membrane-anchored forms of gB that, at heights of 8-10 nm, are significantly shorter than the 16-18-nm postfusion form. Both models also predict extensive conformational changes in domains III and V during the transition from a prefusion to the postfusion conformation.

While comparisons to the prefusion VSV G structure provide a valuable working model for the prefusion gB structure, the prefusion form of gB likely differs from that of VSV G. Alphaherpesvirus entry can occur at neutral pH and requires triggering by other entry glycoproteins, whereas triggering of conformational changes in VSV G is pH-dependent, reversible, and does not require other viral glycoproteins (Ferlin et al., 2014; Roche et al., 2008). Interestingly, exposure to low pH has been shown to cause antigenic changes in HSV-1 and HSV-2 gB, some of which are reversible (Cairns et al., 2011; Dollery et al., 2010a; Dollery et al., 2011; Muggeridge, 2012; Nicola, 2016; Siekavizza-Robles et al., 2010; Weed et al., 2018). Ultimately, a detailed understanding of the prefusion form of gB, including the organization of domains III and V, will require determination of the structure at atomic resolution.

Functional regions across gB

HSV gB elicits neutralizing antibodies (Cairns et al., 2015; Cairns et al., 2014), but all of the antibodies that have been tested can bind to the postfusion conformation (Bender et al., 2007), suggesting that these MAbs do not inhibit fusion by trapping a prefusion conformation. Neutralizing MAbs map to different domains within the gB ectodomain, indicating that multiple regions of gB are critical for proper refolding during fusion and/or interactions with other glycoproteins. Likewise, insertion mutations in each of the gB domains can prevent fusion function (Lin and Spear, 2007). By mapping the epitopes for the neutralizing antibodies, four distinct functional regions (FR) on the surface of HSV gB were defined (Bender et al., 2007). FR1 lies at the base of the postfusion structure and includes the fusion loops and regions of both domain I and V. FR2 maps to domain II and may overlap a gHgL interaction site (see below). FR3 is located on the gB crown and includes regions of domain III and IV. Finally, FR4 maps to the N-terminal region of gB that was not resolved in the structure and includes the binding site for heparan sulfate proteoglycan (HSPG) (Laquerre et al., 1998).

Receptors for gB

In addition to its role as the fusogen, HSV gB binds several host surface molecules. While gB can bind to HSPG, this interaction is not essential for virus entry. HSV gB can also bind to paired immunoglobulin-like type 2 receptor-α (PILRα), an immune regulator (Fan et al., 2009; Satoh et al., 2008; Wang et al., 2009). PILRα is not required for HSV entry into all cells; however, its expression in the absence of gD receptors can mediate virus entry into cells and low levels of cell-cell fusion. Interestingly, gD must be present for PILRα-mediated fusion to occur. The clinical importance of PILRα has yet to be determined; however, it may contribute to entry into retinal pigment epithelial cells (Shukla et al., 2009) and replication on murine cornea (Arii et al., 2010b). Two additional gB receptors that enhance HSV entry have been described, including myelin-associated glycoprotein (MAG) (Suenaga et al., 2010) and myosin-9 (also called non-muscle myosin heavy chain IIA, NMMHC-IIA) (Arii et al., 2010a). MAG also binds to VZV gB and can enhance VZV entry into cells (Suenaga et al., 2010), although it is not required for infection. The ability of receptors to trigger fusion by binding to gB diverges from the current model of entry, in which a signal for fusion is transmitted to gB by gHgL. Future structural studies of gB complexed with these receptors may reveal a mechanism of activation.

Viral accessory proteins for gB

Although transfection of gD-receptor-bearing cells with the core HSV entry glycoproteins gD, gH, gL, and gB is sufficient for cell-cell fusion, other virally encoded proteins influence fusion in the context of the viral entry. For example, while certain mutations in the gB CT impart a syncytial (syn) phenotype (Cooper et al., 2018; Gage et al., 1993; Silverman et al., 2012), syn mutations also map to other viral genes, including gK, UL20, and UL24 (Dolter et al., 1994; Leiva-Torres et al., 2010). gK and UL20 interact both with gB and with each other (Chouljenko et al., 2010). Although gK is not required for cell-cell fusion, deletion of the N terminus of gK inhibits entry into neurons in culture and spread to the mouse trigeminal ganglia after ocular infection (Jambunathan et al., 2015; Musarrat et al., 2018). Additional viral proteins also contribute to fusion in potentially complex ways. For example, UL21 is required for the syncytial phenotype of gB syn mutants, but not gK, UL20, or UL24 syn mutants (Sarfo et al., 2017). Other proteins that also have been reported to influence fusion include gE, gI, gM, gN, UL11, UL16, and UL45 (Dollery et al., 2010b; El Kasmi and Lippe, 2015; Haanes et al., 1994; Han et al., 2012; Kim et al., 2013).

Interactions among entry glycoproteins

Detecting interactions

HSV entry glycoproteins can be chemically crosslinked, indicating their proximity on the viral envelope (Handler et al., 1996); however, formation of a stable entry complex prior to entry is unlikely because individual glycoprotein deletions do not disrupt glycoprotein incorporation into the virion (Rodger et al., 2001). HSV gD is reported to co-immunoprecipitate (coIP) with both gHgL and gB (Gianni et al., 2009), and, in one study, the coIP of gD with gHgL was dependent on the presence of receptor (Perez-Romero et al., 2005). A direct association of purified soluble forms of gHgL and gB in the presence of liposomes has be detected at low pH (Cairns et al., 2011). Lateral interactions among gB trimers also have been observed by EM of soluble gB bound to liposomes (Maurer et al., 2013).

Direct interactions among the essential HSV entry glycoproteins have also been examined using bimolecular fluorescence complementation (BiFC), a technique in which two inactive halves of a fluorescent protein, for example, green fluorescent protein (GFP), are fused onto separate glycoproteins. Interaction between the glycoproteins brings the two halves into proximity such that the active fluorescent protein forms. Using BiFC, interactions between all of the essential HSV entry glycoproteins (gD/gB, gD/gHgL, gHgL/gB) have been detected in two independent studies (Atanasiu et al., 2007; Avitabile et al., 2007). In one study, the interaction of gHgL with gB was dependent on the presence of both gD and its receptor, suggesting a cascade of interactions starting with gD/receptor and ending with gHgL/gB (Atanasiu et al., 2007). Interestingly, although each BiFC construct was functional individually, coexpression of complementary gD and gH BiFC constructs inhibited fusion, suggesting that the gD/gHgL interaction may need to be transient. This was further supported by recent SPR experiments showing that gH/gL binds to gD rapidly and also rapidly dissociates from it (Cairns et al., 2019).

Mapping interactions

Putative interaction sites on the glycoproteins can be mapped by blocking their interactions with neutralizing antibodies that bind to defined epitopes. Using SPR, MAbs that block gD-gHgL binding were used to map the gHgL binding site on gD to a large region adjacent to, but separate from, the receptor binding domain (Cairns et al., 2019). Using BiFC, the gHgL interaction site on gB was mapped to gB domain II, overlapping the epitope for the MAb C226 in FR2 (Figure 5) (Atanasiu et al., 2010b). Similarly, the gB interaction site on gHgL was mapped to a conserved groove in domains H1/H2 of gHgL, overlapping the epitope for the MAb LP11 (Figure 3) (Chowdary et al., 2010).

Interaction sites on glycoproteins can also be mapped by creating chimeric entry glycoproteins that carry regions from non-complementing viral species. By coexpressing various combinations of HSV-1 and saimiriine herpesvirus 1 (SaHV-1) entry glycoproteins, a species-specific functional interaction between gD and gHgL was demonstrated (Fan et al., 2014). By generating gHgL chimeras encoding segments of HSV-1 and SaHV-1 sequence, a gD interaction site within gHgL was mapped to domains H1 and the N terminus of H2, in the membrane-distal region of the gH ectodomain (Fan et al., 2015). Consistent with a model that gD interacts with the N terminus of gHgL, an N-terminally truncated form of HSV gHgL can mediate gD-independent cell-cell fusion (Atanasiu et al., 2013). Taken together, these studies suggest gD may trigger a conformational change in the N-terminus of gHgL.

Species-specific interactions also have been used to map gHgL-gB interactions. Chimeric constructs combining gHgL sequences from HSV-1 and PRV were used to map a species-specific gHgL-gB interaction site to domain H3 at the C terminus of the gHgL ectodomain (Bohm et al., 2016). Interestingly, these chimeric gHgL studies and the BiFC approach identified distinct sites on gHgL that contribute to the gB interaction, in domains H3 and H1/H2, respectively. A potential explanation for this discrepancy is that, in the BiFC studies, MAb binding to H1/H2 may alter the presentation of an interaction site in H3. Conversely, in the chimera studies, domain H3 may influence the presentation of an interaction site in H1/H2.

Defining interactions among the herpesvirus glycoproteins has been challenging, likely due to low-affinity or transient interactions. Using chimeric constructs to map functional interaction sites may reveal domains of a protein that are required for an interaction, but this approach does not exclude the possibility that other regions of the protein also contribute to the interaction. Using neutralizing antibodies to map interactions is complicated by the fact that antibody binding may alter the protein structure and/or block an interaction by steric hindrance. Neither approach can identify the specific residues that mediate glycoprotein interactions. Accurate mapping of HSV glycoprotein interaction sites awaits future studies, in which structures of entry complexes determined by cryoET or x-ray crystallography can be verified using biochemical assays.

Routes of entry

Due, in part, to compelling electron micrographs showing alphaherpesvirus particles fusing directly with the plasma membrane (Granzow et al., 1997), the entry of alphaherpesviruses for a long time was thought to occur at the plasma membrane at a neutral pH. Upon subsequent reassessment, the route of entry into cells for alphaherpesviruses was shown to be dependent on cell type, as for beta- and gammaherpesviruses (Compton et al., 1992; Hutt-Fletcher, 2007; Ryckman et al., 2006). For example, HSV enters keratinocytes, human epithelial HeLa cells, and Chinese hamster ovary (CHO) cells in a pH-dependent manner via endocytosis (Nicola et al., 2005; Nicola et al., 2003; Nicola and Straus, 2004), whereas HSV fuses directly with the plasma membrane during entry into neurons, keratinocytes, human epithelial Hep2 cells, and Vero cells (Fuller and Spear, 1987; Koyama and Uchida, 1987; Nicola et al., 2005; Rahn et al., 2011; Sodeik et al., 1997; Wittels and Spear, 1991). Similarly, equine herpesvirus (EHV) enters CHO cells and endothelial cells via endocytosis but enters rabbit kidney RK13 cells by fusion at the plasma membrane (Frampton et al., 2007; Hasebe et al., 2009; Van de Walle et al., 2008).

Why endocytosis is required for entry into some cells but not into others is unknown. For entry via endocytosis, the low pH environment of the endosome appears to serve as a trigger for fusion, since agents that disrupt endosomal pH prevent entry (Nicola et al., 2005; Nicola et al., 2003). Low pH can affect gB conformation and oligomeric state, in both reversible and irreversible manners (Cairns et al., 2011; Dollery et al., 2010a; Dollery et al., 2011; Muggeridge, 2012; Weed et al., 2018). Interestingly, there is evidence that gC may selectively facilitate low-pH entry by regulating conformational changes in gB (Komala Sari et al., 2020). VSV G, another class III fusion protein, also undergoes reversible pH-triggered conformational changes (Roche et al., 2007), raising the question of whether gB is triggered similarly to VSV G. Nevertheless, endocytosis appears to contribute more to entry than simply an exposure to low pH, because exposure of HSV to acidic pH fails to trigger the fusion of bound virus with the plasma membrane (Walker et al., 2015). In addition, in some cases, HSV-1 can enter cells via endocytosis without a low pH requirement (Milne et al., 2005).

Discussion

Recent work, driven in large part by structural studies of the core entry machinery, has improved our understanding of the stepwise mechanism for alphaherpesvirus entry into cells (Figure 1B). The identification of entry receptors has shown that nectin-1 serves as a common receptor that is critical for infection. The biological relevance of the receptors that bind to gB, the HSV receptor HVEM, and the potential entry receptors for VZV (Chen et al., 2004; Suenaga et al., 2015; Suenaga et al., 2010; Yang et al., 2016) is yet unclear. The comparison of several gD structures indicates that the C terminus of the gD ectodomain shifts upon receptor binding. Future studies will determine how this shift exposes a gHgL interaction site as well as map the precise location of this site, be it within the gD C terminus or within the gD core.

gHgL structures from three different alphaherpesviruses indicated that this essential entry glycoprotein complex is not a viral fusogen and allowed the design of structure-based experiments to dissect its function and sites of interaction. How the gHgL conveys a fusion signal from gD (and/or receptor) to gB has not yet been determined. gHgL could, potentially, adopt multiple conformations, including a native, pre-triggered conformation, a gD-activated conformation, and a post-triggered conformation.

To date, all of the gB variants isolated for crystallographic studies, including those containing intact TM and CT, adopted the postfusion conformation (Cooper et al., 2018; Vitu et al., 2013). CryoET of membrane-associated gB reveals compact forms of gB with trimeric symmetry that may represent a prefusion or an alternative conformation (Fontana et al., 2017; Zeev-Ben-Mordehai et al., 2016). These compact structures cannot accommodate the extended coil of domain III, seen in the postfusion form, indicating that gB undergoes extensive conformational rearrangement to mediate membrane fusion. Structural and mutagenic data suggest that the gB CT acts as a clamp, regulating gB activation.

Future trends

Despite recent advances, many questions regarding the entry of alphaherpesviruses remain. Structural approaches, including both crystallography and cryoET, will continue to prove indispensable for addressing the remaining questions in alphaherpesvirus entry. CryoET in particular is being employed more frequently to examine the structures of individual proteins. Results from these structural studies will provide the basis for hypotheses that can be tested using mutagenesis and functional assays of entry and membrane fusion.

What do the complexes of gD/gHgL and gHgL/gB look like? What surfaces of the proteins serve as contact sites? Stable association of these complexes will be required for structural studies, a notable disadvantage for the alphaherpesviruses compared to other herpesviruses (Ciferri et al., 2015; Sathiyamoorthy et al., 2016). Identification of antibodies and/or mutations that stabilize potentially transient conformations of these proteins may facilitate structural studies of the complexes.

Why is gHgL required for the entry of herpesviruses? Viruses from most other families use a single protein to mediate both receptor-binding and membrane fusion. Paramyxoviruses encode separate receptor-binding and fusion proteins; however, they do not require an additional protein to serve as a "bridge" between the receptor-binding protein and fusion protein. How exactly does gHgL trigger gB? Although a purified form of the gHgL ectodomain can trigger fusion at low levels (Atanasiu et al., 2010a), mutational studies indicate that the gH TM and short CT domains are critical for function (Browne et al., 1996; Harman et al., 2002; Wilson et al., 1994). Do interactions between the gH CT and gB CT influence gB triggering (Cooper et al., 2018; Rogalin and Heldwein, 2015; Silverman and Heldwein, 2013)? In addition, does gHgL have any direct impact on the cellular or viral membranes at the site of fusion? What additional roles does gHgL play during VZV entry?

What conformational changes are required for and/or triggered by complex formation among the entry glycoproteins? Does gHgL adopt multiple distinct conformations as it interacts with gD and gB? What does the prefusion form of gB look like and where are the fusion peptides positioned in this form? What intermediate conformation(s) does gB fold into as it transitions from its prefusion to postfusion form? Can an extended form of gB, embedded in both the cellular and viral membranes, be captured? Defining the conformational changes required for entry may highlight regions susceptible to inhibition by small molecules, offering tools to trap intermediate conformations for further study and potentially providing a basis for the development of antiviral drugs that inhibit virus entry.

Web resources

Structures of the alphaherpesvirus entry glycoproteins can be accessed from the protein data bank (https://www.rcsb.org). These crystal structures include unliganded HSV-1 gD (1L2G, 2C3A, and 2C36) (Carfi et al., 2001; Krummenacher et al., 2005), HSV-1 gD bound to HVEM (1JMA) (Carfi et al., 2001), HSV-1 gD bound to nectin-1 (3SKU and 3U82) (Di Giovine et al., 2011; Zhang et al., 2011), HSV-2 gD (4MYV), HSV-2 gD bound to nectin-1 (4MYW) (Lu et al., 2014), PRV gD (5X5V), PRV gD bound to nectin-1 (5X5W) (Li et al., 2017a), HSV-2 gHgL (3M1C) (Chowdary et al., 2010), PRV gH (2XQY) (Backovic et al., 2010), VZV gHgL (4XHJ) (Xing et al., 2015), postfusion forms of HSV-1 gB (2GUM, 2NWF, and 5V2S) (Cooper et al., 2018; Heldwein et al., 2006; Stampfer et al., 2010), and PRV gB (6ESC and 5YS6) (Li et al., 2017b; Vallbracht et al., 2017). A model of membrane-anchored HSV-1 gB generated using cryoET also is available (5FZ2) (Zeev-Ben-Mordehai et al., 2016).

Acknowledgements

TMC is supported by grants AI-18289, AI-142940, and AI-139618, and SAC is supported by grant AI-148478 from the National Institute of Allergy and Infectious Diseases of the National Institutes of Health.

References

Akkarawongsa, R., Pocaro, N.E., Case, G., Kolb, A.W., and Brandt, C.R. (2009). Multiple peptides homologous to herpes simplex virus type 1 glycoprotein B inhibit viral infection. Antimicrob Agents Chemother *53*, 987-996.

Arii, J., Goto, H., Suenaga, T., Oyama, M., Kozuka-Hata, H., Imai, T., Minowa, A., Akashi, H., Arase, H., Kawaoka, Y., *et al.* (2010a). Non-muscle myosin IIA is a functional entry receptor for herpes simplex virus-1. Nature *467*, 859-862.

Arii, J., Wang, J., Morimoto, T., Suenaga, T., Akashi, H., Arase, H., and Kawaguchi, Y. (2010b). A single-amino-acid substitution in herpes simplex virus 1 envelope glycoprotein B at a site required for binding to the paired immunoglobulin-like type 2 receptor alpha (PILRalpha) abrogates PILRalpha-dependent viral entry and reduces pathogenesis. J Virol *84*, 10773-10783.

Atanasiu, D., Cairns, T.M., Whitbeck, J.C., Saw, W.T., Rao, S., Eisenberg, R.J., and Cohen, G.H. (2013). Regulation of herpes simplex virus gB-induced cell-cell fusion by mutant forms of gH/gL in the absence of gD and cellular receptors. mBio *4*.

Atanasiu, D., Saw, W.T., Cohen, G.H., and Eisenberg, R.J. (2010a). Cascade of events governing cell-cell fusion induced by herpes simplex virus glycoproteins gD, gH/gL, and gB. J Virol *84*, 12292-12299.

Atanasiu, D., Saw, W.T., Eisenberg, R.J., and Cohen, G.H. (2016). Regulation of HSV glycoprotein induced cascade of events governing cell-cell fusion. J Virol.

Atanasiu, D., Saw, W.T., Lazear, E., Whitbeck, J.C., Cairns, T.M., Lou, H., Eisenberg, R.J., and Cohen, G.H. (2018). Using Antibodies and Mutants To Localize the Presumptive gH/gL Binding Site on Herpes Simplex Virus gD. J Virol *92*.

Atanasiu, D., Whitbeck, J.C., Cairns, T.M., Reilly, B., Cohen, G.H., and Eisenberg, R.J. (2007). Bimolecular complementation reveals that glycoproteins gB and gH/gL of herpes simplex virus interact with each other during cell fusion. Proceedings of the National Academy of Sciences of the United States of America *104*, 18718-18723.

Atanasiu, D., Whitbeck, J.C., de Leon, M.P., Lou, H., Hannah, B.P., Cohen, G.H., and Eisenberg, R.J. (2010b). Bimolecular complementation defines functional regions of Herpes simplex virus gB that are involved with gH/gL as a necessary step leading to cell fusion. J Virol *84*, 3825-3834.

Avitabile, E., Forghieri, C., and Campadelli-Fiume, G. (2007). Complexes between herpes simplex virus glycoproteins gD, gB, and gH detected in cells by complementation of split enhanced green fluorescent protein. J Virol *81*, 11532-11537.

Awasthi, S., Belshe, R.B., and Friedman, H.M. (2014). Better neutralization of herpes simplex virus type 1 (HSV-1) than HSV-2 by antibody from recipients of GlaxoSmithKline HSV-2 glycoprotein D2 subunit vaccine. The Journal of infectious diseases *210*, 571-575.

Awasthi, S., Mahairas, G.G., Shaw, C.E., Huang, M.L., Koelle, D.M., Posavad, C., Corey, L., and Friedman, H.M. (2015). A Dual-Modality Herpes Simplex Virus 2 Vaccine for Preventing Genital Herpes by Using Glycoprotein C and D Subunit Antigens To Induce Potent Antibody Responses and Adenovirus Vectors Containing Capsid and Tegument Proteins as T Cell Immunogens. J Virol *89*, 8497-8509.

Babic, N., Klupp, B.G., Makoschey, B., Karger, A., Flamand, A., and Mettenleiter, T.C. (1996). Glycoprotein gH of pseudorabies virus is essential for penetration and propagation in cell culture and in the nervous system of mice. J Gen Virol *77 (Pt 9)*, 2277-2285.

Backovic, M., DuBois, R.M., Cockburn, J.J., Sharff, A.J., Vaney, M.C., Granzow, H., Klupp, B.G., Bricogne, G., Mettenleiter, T.C., and Rey, F.A. (2010). Structure of a core fragment of glycoprotein H from pseudorabies virus in complex with antibody. Proceedings of the National Academy of Sciences of the United States of America *107*, 22635-22640.

Backovic, M., and Jardetzky, T.S. (2009). Class III viral membrane fusion proteins. Current opinion in structural biology *19*, 189-196.

Backovic, M., Jardetzky, T.S., and Longnecker, R. (2007). Hydrophobic residues that form putative fusion loops of Epstein-Barr virus glycoprotein B are critical for fusion activity. J Virol *81*, 9596-9600.

Backovic, M., Longnecker, R., and Jardetzky, T.S. (2009). Structure of a trimeric variant of the Epstein-Barr virus glycoprotein B. Proceedings of the National Academy of Sciences of the United States of America *106*, 2880-2885.

Baghian, A., Huang, L., Newman, S., Jayachandra, S., and Kousoulas, K.G. (1993). Truncation of the carboxy-terminal 28 amino acids of glycoprotein B specified by herpes simplex virus type 1 mutant amb1511-7 causes extensive cell fusion. J Virol *67*, 2396-2401.

Baquero, E., Albertini, A.A., Raux, H., Buonocore, L., Rose, J.K., Bressanelli, S., and Gaudin, Y. (2015). Structure of the low pH conformation of Chandipura virus G reveals important features in the evolution of the vesiculovirus glycoprotein. PLoS Pathog *11*, e1004756.

Belot, L., Ouldali, M., Roche, S., Legrand, P., Gaudin, Y., and Albertini, A.A. (2020). Crystal structure of Mokola virus glycoprotein in its post-fusion conformation. PLoS Pathog *16*, e1008383.

Bender, F.C., Samanta, M., Heldwein, E.E., de Leon, M.P., Bilman, E., Lou, H., Whitbeck, J.C., Eisenberg, R.J., and Cohen, G.H. (2007). Antigenic and mutational analyses of herpes simplex virus glycoprotein B reveal four functional regions. J Virol *81*, 3827-3841.

Birlea, M., Owens, G.P., Eshleman, E.M., Ritchie, A., Traktinskiy, I., Bos, N., Seitz, S., Azarkh, Y., Mahalingam, R., Gilden, D., *et al.* (2013). Human anti-varicella-zoster virus (VZV) recombinant monoclonal antibody produced after Zostavax immunization recognizes the gH/gL complex and neutralizes VZV infection. J Virol *87*, 415-421.

Bohm, S.W., Backovic, M., Klupp, B.G., Rey, F.A., Mettenleiter, T.C., and Fuchs, W. (2016). Functional Characterization of Glycoprotein H Chimeras Composed of Conserved Domains of the Pseudorabies Virus and Herpes Simplex Virus 1 Homologs. J Virol *90*, 421-432.

Browne, H., Bruun, B., and Minson, T. (2001). Plasma membrane requirements for cell fusion induced by herpes simplex virus type 1 glycoproteins gB, gD, gH and gL. J Gen Virol *82*, 1419-1422.

Browne, H.M., Bruun, B.C., and Minson, A.C. (1996). Characterization of herpes simplex virus type 1 recombinants with mutations in the cytoplasmic tail of glycoprotein H. J Gen Virol *77 (Pt 10)*, 2569-2573.

Buckmaster, E.A., Gompels, U., and Minson, A. (1984). Characterisation and physical mapping of an HSV-1 glycoprotein of approximately 115 X 10(3) molecular weight. Virology *139*, 408-413.

Burke, H.G., and Heldwein, E.E. (2015). Crystal Structure of the Human Cytomegalovirus Glycoprotein B. PLoS Pathog *11*, e1005227.

Cai, W.H., Gu, B., and Person, S. (1988). Role of glycoprotein B of herpes simplex virus type 1 in viral entry and cell fusion. J Virol *62*, 2596-2604.

Cairns, T.M., Ditto, N.T., Atanasiu, D., Lou, H., Brooks, B.D., Saw, W.T., Eisenberg, R.J., and Cohen, G.H. (2019). Surface Plasmon Resonance Reveals Direct Binding of Herpes Simplex Virus Glycoproteins gH/gL to gD and Locates a gH/gL Binding Site on gD. J Virol *93*.

Cairns, T.M., Ditto, N.T., Lou, H., Brooks, B.D., Atanasiu, D., Eisenberg, R.J., and Cohen, G.H. (2017). Global sensing of the antigenic structure of herpes simplex virus gD using high-throughput array-based SPR imaging. PLoS Pathog *13*, e1006430.

Cairns, T.M., Friedman, L.S., Lou, H., Whitbeck, J.C., Shaner, M.S., Cohen, G.H., and Eisenberg, R.J. (2007). N-terminal mutants of herpes simplex virus type 2 gH are transported without gL but require gL for function. J Virol *81*, 5102-5111.

Cairns, T.M., Huang, Z.Y., Gallagher, J.R., Lin, Y., Lou, H., Whitbeck, J.C., Wald, A., Cohen, G.H., and Eisenberg, R.J. (2015). Patient-Specific Neutralizing Antibody Responses to Herpes Simplex Virus Are Attributed to Epitopes on gD, gB, or Both and Can Be Type Specific. J Virol *89*, 9213-9231.

Cairns, T.M., Huang, Z.Y., Whitbeck, J.C., Ponce de Leon, M., Lou, H., Wald, A., Krummenacher, C., Eisenberg, R.J., and Cohen, G.H. (2014). Dissection of the antibody response against herpes simplex virus glycoproteins in naturally infected humans. J Virol *88*, 12612-12622.

Cairns, T.M., Landsburg, D.J., Whitbeck, J.C., Eisenberg, R.J., and Cohen, G.H. (2005). Contribution of cysteine residues to the structure and function of herpes simplex virus gH/gL. Virology *332*, 550-562.

Cairns, T.M., Shaner, M.S., Zuo, Y., Ponce-de-Leon, M., Baribaud, I., Eisenberg, R.J., Cohen, G.H., and Whitbeck, J.C. (2006). Epitope mapping of herpes simplex virus type 2 gH/gL defines distinct antigenic sites, including some associated with biological function. J Virol *80*, 2596-2608.

Cairns, T.M., Whitbeck, J.C., Lou, H., Heldwein, E.E., Chowdary, T.K., Eisenberg, R.J., and Cohen, G.H. (2011). Capturing the herpes simplex virus core fusion complex (gB-gH/gL) in an acidic environment. J Virol *85*, 6175-6184.

Campadelli-Fiume, G., Petrovic, B., Leoni, V., Gianni, T., Avitabile, E., Casiraghi, C., and Gatta, V. (2016). Retargeting Strategies for Oncolytic Herpes Simplex Viruses. Viruses *8*, 63.

Carfi, A., Willis, S.H., Whitbeck, J.C., Krummenacher, C., Cohen, G.H., Eisenberg, R.J., and Wiley, D.C. (2001). Herpes simplex virus glycoprotein D bound to the human receptor HveA. Molecular cell *8*, 169-179.

Ch'ng, T.H., Spear, P.G., Struyf, F., and Enquist, L.W. (2007). Glycoprotein D-independent spread of pseudorabies virus infection in cultured peripheral nervous system neurons in a compartmented system. J Virol *81*, 10742-10757.

Chandramouli, S., Ciferri, C., Nikitin, P.A., Calo, S., Gerrein, R., Balabanis, K., Monroe, J., Hebner, C., Lilja, A.E., Settembre, E.C., *et al.* (2015). Structure of HCMV glycoprotein B in the postfusion conformation bound to a neutralizing human antibody. Nature communications *6*, 8176.

Chen, J., Sathiyamoorthy, K., Zhang, X., Schaller, S., Perez White, B.E., Jardetzky, T.S., and Longnecker, R. (2018). Ephrin receptor A2 is a functional entry receptor for Epstein-Barr virus. Nat Microbiol *3*, 172-180.

Chen, J., Zhang, X., Schaller, S., Jardetzky, T.S., and Longnecker, R. (2019). Ephrin Receptor A4 is a New Kaposi's Sarcoma-Associated Herpesvirus Virus Entry Receptor. mBio *10*.

Chen, J.J., Zhu, Z., Gershon, A.A., and Gershon, M.D. (2004). Mannose 6-phosphate receptor dependence of varicella zoster virus infection in vitro and in the epidermis during varicella and zoster. Cell *119*, 915-926.

Chesnokova, L.S., and Hutt-Fletcher, L.M. (2011). Fusion of Epstein-Barr virus with epithelial cells can be triggered by alphavbeta5 in addition to alphavbeta6 and alphavbeta8, and integrin binding triggers a conformational change in glycoproteins gHgL. J Virol *85*, 13214-13223.

Chesnokova, L.S., Nishimura, S.L., and Hutt-Fletcher, L.M. (2009). Fusion of epithelial cells by Epstein-Barr virus proteins is triggered by binding of viral glycoproteins gHgL to integrins alphavbeta6 or alphavbeta8. Proceedings of the National Academy of Sciences of the United States of America *106*, 20464-20469.

Chouljenko, V.N., Iyer, A.V., Chowdhury, S., Kim, J., and Kousoulas, K.G. (2010). The herpes simplex virus type 1 UL20 protein and the amino terminus of glycoprotein K (gK) physically interact with gB. J Virol *84*, 8596-8606.

Chowdary, T.K., Cairns, T.M., Atanasiu, D., Cohen, G.H., Eisenberg, R.J., and Heldwein, E.E. (2010). Crystal structure of the conserved herpesvirus fusion regulator complex gH-gL. Nature structural and molecular biology *17*, 882-888.

Ciferri, C., Chandramouli, S., Donnarumma, D., Nikitin, P.A., Cianfrocco, M.A., Gerrein, R., Feire, A.L., Barnett, S.W., Lilja, A.E., Rappuoli, R., *et al.* (2015). Structural and biochemical studies of HCMV gH/gL/gO and Pentamer reveal mutually exclusive cell entry complexes. Proceedings of the National Academy of Sciences of the United States of America *112*, 1767-1772.

Cocchi, F., Fusco, D., Menotti, L., Gianni, T., Eisenberg, R.J., Cohen, G.H., and Campadelli-Fiume, G. (2004a). The soluble ectodomain of herpes simplex virus gD contains a membrane-proximal pro-fusion domain and suffices to mediate virus entry. Proceedings of the National Academy of Sciences of the United States of America *101*, 7445-7450.

Cocchi, F., Menotti, L., Di Ninni, V., Lopez, M., and Campadelli-Fiume, G. (2004b). The herpes simplex virus JMP mutant enters receptor-negative J cells through a novel pathway independent of the known receptors nectin1, HveA, and nectin2. J Virol *78*, 4720-4729.

Cocchi, F., Menotti, L., Mirandola, P., Lopez, M., and Campadelli-Fiume, G. (1998). The ectodomain of a novel member of the immunoglobulin subfamily related to the poliovirus receptor has the attributes of a bona fide receptor for herpes simplex virus types 1 and 2 in human cells. J Virol *72*, 9992-10002.

Cohen, G.H., Dietzschold, B., Ponce de Leon, M., Long, D., Golub, E., Varrichio, A., Pereira, L., and Eisenberg, R.J. (1984). Localization and synthesis of an antigenic determinant of herpes simplex virus glycoprotein D that stimulates the production of neutralizing antibody. J Virol *49*, 102-108.

Cole, N.L., and Grose, C. (2003). Membrane fusion mediated by herpesvirus glycoproteins: the paradigm of varicella-zoster virus. Rev Med Virol *13*, 207-222.

Compton, T., Nepomuceno, R.R., and Nowlin, D.M. (1992). Human cytomegalovirus penetrates host cells by pH-independent fusion at the cell surface. Virology *191*, 387-395.

Connolly, S.A., Landsburg, D.J., Carfi, A., Wiley, D.C., Eisenberg, R.J., and Cohen, G.H. (2002). Structure-based analysis of the herpes simplex virus glycoprotein D binding site present on herpesvirus entry mediator HveA (HVEM). J Virol *76*, 10894-10904.

Connolly, S.A., and Longnecker, R. (2012). Residues within the C-terminal arm of the herpes simplex virus 1 glycoprotein B ectodomain contribute to its refolding during the fusion step of virus entry. J Virol *86*, 6386-6393.

Connolly, S.A., Whitbeck, J.J., Rux, A.H., Krummenacher, C., van Drunen Littel-van den Hurk, S., Cohen, G.H., and Eisenberg, R.J. (2001). Glycoprotein D homologs in herpes simplex virus type 1, pseudorabies virus, and bovine herpes virus type 1 bind directly to human HveC(nectin-1) with different affinities. Virology *280*, 7-18.

Cooper, R.S., Georgieva, E.R., Borbat, P.P., Freed, J.H., and Heldwein, E.E. (2018). Structural basis for membrane anchoring and fusion regulation of the herpes simplex virus fusogen gB. Nature structural and molecular biology *25*, 416-424.

Cooper, R.S., and Heldwein, E.E. (2015). Herpesvirus gB: A Finely Tuned Fusion Machine. Viruses *7*, 6552-6569.

Desai, P.J., Schaffer, P.A., and Minson, A.C. (1988). Excretion of non-infectious virus particles lacking glycoprotein H by a temperature-sensitive mutant of herpes simplex virus type 1: evidence that gH is essential for virion infectivity. J Gen Virol *69 (Pt 6)*, 1147-1156.

Di Giovine, P., Settembre, E.C., Bhargava, A.K., Luftig, M.A., Lou, H., Cohen, G.H., Eisenberg, R.J., Krummenacher, C., and Carfi, A. (2011). Structure of herpes simplex virus glycoprotein D bound to the human receptor nectin-1. PLoS Pathog *7*, e1002277.

Diakidi-Kosta, A., Michailidou, G., Kontogounis, G., Sivropoulou, A., and Arsenakis, M. (2003). A single amino acid substitution in the cytoplasmic tail of the glycoprotein B of herpes simplex virus 1 affects both syncytium formation and binding to intracellular heparan sulfate. Virus Res *93*, 99-108.

Dollery, S.J., Delboy, M.G., and Nicola, A.V. (2010a). Low pH-induced conformational change in herpes simplex virus glycoprotein B. J Virol *84*, 3759-3766.

Dollery, S.J., Lane, K.D., Delboy, M.G., Roller, D.G., and Nicola, A.V. (2010b). Role of the UL45 protein in herpes simplex virus entry via low pH-dependent endocytosis and its relationship to the conformation and function of glycoprotein B. Virus Res *149*, 115-118.

Dollery, S.J., Wright, C.C., Johnson, D.C., and Nicola, A.V. (2011). Low-pH-dependent changes in the conformation and oligomeric state of the prefusion form of herpes simplex virus glycoprotein B are separable from fusion activity. J Virol *85*, 9964-9973.

Dolter, K.E., Ramaswamy, R., and Holland, T.C. (1994). Syncytial mutations in the herpes simplex virus type 1 gK (UL53) gene occur in two distinct domains. J Virol *68*, 8277-8281.

Du, R., Wang, L., Xu, H., Wang, Z., Zhang, T., Wang, M., Ning, Y., Deng, F., Hu, Z., Wang, H., *et al.* (2017). A novel glycoprotein D-specific monoclonal antibody neutralizes herpes simplex virus. Antiviral research *147*, 131-141.

Dubin, G., and Jiang, H. (1995). Expression of herpes simplex virus type 1 glycoprotein L (gL) in transfected mammalian cells: evidence that gL is not independently anchored to cell membranes. J Virol *69*, 4564-4568.

Duus, K.M., and Grose, C. (1996). Multiple regulatory effects of varicella-zoster virus (VZV) gL on trafficking patterns and fusogenic properties of VZV gH. J Virol *70*, 8961-8971.

Duus, K.M., Hatfield, C., and Grose, C. (1995). Cell surface expression and fusion by the varicella-zoster virus gH:gL glycoprotein complex: analysis by laser scanning confocal microscopy. Virology *210*, 429-440.

Edwards, R.G., Kopp, S.J., Karaba, A.H., Wilcox, D.R., and Longnecker, R. (2015). Herpesvirus entry mediator on radiation-resistant cell lineages promotes ocular herpes simplex virus 1 pathogenesis in an entry-independent manner. mBio *6*, e01532-01515.

Edwards, R.G., and Longnecker, R. (2017). Herpesvirus Entry Mediator and Ocular Herpesvirus Infection: More than Meets the Eye. J Virol *91*.

El Kasmi, I., and Lippe, R. (2015). Herpes simplex virus 1 gN partners with gM to modulate the viral fusion machinery. J Virol *89*, 2313-2323.

Engel, J.P., Boyer, E.P., and Goodman, J.L. (1993). Two novel single amino acid syncytial mutations in the carboxy terminus of glycoprotein B of herpes simplex virus type 1 confer a unique pathogenic phenotype. Virology *192*, 112-120.

Falanga, A., Tarallo, R., Vitiello, G., Vitiello, M., Perillo, E., Cantisani, M., D'Errico, G., Galdiero, M., and Galdiero, S. (2012). Biophysical characterization and membrane interaction of the two fusion loops of glycoprotein B from herpes simplex type I virus. PloS one *7*, e32186.

Fan, Q., Kopp, S.J., Connolly, S.A., and Longnecker, R. (2017). Structure-Based Mutations in the Herpes Simplex Virus 1 Glycoprotein B Ectodomain Arm Impart a Slow-Entry Phenotype. mBio *8*.

Fan, Q., Lin, E., Satoh, T., Arase, H., and Spear, P.G. (2009). Differential effects on cell fusion activity of mutations in herpes simplex virus 1 glycoprotein B (gB) dependent on whether a gD receptor or a gB receptor is overexpressed. J Virol *83*, 7384-7390.

Fan, Q., and Longnecker, R. (2012). Is nectin-1 the "master" receptor for deadly herpes B virus infection? Virulence *3*, 405.

Fan, Q., Longnecker, R., and Connolly, S.A. (2014). Substitution of herpes simplex virus 1 entry glycoproteins with those of saimiriine herpesvirus 1 reveals a gD-gH/gL functional interaction and a region within the gD profusion domain that is critical for fusion. J Virol *88*, 6470-6482.

Fan, Q., Longnecker, R., and Connolly, S.A. (2015). A Functional Interaction between Herpes Simplex Virus 1 Glycoprotein gH/gL Domains I and II and gD Is Defined by Using Alphaherpesvirus gH and gL Chimeras. J Virol *89*, 7159-7169.

Fan, Z., Grantham, M.L., Smith, M.S., Anderson, E.S., Cardelli, J.A., and Muggeridge, M.I. (2002). Truncation of herpes simplex virus type 2 glycoprotein B increases its cell surface expression and activity in cell-cell fusion, but these properties are unrelated. J Virol *76*, 9271-9283.

Ferlin, A., Raux, H., Baquero, E., Lepault, J., and Gaudin, Y. (2014). Characterization of pH-sensitive molecular switches that trigger the structural transition of vesicular stomatitis virus glycoprotein from the postfusion state toward the prefusion state. J Virol *88*, 13396-13409.

Foa-Tomasi, L., Avitabile, E., Boscaro, A., Brandimarti, R., Gualandri, R., Manservigi, R., Dall'Olio, F., Serafini-Cessi, F., and Fiume, G.C. (1991). Herpes simplex virus (HSV) glycoprotein H is partially processed in a cell line that expresses the glycoprotein and fully processed in cells infected with deletion or ts mutants in the known HSV glycoproteins. Virology *180*, 474-482.

Fontana, J., Atanasiu, D., Saw, W.T., Gallagher, J.R., Cox, R.G., Whitbeck, J.C., Brown, L.M., Eisenberg, R.J., and Cohen, G.H. (2017). The Fusion Loops of the Initial Prefusion Conformation of Herpes Simplex Virus 1 Fusion Protein Point Toward the Membrane. mBio *8*.

Forrester, A., Farrell, H., Wilkinson, G., Kaye, J., Davis-Poynter, N., and Minson, T. (1992). Construction and properties of a mutant of herpes simplex virus type 1 with glycoprotein H coding sequences deleted. J Virol *66*, 341-348.

Foster, T.P., Melancon, J.M., and Kousoulas, K.G. (2001). An alpha-helical domain within the carboxyl terminus of herpes simplex virus type 1 (HSV-1) glycoprotein B (gB) is associated with cell fusion and resistance to heparin inhibition of cell fusion. Virology *287*, 18-29.

Frampton, A.R., Jr., Stolz, D.B., Uchida, H., Goins, W.F., Cohen, J.B., and Glorioso, J.C. (2007). Equine herpesvirus 1 enters cells by two different pathways, and infection requires the activation of the cellular kinase ROCK1. J Virol *81*, 10879-10889.

Fuchs, W., Backovic, M., Klupp, B.G., Rey, F.A., and Mettenleiter, T.C. (2012). Structure-based mutational analysis of the highly conserved domain IV of glycoprotein H of pseudorabies virus. J Virol *86*, 8002-8013.

Fuller, A.O., and Lee, W.C. (1992). Herpes simplex virus type 1 entry through a cascade of virus-cell interactions requires different roles of gD and gH in penetration. J Virol *66*, 5002-5012.

Fuller, A.O., and Spear, P.G. (1987). Anti-glycoprotein D antibodies that permit adsorption but block infection by herpes simplex virus 1 prevent virion-cell fusion at the cell surface. Proceedings of the National Academy of Sciences of the United States of America *84*, 5454-5458.

Gage, P.J., Levine, M., and Glorioso, J.C. (1993). Syncytium-inducing mutations localize to two discrete regions within the cytoplasmic domain of herpes simplex virus type 1 glycoprotein B. J Virol *67*, 2191-2201.

Galdiero, M., Whiteley, A., Bruun, B., Bell, S., Minson, T., and Browne, H. (1997). Site-directed and linker insertion mutagenesis of herpes simplex virus type 1 glycoprotein H. J Virol *71*, 2163-2170.

Galdiero, S., Falanga, A., Vitiello, M., Browne, H., Pedone, C., and Galdiero, M. (2005). Fusogenic domains in herpes simplex virus type 1 glycoprotein H. J Biol Chem *280*, 28632-28643.

Galdiero, S., Falanga, A., Vitiello, M., D'Isanto, M., Collins, C., Orrei, V., Browne, H., Pedone, C., and Galdiero, M. (2007). Evidence for a role of the membrane-proximal region of herpes simplex virus Type 1 glycoprotein H in membrane fusion and virus inhibition. ChemBioChem *8*, 885-895.

Galdiero, S., Vitiello, M., D'Isanto, M., Falanga, A., Collins, C., Raieta, K., Pedone, C., Browne, H., and Galdiero, M. (2006). Analysis of synthetic peptides from heptad-repeat domains of herpes simplex virus type 1 glycoproteins H and B. J Gen Virol *87*, 1085-1097.

Galen, B., Cheshenko, N., Tuyama, A., Ramratnam, B., and Herold, B.C. (2006). Access to nectin favors herpes simplex virus infection at the apical surface of polarized human epithelial cells. J Virol *80*, 12209-12218.

Gallagher, J.R., Atanasiu, D., Saw, W.T., Paradisgarten, M.J., Whitbeck, J.C., Eisenberg, R.J., and Cohen, G.H. (2014). Functional fluorescent protein insertions in herpes simplex virus gB report on gB conformation before and after execution of membrane fusion. PLoS Pathog *10*, e1004373.

Gallagher, J.R., Saw, W.T., Atanasiu, D., Lou, H., Eisenberg, R.J., and Cohen, G.H. (2013). Displacement of the C terminus of herpes simplex virus gD is sufficient to expose the fusion-activating interfaces on gD. J Virol *87*, 12656-12666.

Gianni, T., Amasio, M., and Campadelli-Fiume, G. (2009). Herpes simplex virus gD forms distinct complexes with fusion executors gB and gH/gL in part through the C-terminal profusion domain. J Biol Chem *284*, 17370-17382.

Gianni, T., and Campadelli-Fiume, G. (2012). alphaVbeta3-integrin relocalizes nectin1 and routes herpes simplex virus to lipid rafts. J Virol *86*, 2850-2855.

Gianni, T., Martelli, P.L., Casadio, R., and Campadelli-Fiume, G. (2005a). The ectodomain of herpes simplex virus glycoprotein H contains a membrane alpha-helix with attributes of an internal fusion peptide, positionally conserved in the herpesviridae family. J Virol *79*, 2931-2940.

Gianni, T., Menotti, L., and Campadelli-Fiume, G. (2005b). A heptad repeat in herpes simplex virus 1 gH, located downstream of the alpha-helix with attributes of a fusion peptide, is critical for virus entry and fusion. J Virol *79*, 7042-7049.

Gianni, T., Salvioli, S., Chesnokova, L.S., Hutt-Fletcher, L.M., and Campadelli-Fiume, G. (2013). alphavbeta6- and alphavbeta8-integrins serve as interchangeable receptors for HSV gH/gL to promote endocytosis and activation of membrane fusion. PLoS Pathog *9*, e1003806.

Gompels, U.A., Carss, A.L., Saxby, C., Hancock, D.C., Forrester, A., and Minson, A.C. (1991). Characterization and sequence analyses of antibody-selected antigenic variants of herpes simplex virus show a conformationally complex epitope on glycoprotein H. J Virol *65*, 2393-2401.

Granzow, H., Weiland, F., Jons, A., Klupp, B.G., Karger, A., and Mettenleiter, T.C. (1997). Ultrastructural analysis of the replication cycle of pseudorabies virus in cell culture: a reassessment. J Virol *71*, 2072-2082.

Gruenheid, S., Gatzke, L., Meadows, H., and Tufaro, F. (1993). Herpes simplex virus infection and propagation in a mouse L cell mutant lacking heparan sulfate proteoglycans. J Virol *67*, 93-100.

Grunewald, K., Desai, P., Winkler, D.C., Heymann, J.B., Belnap, D.M., Baumeister, W., and Steven, A.C. (2003). Three-dimensional structure of herpes simplex virus from cryo-electron tomography. Science *302*, 1396-1398.

Haanes, E.J., Nelson, C.M., Soule, C.L., and Goodman, J.L. (1994). The UL45 gene product is required for herpes simplex virus type 1 glycoprotein B-induced fusion. J Virol *68*, 5825-5834.

Hahn, A.S., Kaufmann, J.K., Wies, E., Naschberger, E., Panteleev-Ivlev, J., Schmidt, K., Holzer, A., Schmidt, M., Chen, J., Konig, S., *et al.* (2012). The ephrin receptor tyrosine kinase A2 is a cellular receptor for Kaposi's sarcoma-associated herpesvirus. Nature medicine *18*, 961-966.

Han, J., Chadha, P., Starkey, J.L., and Wills, J.W. (2012). Function of glycoprotein E of herpes simplex virus requires coordinated assembly of three tegument proteins on its cytoplasmic tail. Proceedings of the National Academy of Sciences of the United States of America *109*, 19798-19803.

Handler, C.G., Cohen, G.H., and Eisenberg, R.J. (1996). Cross-linking of glycoprotein oligomers during herpes simplex virus type 1 entry. J Virol *70*, 6076-6082.

Hannah, B.P., Cairns, T.M., Bender, F.C., Whitbeck, J.C., Lou, H., Eisenberg, R.J., and Cohen, G.H. (2009). Herpes simplex virus glycoprotein B associates with target membranes via its fusion loops. J Virol *83*, 6825-6836.

Harman, A., Browne, H., and Minson, T. (2002). The transmembrane domain and cytoplasmic tail of herpes simplex virus type 1 glycoprotein H play a role in membrane fusion. J Virol *76*, 10708-10716.

Harrison, S.C. (2015). Viral membrane fusion. Virology *479-480*, 498-507.

Hasebe, R., Sasaki, M., Sawa, H., Wada, R., Umemura, T., and Kimura, T. (2009). Infectious entry of equine herpesvirus-1 into host cells through different endocytic pathways. Virology *393*, 198-209.

Heineman, T.C., and Hall, S.L. (2002). Role of the varicella-zoster virus gB cytoplasmic domain in gB transport and viral egress. J Virol *76*, 591-599.

Heldwein, E.E., Lou, H., Bender, F.C., Cohen, G.H., Eisenberg, R.J., and Harrison, S.C. (2006). Crystal structure of glycoprotein B from herpes simplex virus 1. Science *313*, 217-220.

Herold, B.C., Visalli, R.J., Susmarski, N., Brandt, C.R., and Spear, P.G. (1994). Glycoprotein C-independent binding of herpes simplex virus to cells requires cell surface heparan sulfate and glycoprotein B. J Gen Virol *75*, 1211-1222.

Herold, B.C., WuDunn, D., Soltys, N., and Spear, P.G. (1991). Glycoprotein C of herpes simplex virus type 1 plays a principal role in the adsorption of virus to cells and in infectivity. J Virol *65*, 1090-1098.

Highlander, S.L., Sutherland, S.L., Gage, P.J., Johnson, D.C., Levine, M., and Glorioso, J.C. (1987). Neutralizing monoclonal antibodies specific for herpes simplex virus glycoprotein D inhibit virus penetration. J Virol *61*, 3356-3364.

Hook, L.M., Cairns, T.M., Awasthi, S., Brooks, B.D., Ditto, N.T., Eisenberg, R.J., Cohen, G.H., and Friedman, H.M. (2018). Vaccine-induced antibodies to herpes simplex virus glycoprotein D epitopes involved in virus entry and cell-to-cell spread correlate with protection against genital disease in guinea pigs. PLoS Pathog *14*, e1007095.

Hutchinson, L., Browne, H., Wargent, V., Davis-Poynter, N., Primorac, S., Goldsmith, K., Minson, A.C., and Johnson, D.C. (1992). A novel herpes simplex virus glycoprotein, gL, forms a complex with glycoprotein H (gH) and affects normal folding and surface expression of gH. J Virol *66*, 2240-2250.

Hutt-Fletcher, L.M. (2007). Epstein-Barr virus entry. J Virol *81*, 7825-7832.

Jackson, J.O., Lin, E., Spear, P.G., and Longnecker, R. (2010). Insertion mutations in herpes simplex virus 1 glycoprotein H reduce cell surface expression, slow the rate of cell fusion, or abrogate functions in cell fusion and viral entry. J Virol *84*, 2038-2046.

Jambunathan, N., Charles, A.S., Subramanian, R., Saied, A.A., Naderi, M., Rider, P., Brylinski, M., Chouljenko, V.N., and Kousoulas, K.G. (2015). Deletion of a Predicted beta-Sheet Domain within the Amino Terminus of Herpes Simplex Virus Glycoprotein K Conserved among Alphaherpesviruses Prevents Virus Entry into Neuronal Axons. J Virol *90*, 2230-2239.

Johnson, D.C., Burke, R.L., and Gregory, T. (1990). Soluble forms of herpes simplex virus glycoprotein D bind to a limited number of cell surface receptors and inhibit virus entry into cells. J Virol *64*, 2569-2576.

Jones, N.A., and Geraghty, R.J. (2004). Fusion activity of lipid-anchored envelope glycoproteins of herpes simplex virus type 1. Virology *324*, 213-228.

Kadlec, J., Loureiro, S., Abrescia, N.G., Stuart, D.I., and Jones, I.M. (2008). The postfusion structure of baculovirus gp64 supports a unified view of viral fusion machines. Nature structural and molecular biology *15*, 1024-1030.

Karaba, A.H., Kopp, S.J., and Longnecker, R. (2011). Herpesvirus entry mediator and nectin-1 mediate herpes simplex virus 1 infection of the murine cornea. J Virol *85*, 10041-10047.

Karaba, A.H., Kopp, S.J., and Longnecker, R. (2012). Herpesvirus entry mediator is a serotype specific determinant of pathogenesis in ocular herpes.

Proceedings of the National Academy of Sciences of the United States of America *109*, 20649-20654.

Kim, I.J., Chouljenko, V.N., Walker, J.D., and Kousoulas, K.G. (2013). Herpes simplex virus 1 glycoprotein M and the membrane-associated protein UL11 are required for virus-induced cell fusion and efficient virus entry. J Virol *87*, 8029-8037.

Klupp, B.G., Fuchs, W., Weiland, E., and Mettenleiter, T.C. (1997). Pseudorabies virus glycoprotein L is necessary for virus infectivity but dispensable for virion localization of glycoprotein H. J Virol *71*, 7687-7695.

Klupp, B.G., and Mettenleiter, T.C. (1999). Glycoprotein gL-independent infectivity of pseudorabies virus is mediated by a gD-gH fusion protein. J Virol *73*, 3014-3022.

Klupp, B.G., Nixdorf, R., and Mettenleiter, T.C. (2000). Pseudorabies virus glycoprotein M inhibits membrane fusion. J Virol *74*, 6760-6768.

Klyachkin, Y.M., Stoops, K.D., and Geraghty, R.J. (2006). Herpes simplex virus type 1 glycoprotein L mutants that fail to promote trafficking of glycoprotein H and fail to function in fusion can induce binding of glycoprotein L-dependent anti-glycoprotein H antibodies. J Gen Virol *87*, 759-767.

Komala Sari, T., Gianopulos, K.A., Weed, D.J., Schneider, S.M., Pritchard, S.M., and Nicola, A.V. (2020). Herpes Simplex Virus Glycoprotein C Regulates Low-pH Entry. mSphere *5*.

Kopp, S.J., Banisadr, G., Glajch, K., Maurer, U.E., Grunewald, K., Miller, R.J., Osten, P., and Spear, P.G. (2009). Infection of neurons and encephalitis after intracranial inoculation of herpes simplex virus requires the entry receptor nectin-1. Proceedings of the National Academy of Sciences of the United States of America *106*, 17916-17920.

Koyama, A.H., and Uchida, T. (1987). The mode of entry of herpes simplex virus type 1 into Vero cells. Microbiol Immunol *31*, 123-130.

Krummenacher, C., Baribaud, F., Ponce de Leon, M., Baribaud, I., Whitbeck, J.C., Xu, R., Cohen, G.H., and Eisenberg, R.J. (2004). Comparative usage of herpesvirus entry mediator A and nectin-1 by laboratory strains and clinical isolates of herpes simplex virus. Virology *322*, 286-299.

Krummenacher, C., Baribaud, I., Ponce de Leon, M., Whitbeck, J.C., Lou, H., Cohen, G.H., and Eisenberg, R.J. (2000). Localization of a binding site for herpes simplex virus glycoprotein D on herpesvirus entry mediator C by using antireceptor monoclonal antibodies. J Virol *74*, 10863-10872.

Krummenacher, C., Nicola, A.V., Whitbeck, J.C., Lou, H., Hou, W., Lambris, J.D., Geraghty, R.J., Spear, P.G., Cohen, G.H., and Eisenberg, R.J. (1998). Herpes simplex virus glycoprotein D can bind to poliovirus receptor-related protein 1 or herpesvirus entry mediator, two structurally unrelated mediators of virus entry. J Virol *72*, 7064-7074.

Krummenacher, C., Rux, A.H., Whitbeck, J.C., Ponce-de-Leon, M., Lou, H., Baribaud, I., Hou, W., Zou, C., Geraghty, R.J., Spear, P.G., *et al.* (1999). The first immunoglobulin-like domain of HveC is sufficient to bind herpes simplex virus gD with full affinity, while the third domain is involved in oligomerization of HveC. J Virol *73*, 8127-8137.

Krummenacher, C., Supekar, V.M., Whitbeck, J.C., Lazear, E., Connolly, S.A., Eisenberg, R.J., Cohen, G.H., Wiley, D.C., and Carfi, A. (2005). Structure of unliganded HSV gD reveals a mechanism for receptor-mediated activation of virus entry. The EMBO journal *24*, 4144-4153.

Laquerre, S., Argnani, R., Anderson, D.B., Zucchini, S., Manservigi, R., and Glorioso, J.C. (1998). Heparan sulfate proteoglycan binding by herpes simplex virus type 1 glycoproteins B and C, which differ in their contributions to virus attachment, penetration, and cell-to-cell spread. J Virol *72*, 6119-6130.

Lawrence, R., Yabe, T., Hajmohammadi, S., Rhodes, J., McNeely, M., Liu, J., Lamperti, E.D., Toselli, P.A., Lech, M., Spear, P.G., *et al.* (2007). The principal neuronal gD-type 3-O-sulfotransferases and their products in central and peripheral nervous system tissues. Matrix Biol *26*, 442-455.

Lazear, E., Carfi, A., Whitbeck, J.C., Cairns, T.M., Krummenacher, C., Cohen, G.H., and Eisenberg, R.J. (2008). Engineered disulfide bonds in herpes simplex virus type 1 gD separate receptor binding from fusion initiation and viral entry. J Virol *82*, 700-709.

Lazear, E., Whitbeck, J.C., Ponce-de-Leon, M., Cairns, T.M., Willis, S.H., Zuo, Y., Krummenacher, C., Cohen, G.H., and Eisenberg, R.J. (2012). Antibody-induced conformational changes in herpes simplex virus glycoprotein gD reveal new targets for virus neutralization. J Virol *86*, 1563-1576.

Lazear, E., Whitbeck, J.C., Zuo, Y., Carfi, A., Cohen, G.H., Eisenberg, R.J., and Krummenacher, C. (2014). Induction of conformational changes at the N-terminus of herpes simplex virus glycoprotein D upon binding to HVEM and nectin-1. Virology *448*, 185-195.

Lee, C.C., Lin, L.L., Chan, W.E., Ko, T.P., Lai, J.S., and Wang, A.H. (2013). Structural basis for the antibody neutralization of herpes simplex virus. Acta crystallographica Section D, Biological crystallography *69*, 1935-1945.

Leiva-Torres, G.A., Rochette, P.A., and Pearson, A. (2010). Differential importance of highly conserved residues in UL24 for herpes simplex virus 1 replication in vivo and reactivation. J Gen Virol *91*, 1109-1116.

Li, A., Lu, G., Qi, J., Wu, L., Tian, K., Luo, T., Shi, Y., Yan, J., and Gao, G.F. (2017a). Structural basis of nectin-1 recognition by pseudorabies virus glycoprotein D. PLoS Pathog *13*, e1006314.

Li, X., Yang, F., Hu, X., Tan, F., Qi, J., Peng, R., Wang, M., Chai, Y., Hao, L., Deng, J., *et al.* (2017b). Two classes of protective antibodies against Pseudorabies virus variant glycoprotein B: Implications for vaccine design. PLoS Pathog *13*, e1006777.

Lin, E., and Spear, P.G. (2007). Random linker-insertion mutagenesis to identify functional domains of herpes simplex virus type 1 glycoprotein B. Proceedings of the National Academy of Sciences of the United States of America *104*, 13140-13145.

Locksley, R.M., Killeen, N., and Lenardo, M.J. (2001). The TNF and TNF receptor superfamilies: integrating mammalian biology. Cell *104*, 487-501.

Lopez, M., Cocchi, F., Menotti, L., Avitabile, E., Dubreuil, P., and Campadelli-Fiume, G. (2000). Nectin2alpha (PRR2alpha or HveB) and nectin2delta are low-efficiency mediators for entry of herpes simplex virus mutants carrying the Leu25Pro substitution in glycoprotein D. J Virol *74*, 1267-1274.

Lopper, M., and Compton, T. (2004). Coiled-coil domains in glycoproteins B and H are involved in human cytomegalovirus membrane fusion. J Virol *78*, 8333-8341.

Lu, G., Zhang, N., Qi, J., Li, Y., Chen, Z., Zheng, C., Gao, G.F., and Yan, J. (2014). Crystal structure of herpes simplex virus 2 gD bound to nectin-1 reveals a conserved mode of receptor recognition. J Virol *88*, 13678-13688.

Maresova, L., Pasieka, T.J., and Grose, C. (2001). Varicella-zoster Virus gB and gE coexpression, but not gB or gE alone, leads to abundant fusion and syncytium formation equivalent to those from gH and gL coexpression. J Virol *75*, 9483-9492.

Mata, M., Zhang, M., Hu, X., and Fink, D.J. (2001). HveC (nectin-1) is expressed at high levels in sensory neurons, but not in motor neurons, of the rat peripheral nervous system. Journal of neurovirology *7*, 476-480.

Matsuura, H., Kirschner, A.N., Longnecker, R., and Jardetzky, T.S. (2010). Crystal structure of the Epstein-Barr virus (EBV) glycoprotein H/glycoprotein L (gH/gL) complex. Proceedings of the National Academy of Sciences of the United States of America *107*, 22641-22646.

Maurer, U.E., Zeev-Ben-Mordehai, T., Pandurangan, A.P., Cairns, T.M., Hannah, B.P., Whitbeck, J.C., Eisenberg, R.J., Cohen, G.H., Topf, M., Huiskonen, J.T., *et al.* (2013). The structure of herpesvirus fusion glycoprotein B-bilayer complex reveals the protein-membrane and lateral protein-protein interaction. Structure *21*, 1396-1405.

Melikyan, G.B., Markosyan, R.M., Hemmati, H., Delmedico, M.K., Lambert, D.M., and Cohen, F.S. (2000). Evidence that the transition of HIV-1 gp41 into a six-helix bundle, not the bundle configuration, induces membrane fusion. J Cell Biol *151*, 413-423.

Menotti, L., Cerretani, A., Hengel, H., and Campadelli-Fiume, G. (2008). Construction of a fully retargeted herpes simplex virus 1 recombinant capable of entering cells solely via human epidermal growth factor receptor 2. J Virol *82*, 10153-10161.

Milne, R.S., Connolly, S.A., Krummenacher, C., Eisenberg, R.J., and Cohen, G.H. (2001). Porcine HveC, a member of the highly conserved HveC/nectin 1 family, is a functional alphaherpesvirus receptor. Virology *281*, 315-328.

Milne, R.S., Nicola, A.V., Whitbeck, J.C., Eisenberg, R.J., and Cohen, G.H. (2005). Glycoprotein D receptor-dependent, low-pH-independent endocytic entry of herpes simplex virus type 1. J Virol *79*, 6655-6663.

Mizoguchi, A., Nakanishi, H., Kimura, K., Matsubara, K., Ozaki-Kuroda, K., Katata, T., Honda, T., Kiyohara, Y., Heo, K., Higashi, M., *et al.* (2002). Nectin: an adhesion molecule involved in formation of synapses. J Cell Biol *156*, 555-565.

Montalvo, E.A., and Grose, C. (1986). Neutralization epitope of varicella zoster virus on native viral glycoprotein gp118 (VZV glycoprotein gpIII). Virology *149*, 230-241.

Montgomery, R.I., Warner, M.S., Lum, B.J., and Spear, P.G. (1996). Herpes simplex virus-1 entry into cells mediated by a novel member of the TNF/NGF receptor family. Cell *87*, 427-436.

Muggeridge, M.I. (2000). Characterization of cell-cell fusion mediated by herpes simplex virus 2 glycoproteins gB, gD, gH and gL in transfected cells. J Gen Virol *81*, 2017-2027.

Muggeridge, M.I. (2012). Glycoprotein B of herpes simplex virus 2 has more than one intracellular conformation and is altered by low pH. J Virol *86*, 6444-6456.

Muggeridge, M.I., Grantham, M.L., and Johnson, F.B. (2004). Identification of syncytial mutations in a clinical isolate of herpes simplex virus 2. Virology *328*, 244-253.

Musarrat, F., Jambunathan, N., Rider, P.J.F., Chouljenko, V.N., and Kousoulas, K.G. (2018). The Amino Terminus of Herpes Simplex Virus 1 Glycoprotein K (gK) Is Required for gB Binding to Akt, Release of Intracellular Calcium, and Fusion of the Viral Envelope with Plasma Membranes. J Virol *92*.

Nakanishi, H., and Takai, Y. (2004). Roles of nectins in cell adhesion, migration and polarization. Biol Chem *385*, 885-892.

Nicola, A.V. (2016). Herpesvirus Entry into Host Cells Mediated by Endosomal Low pH. Traffic *17*, 965-975.

Nicola, A.V., Hou, J., Major, E.O., and Straus, S.E. (2005). Herpes simplex virus type 1 enters human epidermal keratinocytes, but not neurons, via a pH-dependent endocytic pathway. J Virol *79*, 7609-7616.

Nicola, A.V., McEvoy, A.M., and Straus, S.E. (2003). Roles for endocytosis and low pH in herpes simplex virus entry into HeLa and Chinese hamster ovary cells. J Virol *77*, 5324-5332.

Nicola, A.V., Peng, C., Lou, H., Cohen, G.H., and Eisenberg, R.J. (1997). Antigenic structure of soluble herpes simplex virus (HSV) glycoprotein D correlates with inhibition of HSV infection. J Virol *71*, 2940-2946.

Nicola, A.V., Ponce de Leon, M., Xu, R., Hou, W., Whitbeck, J.C., Krummenacher, C., Montgomery, R.I., Spear, P.G., Eisenberg, R.J., and Cohen, G.H. (1998). Monoclonal antibodies to distinct sites on herpes simplex virus (HSV) glycoprotein D block HSV binding to HVEM. J Virol *72*, 3595-3601.

Nicola, A.V., and Straus, S.E. (2004). Cellular and viral requirements for rapid endocytic entry of herpes simplex virus. J Virol *78*, 7508-7517.

Nishimura, M., and Mori, Y. (2019). Entry of betaherpesviruses. Adv Virus Res *104*, 283-312.

Novotny, M.J., Parish, M.L., and Spear, P.G. (1996). Variability of herpes simplex virus 1 gL and anti-gL antibodies that inhibit cell fusion but not viral infectivity. Virology *221*, 1-13.

Okubo, Y., Uchida, H., Wakata, A., Suzuki, T., Shibata, T., Ikeda, H., Yamaguchi, M., Cohen, J.B., Glorioso, J.C., Tagaya, M., *et al.* (2016). Syncytial Mutations Do Not Impair the Specificity of Entry and Spread of a Glycoprotein D Receptor-Retargeted Herpes Simplex Virus. J Virol *90*, 11096-11105.

Oliver, S.L., Brady, J.J., Sommer, M.H., Reichelt, M., Sung, P., Blau, H.M., and Arvin, A.M. (2013). An immunoreceptor tyrosine-based inhibition motif in varicella-zoster virus glycoprotein B regulates cell fusion and skin pathogenesis. Proceedings of the National Academy of Sciences of the United States of America *110*, 1911-1916.

Oliver, S.L., Sommer, M., Zerboni, L., Rajamani, J., Grose, C., and Arvin, A.M. (2009). Mutagenesis of varicella-zoster virus glycoprotein B: putative fusion loop residues are essential for viral replication, and the furin cleavage motif contributes to pathogenesis in skin tissue in vivo. J Virol *83*, 7495-7506.

Oliver, S.L., Yang, E., and Arvin, A.M. (2016). Varicella-Zoster Virus Glycoproteins: Entry, Replication, and Pathogenesis. Curr Clin Microbiol Rep *3*, 204-215.

Ono, E., Amagai, K., Taharaguchi, S., Tomioka, Y., Yoshino, S., Watanabe, Y., Cherel, P., Houdebine, L.M., Adam, M., Eloit, M., *et al.* (2004). Transgenic mice expressing a soluble form of porcine nectin-1/herpesvirus entry mediator C as a model for pseudorabies-resistant livestock. Proceedings of the National Academy of Sciences of the United States of America *101*, 16150-16155.

Parry, C., Bell, S., Minson, T., and Browne, H. (2005). Herpes simplex virus type 1 glycoprotein H binds to alphavbeta3 integrins. J Gen Virol *86*, 7-10.

Peng, R., Zhang, S., Cui, Y., Shi, Y., Gao, G.F., and Qi, J. (2017). Structures of human-infecting Thogotovirus fusogens support a common ancestor with insect baculovirus. Proceedings of the National Academy of Sciences of the United States of America *114*, E8905-E8912.

Peng, T., Ponce de Leon, M., Novotny, M.J., Jiang, H., Lambris, J.D., Dubin, G., Spear, P.G., Cohen, G.H., and Eisenberg, R.J. (1998a). Structural and antigenic analysis of a truncated form of the herpes simplex virus glycoprotein gH-gL complex. J Virol *72*, 6092-6103.

Peng, T., Ponce-de-Leon, M., Jiang, H., Dubin, G., Lubinski, J.M., Eisenberg, R.J., and Cohen, G.H. (1998b). The gH-gL complex of herpes simplex virus (HSV) stimulates neutralizing antibody and protects mice against HSV type 1 challenge. J Virol *72*, 65-72.

Perez-Romero, P., Perez, A., Capul, A., Montgomery, R., and Fuller, A.O. (2005). Herpes simplex virus entry mediator associates in infected cells in a complex with viral proteins gD and at least gH. J Virol *79*, 4540-4544.

Pertel, P.E., Fridberg, A., Parish, M.L., and Spear, P.G. (2001). Cell fusion induced by herpes simplex virus glycoproteins gB, gD, and gH-gL requires a gD receptor but not necessarily heparan sulfate. Virology *279*, 313-324.

Petrovic, B., Leoni, V., Gatta, V., Zaghini, A., Vannini, A., and Campadelli-Fiume, G. (2018). Dual Ligand Insertion in gB and gD of Oncolytic Herpes Simplex Viruses for Retargeting to a Producer Vero Cell Line and to Cancer Cells. J Virol *92*.

Raftery, M.J., Behrens, C.K., Muller, A., Krammer, P.H., Walczak, H., and Schonrich, G. (1999). Herpes simplex virus type 1 infection of activated cytotoxic T cells: Induction of fratricide as a mechanism of viral immune evasion. J Exp Med *190*, 1103-1114.

Rahn, E., Petermann, P., Hsu, M.J., Rixon, F.J., and Knebel-Morsdorf, D. (2011). Entry pathways of herpes simplex virus type 1 into human keratinocytes are dynamin- and cholesterol-dependent. PloS one *6*, e25464.

Richart, S.M., Simpson, S.A., Krummenacher, C., Whitbeck, J.C., Pizer, L.I., Cohen, G.H., Eisenberg, R.J., and Wilcox, C.L. (2003). Entry of herpes simplex virus type 1 into primary sensory neurons in vitro is mediated by Nectin-1/HveC. J Virol *77*, 3307-3311.

Roberts, S.R., Ponce de Leon, M., Cohen, G.H., and Eisenberg, R.J. (1991). Analysis of the intracellular maturation of the herpes simplex virus type 1 glycoprotein gH in infected and transfected cells. Virology *184*, 609-624.

Roche, S., Albertini, A.A., Lepault, J., Bressanelli, S., and Gaudin, Y. (2008). Structures of vesicular stomatitis virus glycoprotein: membrane fusion revisited. Cell Mol Life Sci *65*, 1716-1728.

Roche, S., Bressanelli, S., Rey, F.A., and Gaudin, Y. (2006). Crystal structure of the low-pH form of the vesicular stomatitis virus glycoprotein G. Science *313*, 187-191.

Roche, S., Rey, F.A., Gaudin, Y., and Bressanelli, S. (2007). Structure of the prefusion form of the vesicular stomatitis virus glycoprotein G. Science *315*, 843-848.

Rodger, G., Boname, J., Bell, S., and Minson, T. (2001). Assembly and organization of glycoproteins B, C, D, and H in herpes simplex virus type 1 particles lacking individual glycoproteins: No evidence for the formation of a complex of these molecules. J Virol *75*, 710-716.

Rodriguez, J.E., Moninger, T., and Grose, C. (1993). Entry and egress of varicella virus blocked by same anti-gH monoclonal antibody. Virology *196*, 840-844.

Rogalin, H.B., and Heldwein, E.E. (2015). Interplay between the Herpes Simplex Virus 1 gB Cytodomain and the gH Cytotail during Cell-Cell Fusion. J Virol *89*, 12262-12272.

Roop, C., Hutchinson, L., and Johnson, D.C. (1993). A mutant herpes simplex virus type 1 unable to express glycoprotein L cannot enter cells, and its particles lack glycoprotein H. J Virol *67*, 2285-2297.

Rux, A.H., Lou, H., Lambris, J.D., Friedman, H.M., Eisenberg, R.J., and Cohen, G.H. (2002). Kinetic analysis of glycoprotein C of herpes simplex virus types 1 and 2 binding to heparin, heparan sulfate, and complement component C3b. Virology *294*, 324-332.

Rux, A.H., Willis, S.H., Nicola, A.V., Hou, W., Peng, C., Lou, H., Cohen, G.H., and Eisenberg, R.J. (1998). Functional region IV of glycoprotein D from herpes simplex virus modulates glycoprotein binding to the herpesvirus entry mediator. J Virol *72*, 7091-7098.

Ryckman, B.J., Jarvis, M.A., Drummond, D.D., Nelson, J.A., and Johnson, D.C. (2006). Human cytomegalovirus entry into epithelial and endothelial cells depends on genes UL128 to UL150 and occurs by endocytosis and low-pH fusion. J Virol *80*, 710-722.

Santoro, F., Greenstone, H.L., Insinga, A., Liszewski, M.K., Atkinson, J.P., Lusso, P., and Berger, E.A. (2003). Interaction of glycoprotein H of human herpesvirus 6 with the cellular receptor CD46. J Biol Chem *278*, 25964-25969.

Sarfo, A., Starkey, J., Mellinger, E., Zhang, D., Chadha, P., Carmichael, J., and Wills, J.W. (2017). The UL21 Tegument Protein of Herpes Simplex Virus 1 Is Differentially Required for the Syncytial Phenotype. J Virol *91*.

Sathiyamoorthy, K., Hu, Y.X., Mohl, B.S., Chen, J., Longnecker, R., and Jardetzky, T.S. (2016). Structural basis for Epstein-Barr virus host cell tropism mediated by gp42 and gHgL entry glycoproteins. Nature communications *7*, 13557.

Satoh, T., Arii, J., Suenaga, T., Wang, J., Kogure, A., Uehori, J., Arase, N., Shiratori, I., Tanaka, S., Kawaguchi, Y., *et al.* (2008). PILRalpha is a herpes simplex virus-1 entry coreceptor that associates with glycoprotein B. Cell *132*, 935-944.

Saw, W.T., Matsuda, Z., Eisenberg, R.J., Cohen, G.H., and Atanasiu, D. (2015). Using a split luciferase assay (SLA) to measure the kinetics of cell-

cell fusion mediated by herpes simplex virus glycoproteins. Methods *90*, 68-75.

Schmidt, J., Gerdts, V., Beyer, J., Klupp, B.G., and Mettenleiter, T.C. (2001). Glycoprotein D-independent infectivity of pseudorabies virus results in an alteration of in vivo host range and correlates with mutations in glycoproteins B and H. J Virol *75*, 10054-10064.

Schroder, C., and Keil, G.M. (1999). Bovine herpesvirus 1 requires glycoprotein H for infectivity and direct spreading and glycoproteins gH(W450) and gB for glycoprotein D-independent cell-to-cell spread. J Gen Virol *80 (Pt 1)*, 57-61.

Showalter, S.D., Zweig, M., and Hampar, B. (1981). Monoclonal antibodies to herpes simplex virus type 1 proteins, including the immediate-early protein ICP 4. Infection and immunity *34*, 684-692.

Shukla, D., Liu, J., Blaiklock, P., Shworak, N.W., Bai, X., Esko, J.D., Cohen, G.H., Eisenberg, R.J., Rosenberg, R.D., and Spear, P.G. (1999). A novel role for 3-O-sulfated heparan sulfate in herpes simplex virus 1 entry. Cell *99*, 13-22.

Shukla, S.Y., Singh, Y.K., and Shukla, D. (2009). Role of nectin-1, HVEM, and PILR-alpha in HSV-2 entry into human retinal pigment epithelial cells. Investigative ophthalmology and visual science *50*, 2878-2887.

Si, Z., Zhang, J., Shivakoti, S., Atanasov, I., Tao, C.L., Hui, W.H., Zhou, K., Yu, X., Li, W., Luo, M.*, et al.* (2018). Different functional states of fusion protein gB revealed on human cytomegalovirus by cryo electron tomography with Volta phase plate. PLoS Pathog *14*, e1007452.

Siekavizza-Robles, C.R., Dollery, S.J., and Nicola, A.V. (2010). Reversible conformational change in herpes simplex virus glycoprotein B with fusion-from-without activity is triggered by mildly acidic pH. Virology journal *7*, 352.

Silverman, J.L., Greene, N.G., King, D.S., and Heldwein, E.E. (2012). Membrane requirement for folding of the herpes simplex virus 1 gB cytodomain suggests a unique mechanism of fusion regulation. J Virol *86*, 8171-8184.

Silverman, J.L., and Heldwein, E.E. (2013). Mutations in the cytoplasmic tail of herpes simplex virus 1 gH reduce the fusogenicity of gB in transfected cells. J Virol *87*, 10139-10147.

Silverman, J.L., Sharma, S., Cairns, T.M., and Heldwein, E.E. (2010). Fusion-deficient insertion mutants of herpes simplex virus 1 glycoprotein B adopt the trimeric postfusion conformation. J Virol *84*, 2001-2012.

Sodeik, B., Ebersold, M.W., and Helenius, A. (1997). Microtubule-mediated transport of incoming herpes simplex virus 1 capsids to the nucleus. J Cell Biol *136*, 1007-1021.

Spear, P.G., Manoj, S., Yoon, M., Jogger, C.R., Zago, A., and Myscofski, D. (2006). Different receptors binding to distinct interfaces on herpes simplex virus gD can trigger events leading to cell fusion and viral entry. Virology *344*, 17-24.

Stampfer, S.D., Lou, H., Cohen, G.H., Eisenberg, R.J., and Heldwein, E.E. (2010). Structural basis of local, pH-dependent conformational changes in glycoprotein B from herpes simplex virus type 1. J Virol *84*, 12924-12933.

Suenaga, T., Matsumoto, M., Arisawa, F., Kohyama, M., Hirayasu, K., Mori, Y., and Arase, H. (2015). Sialic Acids on Varicella-Zoster Virus Glycoprotein B Are Required for Cell-Cell Fusion. J Biol Chem *290*, 19833-19843.

Suenaga, T., Satoh, T., Somboonthum, P., Kawaguchi, Y., Mori, Y., and Arase, H. (2010). Myelin-associated glycoprotein mediates membrane fusion and entry of neurotropic herpesviruses. Proceedings of the National Academy of Sciences of the United States of America *107*, 866-871.

Takahashi, K., Nakanishi, H., Miyahara, M., Mandai, K., Satoh, K., Satoh, A., Nishioka, H., Aoki, J., Nomoto, A., Mizoguchi, A., *et al.* (1999). Nectin/PRR: an immunoglobulin-like cell adhesion molecule recruited to cadherin-based adherens junctions through interaction with Afadin, a PDZ domain-containing protein. J Cell Biol *145*, 539-549.

Taylor, J.M., Lin, E., Susmarski, N., Yoon, M., Zago, A., Ware, C.F., Pfeffer, K., Miyoshi, J., Takai, Y., and Spear, P.G. (2007). Alternative entry receptors for herpes simplex virus and their roles in disease. Cell Host Microbe *2*, 19-28.

Tiwari, V., O'Donnell, C., Copeland, R.J., Scarlett, T., Liu, J., and Shukla, D. (2007). Soluble 3-O-sulfated heparan sulfate can trigger herpes simplex virus type 1 entry into resistant Chinese hamster ovary (CHO-K1) cells. J Gen Virol *88*, 1075-1079.

Turner, A., Bruun, B., Minson, T., and Browne, H. (1998). Glycoproteins gB, gD, and gHgL of herpes simplex virus type 1 are necessary and sufficient to mediate membrane fusion in a Cos cell transfection system. J Virol *72*, 873-875.

Tuzmen, C., Cairns, T.M., Atanasiu, D., Lou, H., Saw, W.T., Hall, B.L., Cohen, J.B., Cohen, G.H., and Glorioso, J.C. (2020). Point Mutations in Retargeted gD Eliminate the Sensitivity of EGFR/EGFRvIII-Targeted HSV to Key Neutralizing Antibodies. Mol Ther Methods Clin Dev *16*, 145-154.

Uchida, H., Chan, J., Goins, W.F., Grandi, P., Kumagai, I., Cohen, J.B., and Glorioso, J.C. (2010). A double mutation in glycoprotein gB compensates for ineffective gD-dependent initiation of herpes simplex virus type 1 infection. J Virol *84*, 12200-12209.

Vallbracht, M., Backovic, M., Klupp, B.G., Rey, F.A., and Mettenleiter, T.C. (2019). Common characteristics and unique features: A comparison of the fusion machinery of the alphaherpesviruses Pseudorabies virus and Herpes simplex virus. Adv Virus Res *104*, 225-281.

Vallbracht, M., Brun, D., Tassinari, M., Vaney, M.C., Pehau-Arnaudet, G., Guardado-Calvo, P., Haouz, A., Klupp, B.G., Mettenleiter, T.C., Rey, F.A., *et al.* (2017). Structure-function dissection of the Pseudorabies virus glycoprotein B fusion loops. J Virol *92*, e01203-01201.

Vallbracht, M., Fuchs, W., Klupp, B.G., and Mettenleiter, T.C. (2018). Functional Relevance of the Transmembrane Domain and Cytoplasmic Tail of the Pseudorabies Virus Glycoprotein H for Membrane Fusion. J Virol *92*, e00376-00318.

Van de Walle, G.R., Peters, S.T., VanderVen, B.C., O'Callaghan, D.J., and Osterrieder, N. (2008). Equine herpesvirus 1 entry via endocytosis is facilitated by alphaV integrins and an RSD motif in glycoprotein D. J Virol *82*, 11859-11868.

Vitu, E., Sharma, S., Stampfer, S.D., and Heldwein, E.E. (2013). Extensive mutagenesis of the HSV-1 gB ectodomain reveals remarkable stability of its postfusion form. J Mol Biol *425*, 2056-2071.

Vleck, S.E., Oliver, S.L., Brady, J.J., Blau, H.M., Rajamani, J., Sommer, M.H., and Arvin, A.M. (2011). Structure-function analysis of varicella-zoster virus glycoprotein H identifies domain-specific roles for fusion and skin tropism. Proceedings of the National Academy of Sciences of the United States of America *108*, 18412-18417.

Walker, E.B., Pritchard, S.M., Cunha, C.W., Aguilar, H.C., and Nicola, A.V. (2015). Polyethylene glycol-mediated fusion of herpes simplex type 1 virions with the plasma membrane of cells that support endocytic entry. Virology journal *12*, 190.

Wanas, E., Efler, S., Ghosh, K., and Ghosh, H.P. (1999). Mutations in the conserved carboxy-terminal hydrophobic region of glycoprotein gB affect infectivity of herpes simplex virus. J Gen Virol *80 (Pt 12)*, 3189-3198.

Wang, D., and Shenk, T. (2005). Human cytomegalovirus virion protein complex required for epithelial and endothelial cell tropism. Proceedings of the National Academy of Sciences of the United States of America *102*, 18153-18158.

Wang, J., Fan, Q., Satoh, T., Arii, J., Lanier, L.L., Spear, P.G., Kawaguchi, Y., and Arase, H. (2009). Binding of herpes simplex virus glycoprotein B (gB) to paired immunoglobulin-like type 2 receptor alpha depends on specific sialylated O-linked glycans on gB. J Virol *83*, 13042-13045.

Waning, D.L., Russell, C.J., Jardetzky, T.S., and Lamb, R.A. (2004). Activation of a paramyxovirus fusion protein is modulated by inside-out signaling from the cytoplasmic tail. Proceedings of the National Academy of Sciences of the United States of America *101*, 9217-9222.

Warner, M.S., Geraghty, R.J., Martinez, W.M., Montgomery, R.I., Whitbeck, J.C., Xu, R., Eisenberg, R.J., Cohen, G.H., and Spear, P.G. (1998). A cell surface protein with herpesvirus entry activity (HveB) confers susceptibility to infection by mutants of herpes simplex virus type 1, herpes simplex virus type 2, and pseudorabies virus. Virology *246*, 179-189.

Weed, D.J., Dollery, S.J., Komala Sari, T., and Nicola, A.V. (2018). Acidic pH Mediates Changes in Antigenic and Oligomeric Conformation of Herpes Simplex Virus gB and Is a Determinant of Cell-Specific Entry. J Virol *92*.

Whitbeck, J.C., Connolly, S.A., Willis, S.H., Hou, W., Krummenacher, C., Ponce de Leon, M., Lou, H., Baribaud, I., Eisenberg, R.J., and Cohen, G.H. (2001). Localization of the gD-binding region of the human herpes simplex virus receptor, HveA. J Virol *75*, 171-180.

Whitbeck, J.C., Muggeridge, M.I., Rux, A.H., Hou, W., Krummenacher, C., Lou, H., van Geelen, A., Eisenberg, R.J., and Cohen, G.H. (1999). The major neutralizing antigenic site on herpes simplex virus glycoprotein D overlaps a receptor-binding domain. J Virol *73*, 9879-9890.

Whitbeck, J.C., Peng, C., Lou, H., Xu, R., Willis, S.H., Ponce de Leon, M., Peng, T., Nicola, A.V., Montgomery, R.I., Warner, M.S., *et al.* (1997). Glycoprotein D of herpes simplex virus (HSV) binds directly to HVEM, a member of the tumor necrosis factor receptor superfamily and a mediator of HSV entry. J Virol *71*, 6083-6093.

White, J.M., Delos, S.E., Brecher, M., and Schornberg, K. (2008). Structures and mechanisms of viral membrane fusion proteins: multiple variations on a common theme. Crit Rev Biochem Mol Bio *43*, 189-219.

Willis, S.H., Rux, A.H., Peng, C., Whitbeck, J.C., Nicola, A.V., Lou, H., Hou, W., Salvador, L., Eisenberg, R.J., and Cohen, G.H. (1998). Examination of the kinetics of herpes simplex virus glycoprotein D binding to the herpesvirus entry mediator, using surface plasmon resonance. J Virol *72*, 5937-5947.

Wilson, D.W., Davis-Poynter, N., and Minson, A.C. (1994). Mutations in the cytoplasmic tail of herpes simplex virus glycoprotein H suppress cell fusion by a syncytial strain. J Virol *68*, 6985-6993.

Wittels, M., and Spear, P.G. (1991). Penetration of cells by herpes simplex virus does not require a low pH-dependent endocytic pathway. Virus Res *18*, 271-290.

Wu, P., Reed, W.M., and Lee, L.F. (2001). Glycoproteins H and L of Marek's disease virus form a hetero-oligomer essential for translocation and cell surface expression. Arch Virol *146*, 983-992.

Wyss, S., Dimitrov, A.S., Baribaud, F., Edwards, T.G., Blumenthal, R., and Hoxie, J.A. (2005). Regulation of human immunodeficiency virus type 1 envelope glycoprotein fusion by a membrane-interactive domain on the gp41 cytoplasmic tail. JVirol *79*, 12231-12241.

Xing, Y., Oliver, S.L., Nguyen, T., Ciferri, C., Nandi, A., Hickman, J., Giovani, C., Yang, E., Palladino, G., Grose, C., *et al.* (2015). A site of varicella-zoster virus vulnerability identified by structural studies of neutralizing antibodies bound to the glycoprotein complex gHgL. Proceedings of the National Academy of Sciences of the United States of America *112*, 6056-6061.

Yang, E., Arvin, A.M., and Oliver, S.L. (2014). The cytoplasmic domain of varicella-zoster virus glycoprotein H regulates syncytia formation and skin pathogenesis. PLoS Pathog *10*, e1004173.

Yang, E., Arvin, A.M., and Oliver, S.L. (2016). Role for the alphaV Integrin Subunit in Varicella-Zoster Virus-Mediated Fusion and Infection. J Virol *90*, 7567-7578.

Yang, F., Lin, S., Ye, F., Yang, J., Qi, J., Chen, Z., Lin, X., Wang, J., Yue, D., Cheng, Y., *et al.* (2020). Structural Analysis of Rabies Virus Glycoprotein Reveals pH-Dependent Conformational Changes and Interactions with a Neutralizing Antibody. Cell Host Microbe *27*, 441-453 e447.

Yin, H.S., Paterson, R.G., Wen, X., Lamb, R.A., and Jardetzky, T.S. (2005). Structure of the uncleaved ectodomain of the paramyxovirus (hPIV3) fusion protein. PNAS *102*, 9288-9293.

Yin, H.S., Wen, X., Paterson, R.G., Lamb, R.A., and Jardetzky, T.S. (2006). Structure of the parainfluenza virus 5 F protein in its metastable, prefusion conformation. Nature *439*, 38-44.

Yoon, M., Zago, A., Shukla, D., and Spear, P.G. (2003). Mutations in the N termini of herpes simplex virus type 1 and 2 gDs alter functional interactions with the entry/fusion receptors HVEM, nectin-2, and 3-O-sulfated heparan sulfate but not with nectin-1. J Virol *77*, 9221-9231.

Zago, A., Jogger, C.R., and Spear, P.G. (2004). Use of herpes simplex virus and pseudorabies virus chimeric glycoprotein D molecules to identify regions critical for membrane fusion. Proceedings of the National Academy of Sciences of the United States of America *101*, 17498-17503.

Zeev-Ben-Mordehai, T., Vasishtan, D., Hernandez Duran, A., Vollmer, B., White, P., Prasad Pandurangan, A., Siebert, C.A., Topf, M., and Grunewald, K. (2016). Two distinct trimeric conformations of natively membrane-anchored full-length herpes simplex virus 1 glycoprotein B. Proceedings of the National Academy of Sciences of the United States of America *113*, 4176-4181.

Zhang, N., Yan, J., Lu, G., Guo, Z., Fan, Z., Wang, J., Shi, Y., Qi, J., and Gao, G.F. (2011). Binding of herpes simplex virus glycoprotein D to nectin-1 exploits host cell adhesion. Nature communications *2*, 577.

Zhou, G., and Roizman, B. (2006). Construction and properties of a herpes simplex virus 1 designed to enter cells solely via the IL-13alpha2 receptor. Proceedings of the National Academy of Sciences of the United States of America *103*, 5508-5513.

Zhou, G., and Roizman, B. (2007). Separation of receptor-binding and profusogenic domains of glycoprotein D of herpes simplex virus 1 into distinct interacting proteins. Proceedings of the National Academy of Sciences of the United States of America *104*, 4142-4146.

Zhou, G., Ye, G.J., Debinski, W., and Roizman, B. (2002). Engineered herpes simplex virus 1 is dependent on IL13Ralpha 2 receptor for cell entry and independent of glycoprotein D receptor interaction. Proceedings of the National Academy of Sciences of the United States of America *99*, 15124-15129.

Chapter 3

Navigating the Cytoplasm: Delivery of the Alphaherpesvirus Genome to the Nucleus

Gregory A. Smith*

Department of Microbiology-Immunology, Northwestern University Feinberg School of Medicine, Chicago, IL, USA

*g-smith3@northwestern.edu

DOI: https://doi.org/10.21775/9781913652555.03

Abstract

Herpesviruses virions are large and complex structures that deliver their genetic content to nuclei upon entering cells. This property is not unusual as many other viruses including the adenoviruses, orthomyxoviruses, papillomaviruses, polyomaviruses, and retroviruses, do likewise. However, the means by which viruses in the *alphaherpesvirinae* subfamily accomplish this fundamental stage of the infectious cycle is tied to their defining ability to efficiently invade the nervous system. Fusion of the viral envelope with a cell membrane results in the deposition of the capsid, along with an assortment of tegument proteins, into the cytosol. Establishment of infection requires that the capsid traverse the cytosol, dock at a nuclear pore, and inject its genome into the nucleoplasm. Accumulating evidence indicates that the capsid is not the effector of this delivery process, but is instead shepherded by tegument proteins that remain capsid bound. At the same time, tegument proteins that are released from the capsid upon entry act to increase the susceptibility of the cell to the ensuing infection. Mucosal epithelial cells and neurons are both susceptible to alphaherpesvirus infection and, together, provide the niche to which these viruses have adapted. Although much has been revealed about the functions of *de novo* expressed tegument proteins during the late stages of assembly and egress, this chapter will specifically address the roles of tegument proteins brought into the cell with the incoming virion, and our current understanding of alphaherpesvirus genome delivery to nuclei.

Introduction

For many viruses, replication in epithelial cells lining the respiratory or gastrointestinal tracts amplifies and releases viral particles back into the environment during an acute infection. Although neuroinvasive viruses of the *alphaherpesviridae* subfamily similarly replicate in epithelial cells at exposed mucosal sites, they subsequently spread to innervating axons of sensory and autonomic neurons, which is followed by their retrograde axonal transport to neural soma resident in peripheral nervous system (PNS) ganglia (see Chapter 5). Whereas members of the beta and gammaherpesvirus subfamilies transmit between tissues by modulating receptor use (Borza and Hutt-Fletcher, 2002; Scrivano et al., 2011), the neuroinvasive alphaherpesviruses engage host receptors that are ubiquitously expressed (see Chapter 2). Despite this relative lack of selectivity, infections typically progress in a controlled manner whereby viruses productively infect mucosal epithelial cells and generally establish latent infections in the PNS (see Chapter 6). Although the ability to infect the nervous system by an axonal route is uncommon, alphaherpesviruses are uniquely proficient at it, and yet the ensuing infections are usually asymptomatic. In other words, these neurotropic viruses are highly neuroinvasive yet typically not neurovirulent. Nonetheless, these neuroinvasive infections can cause diseases associated with high morbidity and mortality. For example, herpes simplex keratitis and herpes simplex encephalitis in humans caused by herpes simplex virus type 1 (HSV-1), post-herpetic neuralgia in humans caused by varicella zoster virus (VZV), Aujeszky's disease in pigs and severe pruritus in cows caused by pseudorabies virus (PRV), equine herpes myeloencephalopathy in horses caused by equine herpesvirus 1 (EHV-1), and encephalitis in cows caused by bovine herpesvirus 5 (BHV-5).

Multiple infectious agents sporadically gain access to the nervous system by a hematogenous route, whereas poliovirus and rabies virus are examples of viruses that invade the nervous system axonally. However, unlike alphaherpesviruses, retrograde axonal transport of poliovirus and rabies virus often results in severe or fatal neurological disease (Ohka et al., 1998; Ren and Racaniello, 1992). Furthermore, the neuroinvasive ability of the latter two viruses is dependent upon tissue damage, presumably at neuromuscular junctions. Poliovirus is normally an enteric pathogen that rarely becomes neuroinvasive: an event that can be provoked by coincidental tissue damage such as that obtained by a needle injection (aka provocation poliomyelitis) (Gromeier and Wimmer, 1998). In the case of rabies virus, infection is typically initiated by an animal bite, which is an explicit tissue damage provocation. Both poliovirus and rabies virus are endocytosed at motor nerve endings and are transported retrogradely in axons within endosomes (Gluska et al., 2014; Lewis et al., 2000; Ohka et al., 2004; Salinas et al., 2010). Axons up-regulate retrograde axonal trafficking when damaged (Curtis et al., 1993; Hanz et al., 2003), which may inadvertently promote provocation-based invasion of these viruses. In contrast, alphaherpesviruses do not appear to require provocation as evidenced by the remarkably high incidence of these viruses in their respective host populations. In fact, rather than promoting neuroinvasion axon damage hinders alphaherpesvirus invasion of neurons, indicating that the mechanism used by alphaherpesviruses to engage retrograde axonal transport competes with the axon damage response rather than being supported by it (Koyuncu et al., 2013a). Coincident with these observations, alphaherpesvirus axonal trafficking is not reliant on endosomes but rather is mediated by direct viral engagement of the host cytoskeletal transport machinery in the cytosol (Antinone and Smith, 2006, 2010; Granzow et al., 1997; Luxton et al., 2005; Lycke et al., 1984). Therefore, deciphering the proficiency of alphaherpesvirus neuroinvasion requires knowledge of the fundamental process by which these viruses deliver their genomes to nuclei and the transport effectors that reside in a protein matrix located between the capsid and envelope: a structural referred to as the herpesvirus tegument.

Virion structure and ancestry

The virion structure is a defining characteristic of the Herpesvirales order. At 200-250 nm in diameter, the viral particles are slightly too small to be spatially resolved with a light microscope, but their architecture becomes apparent when viewed by electron microscopy (Morgan et al., 1953; Stackpole, 1969). Virions consist of an icosahedral capsid shell containing the viral genome, surrounded by tegument that, in turn, is enveloped by a lipid bilayer covered in glycoprotein spikes. There are roughly 700 spikes in the HSV-1 envelope (Grunewald et al., 2003). Upon contact with a susceptible cell, the viral particle undergoes a series of architectural rearrangements that coordinate capsid entry into the cell and its subsequent transport through the cytoplasm to the nucleus.

Herpesviruses are notable for being the only members of the *Duplodnaviria* that infect animals (Figure 1) (Baker et al., 2005; Duda et al., 2006; Krupovic and Koonin, 2017; Suhanovsky and Teschke, 2015; Twarock and Luque, 2019; Wikoff et al., 2000). This recently classified realm consists of viruses possessing a capsid shell based on a HK97-type major capsid protein (International Committee on Taxonomy of Viruses). Other *Duplodnaviria* include the tailed bacteriophage (Caudovirales) that, like herpesviruses, have large dsDNA genomes packaged into preassembled procapsid shells by an ATP-fueled motor (Heming et al., 2014; Newcomb et al., 1996; Poon and Roizman, 1993; Trus et al., 1996; Wikoff et al., 2000). The capsid shell of HK97-type viruses consists of varying numbers of capsomers that are reinforced at points of three-fold symmetry, a function that in herpesviruses is performed by the capsid triplex proteins (Dai and Zhou, 2018; Yuan et al., 2018). Genome packaging occurs through a portal, located at one of the twelve capsid vertices where a penton capsomer would otherwise reside (Dedeo et al., 2019; Kornfeind and Visalli, 2018; McElwee et al., 2018; Newcomb et al., 2001). In phages, the portal is often accompanied by a tail that extends from this unique capsid vertex. Because of the asymmetry in particle morphology that the tail provides, it is clear that the phage portal vertex also serves as the exit channel during genome release. Although there is evidence that

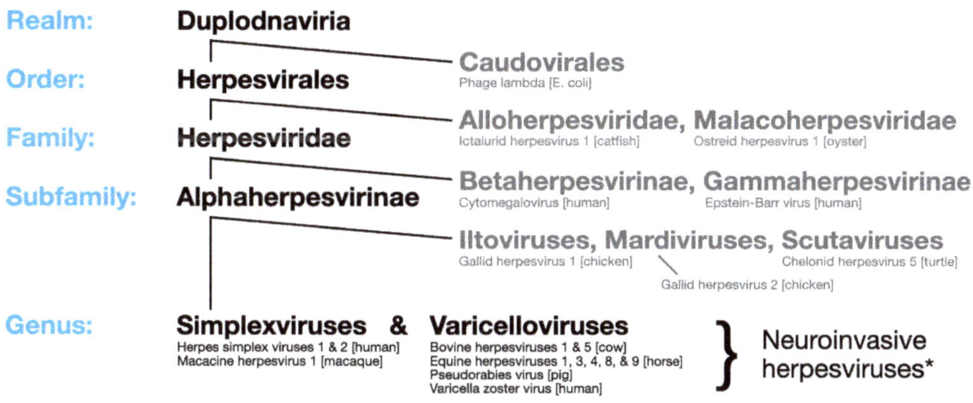

Figure 1. The neuroinvasive herpesviruses consist of members of the simplexvirus and varicellovirus genera. Divergent branches are shown in grey. Example species are shown below each classification with respective hosts in brackets. *There is evidence that iltoviruses may also be neuroinvasive (Williams et al., 1992).

herpesviruses capsids in extracellular virions have tails, such a structure has not been observed on capsids after entry into cells (Schmid et al., 2012).

The capsid interior consists primarily of the genome confined at high pressure, similarly to tailed bacteriophages (Bauer et al., 2013). The maturation protease, VP24, is the only protein known to reside with the genome inside the capsid shell (Kramer et al., 2011; Loret et al., 2008; Person and Desai, 1998; Sheaffer et al., 2001; Spear and Roizman, 1972; Stevenson et al., 1997). VP24 cleaves the protein scaffold that resides inside the capsid prior to genome packaging, allowing the genome to displace the scaffold from the capsid interior (Newcomb et al., 1996; Newcomb et al., 2000; Rixon and McNab, 1999). Despite a suggestion that VP24 could promote genome release from capsids, there is no evidence that VP24 contributes to initial infection (Jovasevic et al., 2008).

The alphaherpesvirus capsid shell consists of the HK97-type major capsid protein, VP5, arranged in 162 capsomers consisting of 150 hexons and 11 pentons, the latter occupying the vertices (triangulation number of 16), with the twelfth vertex consisting of a ring of 12 copies of pUL6 that forms the portal (Cardone et al., 2007). Each copy of VP5 contributing to hexons is capped by the small capsid protein, VP26, such that there are 955 copies of VP5 per capsid and 900 copies of VP26. The capsomers are oriented tip-to-tip instead of side-to-side, with the resulting gaps occupied by the triplex proteins VP19c and VP23 (Newcomb et al., 1993). The pentons provide rigidity to the capsid and are reinforced by pUL17 and pUL25 (Roos et al., 2009; Snijder et al., 2017), which together with the pUL36 tegument protein form five elongated structures that radiate from the pentons to the neighboring triplexes (Fan et al., 2014; Huet et al., 2016; Liu et al., 2017; Toropova et al., 2011; Trus et al., 2007). The pUL17/pUL25/pUL36 structure has been referred to as the C-capsid-specific component (CCSC), the capsid-vertex-specific component (CVSC), and the capsid-associated tegument complex (CATC).

Unlike tailed bacteriophages, herpesvirus virions have a tegument and an envelope in addition to the capsid (Owen et al., 2015). The majority of the tegument mass departs from the icosahedral symmetry of the underlying capsid; therefore, unlike the capsid, much of the tegument and the envelope are not constrained by a rigid architecture (Zhou et al., 1999). The interface between the alphaherpesvirus capsid and tegument resides at the capsid vertices, but the composition of this interface is not yet fully elucidated (Zhou et al., 1999). Part of this interface consists of the interaction between pUL17/pUL25 and the carboxyl terminus of the large tegument protein, pUL36 (Coller et al., 2007; Pasdeloup et al., 2009; Uetz et al., 2006). In a recent advance, the pUL17/pUL25/pUL36 interface was visualized and identified by cryo-electron microscopy at capsid vertices (Dai and Zhou, 2018; Liu et al., 2017; Wang et al., 2018). Although pUL36 is the only tegument protein for which a molecular interaction with the capsid is defined, the pUL16 tegument protein also associates with capsids (Meckes and Wills, 2007). Whether pUL16 binds a capsid protein directly or if its interaction is dependent on pUL36 has not been determined. Other tegument proteins are also reported to bind capsid proteins, including pUL21 (de Wind et al., 1992; Takakuwa et al., 2001), pUL37 (Lee et al., 2008; Uetz et al., 2006), and pUL47 (Scholtes et al., 2010), but it is not clear that these tegument proteins directly bind to assembled capsids. In fact, multiple lines of evidence argue against a direct interaction between pUL37 and capsids, with pUL37 instead tethered to capsids by pUL36 (Cardone et al., 2012; Daniel et al., 2018; Fuchs et al., 2004; Ko et al., 2010; Sandbaumhuter et al., 2013). Nevertheless, there is indirect evidence of additional capsid-tegument interactions, including that virions can form to some degree in the absence of the pUL36-pUL25 interaction (Coller et al., 2007; Schipke et al., 2012), and that the link between pUL36 and the remainder of the tegument is also dispensable for virion formation (Fuchs et al., 2002; Svobodova et al., 2012).

Quantitative fluorescence imaging of individual alphaherpesvirus extracellular particles indicates that pUL36 is present at 5 copies at all 12 vertices along with its binding partner, pUL37 (Bohannon et al., 2013). However, the carboxyl-terminal capsid-binding domain of pUL36 is separated from the bulk of the protein by a linker region that is high in proline content and likely unstructured (Böttcher et al., 2006). Therefore, the majority of pUL36 along with the entirety of pUL37, which binds near the amino terminus of pUL36 (Klupp et al., 2002; Vittone et al., 2005), may have some freedom of movement while anchored to the capsid (Newcomb and Brown, 2010). This is consistent with symmetry-dependent cryo-electron microscopy reconstructions of the capsid that have failed to resolve pUL36 apart from its carboxyl terminus (Dai and Zhou, 2018; Wang et al., 2018). In addition to pUL37, pUL36 tethers the pUS3 protein kinase to the capsid (Daniel et al., 2018) and links to the remainder of the tegument through an interaction with the pUL48 tegument protein (Ko et al., 2010; Svobodova et al., 2012). The pUL36 interaction with pUL48 appears complex, with pUL48 possibly undergoing a substantial conformational change as a result of binding to pUL36 (Svobodova et al., 2012). The significance of this is unknown, but unlike pUS3 and pUL37, pUL48 disassociates from capsids upon cell entry (Aggarwal et al., 2012; Antinone and Smith, 2010;

Granzow et al., 2005; Luxton et al., 2005), hinting that the complexity of the pUL36-pUL48 interaction may be related to its metastability. Based on this accumulated knowledge, pUL36 is a central determinant to tegument disassembly upon cell entry and is currently the only known link between the capsid and the remainder of the tegument (Figure 2).

The envelope is the effector of cell entry, which occurs by its fusion with a cellular membrane (see Chapter 2). There are at least a dozen types of

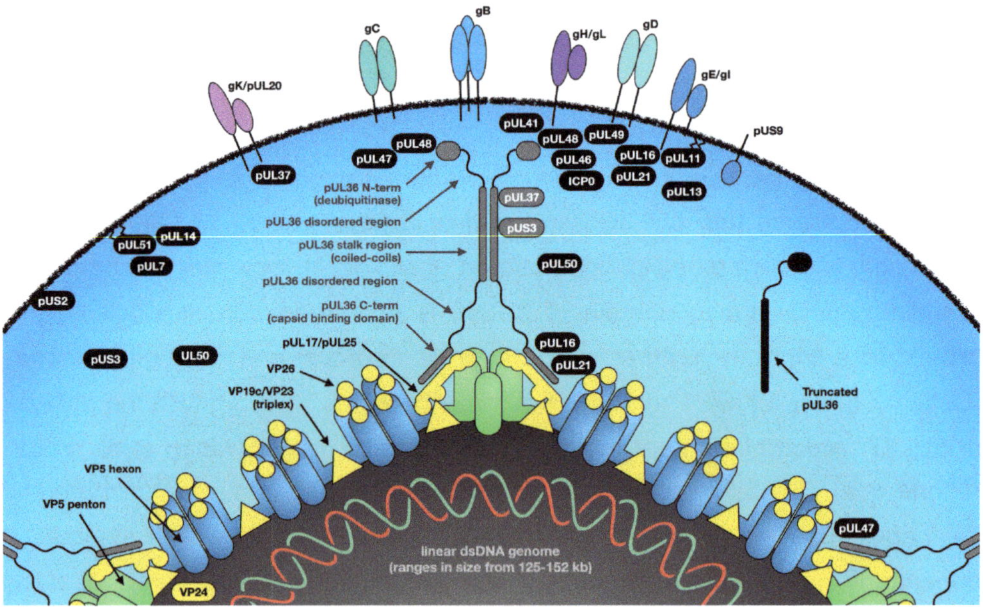

Figure 2. Components of the alphaherpesvirus virion. The icosahedral capsid (cut off at bottom) consists of hexons (blue) and pentons (green) of the major capsid protein, VP5. Other capsid proteins are shown in yellow. The capsid portal is not illustrated. Two of the five pUL17/pUL25/pUL36 copies per vertex are shown for simplicity. The large tegument protein, pUL36, is prominently featured and is shaded gray along with its binding partners, pUL37 and pUS3, which together are retained on the capsid following entry into a cell. Additional tegument proteins (black) are juxtaposed to known binding partners, including envelope proteins, with pUL48 linking the bulk of the tegument to the pUL36 amino terminus. However, unlike the capsid, the precise architectural arrangement of the tegument is largely unknown. The tegument also contains an assortment of cellular proteins and RNA that are not illustrated (Kramer et al., 2011; Loret et al., 2008; Sciortino et al., 2001).

proteins in the alphaherpesvirus envelope. In addition to driving entry, these proteins protect the virion from complement and antibody neutralization (Frank and Friedman, 1989; Hook et al., 2008; Komala Sari et al., 2020a), trigger alterations in the tegument architecture prior to entry (Meckes and Wills, 2008), and modulate the entry process (El Kasmi and Lippe, 2014; Komala Sari et al., 2020b).

Virion disassembly and the post-entry capsid complex

Virion attachment to cells triggers signaling events both into the cell and the virion. Signaling into the cell can be induced by the epidermal growth factor receptor and the virion envelope glycoproteins B (gB) and gD, which activate cellular phosphoinositide 3-kinase (PI3K) (Cheshenko et al., 2013; Zheng et al., 2014). PI3K produces phosphatidylinositol (3,4,5)-trisphosphate (PIP_3) in the cell membrane that recruits pleckstrin-homology domain containing proteins and induces several signaling cascades. Activation of Akt, Cdc42, RhoA>ROCK1, and Erk1/2 are each reported to promote virus entry (Frampton et al., 2007; Hoppe et al., 2006; Oh et al., 2010; Tiwari and Shukla, 2010; Zheng et al., 2014); however, these signaling effects are not consistently observed in all studies (Devadas et al., 2014; Nicola and Straus, 2004). Nevertheless, one consequence of these events is the phosphorylation and inactivation of the actin-depolymerization protein, cofilin, the result of which may cluster receptors in proximity to the virion by promoting actin polymerization underneath the cell membrane (Van den Broeke et al., 2014; Zheng et al., 2014). NF-kB signaling can also result from PI3K/Akt activation. HSV-1, though interestingly not PRV, relies on NF-kB to establish infection and upon entering the cell reinforces NF-kB signaling using a TRAF6 effector domain carried in the HSV-1 pUL37 tegument protein (Liu et al., 2008; Patel et al., 1998; Romero et al., 2020). Why NF-kB is important for HSV-1 but not PRV is unclear, but the absence of an obvious TRAF6 effector domain in PRV pUL37 adds further support that NF-kB signaling is inconsequential to PRV. PI3K/Akt signaling also triggers calcium release from the endoplasmic reticulum, which regulates cytoskeletal dynamics and may prime capsid trafficking to nuclei upon entry (Liu and Cohen, 2015; Nicola and Straus, 2004).

Signaling that propagates into the virion before entry triggers release of pUL16 from the capsid (Meckes and Wills, 2008). The ramifications of this molecular event are not established but the finding demonstrates that the tegument is malleable. The plasticity of the tegument is also evident in ultrastructural studies. Newly formed virions are symmetric, with a tegument layer that is wrapped evenly around the capsid (Chen et al., 2004; Granzow et al., 1997), and which can be extracted from the capsid with nonionic detergent (Newcomb and Brown, 2009). The tegument architecture spontaneously redistributes over a period of 1-2 days such that the tegument mass is displaced predominantly to one side the capsid and becomes resistant to detergent extraction (Newcomb and Brown, 2009). Particles with spontaneous tegument displacement have been described in detail by transmission electron microscopy, cryo-electron tomography, and fluorescence ensemble mapping (Bohannon et al., 2013; Grunewald et al., 2003; Newcomb and Brown, 2009). Notably, the displaced tegument architecture is consistently observed when virions are in contact with uninfected cells, suggesting that cell contact can trigger the architectural change if it has not already occurred. Furthermore, the bulk of the tegument mass is specifically displaced to the side of the capsid facing away from the contacted cell (Fuller et al., 1989; Maurer et al., 2008). These events are hypothesized to prepare the virion for fusion and release of the capsid into the cytosol (Smith, 2017). Therefore, signals propagated into the virion and cell may act cooperatively to prime the ensuing infection.

Fusion of the herpesvirus envelope with a target cellular membrane, either the plasma membrane or the membrane of an endocytic vesicle, deposits the capsid and the tegument into the cell cytosol. What happens next was long shrouded in mystery. Early electron microscopy studies revealed that capsids deposited into the cytosol relocalize to the outer surface of nuclei where they release their genetic content: the electron-dense genome core notably absent from capsids docked on nuclear pores (Batterson et al., 1983; Granzow et al., 1997; Lycke et al., 1988). However, a series of events must occur before the ~125-nm capsid, which is too large to passively diffuse to the nucleus, can reach its destination (Luby-Phelps, 1994, 2000). This begins with the release of the incoming capsid from the membrane and the disassembly of the majority of the tegument. As noted earlier, the latter begins with release of pUL16 from the capsid prior to entry into the cell (Meckes and Wills, 2008). Subsequent exposure to the cytosol likely allows the pUS3 and pUL13 tegument protein kinases to further drive this process, as has been inferred from *in vitro* studies (Morrison et al., 1998). The liberated tegument proteins rapidly manipulate cellular processes prior to nuclear genome delivery and viral gene expression. The first recognized effector functions of virion-delivered tegument proteins were the virion host shutoff (VHS) function of pUL41 and the transcriptional activation function of pUL48 (aka VP16, alphaTIF, Vmw65). This recognition was achieved by several experimental approaches including demonstration that UV-inactivated virions, which enter cells but do not express viral gene products, delivered these activities into cells (Ace et al., 1989; Batterson and Roizman, 1983; Fenwick and Walker, 1978; Nishioka and Silverstein, 1978).

Not all tegument proteins disassociate from capsids upon entry. The cytosolic capsid along with the subset of tegument proteins that remain capsid bound collectively make up what will be referred to here as the post-entry capsid complex (PECC). The tegument proteins of the PECC likely contribute to electron-dense material seen on capsid vertices following entry into cells (Sodeik et al., 1997). At least three tegument proteins are constituents of the PECC: pUL36, pUL37, and pUS3 (shown in gray in Figure 2). Retention of each of these tegument proteins on capsids was demonstrated by immuno-gold electron microscopy and by live-cell imaging of recombinant fluorescent viruses for HSV-1 and PRV (Aggarwal et al., 2012; Antinone and Smith, 2010; Coller and Smith, 2008; Granzow et al., 2005; Luxton et al., 2005). Thus, the composition of the PECC is conserved in these representative members of the simplexvirus and varicellovirus genera (Figure 1). Each of the known tegument constituents of the PECC has enzymatic activity: pUL36 has a deubiquitinase (DUB) domain in its amino terminus (Kattenhorn et al., 2005), pUL37 has a deamidase activity (Zhao et al., 2016), and pUS3 is a serine/threonine protein kinase (Frame et al., 1987; Purves et al., 1987). While these activities are noteworthy, they do not account for all of the functions of these proteins (Lee et al., 2006; Stults and Smith, 2019; Tian et al., 2018). The isolation of a temperature-sensitive mutant of HSV-1, which was subsequently mapped to the gene encoding pUL36, provided an early insight into one aspect of the PECC: at the non-permissive temperature, genomes were not released from capsids upon arriving at nuclear pores (Abaitua et al., 2011; Abaitua et al., 2009; Batterson et al., 1983).

Subsequent studies of the PECC were aided by producing recombinant viruses expressing fluorescent proteins fused to capsid components, which allowed for monitoring capsid dynamics and composition in living cells. The first such reporter virus encoded the green-fluorescent protein (GFP) fused to VP26 (Desai and Person, 1998). GFP-VP26 fusions are useful for studying initial infection because they offer bright fluorescence; however, they impair virus assembly and egress from nuclei during late infection (Bohannon et al., 2012; Krautwald et al., 2008; Nagel et al., 2012). A more recent deign that places GFP on the carboxyl-terminus of VP26 offers improved preservation of viral activity (Hogue et al., 2018). A subsequent design fused a fluorescent protein to pUL25, which offered a good balance between fluorescent emissions and low impact on virus replication (Bohannon et al., 2012; Conway et al., 2010). Whereas pUL25 and VP26 are surface-exposed capsid components, a more recent approach fused a fluorescent protein to the capsid maturation protease, VP24, which resides inside the capsid shell (Maier et al., 2016). Although VP24 fusions are 3-fold dimmer than pUL25 fusions, which are in turn 4-fold dimmer than VP26 fusions, they mitigate concerns that the fluorescent protein may interfere with capsid-tegument and capsid-host interactions. While fluorescence imaging has been used to gain structural insights into alphaherpesvirus virions (Beilstein et al., 2019; Bohannon et al., 2013; Laine et al., 2015), the primary application of recombinant viruses expressing fluorescent protein fusions is to monitor capsid transport in living cells (Diwaker and Wilson, 2019; Smith, 2012).

Navigating the cellular landscape

Upon fusion-based entry at the plasma membrane, the PECC encounters cortical filamentous actin (F-actin) that impedes intracellular capsid trafficking (Jacob et al., 2015; Maurer et al., 2008). As already noted, Akt signaling initiated prior to entry may help dismantle this intrinsic hurdle to infection (Liu and Cohen, 2015; Nicola and Straus, 2004). In addition, actin depolymerization by cofilin, which was suppressed prior to entry, is now promoted (Zheng et al., 2014). Cofilin activation is driven by the pUS3 component of the PECC (Jacob et al., 2013). In the absence of pUS3, genome delivery to the nucleus is less efficient, but can be restored by depolymerizing F-actin with cytochalasin D (Jacob et al., 2015). Thus, pUS3 is directly involved in breeching the cortical actin and allowing the capsid to gain deeper access to the cytoplasm (Figure 3). ROCK1 is also postulated to contribute to initial infection, but its precise role is unclear given that it inactivates cofilin (Frampton et al., 2010; Maekawa et al., 1999). Escape from the cortical actin also requires the microtubule plus-tip proteins: EB1, CLIP-170, and dynactin (Jovasevic et al., 2015). These proteins concentrate at the plus-ends of growing microtubules that typically reside at the cell periphery, and transient knock-down of any of these proteins restricts trafficking of the PECC. One explanation of this effect is that the plus-tip proteins may promote virus trafficking indirectly by maintaining proper cytoskeletal homeostasis. Alternatively, these cellular proteins may work together to capture the PECC and load it onto microtubules for subsequent retrograde transport by the cytoplasmic dynein microtubule motor, similar to endosome loading (Schuster et al., 2011). In support of the latter, the pUL36 component of the PECC associates with dynactin (Radtke et al., 2010; Zaichick et al., 2013). Given that depolymerization of actin by cytochalasin D rescues the escape of a US3-null virus from the cortical actin (Jacob et al., 2015), it would be interesting to determine whether the requirement of the plus-tip proteins would be reduced under this condition as well. More recently, F-actin was demonstrated to compete with microtubules for EB1 (Alberico et al., 2016). This finding hints at further intricacy to the interplay between the virus particle, cortical actin, and microtubules. Possibly adding to this complexity, there are mixed reports regarding whether the proteasome may also participate in this initial step of infection. In one study, inhibition of the proteasome with MG132 caused viral particles to remain at the cell periphery following entry into cells (Delboy et al., 2008). The same group also reported that the HSV-1 E3 ubiquitin ligase, ICP0, works in tandem with proteasomes to facilitate trafficking from the periphery to the nucleus (Delboy and Nicola, 2011). However, the effect of MG132 could not be reproduced by another group (Sekine et al., 2017). Also of interest, local protein synthesis must occur upon PRV entry into axon terminals for retrograde axonal transport to ensue

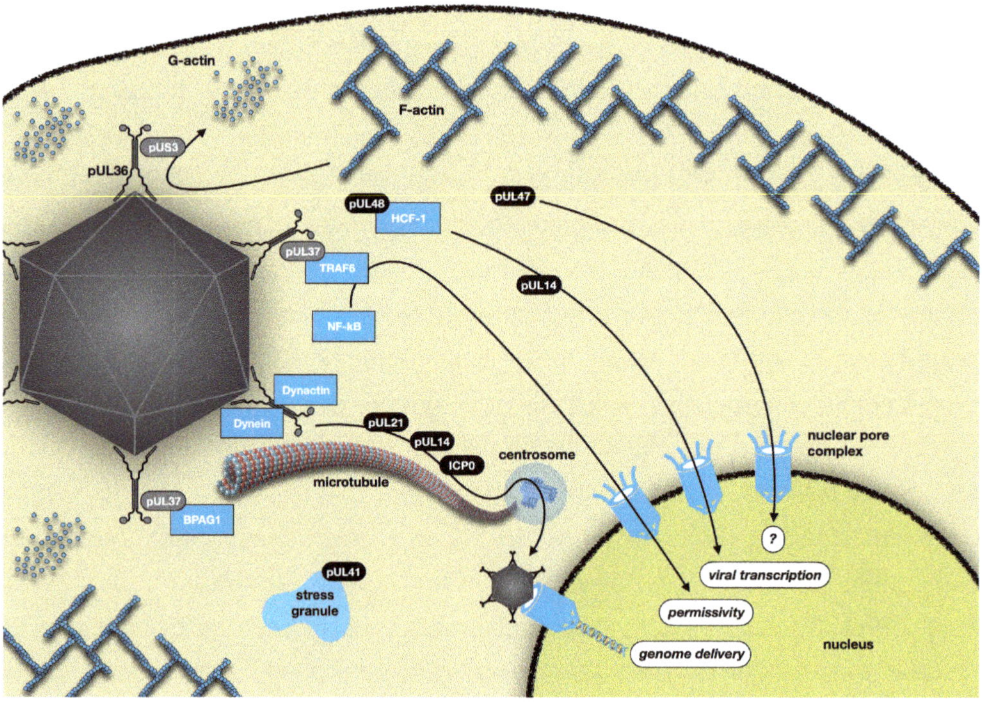

Figure 3. Post-entry events contributing to the delivery of viral genome to the cell nucleus. The post-entry capsid complex (consisting of the capsid, pUL36, pUL37, and pUS3) is shown in gray in the foreground at left. Tegument proteins that disassociate from the capsid are shown in black. Although the ICP0, pUL14 and pUL21 tegument proteins are shown apart from the capsid, their fates have not been reported and are therefore shown in black and gray. Cellular proteins are blue. Additional descriptions of viral effector functions are provided in Table 1.

(Koyuncu et al., 2013b). One of the proteins synthesized upon infection is Lis1: a protein that can facilitate activation of dynein (Elshenawy et al., 2020; Htet et al., 2020; Marzo et al., 2020). Whether new protein synthesis is also needed in non-neuronal cells is unknown.

The PECC departs from the cell periphery by microtubule-based transport, and the microtubule depolymerizing drug, nocodazole, compromises this step of infection in non-polarized cells, polarized epithelia, and neurons (Kristensson et al., 1986b; Sodeik et al., 1997; Topp et al., 1996). Whereas microtubule transport is essential for infection of neurons via axons, it is dispensable in non-polarized cells although it substantially increases the rate of nuclear genome delivery (Richards et al., 2017; Sekine et al., 2017; Sodeik et al., 1997). Dynein is responsible for minus-end-directed microtubule-based transport in mammalian cells including retrograde axonal transport in neurons, whereas multiple kinesin family motors are responsible for plus-end-directed transport (Hirokawa and Noda, 2008; Reck-Peterson et al., 2018). Dynein inhibition also impairs trafficking of viral particles to nuclei in neurons and non-neuronal cells (Frampton et al., 2010; Jovasevic et al., 2015; Kristensson et al., 1986a). Close inspection of PECCs by transmission electron microscopy hints at dynein-like structures attached at capsid vertices, which is supported by immunogold detection (Sodeik et al., 1997). There is also indication that an opposing kinesin motor is present on the PECC. Inhibition of dynein-based transport results in accumulation of the particles at the cell periphery, which was interpreted as evidence of a kinesin motor (Dohner et al., 2002). In addition, time-lapse imaging of the PECC in neurons reveals a subtle bi-directionality to their movement with short-transient reversals in motion likely due to the presence of a kinesin motor (Smith et al., 2004). It is unknown whether a kinesin motor is of functional significance at this stage of infection or if the virus is actively suppressing a kinesin activity to afford preferential retrograde transport by dynein. The latter scenario may explain how alphaherpesviruses sustain retrograde transport over long distances, which is critical for their neuroinvasive phenotype. In contrast, adenoviruses, which are generally not neuroinvasive, also bind dynein and a kinesin but do not favor the activity of one motor over the other and use the motors to facilitate diffusion in the cytosol (Bremner et al., 2009; Gazzola et al., 2009; Suomalainen et al., 1999; Zhou et al., 2018). The ability of alphaherpesviruses to directly engage the microtubule transport machinery and sustain retrograde transport in axons also enables rapid genome delivery to the nucleus in non-neuronal cells (Richards et al., 2017). Although comparisons between viruses are challenging, there is a tentative indication that HSV-1 outpaces adenovirus genome delivery to the nucleus (Sekine et al., 2017). The PECC averages a speed of 1.2 microns/s and can sustain speeds in excess of 3 microns/s as it is transported towards the minus end of microtubules that are located at the centrosome (Smith et al., 2004).

The centrosome presents an interesting quandary for incoming viral particles. Because the minus-ends of microtubules reside at the centrosome, dynein cannot carry the PECC beyond this organelle, and capsids have been observed to transiently accumulate here on their way to the nucleus (Sodeik et al., 1997). Whereas adenoviruses use CRM1-dependent microtubule-based transport to reach the nucleus (Greber and Flatt, 2019; Strunze et al., 2005; Wang et al., 2017), this does not appear to be the case for HSV-1 (Sekine et al., 2017). Alphaherpesvirus transport from the centrosome to the nucleus was hypothesized to occur by a kinesin motor, but data supporting this model are not available. However, there is evidence that the cellular protein, dystonin/ BPAG1, contributes to this step of infection (McElwee et al., 2013). Dystonin/ BPAG1 is a member of the plakin family of proteins (Kunzli et al., 2016). There are three main isoforms of the protein with distinct patterns of expression: BPAG1a is expressed in multiple tissues including the brain, BPAG1b is expressed in muscle, and BPAG1e is expressed in epithelia. These proteins range in size from 300-800 kDa and consist of a plakin domain and an assortment of additional domains that vary between isoforms. Dystonin/BPAG1 binds to dynactin and EB1, which accounts for its plus-tip distribution in cells, and supports retrograde axonal transport of endogenous cargoes in neurons (Kapur et al., 2014; Liu et al., 2003; Pasdeloup et al., 2013; Poliakova et al., 2014). These properties suggest that Dystonin/BPAG1 may promote transport of the PECC from the periphery to the centrosome. However, knock-down of dystonin/BPAG1 has only a modest effect on minus-end-directed microtubule transport to the centrosome and, instead, impedes subsequent transport from the centrosome to the nucleus (McElwee et al., 2013). Centrosome-to-nucleus trafficking is enigmatic, and the dystonin/BPAG phenotype provides an insight into this transport mechanism. Dystonin/BPAG may actively participate in the transport process or may serve as a release factor that frees the PECC from dynein upon arrival at the centrosome. Further insights into this mechanism will be aided by the findings that that pUL37 binds dystonin/BPAG1, and that pUL36 has a nuclear localization signal (NLS) that also contributes to this final transport step to the nucleus (Abaitua et al., 2012; Pasdeloup et al., 2013).

Capsids arriving at the nucleus preferentially bind to nuclear pore complexes (NPCs) (Anderson et al., 2014). The FG-repeat-containing nucleoporins that extend from the cytosolic surface of the NPC, Nup214 and Nup358/RanBP2, along with importin-beta, participate in capsid docking (Copeland et al., 2009; Ojala et al., 2000; Pasdeloup et al., 2009). Although importin-alphas function in a complex with importin-beta and bind to nuclear localization signals in substrate proteins, no individual importin-alpha has been identified as essential for capsid docking (Dohner et al., 2018). Either HSV-1 can use more than one importin-alpha or importin-alphas are not required during initial infection. HSV-1 capsids bind to Nup214, as does the pUL25 capsid protein of HSV-1 and PRV when transiently expressed; whether importin-beta or Nup358/RanBP2 participates in these interactions is not known (Pasdeloup et al., 2009). The role of pUL25's binding partner, pUL36, in the NPC interactions is also unclear. Antibodies reactive to pUL36 block capsid accumulation at the nuclear membrane when loaded into cells prior to infection, but this could be an indirect result of interfering with the role of pUL36 in capsid trafficking to nuclei (Copeland et al., 2009).

Although the role of pUL36 during capsid docking at NPCs requires further study, pUL36 is critical for the subsequent release of the viral genome from the capsid (Abaitua et al., 2011; Abaitua et al., 2009; Batterson et al., 1983;

Roberts et al., 2009), as is pUL25 (Huffman et al., 2017; Preston et al., 2008; Rode et al., 2011). In contrast, pUL37 is dispensable for genome release into nuclei (Roberts et al., 2009). A proteolytic event is also necessary for genome release at NPCs and cleavage of pUL36 is coincident with this step, suggesting the role of pUL36 in genome release involves its processing (Jovasevic et al., 2008).

Alphaherpesvirus capsids orient at nuclear pore complexes such that a vertex faces into the pore (Granzow et al., 1997; Ojala et al., 2000). Given the similarities between the tailed bacteriophage and the herpesvirus capsid, it seems likely that the inward facing vertex is the portal, which would indicate that there are mechanisms at play to correctly orient the capsid and select for the portal vortex to face the NPC. Such an orientation may also be consistent with the high pressure within the capsid contributing to nuclear genome delivery by releasing the genome directly into the pore (Brandariz-Nunez et al., 2019), but no specific mechanism has yet been proposed. Following genome injection, the residual capsid shell remains intact and does not undergo any obvious morphological change (Granzow et al., 1997; Ojala et al., 2000). The empty capsid relic at the NPC is a testament to the stability of the alphaherpesvirus capsid shell, which may be an important factor to prevent detection of incoming viral DNA by nucleic acid sensors in the cytosol prior to arrival at the nucleus (Huffman et al., 2017).

Viral effectors of genome delivery

Investigations into the molecular mechanisms governing intracellular viral trafficking and nuclear genome delivery are complicated by the fact that the capsid and tegument proteins are structural elements of the virion, and their mutagenesis can, in some cases, prevent virion production. This is a significant technical dilemma. Although viruses lacking genes encoding essential proteins can be produced by trans-complementation, the presence of the trans-complemented protein in the virion precludes study of the deletion phenotype during initial infection. Many virion proteins that are non-essential for virus propagation in culture have been ruled out as essential for capsid trafficking to nuclei, including the pUS3 component of the PECC (Antinone et al., 2006). However, alphaherpesviruses cannot be produced in the absence of pUL36 (Desai, 2000; Smith and Enquist, 1999; Zhang et al., 2010) as well as the majority of the capsid proteins, and their study has required a more nuanced approach.

Capsid

Despite the grandiose assembly of the herpesvirus capsid, there is little evidence that it actively contributes to intracellular trafficking. VP26, which decorates the tips of hexons, is a compelling candidate for an effector protein during initial trafficking given its presence on the capsid surface at a high-copy number and its location away from pentons where it could otherwise be buried by tegument proteins (Figure 2). Yeast-2-hybrid studies identified dynein light chains as binding partners for VP26 (Douglas et al., 2004), which was unexpected given that dynein associates with the vertices of the PECC where VP26 is conspicuously absent (Sodeik et al., 1997). Nevertheless, tentative evidence that VP26 facilitates capsid transport was obtained by using a baculovirus expression system to assemble capsids with or without VP26, and then injecting the capsids into mammalian cells. The VP26-containing capsids tended to localize near the nucleus whereas capsids lacking VP26 remained dispersed throughout the cytoplasm (Douglas et al., 2004). However, follow up studies could not identify a role for VP26 in binding dynein to capsids (Wolfstein et al., 2006). Furthermore, viruses lacking VP26 engaged in dynein-based transport indistinguishably from wild-type viruses (Antinone et al., 2006; Desai et al., 1998; Dohner et al., 2006). While these studies were completed over a decade ago, they bear witness to an important consideration when studying components of the PECC as potential effectors of intracellular transport: findings from reductionist models require confirmation in the context of infection.

The major capsid protein, VP5, is essential for virus propagation. While much of VP5 is buried within the capsid shell, the apical domain of the protein is exposed on the outer capsid surface where it could interact with cellular proteins. Mutations of the apical region of VP5 are not lethal but hinder capsid maturation late during infection. However, the apical mutant viruses have no measurable reduction in retrograde axonal transport, indicating that dynein recruitment is unperturbed (Ruhge et al., 2018).

The pUL25 protein is also of interest due to its exposure at vertices and because it adopts a unique conformation at the portal vertex (McElwee et al., 2018). A partial structure is available for HSV-1 pUL25 that reveals a box-like core with multiple exposed peptide loops (Bowman et al., 2006). Although pUL25 is essential for virus propagation (Klupp et al., 2006; Kuhn et al., 2008; McNab et al., 1998), a series of pUL25 mutants were produced based on the protein structure and found to impair genome encapsidation and nuclear egress during late infection (Cockrell et al., 2009; O'Hara et al., 2010). One exception was mutation of the HSV-1 pUL25 carboxyl terminus, which notably prevented release of the genome at the NPC (Huffman et al., 2017).

Beyond these targeted mutations, capsids isolated from infected cell nuclei, which are mostly devoid of tegument proteins, do not bind to dynein, kinesin-1, or kinesin-2. Capsid and tegument exposed from extracellular virions by detergent treatment are also incompetent for motor binding but, when extracted to remove the bulk of the tegument, the remaining capsid-tegument complexes bind all three motors and engage in microtubule transport in an *in vitro* assay (Radtke et al., 2010; Wolfstein et al., 2006). The pUL36 and pUL37 tegument proteins were considered candidates for this motor binding, and additional data supports that these proteins are effectors of PECC intracellular trafficking and genome delivery to nuclei.

pUL36

The pUL36 tegument protein is the largest protein encoded by the alphaherpesviruses, with a predicted molecular weight in excess of 300 kDa. The protein has a deubiquitinase (DUB) domain at the amino terminus and a capsid-binding domain (CBD) at the carboxy terminus (Coller et al., 2007; Kattenhorn et al., 2005). The domain organization of pUL36 can be simplified as DUB-linker-stalk-linker-CBD. The structures of two fragments of pUL36 have been determined. The first is the DUB from a non-neuroinvasive betaherpesvirus in complex with its ubiquitin substrate (Schlieker et al., 2007). The pUL36 DUB can be heterologously expressed by itself and has been adopted as a tool to keep proteins in a hypo-ubiquitinated state due to its potent activity (Stringer and Piper, 2011). During infection, the DUB is essential for transmission of PRV from epithelia to neurons *in vivo* but is dispensable for the delivery of genomes to nuclei of isolated epithelial and neuronal cells in culture; therefore, the DUB is not a core component of the genome delivery machinery (Huffmaster et al., 2015; Lee et al., 2009). The DUB is separated from the bulk of pUL36 by a divergent linker region that has one conserved motif embedded inside of it: a NLS (Abaitua and O'Hare, 2008). HSV-1 mutated in the pUL36 NLS are unable to propagate in culture yet produce fully formed extracellular viral particles (Abaitua et al., 2012). These particles enter cells but subsequently fail to traffic to nuclei and, instead, accumulate at the centrosome. Although more detailed studies are needed, these results indicate that dynein-based transport to the centrosome is not dependent on the NLS, whereas subsequent transport from the centrosome to the nucleus relies on it. This phenotype is reminiscent to wild-type HSV-1 infections in cells knocked-down for dystonin/BPAG1 (McElwee et al., 2013). Although pUL37 binds to dystonin/BPAG1 there is no indication of whether pUL36 does as well. The specific contributions of the pUL36 NLS and dystonin/BPAG1 to centrosome-to-nucleus trafficking of the PECC will require further investigation.

The second structurally resolved region of pUL36 forms an elongated fiber that is predicted to be part of a larger helical region near the center of the protein (Scrima et al., 2015). This region was proposed to form a long coiled-coil stalk that could form from multiple copies of the protein. If pUL36 were a component of the portal tail, the full extent of such an elongated and flexible structure may not be easily identified by cryo-electron microscopy (McElwee et al., 2018; Schmid et al., 2012). The formation of an elongated structure could also explain why the protein is sensitive to sonication (Newcomb and Brown, 2010).

In addition to full-length pUL36 that is anchored to capsid vertices, HSV-1 and PRV virions package naturally occurring truncated forms of pUL36 that lack the CBD (Michael et al., 2006; Radtke et al., 2010). These truncated forms of pUL36 are of unknown significance but presumably dissociate from the capsid upon entry into a cell. This could allow for the dispersal of the DUB and its access to a greater spectrum of host substrates. Unexpectedly, recombinant HSV-1 lacking the CBD of pUL36 remains competent to produce extracellular viral particles (Schipke et al., 2012). Upon entering a cell, the PECC loses the truncated pUL36 and fails to traffic to the nucleus. Thus, the retention of pUL36 on the capsid is essential because it mediates minus-end-directed microtubule transport by tethering dynein to the PECC (Zaichick et al., 2013). In fact, pUL36 interacts with dynein and dynactin in the absence of other viral proteins, and engages in robust microtubule-based transport when transiently expressed as a truncated form that lacks the carboxyl-terminal CBD. This finding has enabled studies of pUL36 function in a simplified system (Zaichick et al., 2013). Several observations made from this model require further study. Similar to capsids following entry, truncated pUL36 engages in processive microtubule-based transport, but unlike capsids this motion is highly bidirectional. As already discussed, the PECC may down-regulate an associated kinesin motor, which based on this latter observation appears to require viral proteins in addition to pUL36. These observations also indicate that the CBD may be autoinhibitory with regard to trafficking such that pUL36 expressed during infection may be inactive until the CBD is sequestered by binding to a capsid. In addition to these activities, capsid-retained pUL36 is also essential because it tethers pUL37 to the PECC (Daniel et al., 2018).

pUL37

Like pUL36, pUL37 supports PECC trafficking (Krautwald et al., 2009). Although there is currently no indication that pUL37 binds to microtubule motors, it modulates their activities (Richards et al., 2017). The amino-terminal half of pUL37 is formally dispensable for propagation in culture (Grzesik et al., 2019) but contains the region 2 (R2) effector that is necessary for full infectivity (Koenigsberg and Heldwein, 2017; Pitts et al., 2014). R2 sustains minus-end directed microtubule motion that is fundamental to retrograde axonal transport and the alphaherpesvirus neuroinvasive mechanism (Richards et al., 2017). When R2 is mutated, incoming HSV-1 and PRV display balanced plus- and minus-end-directed motion, reminiscent of the native state of adenovirus. Whereas R2 enhances trafficking to the cell nucleus in epithelial cells, R2 is essential for sustained long-distance trafficking to neuronal nuclei via axons. Thus, R2 mutant viruses are viable but are not neuroinvasive. The mechanism of action for R2 has not been determined but given that pUL37 binds dystonin/BPAG1, which is necessary for centrosome-to-NPC transport, it seems reasonable to speculate that R2 may cooperate with dystonin/BPAG1 to mediate a switch to kinesin-based transport upon arriving at the centrosome. In this scenario, viruses mutated for R2 would be incapable of suppressing kinesin and thereby cause kinesin and dynein to compete with one another on the way to the centrosome. Because R2 mutants are non-neuroinvasive, they are compelling candidates as live-attenuated vaccines for veterinary and clinical alphaherpesviruses (see Chapter 10) (Pickard et al., 2020; Richards et al., 2017).

The carboxyl-terminal half of pUL37 is a flexible elongated structure that is essential for viral propagation (Grzesik et al., 2019; Koenigsberg and Heldwein, 2018). This part of the protein binds pUL36 and also has deamidase activity (Bucks et al., 2011; Zhao et al., 2016). The latter antagonizes antiviral sensors but is dispensable for nuclear genome delivery during initial infection (Stults and Smith, 2019; Zhang et al., 2018; Zhao et al., 2016). As mentioned above, HSV-1 pUL37 has a TRAF6-binding motif in the carboxyl-terminal tail that activates NF-kB signaling (Liu et al., 2008). Both the deamidase and the TRAF6-binding site are absent from varicelloviruses, including PRV, indicating that these effectors are not part of the core genome delivery mechanism that is fundamental to the neuroinvasive phenotype of the alphaherpesviruses (Liu et al., 2008; Stults and Smith, 2019).

Other tegument proteins

In addition to pUL36 and pUL37, the pUS3 protein kinase is also a component of the PECC, but either it does not contribute to genome delivery or it is functionally redundant with another PECC component (Antinone et al., 2006; Coller and Smith, 2008; Olsen et al., 2006). The PECC may also contain additional tegument components that have yet to be identified either because reagents are not available for their study, or because their copy number in the PECC is below detection levels. For example, emissions from pUL13-GFP fusions are not easily detectable in extracellular PRV virions, precluding their use to examine the PECC (Coller and Smith, 2008). Whether tegument proteins that disassociate from the capsid upon entry contribute indirectly to PECC trafficking is yet to be seen.

The pUL14 tegument protein is tightly associated with capsids extracted from extracellular virions (Radtke et al., 2010), but is not established as a component of the PECC. HSV-1 lacking pUL14 propagates slowly but produces near wild-type levels of infections virus (Cunningham et al., 2000). The decreased rate of infection can be accounted for by defects in trafficking the PECC and the pUL48 tegument protein to the nucleus (Yamauchi et al., 2008). A screen for cellular binding partners of pUL14 did not identify proteins obviously involved in intracellular trafficking (Wu et al., 2011), which may indicate that pUL14 acts on viral proteins. Consistent with this possibility, pUL14 redistributes VP26 to the nucleus during transient co-expression (Yamauchi et al., 2001). However, VP26 does not appear to anchor tegument on assembled capsids based on cryo-electron microscopy reconstructions (Chen et al., 2001; Dai and Zhou, 2018; Zhou et al., 1999), making it unclear if the pUL14-VP26 interaction is relevant during initial infection. These observations do not rule out pUL14 as a PECC constituent; in fact, the finding that nuclear delivery of incoming HSV-1 lacking pUL14 cannot be rescued by expression of pUL14 in trans argues that the activity of pUL14 during initial infection may be conveyed as part of the PECC (Yamauchi et al., 2008).

The pUL21 tegument protein interacts with a dynein light chain (Yan et al., 2019), analogous to VP26 and several other viral proteins (Douglas et al., 2004). While the significance of this interaction is unclear, recombinant HSV-1 and HSV-2 lacking pUL21 infect cells with delayed kinetics due to a defect in early trafficking (Le Sage et al., 2013; Mbong et al., 2012). The null mutant also invades the nervous system of mice with delayed kinetics (Yan et al., 2019). Similarly, axonal infection of primary neurons in culture is delayed, but it is unknown whether the minus-end transport velocity and processivity are reduced or if transport following entry is initiated at reduced efficiency. As is the case for pUL14, pUL21 is tightly associated with capsids extracted from extracellular virions (Yan et al., 2019), but is not an established PECC component.

The pUL41 tegument protein is responsible for the virion-host shutoff activity. Its endoribonuclease activity degrades host mRNA and disrupts stress granules (Finnen et al., 2014; Finnen et al., 2016). Although pUL41 is dispensable in culture (Read and Frenkel, 1983; Strelow and Leib, 1995), its ability to rapidly antagonize translation and stress granule function upon entry likely supports its importance *in vivo* as a means to limit anti-viral responses (Onomoto et al., 2014). However, it is unclear if antiviral responses antagonize trafficking of the PECCs and the deliver genomes to nuclei *in vivo*.

ICP0 (intracellular protein zero) is so named because it was not originally recognized as a structural component of virions. Nevertheless, ICP0 is a low-copy tegument protein (Yao and Courtney, 1992) that is tightly associated with capsids extracted from extracellular virions (Delboy et al., 2010; Maringer and Elliott, 2010). Although ICP0 is not an established PECC component, the PECC of ICP0-null HSV-1 traffics to nuclei inefficiently, and ICP0 was hypothesized to promote nuclear delivery by targeting a cellular or a viral protein for degradation by virtue of its E3 ubiquitin ligase activity (Delboy and Nicola, 2011). More recently, ICP0 was reported to down regulate autophagy early during infection (Waisner and Kalamvoki, 2019). The 3-hour time point examined in these experiments is too late to conclude whether virion-delivered, or *de novo* expressed, ICP0 was responsible for the effect on autophagy. If virion-delivered ICP0 antagonizes autophagy, this could account for the nuclear trafficking defect seen with the null virus.

Each of the proteins discussed above requires further study to delineate their specific contributions to initial infection, as do other tegument proteins that have yet to be examined in this context. A summary of the contributions of tegument proteins during initial infection is presented in Table 1.

Conclusions and outlook

Delivery of viral genomes to the nucleus is not a trivial problem, especially for the alphaherpesviruses, which accomplish this in two cell types with distinct polarized architectures: epithelial cells and neurons. There are multiple obstacles that these viruses overcome to traffic to nuclei including escaping the cortical actin network upon entering cells, loading onto microtubules, suppressing kinesin motors in favor of sustaining dynein-based transport to the centrosome, trafficking from the centrosome to the nucleus, docking at nuclear pore complexes, and releasing the genome into the nucleus. The specific tactics used by these viruses to overcome these obstacles contribute to their broader neuroinvasive phenotype. The complexity of the tegument provides for a rich assortment of viral effector proteins that enter the cytosol upon entry and rapidly increase the permissivity of a cell to infection. Of these, pUL36 and pUL37 remain associated with the incoming capsid and are essential for its trafficking. How these proteins function is beginning to be delineated, but our working understanding of their contributions to initial infection is rudimentary. There are many exciting questions remaining to be addressed, and likely many more that have yet to be realized. A full accounting of PECC components is needed and, ultimately, a molecular understanding of the contributions of each component to initial infection. An understanding of the alphaherpesvirus genome delivery mechanism will not only advance the fields of virology, neuroscience, and cell biology but also provide a foundation for next generation gene delivery vectors, trans-synaptic tracers, oncolytic vectors, and vaccines.

Table 1. Activities of virion-delivered tegument proteins during initial infection. Tegument proteins are multi-functional proteins that have roles throughout the infectious cycle, as well as serving as structural components of virions. In many instances, studies do not specifically examine the effects of these proteins during initial infection, prior to their *de novo* expression. The information summarizes the reported functions of individual tegument proteins during initial infection. The listed functions are not necessarily dependent on the indicated enzymatic activities (see text). Protein names are based on HSV/PRV nomenclature.

	Enzymatic activity	Capsid retained	Functions that are reported or implicated during ingress	Reference
pUL7	-	?	?	
pUL13	Protein kinase	?	Promotes tegument disassembly by phosphorylating VP22.	Morrison 1998a
pUL14	-	?	Promotes nuclear trafficking of capsids and pUL48 by unknown mechanisms.	Yamauchi 2008
pUL16	-	-	Promotes tegument disassembly by detaching from the capsid.	Meckes 2008
pUL21	-	?	Enhances infection of neurons via retrograde axonal transport, possibly by binding to a dynein light chain.	Yan 2019
pUL36 (VP1/2)	Deubiquitinase	+	Retains pUL37 and pUS3 on the capsid. Recruits dynein microtubule motor to promote capsid trafficking to the centrosome. NLS supports capsid transport from centrosome to nucleus. Cleavage is coincident with genome release from the capsid.	Jovasevic 2008 Abaitua 2012 Zaichick 2013 Daniel 2018
pUL37	Deamidase	+	Sustains dynein-based microtubule transport by suppressing opposing transport. Binds dystonin/BPAG1, which enhances capsid transport from centrosome to nucleus. Activates NF-kB signaling.	Liu 2008 McElwee 2013 Richards 2017
pUL41 (VHS)	Endoribonuclease	?	Disrupts pre-existing stress granules and antagonizes translation of host mRNA.	Kwong 1988 Finnen 2014, 2016
pUL46 (VP11/12)	-	-	?	
pUL47 (VP13/14)	-	-	Traffics into nucleus, but significance unknown.	Morrison 1998b
pUL48 (VP16)	-	-	Binds host HCF-1 protein to import into nucleus and promote viral gene expression.	La Boissiere 1999
pUL49 (VP22)	-	-	?	
pUL50	dUTPase	?	?	
pUS3	Protein kinase	+	Disassembles filamentous actin to promote capsid trafficking to nucleus.	Jacob 2015
ICP0	E3 ubiquitin ligase	?	Promotes nuclear trafficking of capsids possibly by targeting anti-viral host proteins for proteasomal degradation.	Delboy 2011
ICP4	-	?	?	

References

Abaitua, F., Daikoku, T., Crump, C.M., Bolstad, M., and O'Hare, P. (2011). A single mutation responsible for temperature-sensitive entry and assembly defects in the VP1-2 protein of herpes simplex virus. J Virol *85*, 2024-2036.

Abaitua, F., Hollinshead, M., Bolstad, M., Crump, C.M., and O'Hare, P. (2012). A nuclear localization signal in herpesvirus protein VP1-2 is essential for infection via capsid routing to the nuclear pore. J Virol *86*, 8998-9014.

Abaitua, F., and O'Hare, P. (2008). Identification of a highly conserved, functional nuclear localization signal within the N-terminal region of herpes simplex virus type 1 VP1-2 tegument protein. J Virol *82*, 5234-5244.

Abaitua, F., Souto, R.N., Browne, H., Daikoku, T., and O'Hare, P. (2009). Characterization of the herpes simplex virus (HSV)-1 tegument protein VP1-2 during infection with the HSV temperature-sensitive mutant tsB7. J Gen Virol *90*, 2353-2363.

Ace, C.I., McKee, T.A., Ryan, J.M., Cameron, J.M., and Preston, C.M. (1989). Construction and characterization of a herpes simplex virus type 1 mutant unable to transinduce immediate-early gene expression. J Virol *63*, 2260-2269.

Aggarwal, A., Miranda-Saksena, M., Boadle, R.A., Kelly, B.J., Diefenbach, R.J., Alam, W., and Cunningham, A.L. (2012). Ultrastructural visualization of individual tegument protein dissociation during entry of herpes simplex virus 1 into human and rat dorsal root ganglion neurons. J Virol *86*, 6123-6137.

Alberico, E.O., Zhu, Z.C., Wu, Y.O., Gardner, M.K., Kovar, D.R., and Goodson, H.V. (2016). Interactions between the Microtubule Binding Protein EB1 and F-Actin. J Mol Biol *428*, 1304-1314.

Anderson, F., Savulescu, A.F., Rudolph, K., Schipke, J., Cohen, I., Ibiricu, I., Rotem, A., Grunewald, K., Sodeik, B., and Harel, A. (2014). Targeting of viral capsids to nuclear pores in a cell-free reconstitution system. Traffic *15*, 1266-1281.

Antinone, S.E., Shubeita, G.T., Coller, K.E., Lee, J.I., Haverlock-Moyns, S., Gross, S.P., and Smith, G.A. (2006). The Herpesvirus capsid surface protein, VP26, and the majority of the tegument proteins are dispensable for capsid transport toward the nucleus. J Virol *80*, 5494-5498.

Antinone, S.E., and Smith, G.A. (2006). Two modes of herpesvirus trafficking in neurons: membrane acquisition directs motion. J Virol *80*, 11235-11240.

Antinone, S.E., and Smith, G.A. (2010). Retrograde axon transport of herpes simplex virus and pseudorabies virus: a live-cell comparative analysis. J Virol *84*, 1504-1512.

Baker, M.L., Jiang, W., Rixon, F.J., and Chiu, W. (2005). Common ancestry of herpesviruses and tailed DNA bacteriophages. J Virol *79*, 14967-14970.

Batterson, W., Furlong, D., and Roizman, B. (1983). Molecular genetics of herpes simplex virus. VIII. further characterization of a temperature-sensitive mutant defective in release of viral DNA and in other stages of the viral reproductive cycle. J Virol *45*, 397-407.

Batterson, W., and Roizman, B. (1983). Characterization of the herpes simplex virion-associated factor responsible for the induction of alpha genes. J Virol *46*, 371-377.

Bauer, D.W., Huffman, J.B., Homa, F.L., and Evilevitch, A. (2013). Herpes virus genome, the pressure is on. J Am Chem Soc *135*, 11216-11221.

Beilstein, F., Cohen, G.H., Eisenberg, R.J., Nicolas, V., Esclatine, A., and Pasdeloup, D. (2019). Dynamic organization of Herpesvirus glycoproteins on the viral envelope revealed by super-resolution microscopy. PLoS Pathog *15*, e1008209.

Bohannon, K.P., Jun, Y., Gross, S.P., and Smith, G.A. (2013). Differential protein partitioning within the herpesvirus tegument and envelope underlies a complex and variable virion architecture. Proc Natl Acad Sci U S A *110*, E1613-1620.

Bohannon, K.P., Sollars, P.J., Pickard, G.E., and Smith, G.A. (2012). Fusion of a fluorescent protein to the pUL25 minor capsid protein of pseudorabies virus allows live-cell capsid imaging with negligible impact on infection. J Gen Virol *93*, 124-129.

Borza, C.M., and Hutt-Fletcher, L.M. (2002). Alternate replication in B cells and epithelial cells switches tropism of Epstein-Barr virus. Nat Med *8*, 594-599.

Böttcher, S., Klupp, B.G., Granzow, H., Fuchs, W., Michael, K., and Mettenleiter, T.C. (2006). Identification of a 709-amino-acid internal non-essential region within the essential conserved tegument protein (p)UL36 of pseudorabies virus. J Virol *80*, 9910-9915.

Bowman, B.R., Welschhans, R.L., Jayaram, H., Stow, N.D., Preston, V.G., and Quiocho, F.A. (2006). Structural characterization of the UL25 DNA-packaging protein from herpes simplex virus type 1. J Virol *80*, 2309-2317.

Brandariz-Nunez, A., Liu, T., Du, T., and Evilevitch, A. (2019). Pressure-driven release of viral genome into a host nucleus is a mechanism leading to herpes infection. Elife *8*.

Bremner, K.H., Scherer, J., Yi, J., Vershinin, M., Gross, S.P., and Vallee, R.B. (2009). Adenovirus transport via direct interaction of cytoplasmic dynein with the viral capsid hexon subunit. Cell Host Microbe *6*, 523-535.

Bucks, M.A., Murphy, M.A., O'Regan, K.J., and Courtney, R.J. (2011). Identification of interaction domains within the UL37 tegument protein of herpes simplex virus type 1. Virology *416*, 42-53.

Cardone, G., Newcomb, W.W., Cheng, N., Wingfield, P.T., Trus, B.L., Brown, J.C., and Steven, A.C. (2012). The UL36 tegument protein of herpes simplex virus 1 has a composite binding site at the capsid vertices. J Virol *86*, 4058-4064.

Cardone, G., Winkler, D.C., Trus, B.L., Cheng, N., Heuser, J.E., Newcomb, W.W., Brown, J.C., and Steven, A.C. (2007). Visualization of the herpes simplex virus portal in situ by cryo-electron tomography. Virology *361*, 426-434.

Chen, D.H., Jakana, J., McNab, D., Mitchell, J., Zhou, Z.H., Dougherty, M., Chiu, W., and Rixon, F.J. (2001). The pattern of tegument-capsid interaction in the herpes simplex virus type 1 virion is not influenced by the small hexon-associated protein VP26. J Virol *75*, 11863-11867.

Chen, J.J., Zhu, Z., Gershon, A.A., and Gershon, M.D. (2004). Mannose 6-phosphate receptor dependence of varicella zoster virus infection in vitro and in the epidermis during varicella and zoster. Cell *119*, 915-926.

Cheshenko, N., Trepanier, J.B., Stefanidou, M., Buckley, N., Gonzalez, P., Jacobs, W., and Herold, B.C. (2013). HSV activates Akt to trigger calcium release and promote viral entry: novel candidate target for treatment and suppression. FASEB J *27*, 2584-2599.

Cockrell, S.K., Sanchez, M.E., Erazo, A., and Homa, F.L. (2009). Role of the UL25 protein in herpes simplex virus DNA encapsidation. J Virol *83*, 47-57.

Coller, K.E., Lee, J.I., Ueda, A., and Smith, G.A. (2007). The capsid and tegument of the alpha herpesviruses are linked by an interaction between the UL25 and VP1/2 proteins. J Virol *81*, 11790-11797.

Coller, K.E., and Smith, G.A. (2008). Two viral kinases are required for sustained long distance axon transport of a neuroinvasive herpesvirus. Traffic *9*, 1458-1470.

Conway, J.F., Cockrell, S.K., Copeland, A.M., Newcomb, W.W., Brown, J.C., and Homa, F.L. (2010). Labeling and localization of the herpes simplex virus capsid protein UL25 and its interaction with the two triplexes closest to the penton. J Mol Biol *397*, 575-586.

Copeland, A.M., Newcomb, W.W., and Brown, J.C. (2009). Herpes simplex virus replication: roles of viral proteins and nucleoporins in capsid-nucleus attachment. J Virol *83*, 1660-1668.

Cunningham, C., Davison, A.J., MacLean, A.R., Taus, N.S., and Baines, J.D. (2000). Herpes simplex virus type 1 gene UL14: phenotype of a null mutant and identification of the encoded protein. J Virol *74*, 33-41.

Curtis, R., Adryan, K.M., Zhu, Y., Harkness, P.J., Lindsay, R.M., and DiStefano, P.S. (1993). Retrograde axonal transport of ciliary neurotrophic factor is increased by peripheral nerve injury. Nature *365*, 253-255.

Dai, X., and Zhou, Z.H. (2018). Structure of the herpes simplex virus 1 capsid with associated tegument protein complexes. Science *360*, eaao7298.

Daniel, G.R., Pegg, C.E., and Smith, G.A. (2018). Dissecting the herpesvirus architecture by targeted proteolysis. J Virol *92*, e00738-00718.

de Wind, N., Wagenaar, F., Pol, J., Kimman, T., and Berns, A. (1992). The pseudorabies virus homology of the herpes simplex virus UL21 gene product is a capsid protein which is involved in capsid maturation. J Virol *66*, 7096-7103.

Dedeo, C.L., Cingolani, G., and Teschke, C.M. (2019). Portal Protein: The Orchestrator of Capsid Assembly for the dsDNA Tailed Bacteriophages and Herpesviruses. Annu Rev Virol *6*, 141-160.

Delboy, M.G., and Nicola, A.V. (2011). A pre-immediate-early role for tegument ICP0 in the proteasome-dependent entry of herpes simplex virus. J Virol *85*, 5910-5918.

Delboy, M.G., Roller, D.G., and Nicola, A.V. (2008). Cellular proteasome activity facilitates herpes simplex virus entry at a postpenetration step. J Virol *82*, 3381-3390.

Delboy, M.G., Siekavizza-Robles, C.R., and Nicola, A.V. (2010). Herpes simplex virus tegument ICP0 is capsid associated, and its E3 ubiquitin ligase domain is important for incorporation into virions. J Virol *84*, 1637-1640.

Desai, P., DeLuca, N.A., and Person, S. (1998). Herpes simplex virus type 1 VP26 is not essential for replication in cell culture but influences production of infectious virus in the nervous system of infected mice. Virology *247*, 115-124.

Desai, P., and Person, S. (1998). Incorporation of the green fluorescent protein into the herpes simplex virus type 1 capsid. J Virol *72*, 7563-7568.

Desai, P.J. (2000). A null mutation in the UL36 gene of herpes simplex virus type 1 results in accumulation of unenveloped DNA-filled capsids in the cytoplasm of infected cells. J Virol *74*, 11608-11618.

Devadas, D., Koithan, T., Diestel, R., Prank, U., Sodeik, B., and Dohner, K. (2014). Herpes simplex virus internalization into epithelial cells requires Na+/H+ exchangers and p21-activated kinases but neither clathrin- nor caveolin-mediated endocytosis. J Virol *88*, 13378-13395.

Diwaker, D., and Wilson, D.W. (2019). Microtubule-Dependent Trafficking of Alphaherpesviruses in the Nervous System: The Ins and Outs. Viruses *11*.

Dohner, K., Radtke, K., Schmidt, S., and Sodeik, B. (2006). Eclipse phase of herpes simplex virus type 1 infection: Efficient dynein-mediated capsid transport without the small capsid protein VP26. J Virol *80*, 8211-8224.

Dohner, K., Ramos-Nascimento, A., Bialy, D., Anderson, F., Hickford-Martinez, A., Rother, F., Koithan, T., Rudolph, K., Buch, A., Prank, U., *et al.* (2018). Importin alpha1 is required for nuclear import of herpes simplex virus proteins and capsid assembly in fibroblasts and neurons. PLoS Pathog *14*, e1006823.

Dohner, K., Wolfstein, A., Prank, U., Echeverri, C., Dujardin, D., Vallee, R., and Sodeik, B. (2002). Function of Dynein and dynactin in herpes simplex virus capsid transport. Mol Biol Cell *13*, 2795-2809.

Douglas, M.W., Diefenbach, R.J., Homa, F.L., Miranda-Saksena, M., Rixon, F.J., Vittone, V., Byth, K., and Cunningham, A.L. (2004). Herpes simplex virus type 1 capsid protein VP26 interacts with dynein light chains RP3 and Tctex1 and plays a role in retrograde cellular transport. J Biol Chem *279*, 28522-28530.

Duda, R.L., Hendrix, R.W., Huang, W.M., and Conway, J.F. (2006). Shared architecture of bacteriophage SPO1 and herpesvirus capsids. Curr Biol *16*, R11-13.

El Kasmi, I., and Lippe, R. (2014). HSV-1 gN partners with gM to modulate the viral fusion machinery. J Virol *89*, 2313-2323.

Elshenawy, M.M., Kusakci, E., Volz, S., Baumbach, J., Bullock, S.L., and Yildiz, A. (2020). Lis1 activates dynein motility by modulating its pairing with dynactin. Nat Cell Biol *22*, 570-578.

Fan, W.H., Roberts, A.P., McElwee, M., Bhella, D., Rixon, F.J., and Lauder, R. (2014). The Large Tegument Protein pUL36 is Essential for Formation of the Capsid Vertex Specific Component at the Capsid-Tegument Interface of HSV-1. J Virol *89*, 1502-1511.

Fenwick, M.L., and Walker, M.J. (1978). Suppression of the synthesis of cellular macromolecules by herpes simplex virus. J Gen Virol *41*, 37-51.

Finnen, R.L., Hay, T.J., Dauber, B., Smiley, J.R., and Banfield, B.W. (2014). The herpes simplex virus 2 virion-associated ribonuclease vhs interferes with stress granule formation. J Virol *88*, 12727-12739.

Finnen, R.L., Zhu, M., Li, J., Romo, D., and Banfield, B.W. (2016). Herpes Simplex Virus 2 Virion Host Shutoff Endoribonuclease Activity Is Required To Disrupt Stress Granule Formation. J Virol *90*, 7943-7955.

Frame, M.C., Purves, F.C., McGeoch, D.J., Marsden, H.S., and Leader, D.P. (1987). Identification of the herpes simplex virus protein kinase as the product of viral gene US3. J Gen Virol *68 (Pt 10)*, 2699-2704.

Frampton, A.R., Jr., Stolz, D.B., Uchida, H., Goins, W.F., Cohen, J.B., and Glorioso, J.C. (2007). Equine herpesvirus 1 enters cells by two different pathways, and infection requires the activation of the cellular kinase ROCK1. J Virol *81*, 10879-10889.

Frampton, A.R., Jr., Uchida, H., von Einem, J., Goins, W.F., Grandi, P., Cohen, J.B., Osterrieder, N., and Glorioso, J.C. (2010). Equine herpesvirus type 1 (EHV-1) utilizes microtubules, dynein, and ROCK1 to productively infect cells. Vet Microbiol *141*, 12-21.

Frank, I., and Friedman, H.M. (1989). A novel function of the herpes simplex virus type 1 Fc receptor: participation in bipolar bridging of antiviral immunoglobulin G. J Virol *63*, 4479-4488.

Fuchs, W., Granzow, H., Klupp, B.G., Kopp, M., and Mettenleiter, T.C. (2002). The UL48 tegument protein of pseudorabies virus is critical for intracytoplasmic assembly of infectious virions. J Virol *76*, 6729-6742.

Fuchs, W., Klupp, B.G., Granzow, H., and Mettenleiter, T.C. (2004). Essential function of the pseudorabies virus UL36 gene product is independent of its interaction with the UL37 protein. J Virol *78*, 11879-11889.

Fuller, A.O., Santos, R.E., and Spear, P.G. (1989). Neutralizing antibodies specific for glycoprotein H of herpes simplex virus permit viral attachment to cells but prevent penetration. J Virol *63*, 3435-3443.

Gazzola, M., Burckhardt, C.J., Bayati, B., Engelke, M., Greber, U.F., and Koumoutsakos, P. (2009). A stochastic model for microtubule motors describes the in vivo cytoplasmic transport of human adenovirus. PLoS Comput Biol *5*, e1000623.

Gluska, S., Zahavi, E.E., Chein, M., Gradus, T., Bauer, A., Finke, S., and Perlson, E. (2014). Rabies Virus Hijacks and accelerates the p75NTR retrograde axonal transport machinery. PLoS Pathog *10*, e1004348.

Granzow, H., Klupp, B.G., and Mettenleiter, T.C. (2005). Entry of pseudorabies virus: an immunogold-labeling study. J Virol *79*, 3200-3205.

Granzow, H., Weiland, F., Jons, A., Klupp, B.G., Karger, A., and Mettenleiter, T.C. (1997). Ultrastructural analysis of the replication cycle of pseudorabies virus in cell culture: a reassessment. J Virol *71*, 2072-2082.

Greber, U.F., and Flatt, J.W. (2019). Adenovirus Entry: From Infection to Immunity. Annu Rev Virol *6*, 177-197.

Gromeier, M., and Wimmer, E. (1998). Mechanism of injury-provoked poliomyelitis. J Virol *72*, 5056-5060.

Grunewald, K., Desai, P., Winkler, D.C., Heymann, J.B., Belnap, D.M., Baumeister, W., and Steven, A.C. (2003). Three-dimensional structure of herpes simplex virus from cryo-electron tomography. Science *302*, 1396-1398.

Grzesik, P., Pryce, E.N., Bhalala, A., Vij, M., Ahmed, R., Etienne, L., Perez, P., McCaffery, J.M., and Desai, A.P.J. (2019). Functional Domains of the Herpes Simplex Virus Type 1 Tegument Protein pUL37: The Amino Terminus is Dispensable for Virus Replication in Tissue Culture. Viruses *11*.

Hanz, S., Perlson, E., Willis, D., Zheng, J.Q., Massarwa, R., Huerta, J.J., Koltzenburg, M., Kohler, M., van-Minnen, J., Twiss, J.L., *et al.* (2003). Axoplasmic importins enable retrograde injury signaling in lesioned nerve. Neuron *40*, 1095-1104.

Heming, J.D., Huffman, J.B., Jones, L.M., and Homa, F.L. (2014). Isolation and characterization of the herpes simplex virus 1 terminase complex. J Virol *88*, 225-236.

Hirokawa, N., and Noda, Y. (2008). Intracellular transport and kinesin superfamily proteins, KIFs: structure, function, and dynamics. Physiol Rev *88*, 1089-1118.

Hogue, I.B., Jean, J., Esteves, A.D., Tanneti, N.S., Scherer, J., and Enquist, L.W. (2018). Functional Carboxy-Terminal Fluorescent Protein Fusion to Pseudorabies Virus Small Capsid Protein VP26. J Virol *92*.

Hook, L.M., Huang, J., Jiang, M., Hodinka, R., and Friedman, H.M. (2008). Blocking antibody access to neutralizing domains on glycoproteins involved in entry as a novel mechanism of immune evasion by herpes simplex virus type 1 glycoproteins C and E. J Virol *82*, 6935-6941.

Hoppe, S., Schelhaas, M., Jaeger, V., Liebig, T., Petermann, P., and Knebel-Morsdorf, D. (2006). Early herpes simplex virus type 1 infection is dependent on regulated Rac1/Cdc42 signalling in epithelial MDCKII cells. J Gen Virol *87*, 3483-3494.

Htet, Z.M., Gillies, J.P., Baker, R.W., Leschziner, A.E., DeSantis, M.E., and Reck-Peterson, S.L. (2020). LIS1 promotes the formation of activated cytoplasmic dynein-1 complexes. Nat Cell Biol *22*, 518-525.

Huet, A., Makhov, A.M., Huffman, J.B., Vos, M., Homa, F.L., and Conway, J.F. (2016). Extensive subunit contacts underpin herpesvirus capsid stability and interior-to-exterior allostery. Nature structural and molecular biology.

Huffman, J.B., Daniel, G.R., Falck-Pedersen, E., Huet, A., Smith, G.A., Conway, J.F., and Homa, F.L. (2017). The C-terminus of the herpes simplex virus pUL25 protein is required for release of viral genomes from capsids bound to nuclear pores. J Virol *91*.

Huffmaster, N.J., Sollars, P.J., Richards, A.L., Pickard, G.E., and Smith, G.A. (2015). Dynamic ubiquitination drives herpesvirus neuroinvasion. Proc Natl Acad Sci U S A *112*, 12818-12823.

Jacob, T., Van den Broeke, C., Grauwet, K., Baert, K., Claessen, C., De Pelsmaeker, S., Van Waesberghe, C., and Favoreel, H.W. (2015). Pseudorabies virus US3 leads to filamentous actin disassembly and contributes to viral genome delivery to the nucleus. Vet Microbiol *177*, 379-385.

Jacob, T., Van den Broeke, C., van Troys, M., Waterschoot, D., Ampe, C., and Favoreel, H.W. (2013). Alphaherpesviral US3 kinase induces cofilin dephosphorylation to reorganize the actin cytoskeleton. J Virol *87*, 4121-4126.

Jovasevic, V., Liang, L., and Roizman, B. (2008). Proteolytic cleavage of VP1-2 is required for release of herpes simplex virus 1 DNA into the nucleus. J Virol *82*, 3311-3319.

Jovasevic, V., Naghavi, M.H., and Walsh, D. (2015). Microtubule plus end-associated CLIP-170 initiates HSV-1 retrograde transport in primary human cells. J Cell Biol *211*, 323-337.

Kapur, M., Maloney, M.T., Wang, W., Chen, X., Millan, I., Mooney, T., Yang, J., and Yang, Y. (2014). A SxIP motif interaction at the microtubule plus end is important for processive retrograde axonal transport. Cell Mol Life Sci *71*, 4043-4054.

Kattenhorn, L.M., Korbel, G.A., Kessler, B.M., Spooner, E., and Ploegh, H.L. (2005). A deubiquitinating enzyme encoded by HSV-1 belongs to a family of cysteine proteases that is conserved across the family Herpesviridae. Mol Cell *19*, 547-557.

Klupp, B.G., Fuchs, W., Granzow, H., Nixdorf, R., and Mettenleiter, T.C. (2002). Pseudorabies Virus UL36 Tegument Protein Physically Interacts with the UL37 Protein. J Virol *76*, 3065-3071.

Klupp, B.G., Granzow, H., Keil, G.M., and Mettenleiter, T.C. (2006). The capsid-associated UL25 protein of the alphaherpesvirus pseudorabies virus is nonessential for cleavage and encapsidation of genomic DNA but is required for nuclear egress of capsids. J Virol *80*, 6235-6246.

Ko, D.H., Cunningham, A.L., and Diefenbach, R.J. (2010). The major determinant for addition of tegument protein pUL48 (VP16) to capsids in herpes simplex virus type 1 is the presence of the major tegument protein pUL36 (VP1/2). J Virol *84*, 1397-1405.

Koenigsberg, A.L., and Heldwein, E. (2018). The dynamic nature of the conserved tegument protein UL37 of herpesviruses. J Biol Chem *293*, 15827-15839.

Koenigsberg, A.L., and Heldwein, E.E. (2017). Crystal structure of the N-terminal half of the traffic controller UL37 from Herpes Simplex virus Type 1. J Virol *91*, e01244-01217.

Komala Sari, T., Gianopulos, K.A., and Nicola, A.V. (2020a). Glycoprotein C of Herpes Simplex Virus 1 Shields Glycoprotein B from Antibody Neutralization. J Virol *94*.

Komala Sari, T., Gianopulos, K.A., Weed, D.J., Schneider, S.M., Pritchard, S.M., and Nicola, A.V. (2020b). Herpes Simplex Virus Glycoprotein C Regulates Low-pH Entry. mSphere *5*.

Kornfeind, E.M., and Visalli, R.J. (2018). Human herpesvirus portal proteins: Structure, function, and antiviral prospects. Rev Med Virol *28*, e1972.

Koyuncu, O.O., Hogue, I.B., and Enquist, L.W. (2013a). Virus infections in the nervous system. Cell Host Microbe *13*, 379-393.

Koyuncu, O.O., Perlman, D.H., and Enquist, L.W. (2013b). Efficient retrograde transport of pseudorabies virus within neurons requires local protein synthesis in axons. Cell Host Microbe *13*, 54-66.

Kramer, T., Greco, T.M., Enquist, L.W., and Cristea, I.M. (2011). Proteomic characterization of pseudorabies virus extracellular virions. J Virol *85*, 6427-6441.

Krautwald, M., Fuchs, W., Klupp, B.G., and Mettenleiter, T.C. (2009). Translocation of incoming pseudorabies virus capsids to the cell nucleus is delayed in the absence of tegument protein pUL37. J Virol *83*, 3389-3396.

Krautwald, M., Maresch, C., Klupp, B.G., Fuchs, W., and Mettenleiter, T.C. (2008). Deletion or green fluorescent protein tagging of the pUL35 capsid component of pseudorabies virus impairs virus replication in cell culture and neuroinvasion in mice. J Gen Virol *89*, 1346-1351.

Kristensson, K., Lycke, E., Röyttä, M., Svennerholm, B., and Vahlne, A. (1986a). Neuritic transport of herpes simplex virus in rat sensory neurons in vitro. Effects of substances interacting with microtubular function and axonal flow [nocodazole, taxol and erythro-9-3-(2-hydroxynonyl)adenine]. J Gen Virol *67*, 2023-2028.

Kristensson, K., Lycke, E., Röyttä, M., Svennerholm, B., and Vahlne, A. (1986b). Neuritic transport of herpes simplex virus in rat sensory neurons in vitro. Effects of substances interacting with microtubular function and axonal flow [nocodazole, taxol and erythro-9-3-(2-hydroxynonyl)adenine]. J Gen Virol *67*, 2023-2028.

Krupovic, M., and Koonin, E.V. (2017). Multiple origins of viral capsid proteins from cellular ancestors. Proc Natl Acad Sci U S A *114*, E2401-E2410.

Kuhn, J., Leege, T., Klupp, B.G., Granzow, H., Fuchs, W., and Mettenleiter, T.C. (2008). Partial functional complementation of a pseudorabies virus UL25 deletion mutant by herpes simplex virus type 1 pUL25 indicates overlapping functions of alphaherpesvirus pUL25 proteins. J Virol *82*, 5725-5734.

Kunzli, K., Favre, B., Chofflon, M., and Borradori, L. (2016). One gene but different proteins and diseases: the complexity of dystonin and bullous pemphigoid antigen 1. Exp Dermatol *25*, 10-16.

Laine, R.F., Albecka, A., van de Linde, S., Rees, E.J., Crump, C.M., and Kaminski, C.F. (2015). Structural analysis of herpes simplex virus by optical super-resolution imaging. Nat Commun *6*, 5980.

Le Sage, V., Jung, M., Alter, J.D., Wills, E.G., Johnston, S.M., Kawaguchi, Y., Baines, J.D., and Banfield, B.W. (2013). The Herpes Simplex Virus Type 2 UL21 Protein is Essential for Virus Propagation. J Virol.

Lee, J.H., Vittone, V., Diefenbach, E., Cunningham, A.L., and Diefenbach, R.J. (2008). Identification of structural protein-protein interactions of herpes simplex virus type 1. Virology *378*, 347-354.

Lee, J.I., Luxton, G.W., and Smith, G.A. (2006). Identification of an essential domain in the herpesvirus VP1/2 tegument protein: the carboxy terminus directs incorporation into capsid assemblons. J Virol *80*, 12086-12094.

Lee, J.I., Sollars, P.J., Baver, S.B., Pickard, G.E., Leelawong, M., and Smith, G.A. (2009). A herpesvirus encoded deubiquitinase is a novel neuroinvasive determinant. PLoS Pathog *5*, e1000387.

Lewis, P., Fu, Y., and Lentz, T.L. (2000). Rabies virus entry at the neuromuscular junction in nerve-muscle cocultures. Muscle Nerve *23*, 720-730.

Liu, J.J., Ding, J., Kowal, A.S., Nardine, T., Allen, E., Delcroix, J.D., Wu, C., Mobley, W., Fuchs, E., and Yang, Y. (2003). BPAG1n4 is essential for retrograde axonal transport in sensory neurons. J Cell Biol *163*, 223-229.

Liu, X., and Cohen, J.I. (2015). The role of PI3K/Akt in human herpesvirus infection: From the bench to the bedside. Virology *479-480*, 568-577.

Liu, X., Fitzgerald, K., Kurt-Jones, E., Finberg, R., and Knipe, D.M. (2008). Herpesvirus tegument protein activates NF-kappaB signaling through the TRAF6 adaptor protein. Proc Natl Acad Sci U S A *105*, 11335-11339.

Liu, Y.-T., Jiang, J., Bohannon, K.P., Dai, X., Luxton, G.W., Hui, W.H., Bi, G.-Q., Smith, G.A., and Zhou, Z.H. (2017). A pUL25 dimer interfaces the pseudorabies virus capsid and tegument. J Gen Virol *98*, 2837-2849.

Loret, S., Guay, G., and Lippe, R. (2008). Comprehensive characterization of extracellular herpes simplex virus type 1 virions. J Virol *82*, 8605-8618.

Luby-Phelps, K. (1994). Physical properties of cytoplasm. Curr Opin Cell Biol *6*, 3-9.

Luby-Phelps, K. (2000). Cytoarchitecture and physical properties of cytoplasm: volume, viscosity, diffusion, intracellular surface area. Int Rev Cytol *192*, 189-221.

Luxton, G.W., Haverlock, S., Coller, K.E., Antinone, S.E., Pincetic, A., and Smith, G.A. (2005). Targeting of herpesvirus capsid transport in axons is coupled to association with specific sets of tegument proteins. Proc Natl Acad Sci U S A *102*, 5832-5837.

Lycke, E., Hamark, B., Johansson, M., Krotochwil, A., Lycke, J., and Svennerholm, B. (1988). Herpes simplex virus infection of the human sensory neuron. An electron microscopy study. Arch Virol *101*, 87-104.

Lycke, E., Kristensson, K., Svennerholm, B., Vahlne, A., and Ziegler, R. (1984). Uptake and transport of herpes simplex virus in neurites of rat dorsal root ganglia cells in culture. J Gen Virol *65*, 55-64.

Maekawa, M., Ishizaki, T., Boku, S., Watanabe, N., Fujita, A., Iwamatsu, A., Obinata, T., Ohashi, K., Mizuno, K., and Narumiya, S. (1999). Signaling from Rho to the actin cytoskeleton through protein kinases ROCK and LIM-kinase. Science *285*, 895-898.

Maier, O., Sollars, P.J., Pickard, G.E., and Smith, G.A. (2016). Visualizing herpesvirus procapsids in living cells. J Virol *90*, 10182-10192.

Maringer, K., and Elliott, G. (2010). Recruitment of herpes simplex virus type 1 immediate-early protein ICP0 to the virus particle. J Virol *84*, 4682-4696.

Marzo, M.G., Griswold, J.M., and Markus, S.M. (2020). Pac1/LIS1 stabilizes an uninhibited conformation of dynein to coordinate its localization and activity. Nat Cell Biol *22*, 559-569.

Maurer, U.E., Sodeik, B., and Grunewald, K. (2008). Native 3D intermediates of membrane fusion in herpes simplex virus 1 entry. Proc Natl Acad Sci U S A *105*, 10559-10564.

Mbong, E.F., Woodley, L., Frost, E., Baines, J.D., and Duffy, C. (2012). Deletion of UL21 causes a delay in early stages of the herpes simplex virus type 1 replication cycle. J Virol.

McElwee, M., Beilstein, F., Labetoulle, M., Rixon, F.J., and Pasdeloup, D. (2013). Dystonin/BPAG1 Promotes Plus-End-Directed Transport of Herpes Simplex Virus 1 Capsids on Microtubules during Entry. J Virol *87*, 11008-11018.

McElwee, M., Vijayakrishnan, S., Rixon, F., and Bhella, D. (2018). Structure of the herpes simplex virus portal-vertex. PLoS Biol *16*, e2006191.

McNab, A.R., Desai, P., Person, S., Roof, L.L., Thomsen, D.R., Newcomb, W.W., Brown, J.C., and Homa, F.L. (1998). The product of the herpes simplex virus type 1 UL25 gene is required for encapsidation but not for cleavage of replicated viral DNA. J Virol *72*, 1060-1070.

Meckes, D.G., Jr., and Wills, J.W. (2007). Dynamic interactions of the UL16 tegument protein with the capsid of herpes simplex virus. J Virol *81*, 13028-13036.

Meckes, D.G., Jr., and Wills, J.W. (2008). Structural rearrangement within an enveloped virus upon binding to the host cell. J Virol *82*, 10429-10435.

Michael, K., Bottcher, S., Klupp, B.G., Karger, A., and Mettenleiter, T.C. (2006). Pseudorabies virus particles lacking tegument proteins pUL11 or pUL16 incorporate less full-length pUL36 than wild-type virus, but specifically accumulate a pUL36 N-terminal fragment. J Gen Virol *87*, 3503-3507.

Morgan, C., Ellison, S.A., Rose, H.M., and Moore, D.H. (1953). Electron microscopic examination of inclusion bodies of herpes simplex virus. Proc Soc Exp Biol Med *82*, 454-457.

Morrison, E.E., Wang, Y.F., and Meredith, D.M. (1998). Phosphorylation of structural components promotes dissociation of the herpes simplex virus type 1 tegument. J Virol *72*, 7108-7114.

Nagel, C.H., Dohner, K., Binz, A., Bauerfeind, R., and Sodeik, B. (2012). Improper tagging of the non-essential small capsid protein VP26 impairs nuclear capsid egress of herpes simplex virus. PLoS One *7*, e44177.

Newcomb, W.W., and Brown, J.C. (2009). Time-dependent transformation of the herpesvirus tegument. J Virol *83*, 8082-8089.

Newcomb, W.W., and Brown, J.C. (2010). Structure and capsid association of the herpesvirus large tegument protein UL36. J Virol *84*, 9408-9414.

Newcomb, W.W., Homa, F.L., Thomsen, D.R., Booy, F.P., Trus, B.L., Steven, A.C., Spencer, J.V., and Brown, J.C. (1996). Assembly of the herpes simplex virus capsid: characterization of intermediates observed during cell-free capsid formation. J Mol Biol *263*, 432-446.

Newcomb, W.W., Juhas, R.M., Thomsen, D.R., Homa, F.L., Burch, A.D., Weller, S.K., and Brown, J.C. (2001). The UL6 gene product forms the portal for entry of DNA into the herpes simplex virus capsid. J Virol *75*, 10923-10932.

Newcomb, W.W., Trus, B.L., Booy, F.P., Steven, A.C., Wall, J.S., and Brown, J.C. (1993). Structure of the herpes simplex virus capsid. Molecular composition of the pentons and the triplexes. J Mol Biol *232*, 499-511.

Newcomb, W.W., Trus, B.L., Cheng, N., Steven, A.C., Sheaffer, A.K., Tenney, D.J., Weller, S.K., and Brown, J.C. (2000). Isolation of herpes simplex virus procapsids from cells infected with a protease-deficient mutant virus. J Virol *74*, 1663-1673.

Nicola, A.V., and Straus, S.E. (2004). Cellular and viral requirements for rapid endocytic entry of herpes simplex virus. J Virol *78*, 7508-7517.

Nishioka, Y., and Silverstein, S. (1978). Requirement of protein synthesis for the degradation of host mRNA in Friend erythroleukemia cells infected wtih herpes simplex virus type 1. J Virol *27*, 619-627.

O'Hara, M., Rixon, F.J., Stow, N.D., Murray, J., Murphy, M., and Preston, V.G. (2010). Mutational analysis of the herpes simplex virus type 1 UL25 DNA packaging protein reveals regions that are important after the viral DNA has been packaged. J Virol *84*, 4252-4263.

Oh, M.J., Akhtar, J., Desai, P., and Shukla, D. (2010). A role for heparan sulfate in viral surfing. Biochem Biophys Res Commun *391*, 176-181.

Ohka, S., Matsuda, N., Tohyama, K., Oda, T., Morikawa, M., Kuge, S., and Nomoto, A. (2004). Receptor (CD155)-dependent endocytosis of poliovirus and retrograde axonal transport of the endosome. J Virol *78*, 7186-7198.

Ohka, S., Yang, W.X., Terada, E., Iwasaki, K., and Nomoto, A. (1998). Retrograde transport of intact poliovirus through the axon via the fast transport system. Virology *250*, 67-75.

Ojala, P.M., Sodeik, B., Ebersold, M.W., Kutay, U., and Helenius, A. (2000). Herpes simplex virus type 1 entry into host cells: reconstitution of capsid binding and uncoating at the nuclear pore complex in vitro. Mol Cell Biol *20*, 4922-4931.

Onomoto, K., Yoneyama, M., Fung, G., Kato, H., and Fujita, T. (2014). Antiviral innate immunity and stress granule responses. Trends Immunol *35*, 420-428.

Owen, D.J., Crump, C.M., and Graham, S.C. (2015). Tegument assembly and secondary envelopment of alphaherpesviruses. Viruses *7*, 5084-5114.

Pasdeloup, D., Blondel, D., Isidro, A.L., and Rixon, F.J. (2009). Herpesvirus Capsid Association To The Nuclear Pore Complex And Viral DNA Release Involve The Nucleoporin CAN/Nup214 And The Capsid Protein pUL25. J Virol *83*, 6610-6623.

Pasdeloup, D., McElwee, M., Beilstein, F., Labetoulle, M., and Rixon, F.J. (2013). Herpesvirus tegument protein pUL37 interacts with dystonin/BPAG1 to promote capsid transport on microtubules during egress. J Virol *87*, 2857-2867.

Patel, A., Hanson, J., McLean, T.I., Olgiate, J., Hilton, M., Miller, W.E., and Bachenheimer, S.L. (1998). Herpes simplex type 1 induction of persistent NF-kappa B nuclear translocation increases the efficiency of virus replication. Virology *247*, 212-222.

Person, S., and Desai, P. (1998). Capsids are formed in a mutant virus blocked at the maturation site of the UL26 and UL26.5 open reading frames of herpes simplex virus type 1 but are not formed in a null mutant of UL38 (VP19C). Virology *242*, 193-203.

Pickard, G.E., Brodersen, B., Sollars, P.J., and Smith, G.A. (2020). The pseudorabies virus R2 non-neuroinvasive vaccine: A proof-of-concept study in pigs. Vaccine *38*, 4524-4528.

Pitts, J.D., Klabis, J., Richards, A.L., Smith, G.A., and Heldwein, E.E. (2014). Crystal structure of the herpesvirus inner tegument protein UL37 supports its essential role in control of viral trafficking. J Virol *88*, 5462-5473.

Poliakova, K., Adebola, A., Leung, C.L., Favre, B., Liem, R.K., Schepens, I., and Borradori, L. (2014). BPAG1a and b associate with EB1 and EB3 and modulate vesicular transport, Golgi apparatus structure, and cell migration in C2.7 myoblasts. PLoS One *9*, e107535.

Poon, A.P., and Roizman, B. (1993). Characterization of a temperature-sensitive mutant of the UL15 open reading frame of herpes simplex virus 1. J Virol *67*, 4497-4503.

Preston, V.G., Murray, J., Preston, C.M., McDougall, I.M., and Stow, N.D. (2008). The UL25 gene product of herpes simplex virus type 1 is involved in uncoating of the viral genome. J Virol *82*, 6654-6666.

Purves, F.C., Longnecker, R.M., Leader, D.P., and Roizman, B. (1987). Herpes simplex virus 1 protein kinase is encoded by open reading frame US3 which is not essential for virus growth in cell culture. J Virol *61*, 2896-2901.

Radtke, K., Kieneke, D., Wolfstein, A., Michael, K., Steffen, W., Scholz, T., Karger, A., and Sodeik, B. (2010). Plus- and minus-end directed microtubule motors bind simultaneously to herpes simplex virus capsids using different inner tegument structures. PLoS Pathog *6*, e1000991.

Read, G.S., and Frenkel, N. (1983). Herpes simplex virus mutants defective in the virion-associated shutoff of host polypeptide synthesis and exhibiting abnormal synthesis of alpha (immediate early) viral polypeptides. J Virol *46*, 498-512.

Reck-Peterson, S.L., Redwine, W.B., Vale, R.D., and Carter, A.P. (2018). The cytoplasmic dynein transport machinery and its many cargoes. Nat Rev Mol Cell Biol *19*, 382-398.

Ren, R., and Racaniello, V.R. (1992). Poliovirus spreads from muscle to the central nervous system by neural pathways. J Infect Dis *166*, 747-752.

Richards, A.L., Sollars, P.J., Pitts, J.D., Stults, A.M., Heldwein, E.E., Pickard, G.E., and Smith, G.A. (2017). The pUL37 tegument protein guides alpha-herpesvirus retrograde axonal transport to promote neuroinvasion. PLoS Pathog *13*, e1006741.

Rixon, F.J., and McNab, D. (1999). Packaging-competent capsids of a herpes simplex virus temperature-sensitive mutant have properties similar to those of in vitro-assembled procapsids. J Virol *73*, 5714-5721.

Roberts, A.P., Abaitua, F., O'Hare, P., McNab, D., Rixon, F.J., and Pasdeloup, D. (2009). Differing roles of inner tegument proteins pUL36 and pUL37 during entry of herpes simplex virus type 1. J Virol *83*, 105-116.

Rode, K., Dohner, K., Binz, A., Glass, M., Strive, T., Bauerfeind, R., and Sodeik, B. (2011). Uncoupling uncoating of herpes simplex virus genomes from their nuclear import and gene expression. J Virol *85*, 4271-4283.

Romero, N., Van Waesberghe, C., and Favoreel, H.W. (2020). Pseudorabies virus infection of epithelial cells leads to persistent but aberrant activation of the NF-kappaB pathway, inhibiting hallmark NF-kappaB-induced pro-inflammatory gene expression. J Virol.

Roos, W.H., Radtke, K., Kniesmeijer, E., Geertsema, H., Sodeik, B., and Wuite, G.J. (2009). Scaffold expulsion and genome packaging trigger stabilization of herpes simplex virus capsids. Proc Natl Acad Sci U S A *106*, 9673-9678.

Ruhge, L.L., Huet, A.G.E., Conway, J.F., and Smith, G.A. (2018). The apical region of the herpes simplex virus major capsid protein promotes capsid maturation. J Virol *92*, e00821-00818.

Salinas, S., Schiavo, G., and Kremer, E.J. (2010). A hitchhiker's guide to the nervous system: the complex journey of viruses and toxins. Nat Rev Microbiol *8*, 645-655.

Sandbaumhuter, M., Dohner, K., Schipke, J., Binz, A., Pohlmann, A., Sodeik, B., and Bauerfeind, R. (2013). Cytosolic herpes simplex virus capsids not only require binding inner tegument protein pUL36 but also pUL37 for active transport prior to secondary envelopment. Cell Microbiol *15*, 248-269.

Schipke, J., Pohlmann, A., Diestel, R., Binz, A., Rudolph, K., Nagel, C.H., Bauerfeind, R., and Sodeik, B. (2012). The C-terminus of the large tegument protein pUL36 contains multiple capsid binding sites that function differently during assembly and cell entry of herpes simplex virus. J Virol *86*, 3682-3700.

Schlieker, C., Weihofen, W.A., Frijns, E., Kattenhorn, L.M., Gaudet, R., and Ploegh, H.L. (2007). Structure of a herpesvirus-encoded cysteine protease reveals a unique class of deubiquitinating enzymes. Mol Cell *25*, 677-687.

Schmid, M.F., Hecksel, C.W., Rochat, R.H., Bhella, D., Chiu, W., and Rixon, F.J. (2012). A Tail-like assembly at the portal vertex in intact herpes simplex type-1 virions. PLoS Pathog *8*, e1002961.

Scholtes, L.D., Yang, K., Li, L.X., and Baines, J.D. (2010). The capsid protein encoded by UL17 of herpes simplex virus 1 interacts with the tegument protein VP13/14. J Virol.

Schuster, M., Kilaru, S., Ashwin, P., Lin, C., Severs, N.J., and Steinberg, G. (2011). Controlled and stochastic retention concentrates dynein at microtubule ends to keep endosomes on track. EMBO J *30*, 652-664.

Scrima, N., Lepault, J., Boulard, Y., Pasdeloup, D., Bressanelli, S., and Roche, S. (2015). Insights into herpesvirus tegument organization from structural analyses of the 970 central residues of HSV-1 UL36 protein. J Biol Chem *290*, 8820-8833.

Scrivano, L., Sinzger, C., Nitschko, H., Koszinowski, U.H., and Adler, B. (2011). HCMV spread and cell tropism are determined by distinct virus populations. PLoS Pathog *7*, e1001256.

Sekine, E., Schmidt, N., Gaboriau, D., and O'Hare, P. (2017). Spatiotemporal dynamics of HSV genome nuclear entry and compaction state transitions using bioorthogonal chemistry and super-resolution microscopy. PLoS Pathog *13*, e1006721.

Sheaffer, A.K., Newcomb, W.W., Gao, M., Yu, D., Weller, S.K., Brown, J.C., and Tenney, D.J. (2001). Herpes simplex virus DNA cleavage and packaging proteins associate with the procapsid prior to its maturation. J Virol *75*, 687-698.

Smith, G. (2012). Herpesvirus Transport to the Nervous System and Back Again. Annu Rev Microbiol *66*, 153-176.

Smith, G.A. (2017). Assembly and egress of an alphaherpesvirus clockwork. Adv Anat Embryol Cell Biol *223*, 171-193.

Smith, G.A., and Enquist, L.W. (1999). Construction and transposon mutagenesis in Escherichia coli of a full-length infectious clone of pseudorabies virus, an alphaherpesvirus. J Virol *73*, 6405-6414.

Smith, G.A., Pomeranz, L., Gross, S.P., and Enquist, L.W. (2004). Local modulation of plus-end transport targets herpesvirus entry and egress in sensory axons. Proc Natl Acad Sci U S A *101*, 16034-16039.

Snijder, J., Radtke, K., Anderson, F., Scholtes, L., Corradini, E., Baines, J., Heck, A.J.R., Wuite, G.J.L., Sodeik, B., and Roos, W.H. (2017). Vertex-Specific Proteins pUL17 and pUL25 Mechanically Reinforce Herpes Simplex Virus Capsids. J Virol *91*.

Sodeik, B., Ebersold, M.W., and Helenius, A. (1997). Microtubule-mediated transport of incoming herpes simplex virus 1 capsids to the nucleus. J Cell Biol *136*, 1007-1021.

Spear, P.G., and Roizman, B. (1972). Proteins specified by herpes simplex virus. V. Purification and structural proteins of the herpesvirion. J Virol *9*, 143-159.

Stackpole, C.W. (1969). Herpes-type virus of the frog renal adenocarcinoma. I. Virus development in tumor transplants maintained at low temperature. J Virol *4*, 75-93.

Stevenson, A.J., Morrison, E.E., Chaudhari, R., Yang, C.C., and Meredith, D.M. (1997). Processing and intracellular localization of the herpes simplex virus type 1 proteinase. J Gen Virol *78 (Pt 3)*, 671-675.

Strelow, L.I., and Leib, D.A. (1995). Role of the virion host shutoff (vhs) of herpes simplex virus type 1 in latency and pathogenesis. J Virol *69*, 6779-6786.

Stringer, D.K., and Piper, R.C. (2011). A single ubiquitin is sufficient for cargo protein entry into MVBs in the absence of ESCRT ubiquitination. J Cell Biol *192*, 229-242.

Strunze, S., Trotman, L.C., Boucke, K., and Greber, U.F. (2005). Nuclear targeting of adenovirus type 2 requires CRM1-mediated nuclear export. Mol Biol Cell *16*, 2999-3009.

Stults, A.M., and Smith, G.A. (2019). The Herpes Simplex Virus 1 Deamidase Enhances Propagation but Is Dispensable for Retrograde Axonal Transport into the Nervous System. J Virol *93*.

Suhanovsky, M.M., and Teschke, C.M. (2015). Nature's favorite building block: Deciphering folding and capsid assembly of proteins with the HK97-fold. Virology *479-480*, 487-497.

Suomalainen, M., Nakano, M.Y., Keller, S., Boucke, K., Stidwill, R.P., and Greber, U.F. (1999). Microtubule-dependent plus- and minus end-directed motilities are competing processes for nuclear targeting of adenovirus. J Cell Biol *144*, 657-672.

Svobodova, S., Bell, S., and Crump, C.M. (2012). Analysis of the interaction between the essential herpes simplex virus 1 tegument proteins VP16 and VP1/2. J Virol *86*, 473-483.

Takakuwa, H., Goshima, F., Koshizuka, T., Murata, T., Daikoku, T., and Nishiyama, Y. (2001). Herpes simplex virus encodes a virion-associated protein which promotes long cellular processes in over-expressing cells. Genes Cells *6*, 955-966.

Tian, J., Liu, Y., Liu, X., Sun, X., Zhang, J., and Qu, L. (2018). Feline Herpesvirus 1 US3 Blocks the Type I Interferon Signal Pathway by Targeting Interferon Regulatory Factor 3 Dimerization in a Kinase-Independent Manner. J Virol *92*.

Tiwari, V., and Shukla, D. (2010). Phosphoinositide 3 kinase signalling may affect multiple steps during herpes simplex virus type-1 entry. J Gen Virol *91*, 3002-3009.

Topp, K.S., Bisla, K., Saks, N.D., and Lavail, J.H. (1996). Centripetal transport of herpes simplex virus in human retinal pigment epithelial cells in vitro. Neuroscience *71*, 1133-1144.

Toropova, K., Huffman, J.B., Homa, F.L., and Conway, J.F. (2011). The Herpes Simplex Virus 1 UL17 Protein Is the Second Constituent of the Capsid Vertex-Specific Component Required for DNA Packaging and Retention. J Virol *85*, 7513-7522.

Trus, B.L., Booy, F.P., Newcomb, W.W., Brown, J.C., Homa, F.L., Thomsen, D.R., and Steven, A.C. (1996). The herpes simplex virus procapsid: structure, conformational changes upon maturation, and roles of the triplex proteins VP19c and VP23 in assembly. J Mol Biol *263*, 447-462.

Trus, B.L., Newcomb, W.W., Cheng, N., Cardone, G., Marekov, L., Homa, F.L., Brown, J.C., and Steven, A.C. (2007). Allosteric signaling and a nuclear exit strategy: binding of UL25/UL17 heterodimers to DNA-filled HSV-1 capsids. Mol Cell *26*, 479-489.

Twarock, R., and Luque, A. (2019). Structural puzzles in virology solved with an overarching icosahedral design principle. Nat Commun *10*, 4414.

Uetz, P., Dong, Y.A., Zeretzke, C., Atzler, C., Baiker, A., Berger, B., Rajagopala, S.V., Roupelieva, M., Rose, D., Fossum, E., *et al.* (2006). Herpesviral protein networks and their interaction with the human proteome. Science *311*, 239-242.

Van den Broeke, C., Jacob, T., and Favoreel, H.W. (2014). Rho'ing in and out of cells: viral interactions with Rho GTPase signaling. Small GTPases *5*, e28318.

Vittone, V., Diefenbach, E., Triffett, D., Douglas, M.W., Cunningham, A.L., and Diefenbach, R.J. (2005). Determination of interactions between tegument proteins of herpes simplex virus type 1. J Virol *79*, 9566-9571.

Waisner, H., and Kalamvoki, M. (2019). The ICP0 protein of herpes simplex virus 1 (HSV-1) down-regulates major autophagy adaptor proteins, sequestosome 1 (SQSTM1/p62) and optineurin (OPTN), during the early stages of HSV-1 infection. J Virol.

Wang, I.H., Burckhardt, C.J., Yakimovich, A., Morf, M.K., and Greber, U.F. (2017). The nuclear export factor CRM1 controls juxta-nuclear microtubule-dependent virus transport. J Cell Sci *130*, 2185-2195.

Wang, J., Yuan, S., Zhu, D., Tang, H., Wang, N., Chen, W., Gao, Q., Li, Y., Wang, J., Liu, H., *et al.* (2018). Structure of the herpes simplex virus type 2 C-capsid with capsid-vertex-specific component. Nat Commun *9*, 3668.

Wikoff, W.R., Liljas, L., Duda, R.L., Tsuruta, H., Hendrix, R.W., and Johnson, J.E. (2000). Topologically linked protein rings in the bacteriophage HK97 capsid. Science *289*, 2129-2133.

Wolfstein, A., Nagel, C.H., Radtke, K., Dohner, K., Allan, V.J., and Sodeik, B. (2006). The inner tegument promotes herpes simplex virus capsid motility along microtubules in vitro. Traffic *7*, 227-237.

Wu, F., Xing, J., Wang, S., Li, M., and Zheng, C. (2011). Screening and identification of host factors interacting with UL14 of herpes simplex virus 1. Medical microbiology and immunology *200*, 203-208.

Yamauchi, Y., Kiriyama, K., Kubota, N., Kimura, H., Usukura, J., and Nishiyama, Y. (2008). The UL14 tegument protein of herpes simplex virus type 1 is required for efficient nuclear transport of the alpha transinducing factor VP16 and viral capsids. J Virol *82*, 1094-1106.

Yamauchi, Y., Wada, K., Goshima, F., Takakuwa, H., Daikoku, T., Yamada, M., and Nishiyama, Y. (2001). The UL14 protein of herpes simplex virus type 2 translocates the minor capsid protein VP26 and the DNA cleavage and packaging UL33 protein into the nucleus of coexpressing cells. J Gen Virol *82*, 321-330.

Yan, K., Liu, J., Guan, X., Yin, Y.X., Peng, H., Chen, H.C., and Liu, Z.F. (2019). The Carboxyl Terminus of Tegument Protein pUL21 Contributes to Pseudorabies Virus Neuroinvasion. J Virol *93*.

Yao, F., and Courtney, R.J. (1992). Association of ICP0 but not ICP27 with purified virions of herpes simplex virus type 1. J Virol *66*, 2709-2716.

Yuan, S., Wang, J., Zhu, D., Wang, N., Gao, Q., Chen, W., Tang, H., Wang, J., Zhang, X., Liu, H., *et al.* (2018). Cryo-EM structure of a herpesvirus capsid at 3.1 A. Science *360*.

Zaichick, S.V., Bohannon, K.P., Hughes, A., Sollars, P.J., Pickard, G.E., and Smith, G.A. (2013). The herpesvirus VP1/2 protein is an effector of dynein-mediated capsid transport and neuroinvasion. Cell Host Microbe *13*, 193-203.

Zhang, J., Zhao, J., Xu, S., Li, J., He, S., Zeng, Y., Xie, L., Xie, N., Liu, T., Lee, K., *et al.* (2018). Species-Specific Deamidation of cGAS by Herpes Simplex Virus UL37 Protein Facilitates Viral Replication. Cell Host Microbe *24*, 234-248 e235.

Zhang, Z., Selariu, A., Warden, C., Huang, G., Huang, Y., Zaccheus, O., Cheng, T., Xia, N., and Zhu, H. (2010). Genome-wide mutagenesis reveals that ORF7 is a novel VZV skin-tropic factor. PLoS Pathog *6*, e1000971.

Zhao, J., Zeng, Y., Xu, S., Chen, J., Shen, G., Yu, C., Knipe, D., Yuan, W., Peng, J., Xu, W., *et al.* (2016). A viral deamidase targets the helicase domain of RIG-I to block RNA-induced activation. Cell Host Microbe *20*, 770-784.

Zheng, K., Xiang, Y., Wang, X., Wang, Q., Zhong, M., Wang, S., Wang, X., Fan, J., Kitazato, K., and Wang, Y. (2014). Epidermal growth factor receptor-PI3K signaling controls cofilin activity to facilitate herpes simplex virus 1 entry into neuronal cells. mBio *5*, e00958-00913.

Zhou, J., Scherer, J., Yi, J., and Vallee, R.B. (2018). Role of kinesins in directed adenovirus transport and cytoplasmic exploration. PLoS Pathog *14*, e1007055.

Zhou, Z.H., Chen, D.H., Jakana, J., Rixon, F.J., and Chiu, W. (1999). Visualization of tegument-capsid interactions and DNA in intact herpes simplex virus type 1 virions. J Virol *73*, 3210-3218.

Chapter 4

The Fate of Incoming HSV-1 Genomes Entering the Nucleus

Oren Kobiler* and Amichay Afriat

Department of Clinical Microbiology and Immunology, Sackler School of Medicine, Tel Aviv University, Tel Aviv, Israel

*okobiler@tauex.tau.ac.il

DOI: https://doi.org/10.21775/9781913652555.04

Abstract

Herpesvirus genomes enter the eukaryotic nucleus as large linear double stranded DNA molecules that are free of any proteins (naked DNA). Once inside the nucleus, the HSV-1 genomes immediately associate with proteins that will be instrumental in the organization and regulation of these genomes. These initial interactions are thought to determine the fate of the infecting genomes. In general, the host cell has evolved several mechanisms to suppress viral genomes and induce latent or abortive infections. On the other hand, the virus has evolved to use viral and cellular factors to promote lytic infection. Recent findings suggest that not all viral genomes in the infected nucleus will develop progeny and that not all genetically identical cells will support successful virus propagation. Thus, the decision between different fates of infection is determined at both single-cell and single-genome levels. Here we summarize current knowledge on the conditions and interactions that lead to each outcome and discuss the unknown determinants.

Introduction

The entry of herpesviruses into cells can be divided into two major steps: capsid release into the cytoplasm (see Chapter 2) and injection of viral genomes into the nucleoplasm. The fusion of the viral membrane with the host membrane releases the viral capsid into the cytoplasm together with a set of proteins found in the viral tegument layer. The inner tegument proteins remain associated with the capsid and mediate its movement in the cytoplasm (see Chapter 3). Viral outer tegument proteins (like VHS and VP16) as well as host proteins that are found in the tegument (like DDX3X and MIF) are thought to facilitate viral infection by modulating host signaling and host and viral gene expression (Kwong and Frenkel, 1987; Kwong et al., 1988; Weinheimer et al., 1992; Kelly et al., 2009; Stegen et al., 2013). Viral genomes are released from the viral capsids into the nucleus through nuclear pore complexes as large linear double stranded DNA molecules (Ojala et al., 2000; Pasdeloup et al., 2009). The viral genome enters the nucleus as a naked DNA molecule and starts to associate with proteins once within the nucleoplasm.

The eukaryotic nucleus has a highly ordered structure, with chromosomal territories, nucleolus and other specific sites for different tasks (Lanctot et al., 2007). Furthermore, the eukaryotic genome is bound to histones and is condensed into an ordered, compact structure. The positioning and condensation level of the genome correlate with gene activity (Sexton and Cavalli, 2015). During herpesviruses infection, insertion of large naked viral DNA (vDNA) into the host nucleus results in significant disturbance to this order. As all other mammalian dsDNA viruses enter the nucleus associated with histones or histone like proteins, herpesviruses nuclear entry is probably the only natural scenario in which mammalian nuclei encounter such a major event. Since herpesviruses are ubiquitous and ancient viruses that have probably co-evolved with their hosts before mammalian lineages appeared (McGeoch et al., 2006), it is not surprising that the host nucleus has developed many mechanisms to counteract invasion of these large genomes. Thus, an arms race between these viruses and their hosts has been going on for millions of years, and a major battleground of this conflict is within the host cell nucleus.

The host defense against viral pathogens can be divided to three major immune arms: intrinsic, innate and adaptive. These arms cooperate and overlap to fight off invading pathogens. The first line of intra-nuclear defense against DNA virus infection (i.e., the intrinsic immunity) is coordinated by a set of constitutively expressed host-cell restriction factors (reviewed in (Everett, 2013; Komatsu et al., 2016)). In this chapter, we will focus on mechanisms of the first line of defense while other antiviral innate immune defenses are covered in Chapter 7. Herpesviruses have evolved mechanisms to avoid these host defenses. Since these viral strategies are highly efficient, mutation of the viral counteracting factors is often required to reveal the host defense mechanisms. In the case of HSV-1, many of the host intrinsic immune mechanisms were identified by using viral strains deleted or mutated in the ICP0 gene (Boutell and Everett, 2013).

Herpesvirus/host interactions that take place in the nucleus have been studied for several decades and were recently expanded upon with the improvement of available technology. For example, combining the ability to detect incoming viral genomes by incorporating biorthogonal traceable precursors into the viral genomes [e.g., ethynyl-deoxyuridine (EdU) or ethynyl-deoxycytidine (EdC)], together with high-resolution fluorescent microscopy and live cell imaging, allowed better characterization of viral host interactions during critical early stages of the viral genomes in the nucleus (Diner et al., 2016; Sekine et al., 2017; Alandijany et al., 2018; Dembowski and DeLuca, 2018). Furthermore, recent studies at single-cell resolution have identified differences in the infection outcomes among genetically identical cells (Drayman et al., 2017; Sekine et al., 2017; Cohen et al., 2018; Drayman et al., 2019; Wyler et al., 2019; Cohen et al., 2020) shedding further light on these complex interactions.

In this chapter, we focus on the initial interactions of viral genomes with host and viral factors inside the nucleus that ultimately determine the outcome of infection. Since these interactions are highly complex and usually begin a cascade of downstream processes, we try to highlight concepts rather than specific interactions. We have limited our discussion to events prior to formation of replication compartments (RCs), as we assume that once RCs are formed the cell is committed to the lytic cycle.

Possible fates of viral genomes

Entry of a virus into a permissive and susceptible cell leads to either a productive lytic infection that results in progeny release and cell death, or to an abortive infection, in which the cell survives viral infection without generating new viral particles. In the case of herpesviruses (as well as other viruses), establishment of a latent infection is a third possibility, in which viral genomes remain within the host cell without major changes in cell viability for a relatively long time (see Chapter 6). From this condition, the virus can reactivate and produce progeny. Recent evidence suggests that latent viral genomes could be detected in cells that survived abortive infection (Cohen et al., 2020). The decision between the different outcomes depends partially on viral factors since some viral mutants predominately favor one outcome (Samaniego et al., 1998; Marshall et al., 2000). But, the decision is also determined by host factors. For example, infection with HSV-1 results in lytic infection for most cell types; however, among myeloid cells, only specific types support lytic infection (Linnavuori and Hovi, 1983; Leary et al., 1985; Shivkumar et al., 2016).

During lytic infection viral genomes enter the nucleus and initiate a cascade of viral gene expression. First, the tegument protein VP16 (encoded by the UL48 gene) forms a complex with host transcription activators and induces expression of the immediate early (IE) viral genes (Wysocka and Herr, 2003), which are defined as genes expressed prior to viral protein synthesis (Honess and Roizman, 1974, 1975). The IE proteins regulate expression of early and late genes. Early genes are mostly involved in viral replication (Wu et al., 1988; Muylaert et al., 2011), which takes place in specific sites in the nucleus known as replication compartments (Quinlan et al., 1984; de Bruyn Kops and Knipe, 1988; Kobiler and Weitzman, 2019). The late genes, which are expressed following viral replication (Dremel and DeLuca, 2019a), are mostly structural genes required for assembly of new progeny viruses (Honess and Roizman, 1974, 1975). The release of progeny viruses from infected cells eventually leads to cell death. This is thought to be the default process in most infected cells, especially in non-neuronal cells and in tissue culture.

For alphaherpesviruses, clinical and experimental evidence suggested that latency occurs spontaneously only in peripheral sensory neurons or neurons of the autonomic sympathetic ganglia. Moreover, latency is also host specific, i.e., spontaneous latency occurs almost exclusively in the natural host of the virus (Koyuncu et al., 2018). Thus, it seems that latent infection can only take place under special circumstances. However, quiescent infections that resemble latency can be observed in abortively infected non-neuronal cells (Cohen et al., 2020). *In vitro*, neuronal and non-neuronal cells can be induced into a latency-like state by different methods that inhibit viral replication (O'Neill, 1977; Russell and Preston, 1986; Scheck et al., 1986; Shiraki and Rapp, 1986; Harris and Preston, 1991; McMahon and Walsh, 2008; Edwards and Bloom, 2019; Hu et al., 2020). During latency, the viral genomes are found as episomes (independently maintained, circular vDNA molecules) inside the host nucleus. These genomes are tightly associated with host histones that maintain genomes in the heterochromatin state, which prevents formation of new viral particles (Deshmane and Fraser, 1989; Wang et al., 2005; Cliffe et al., 2009; Kwiatkowski et al., 2009). Limited viral gene expression as well as expression of latency-associated RNAs (LAT RNAs) can be detected during latency (Ma et al., 2014). The most important characteristic of this condition is the ability to reactivate and generate new viral particles. Many stress signals can induce these viral genomes to reactivate (Suzich and Cliffe, 2018).

Abortive infection is defined as an infection in which viruses enter the cell and some viral elements are generated; however, no progeny virus is produced (Aurelian and Roizman, 1964, 1965; Roizman and Aurelian, 1965). The abortive infection can result from blocking the lytic process either before or after vDNA replication takes place (Leary et al., 1985). Thus, several distinct mechanisms can lead to abortive infections. Abortive viral genomes may be indistinguishable from latent genomes (Cohen et al., 2020), especially in the early stages post infection, and differ only retrospectively by their ability to reactivate. While the mechanism of abortive infection is not defined, in this chapter we will consider abortive/latent viral genomes as those that enter the nucleus and do not initiate lytic infection.

Early steps in the nucleus

Initial interactions of viral genomes with viral and host proteins in the nucleus have an important role in determining the fate of infection (Figure 1). Most host proteins that repress gene expression from incoming viral genomes are considered part of nuclear intrinsic immunity. Single-cell analysis suggested that in cells with higher levels of host housekeeping genes, fewer viral genomes can initiate gene expression and replication (Cohen and Kobiler, 2016). Similarly, the number of viral genomes initiating expression is cell-type dependent and is reduced under conditions that increase the levels of intrinsic factors, such as exposing the cells to histone deacetylases inhibitors (Shapira et al., 2016). These results highlight the role of host factors in preventing incoming viral genomes to initiate lytic infection. Viral proteins counteract these intrinsic immunity proteins and stimulate viral gene expression. Viral gene expression correlates with the number of viral genomes that can replicate in individual cells (Cohen and Kobiler, 2016). However, it has not yet been established whether viral gene expression allows more viral genomes to initiate replication. An alternative hypothesis is that replication of multiple viral genomes results in higher levels of gene expression.

In an attempt to characterize the initial virus/host interactions after nuclear entry, there are two major challenges: first, to develop an unbiased method to detect all interactions, and second to analyze these interactions as early as possible after nuclear entry. The first challenge was overcome by pulling down the viral genomes and using mass spectrometry to identify all associated proteins. Most of these studies have been limited to replicating viral genomes (Taylor and Knipe, 2004; Dembowski and DeLuca, 2015; Dembowski et al.,

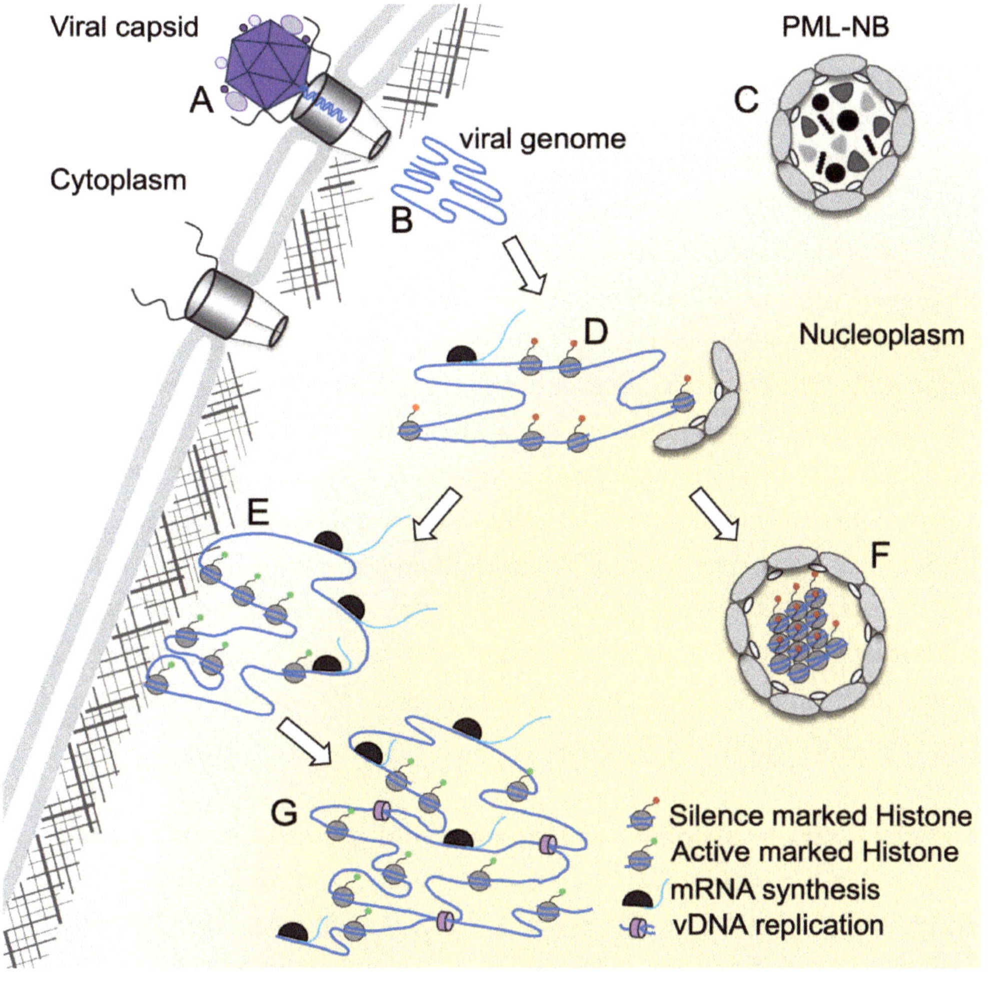

Figure 1. The fates of incoming viral genomes. A schematic illustration of the steps of incoming viral genomes in the nucleus. (A) Capsid docks at the nuclear pore complex (NPC), and viral genome is translocated into the nucleus through the NPC. (B) The naked viral genome enters the nucleus and remains condensed. (C) Promyelocytic Leukemia Nuclear Bodies (PML-NB's) are present near sites of entering genomes. (D) The entering genome becomes circular, binds histones, and initiates IE gene expression while host factors try to repress gene expression. (E) The viral DNA (vDNA) associates with the nuclear lamina where enhanced viral gene expression takes place. (F) vDNA is repressed both by the formation of heterochromatin and by binding of PML proteins. (G) Replication of the vDNA in early RCs. The viral inner tegument and capsid is colored purple, the vDNA is in blue, viral mRNA in light blue and host histones are gray round particles with tails labeled with active (green) and silencing (red) marks. All other host proteins and structures are in different shades of gray.

2017; Reyes et al., 2017) as pulldown was carried out by incorporation of EdC to label the vDNA during the DNA replication process. The second challenge was addressed with the ability to generate infectious viral particles containing viral genomes that incorporated EdC/EdU. Once available, this method was used to visualize the incoming genomes with fluorescent microscopy. Sekine et al. detected the first genomes arriving at the nucleus following incubation of the virus at 4°C for 45 minutes (to prevent membrane fusion) and then shifting to 37°C for 30 minutes (Sekine et al., 2017). Similar labeling method allowed to detect viral genomes in the nucleus as early as 30 minutes after adding virus to cells at 37°C (Alandijany et al., 2018). Dembowski et al. identified host proteins associated with incoming viral genomes at different time points prior to replication using mass spectrometry (Dembowski and DeLuca, 2018). In this study, the first time-point analyzed was one hour post infection (hpi), which corresponds to two hours after the virus was first added to the cells at 37°C. These three papers (Sekine et al., 2017; Alandijany et al., 2018; Dembowski and DeLuca, 2018) used three different methods to initiate infection and determine the initial time of infection. This emphasizes the need for an accepted synchronization method for recognizing initial events of incoming viral genomes (Ralph et al., 2017).

Antiviral defense responses to incoming viral genomes in the nucleus

The nucleus has evolved several mechanisms to cope with foreign DNA. Since vDNA enters naked into the nucleus, it has been assumed that the first interactors are components of these host defense mechanisms. In the analysis of Dembowski et al., several classes of host proteins were found to be associated with the vDNA, including proteins involved in DNA damage response (DDR), proteins from the Promyelocytic Leukemia Nuclear Body (PML-NB), histones, histone modifiers, transcription factors, and RNA processing proteins (Dembowski and DeLuca, 2018). The DDR, PML-NB and histone modifier proteins, alongside the innate immunity sensors IFI16, cGAS and STING (see Chapter 7), are considered to be the first line of host defense response to viral genomes (Alandijany, 2019).

DNA damage response

The viral genome found inside the capsid typically have several nicks and gaps (Kieff et al., 1971; Frenkel and Roizman, 1972; Sheldrick et al., 1973; Smith et al., 2014). Once the genome is released from the capsid into the nucleus, these DNA breaks and the open ends of the linear naked genome trigger nuclear mechanisms targeting foreign DNA that include components of the DDR pathways and the PML-NB (Smith et al., 2014). Therefore, it is not surprising that the proteins associated with DDR and the PML-NB are among the first host factors to be recruited to the incoming genomes (Lilley et al., 2011; Dembowski and DeLuca, 2018).

Many components of the antiviral DDR pathways are degraded or inhibited by viral infection (reviewed in (Smith and Weller, 2015)). These include proteins involved in classic nonhomologous end joining (Lees-Miller et al., 1996; Parkinson et al., 1999; Trigg et al., 2017), homologous recombination signaling (Lilley et al., 2010), and ATR signaling involved in single strand DNA break repair (Mohni et al., 2010). Sensing of damaged DNA by the DDR system can detect the foreign DNA (especially open-ended DNA containing nicks and gaps) and may initiate a cascade of events to prevent viral gene expression before replication initiates (Lilley et al., 2011). This DNA sensing can stimulate both intrinsic and innate antiviral responses (Trigg and Ferguson, 2015). On the other hand, several of the key regulators of DDR are induced during HSV infection and are required for efficient productive lytic infection (Lilley et al., 2005; Mohni et al., 2011; Mohni et al., 2013; Karttunen et al., 2014; Edwards et al., 2018) likely because viral replication is associated with recombination (Wilkinson and Weller, 2003) and many of DDR proteins can be found in viral RCs (Taylor and Knipe, 2004; Dembowski and DeLuca, 2015; Dembowski et al., 2017; Reyes et al., 2017). Recently, it was suggested that the DDR may have an important role in determining the outcome of the infection (lytic vs. latency) since DDR proteins are not activated in HSV1 infected neurons (Brown, 2017). Thus, the DDR response can both positively and negatively regulate viral infection.

Promyelocytic Leukemia Nuclear Body

PML-NBs, also known as nuclear domains 10 (ND10), are proteinaceous nuclear structures that are hubs for many nuclear activities. The PML protein is the key structural protein in the PML-NB assemblies. Several other proteins are also permanent residents of these structures whereas other proteins can be temporarily present under different conditions including different cell types or stresses (Lallemand-Breitenbach and de The, 2018). PML-NBs have been associated with a broad range of cellular functions including apoptosis, DNA repair, epigenetic control, oncogenesis and response to viral infection. It has been suggested that most of these functions are regulated by post-transcriptional modifications, especially by SUMOylation (small ubiquitin-like modifier (Sternsdorf et al., 1997; Muller et al., 1998)).

Incoming HSV-1 genomes rapidly associate with PML-NB (Maul et al., 1996; Everett and Murray, 2005; Alandijany et al., 2018). During wild-type HSV-1 infection, this interaction is transient since the E3 ubiquitin ligase activity of the viral ICP0 protein induces degradation and disassembly of PML-NBs (see below). The PML protein itself entraps viral genomes as early as 30 minutes post incubation with cells at 37°C (Alandijany et al., 2018). Viral genomes entrapped within PML-NBs can also be found in latent neuronal cells (Catez et al., 2012; Maroui et al., 2016). Therefore, it has been suggested that PML entrapment itself has an inhibitory effect on viral genomes. Other PML-NB proteins that colocalize with the PML and viral genomes are Daxx (Death-Domain Associated Protein), Sp100 (Speckled Protein of 100 kDa), and ATRX (α-Thalassemia/Mental Retardation Syndrome X-Linked) (Everett and Murray, 2005; Glass and Everett, 2013; Alandijany et al., 2018). Depletion of each of these proteins by shRNA constructs partially rescues growth defects of HSV-1 deleted for the ICP0 gene but has no effect on wild-type HSV-1 replication (Everett et al., 2008; Lukashchuk and Everett, 2010), suggesting a host defense mechanism that is inhibited by the wild-type HSV-1. Daxx and ATRX are known to promote nucleosome assembly and repressive heterochromatin state (Lukashchuk and Everett, 2010). It was further shown that SUMO modification pathways are required for recruitment of PML-NB proteins to viral genomes (Cuchet-Lourenco et al., 2011).

While the molecular mechanisms by which PML-NBs regulate expression from viral genomes are not fully understood, it was suggested that the main role of PML-NBs is to enhance chromatin repression of HSV-1 genomes (Gu and Zheng, 2016). One of the mechanisms that repress viral genomes is the histone H3.3 chaperone HIRA, which is relocalized to PML-NBs following viral infection. HIRA deposits variant histone H3.3 on viral genomes to promote epigenetic silencing of the incoming viral genomes (Rai et al., 2017; Cohen et al., 2018) and also induce the innate immune defenses (McFarlane et al., 2019).

In some cell lines, depletion of the PML protein result in decreased HSV-1 replication (Xu et al., 2016; Xu and Roizman, 2017; Merkl et al., 2018). Thus, similar to the DDR response, PML protein itself or other proteins found in the PML-NBs may also have pro-viral functions (Xu et al., 2016; Xu and Roizman, 2017; Merkl et al., 2018).

Histones and histone modifiers

Histones do not only function to condense DNA into packaged chromatin but are also master facilitators of transcription and DNA replication. This is achieved by a large array of post-translational modifications on histones regulated by proteins that function as writers, erasers and readers (recently reviewed (Biswas and Rao, 2018)). Writers are proteins that add modifications on the histones (for example, P300 is an histone acetyltransferase), erasers are proteins that remove the modification (such as histone deacetylases, HDACs) and readers are the proteins that recognize the modifications and bind to them (for example, BRD4 that recognize acetylated histones).

Although HSV-1 genomes enter cells as naked DNA molecules, they rapidly associate with histones. These interactions can be detected in cells entering either lytic or latent infection (Deshmane and Fraser, 1989; Herrera and

Triezenberg, 2004; Kent et al., 2004; Kubat et al., 2004; Wang et al., 2005; Huang et al., 2006; Oh and Fraser, 2008; Silva et al., 2008; Cliffe et al., 2009; Lacasse and Schang, 2012). Initially, viral genomes are associated with histones marked with known post-translational modifications correlating with genome silencing, such as histone 3 methylation on lysine 9 and 27 (H3K9me3 and H3K27me3). This suggests that the host utilizes these mechanisms to silence the incoming genomes (Narayanan et al., 2007; Silva et al., 2008; Liang et al., 2009; Lee et al., 2016). Indeed, during latency, viral genomes are maintained in a heterochromatin state with repressed viral gene expression (Deshmane and Fraser, 1989; Wang et al., 2005; Cliffe et al., 2009; Kwiatkowski et al., 2009).

To promote lytic infection, host enzymes directed by viral proteins rapidly remove these silencing marks and install new modification marks that enhance viral gene expression (Liang et al., 2009; Lee et al., 2016). Throughout lytic infection, the nucleosomes are unstable on viral genomes (Lacasse and Schang, 2010, 2012). Recent findings showed that early during lytic infection the majority of viral genomes are found in condensed chromatin structures that are likely to be inaccessible to transcription and replication (Hu et al., 2019). These results suggest that most incoming viral genomes remain silenced and condensed by nucleosome structures throughout the lytic infection, and that only few genomes need to escape the silencing to initiate active replication.

Viral proteins overcome host defenses

Many viral genes are required to complete the replication cycle of HSV-1. However, only three proteins (VP16, ICP0 and ICP4) are thought of as the master regulators that determine the fate of infection (Figure 2). All three proteins have been shown to associate with viral genomes at early time points post-nuclear entry and contribute to lytic infection (Gibeault et al., 2016; Dembowski and DeLuca, 2018).

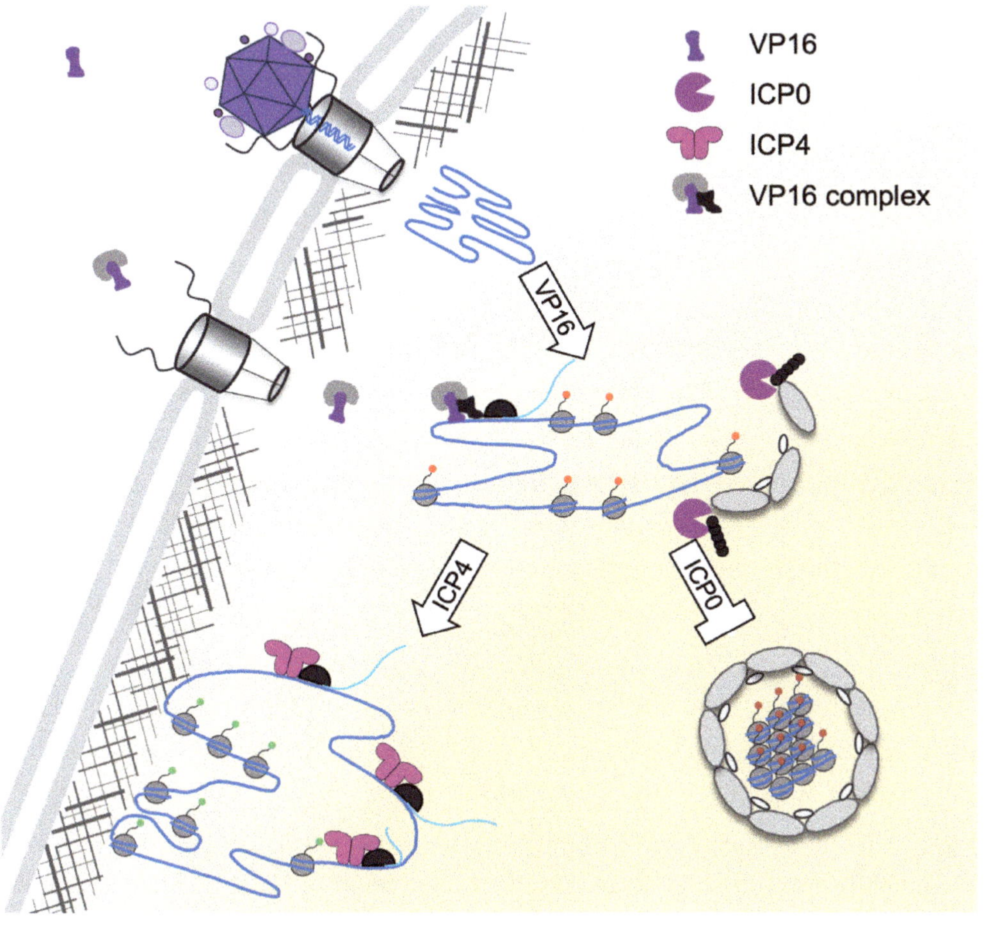

Figure 2. Role of VP16, ICP0 and ICP4 in initial steps of infection. The viral proteins VP16, ICP0 and ICP4 are indicated. VP16 is released from the outer tegument into the cytoplasm, where it interacts with the host transcription factor HCF-1 that facilitates entry into the nucleus. VP16 in complex with the host transcription factors Oct-1 and HCF-1 binds to the promoter regions of the IE genes on viral genomes and initiates their transcription. ICP0 prevents repression of viral genomes mainly by ubiquitinating host proteins involved in silencing and targeting these proteins to degradation. ICP4 induces expression of early genes that will allow formation of active RCs. All host and viral structures are presented in Figure 1.

Viral Protein 16 (VP16)

VP16, encoded by the UL48 gene, is a 490-amino-acids protein located in the outer layer of the viral tegument. The protein is conserved across all alphaherpesviruses and has a central core domain and a C-terminal transcriptional activation domain (TAD) (Hirai et al., 2010) (Figure 3A). VP16 binds both HCF-1 and Oct-1 through its core domain to form the VP16 complex which binds vDNA and induce transcription. An individual virion has 500-1000 copies of VP16 (Wysocka and Herr, 2003). Interestingly, the number of VP16 copies per virion varies significantly compared to other tegument proteins, and VP16 levels correlate with infectivity (El Bilali et al., 2017).

When the viral membrane fuses with the cellular membrane during entry, VP16, together with other tegument proteins, is released to the cytoplasm. In the cytoplasm, VP16 binds the host cell factor 1 (HCF-1). HCF-1, is a modular recruiter of chromatin remodeling proteins from different families. It is known to suppress or promote transcription of various genes related to cell-cycle and proliferation depending on the identity of the histone modifier it recruits (Hancock et al., 2010; Zargar and Tyagi, 2012; Oh et al., 2014).

VP16 translocation into the nucleus is facilitated by the nuclear localization signal (NLS) of HCF-1 (La Boissiere et al., 1999). Once in the nucleus, VP16 and HCF-1 bind Oct-1 and form the VP16 complex. Oct-1 is a versatile transcription factor, with promiscuous DNA binding and a plethora of co-factors. Its multifunctional nature owes to the presence of a POU domain, a highly conserved bipartite DNA-binding domain (Herr and Cleary, 1995). VP16 binds the POU domain of Oct-1 in a manner that suggests species-specific host virus recognition (Kristie et al., 1989; Stern et al., 1989).

The VP16 complex, via Oct-1, binds the regulatory element of the viral IE promoter TAATGARAT (where R=purine) (Misra et al., 1996). Additional binding of VP16 to the adjacent DNA near the Oct-1 binding site was also shown to be important for the formation of the VP16 complex (Babb et al., 2001). In addition, VP16 complex binds various gene regulators via the VP16's

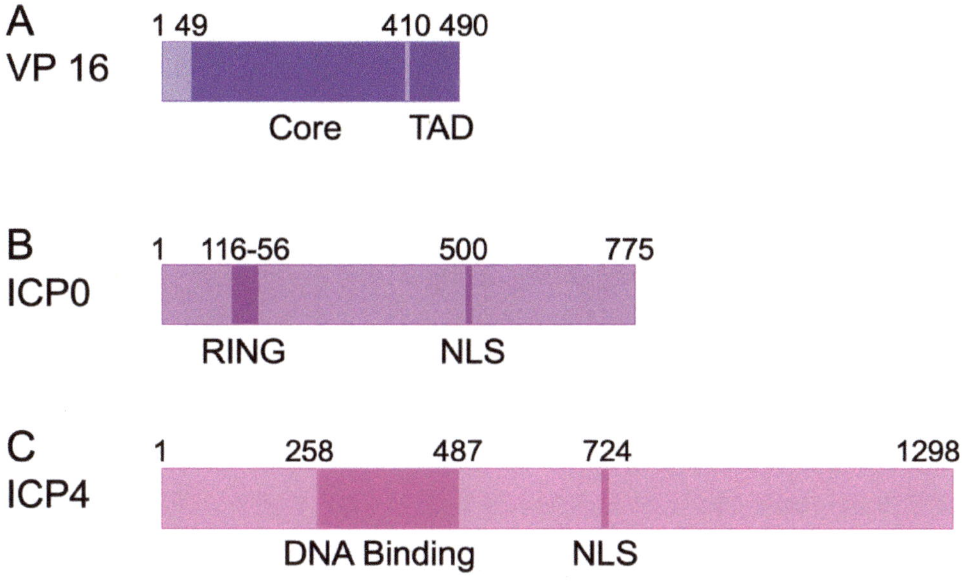

Figure 3. VP16, ICP0 and ICP4 functional domains. Schematic representation of the viral proteins: **(A)** VP16, **(B)** ICP0 and **(C)** ICP4. Functional domains are marked in darker shades and labeled. Sites are numbered according to the amino acid residues.

TAD. The TAD is known to bind histone acetyltransferases, basal transcription factors, histone modifiers and the Mediator complex (Suk and Knipe, 2015). Thus, the VP16 complex and its co-activators are a major driving force in the initiation of viral DNA transcription. Deletion of the VP16 reading frame in HSV-1 inhibits the virus ability to produce viable progeny and causes major defects in the assembly of viral particles (Weinheimer et al., 1992). Mutations in either Oct-1 or HCF-1 binding sites of VP16 or genetic tempering with its TAD results in reduced IE transcription (Smiley and Duncan, 1997; Ottosen et al., 2006).

Successful activation and transcription of IE genes is crucial in determining the genome fate. Decreased activity of VP16 complex often precludes productive infection (Steiner et al., 1990). Initiation of viral gene expression by VP16 is widely viewed as a crucial step in the establishment of lytic infection. Furthermore, since VP16 enhances IE gene expression, it regulates expression of the other major viral regulators: ICP0 and ICP4.

Infected Cell Protein 0 (ICP0)

The 775-amino-acid long IE protein ICP0 is encoded by two copies of the RL2 gene found in the long repeat region of the HSV genome (Wadsworth et al., 1975). The ICP0 gene is one of the very few HSV genes that contain introns. The introns of ICP0, 765 and 136 nucleotides in size, have not been shown to have any significant role in productive lytic infection or during reactivation (Perry et al., 1986; Natarajan et al., 1991; Lium et al., 1998).

ICP0 is not essential for viral replication. Viral strains without functional ICP0 protein can initiate replication only at high multiplicity of infection (MOI) (Stow and Stow, 1986; Sacks and Schaffer, 1987). Interestingly, this effect is cell-type dependent, i.e., the relative efficiency of plating between the mutant and wild type strain can range from 1:1000 in human fibroblast to 1:1 in the U2OS cell line derived from osteosarcoma cells (Boutell and Everett, 2013).

One of the major features of ICP0 is the zinc-containing RING (really interesting new gene) finger domain positioned in the middle exon (Everett, 1988; Everett et al., 1993; Lium and Silverstein, 1997) (Figure 3B). This domain functions as a E3 ubiquitin ligase that targets several host defense proteins for degradation (reviewed at (Boutell and Everett, 2013)). Since ICP0 was shown to autoubiquitinate *in vitro* and is subject to proteasome-dependent degradation (albeit not exclusively), it may suggest that ICP0 has functional ubiquitination sites (Canning et al., 2004).

ICP0 contains a short NLS of seven amino acids within its third exon. Following nuclear entry, ICP0 rapidly localizes to PML-NBs. This localization requires the RING finger domain. The close interaction with PML-NBs allows for a swift and effective proteasomal degradation of several proteins that are located in the PML-NBs, including all PML isoforms and some sp100 isoforms (e.g., SUMO modified) (Boutell et al., 2011). This degradation is mediated by ubiquitination through the E3 ligase activity of ICP0 (Chelbi-Alix and de The, 1999). This ICP0 activity leads to the dispersal of PML-NBs and minimizes the intrinsic antiviral response. Depletion of PML can allow a null-ICP0 mutant to achieve high progeny titers, similar to those of a WT virus, a phenotype that can be reversed with reintroduction of some PML isoforms (Cuchet et al., 2011).

Most targets of the E3 activity of ICP0 are multifunctional host proteins involved in other defense mechanisms, including DNA damage response (DNA-PK catalytic subunit (Lees-Miller et al., 1996), Ring Finger Protein 8 (RNF8) and RNF168 (Lilley et al., 2010)), innate immunity (Myeloid differentiation primary response 88 (MyD88) (van Lint et al., 2010)), autophagy (sequestosome 1 and optineurin (Waisner and Kalamvoki, 2019)) and apoptosis (p53 (Boutell and Everett, 2003)). In addition, it has been shown that the E3 activity of ICP0 leads to an overall depletion of SUMOylated proteins (Muller and Dejean, 1999; Boutell et al., 2011).

It has been suggested that ICP0 also interacts with RE1-Silencing Transcription factor (REST) corepressor (CoREST) (Gu and Roizman, 2007, 2009). Together with histone deacetylases (HDACs) and lysine-specific-demethylase-1 (LSD1), CoREST/REST forms a regulatory complex that represses genes by chromatin condensation (Chong et al., 1995; You et al., 2001; Lakowski et al., 2006). The effect of CoREST/REST on HSV-1 infection could be cell-specific because in some cell types, depletion of CoREST showed no improvement in the replication of ICP0-null mutant viruses compared to the wild-type strain (Everett, 2010).

ICP0 activities are thought to keep viral genomes open and uncondensed at early time points post nuclear entry. Interestingly, while ICP0 homologues in other alphaherpesviruses are all capable of dismantling PML-NBs and can at least partially complement ICP0 activity (Everett et al., 2010), not all are expressed with IE kinetics. This could suggest that either the levels of ICP0 homologous are more abundant in the tegument compared to HSV-1 ICP0 or that some alphaherpesviruses do not require ICP0 activity in very early times post nuclear entry. ICP0 continues to regulate host responses at later times of infection and during latency and reactivation (Lee et al., 2016; Raja et al., 2016). These results indicate that there are probably more direct and indirect functions of this multitasking protein.

Infected Cell Protein 4 (ICP4)

The 1298 amino-acid IE protein ICP4 is encoded by the two copies of the RS1 gene found in the short repeat region of the HSV genome (Wadsworth et al., 1975). As many DNA viruses, HSV genome is transcribed by RNA polymerase II (RNA pol II). Via its DNA binding domain (Figure 3C), ICP4 binds the vDNA as a homodimer and functions as either a transcription activator or a repressor of the RNA pol II system (Metzler and Wilcox, 1985; Tunnicliffe et al., 2017). ICP4 can activate most viral promoters, and its presence and proper activation of E and L genes is essential for productive infection. ICP4 was shown to associate with specific DNA sequences found both in the host and viral genomes (Faber and Wilcox, 1986). Recently, it was shown that ICP4 preferentially binds to the viral genomes compared to the host genomes resulting in induction of viral transcription and reduction in host transcripts (Dremel and DeLuca, 2019b).

Activation of genes by ICP4 requires recruitment of TFIID (Transcription factor II D) and the Mediator complex. TFIID is a general TF of the RNA Pol II preinitiation complex. ICP4 interacts with TFIID through its TAF1 subunit (Carrozza and DeLuca, 1996). The Mediator complex subunit involved in the interaction with ICP4 is yet to be determined, but several candidates have been proposed (Lester and DeLuca, 2011).

Repression by ICP4 is less common and limited to specific viral genes (including ICP4 itself) (Faber and Wilcox, 1988; Lium et al., 1996; Liu et al., 2010). In comparison to its activation capacity, repression by ICP4 is more rigid and requires a very specific binding of a tripartite complex (comprised of ICP4,

TFIID and TFIIB) (Smith et al., 1993) to the gene promoter to block other potential activators (such as the Mediator complex and VP16) (Gu et al., 1995).

The major role of ICP4 is to activate expression of early and late viral genes. ICP4 also binds and activates host genes that are beneficial for viral replication, for example miR-101 (Wang et al., 2016). The overall function of ICP4 is to promote conditions for lytic infection. Live imaging revealed distinct ICP4 foci adjacent to PML-NBs and ICP0 colocalization sites (Everett et al., 2003), suggesting an early association of ICP4 with viral genomes, later to be developed into replication compartments. Spatiotemporal study of entering HSV showed heterogeneity amongst those genomes. While some started accumulating ICP4 (initially as distinct foci and later accumulated along the majority of the genome), other genomes showed little to no localization of ICP4 (Sekine et al., 2017). Viral genomes accumulating ICP4 start to decondense and initiate transcription, and ICP4 may have a role in maintaining these genomes relaxed and accessible by coating the viral genomes (Dremel and DeLuca, 2019b). Genomes that did not accumulate ICP4 remained condensed. These results suggest that once ICP4 binds to viral genomes, they commit to lytic infection.

Host factors supporting HSV-1 lytic infection

Since all viruses are obligate intracellular parasites, they require numerous host factors to complete their life cycle. Many of these host factors are required for viral replication and assembly. Unexpectedly, specific proteins that are part of the host defense response mentioned above are beneficial for the viral replication process. Several virus/host interactions occur early after nuclear entry and influence the outcome of infection.

Most incoming viral genomes remain located near their entry sites, at the nuclear periphery, where they interact with the nuclear lamina (Silva et al., 2008; Silva et al., 2012). Fibroblasts derived from either lamA-/- or lamB1-/- knockout mice have reduced viral progeny compared to wild-type murine fibroblasts suggesting that interactions with lamin proteins enhance expression and replication from the viral genomes (Mou et al., 2008; Silva et al., 2008). It has been suggested that interactions with the nuclear lamina facilitate assembly of VP16 complexes on the viral genomes (Silva et al., 2012). Despite proximity of the lamina layer to viral genome entry sites, in the analysis by Dembowski et al., lamins associated with the vDNA only at 2 hpi, indicating that these lamins are not the first host proteins to interact with viral genomes (Dembowski and DeLuca, 2018). Viral infections that take place a short time before the cell reaches mitosis are less likely to result in a successful infection (Drayman et al., 2017), indicating the importance of intact nucleus for the initiation of the infection.

The single-genome hypothesis

HSV-1 infection of a given cell is currently thought to result in lytic, latent, or abortive infection. But the observations that permissive and susceptible cells can maintain quiescent viral genomes during lytic infection (Hu et al., 2019) or during abortive/latency like infection (Cohen et al., 2020) suggest that there may be a dynamic continuum of possible outcomes rather than a three-solution decision model. This dynamic continuum could be explained by the single-genome hypothesis which suggests that each individual incoming viral genome may separately and stochastically result in a different outcome of infection, as has been suggested for bacteriophage lambda (Golding, 2016).

Support for the single-genome hypothesis is provided by the recent results visualizing single incoming genomes that remain condensed in the nucleus while other viral genomes are being replicated (Sekine et al., 2017; Dembowski and DeLuca, 2018). Small undeveloped RCs were detected within larger RCs using dual-color fluorescent *in-situ* hybridization (Tomer et al., 2019). Even in latently infected neurons, several patterns of quiescent viral genomes have been observed, and in some cells combination of these patterns are detected (Maroui et al., 2016). The identification of condensed inaccessible viral genomes throughout the lytic infection further support this model (Hu et al., 2019).

The concept that not all incoming genomes will share the same fate suggests that productive lytic infection can initiate from only one or few incoming genomes whereas other genomes in the same cell do not contribute to the lytic process or to the genetic pool of progeny viruses (Figure 4). Indeed, each entering genome can form its own RC (Sourvinos and Everett, 2002; Kobiler et al., 2011; Tomer et al., 2019) raising the possibility that there is partial autonomous activity of each incoming genome. Furthermore, only a limited number of incoming alphaherpesviral genomes can initiate expression and replication per cell (Phelan et al., 1997; Kobiler et al., 2010; Kobiler et al., 2011; Taylor et al., 2012; Cohen and Kobiler, 2016; Shapira et al., 2016; Sekine et al., 2017). Since this limited number does not correlate linearly with MOI (i.e., raising the MOI from 10 to 100 results in the increase of the average number of genomes from ~4 to ~7), not all incoming genomes likely initiate expression and replication.

Current paradigm is that host factors try to suppress the incoming viral genomes whereas viral proteins (VP16, ICP0 and ICP4) are required to overcome this suppression and initiate lytic infection. However, this is an

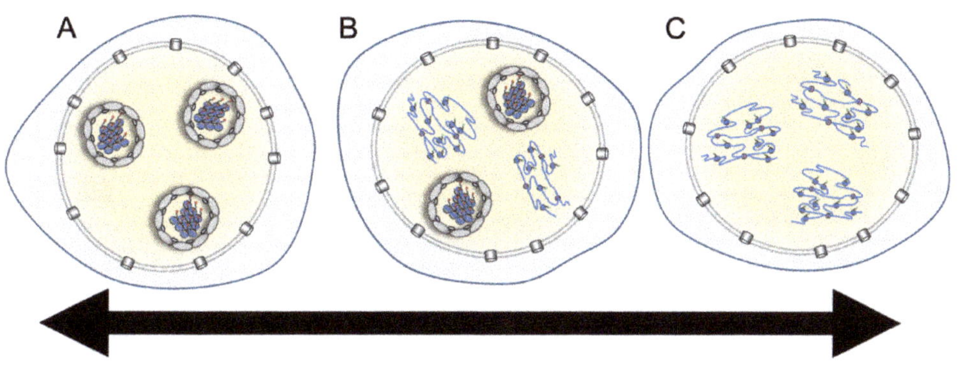

Figure 4. The single-genome hypothesis. A schematic illustration of dynamic continuum of outcomes of infection according to the single genome concept. (A) represent a cell that will follow the abortive or latent cycle as all viral genomes are repressed. (B-C) represent cells that will follow the lytic infection as not all entering genomes will be silenced.

oversimplified view of a complex molecular system that has evolved to allow both the lytic infection as well as the latent infection. This intricate system provides the virus with the opportunity to select either outcome and allows for specific genetic, structural or environmental clues to influence the decision between different outcomes. The single-genome hypothesis suggests that local interactions between viral genomes and viral or host proteins are key for determining the fate of each incoming genome. The probability of any viral genome to encounter either components of the host antiviral defenses or the VP16 complex would then determine the specific outcome for a particular genome. Thus, a rise in the level of antiviral proteins would be expected to increase the likelihood that they would bind the entering genomes and vice versa. Similarly, when VP16 needs to diffuse for a long distance from the viral cellular entry point towards the nucleus, which happens during infection of peripheral neurons where axon ending can be at least few centimeters away from cell body (even for trigeminal neurons in young kids), the likelihood of viral genomes moving inside the capsid on microtubules towards the nucleus to interact with the freely diffusing VP16 is low, increasing the chance for latency (Knipe and Cliffe, 2008). High levels of VP16 complexes (as expected at high MOI) would then be expected to increase the likelihood for lytic infection. Indeed, ICP0-null viruses can complete infection in most cell lines only at high MOI (Stow and Stow, 1986; Sacks and Schaffer, 1987), suggesting that the high probability to interact with VP16 complexes can overcome host defense silencing mechanisms.

Conclusions

Here, we have summarized current knowledge on the decision processes of incoming HSV-1 genomes. We highlighted three major intrinsic host mechanisms that inhibit incoming genomes, and three viral proteins that counteract these mechanisms. We suggested that instead of looking on the outcome of the entire cell to study the viral decision, one should focus on individual viral genomes.

Three major intrinsic nuclear mechanisms are thought to promote HSV-1 genomes silencing: PML-NBs, DDR and histone modifications. As described above, each of these nuclear mechanisms is a multi-protein response, and furthermore, there is crosstalk between them. Therefore, elucidating the roles of specific proteins is challenging. Other challenges result from the finding that inhibition of specific posttranslational modification of histones during viral infection may cause greater effects on the host genome than on viral genomes, adding to the complexity of experimental analysis (Shapira et al., 2016; Arbuckle et al., 2017; Kulej et al., 2017).

Some specific proteins from each host defense response, including PML-NBs, DDR, and histone modifications, can be beneficial for viral replication process (Mohni et al., 2011; Mohni et al., 2013; Kristie, 2015; Xu et al., 2016; Edwards et al., 2018). This might be part of a viral strategy to dismantle host defense mechanisms by reusing parts of these system for its own replication. Alternatively, since these proteins are recruited by the host to inhibit viral replication and they became closely associated with the viral genomes, the virus may have evolved to utilize these nearby specific proteins.

Levels of the viral proteins VP16, ICP0 and ICP4 are critical to promoting lytic infection. VP16 levels are determined by the amount of VP16 arriving within the tegument (differences can be due to MOI and levels of VP16 within the viral particle) and diffusion from the entry site to the nucleus, allowing fine-tuning of outcome of the infection process to these conditions. While ICP0 and ICP4 are also in the tegument (Yao and Courtney, 1989, 1992; Loret et al., 2008) the functional role of these proteins within viral particles is not fully understood (Delboy and Nicola, 2011; Pritchard et al., 2013). But, unlike VP16, which is transcribed from a late gene, ICP0 and ICP4 are transcribed from IE genes and would become available very early in infection. The genes encoding ICP0 and ICP4 are among very few genes that are found in two copies. Two copies for a single gene is a rare phenomenon in virology even among large DNA viruses. One possible explanation for this phenomenon can be the importance of these genes for the infection process requiring that at least one copy should be intact. Alternatively, duplicate copies would allow for a more rapid synthesis of mRNA to meet the need for rapid accumulation of a viral protein.

Future trends

Elucidation of the rapid events that occur following entry of the viral genomes into the nucleus remains a major challenge for the ongoing research. One of the key problems is the lack of synchrony in the entry process. Better synchronization is required to dissect very early events. Currently, one of the best way to synchronize nuclear entry is using the tsB7 mutant (or its derivatives), which permits the capsids to bind the nuclear pore complex at high temperature but allows rapid injection of viral genome into the nucleus only at low temperature (Batterson et al., 1983; Abaitua et al., 2009; Abaitua et al., 2011; Ralph et al., 2017). However, these temperature changes might affect other viral or host processes (Ralph et al., 2017). Alternative approaches for synchronizing nuclear entry are needed.

Even if synchronous nuclear entry could be achieved, according to the single-genome hypothesis, it is likely that each genome has a different fate. Can different early events be detected at a single-genome resolution? Can these early events determine the outcomes for specific genomes?

Finally, the single-genome hypothesis suggests that even a single genome is sufficient to initiate a lytic infection. The virus might have evolved a mechanism to devote most of its efforts to a single or very few genomes, i.e., the viral proteins (VP16 and the immediate early genes) interact with only a few of the incoming viral genomes while the cells suppress the rest of the incoming genomes. Can such a mechanism exist? Can this be the reason for the limited number of genomes expressed in individual cells?

Many questions regarding the early nuclear events remain unanswered. Here, we have discussed only a few of those. While some of these questions are currently too challenging to address due to technological limitations, with rapid improvements in technology, it is only a matter of time until these and other questions are deciphered.

References

Abaitua, F., Daikoku, T., Crump, C.M., Bolstad, M., and O'Hare, P. (2011). A single mutation responsible for temperature-sensitive entry and assembly defects in the VP1-2 protein of herpes simplex virus. J Virol *85*, 2024-2036. https://dx.doi.org/10.1128/JVI.01895-10

Abaitua, F., Souto, R.N., Browne, H., Daikoku, T., and O'Hare, P. (2009). Characterization of the herpes simplex virus (HSV)-1 tegument protein VP1-2 during infection with the HSV temperature-sensitive mutant tsB7. J Gen Virol *90*, 2353-2363. https://dx.doi.org/10.1099/vir.0.012492-0

Alandijany, T. (2019). Host Intrinsic and Innate Intracellular Immunity During Herpes Simplex Virus Type 1 (HSV-1) Infection. Front Microbiol *10*, 2611. https://dx.doi.org/10.3389/fmicb.2019.02611

Alandijany, T., Roberts, A.P.E., Conn, K.L., Loney, C., McFarlane, S., Orr, A., and Boutell, C. (2018). Distinct temporal roles for the promyelocytic leukaemia (PML) protein in the sequential regulation of intracellular host immunity to HSV-1 infection. PLoS Pathog *14*, e1006769. https://dx.doi.org/10.1371/journal.ppat.1006769

Arbuckle, J.H., Gardina, P.J., Gordon, D.N., Hickman, H.D., Yewdell, J.W., Pierson, T.C., Myers, T.G., and Kristie, T.M. (2017). Inhibitors of the Histone Methyltransferases EZH2/1 Induce a Potent Antiviral State and Suppress Infection by Diverse Viral Pathogens. MBio *8*. https://dx.doi.org/10.1128/mBio.01141-17

Aurelian, L., and Roizman, B. (1964). The Host Range of Herpes Simplex Virus; Interferon, Viral DNA, and Antigen Synthesis in Abortive Infection of Dog Kidney Cells. Virology *22*, 452-461

Aurelian, L., and Roizman, B. (1965). Abortive Infection of Canine Cells by Herpes Simplex Virus. Ii. Alternative Suppression of Synthesis of Interferon and Viral Constituents. J Mol Biol *11*, 539-548

Babb, R., Huang, C.C., Aufiero, D.J., and Herr, W. (2001). DNA recognition by the herpes simplex virus transactivator VP16: a novel DNA-binding structure. Mol Cell Biol *21*, 4700-4712. https://dx.doi.org/10.1128/MCB.21.14.4700-4712.2001

Batterson, W., Furlong, D., and Roizman, B. (1983). Molecular genetics of herpes simplex virus. VIII. further characterization of a temperature-sensitive mutant defective in release of viral DNA and in other stages of the viral reproductive cycle. J Virol *45*, 397-407

Biswas, S., and Rao, C.M. (2018). Epigenetic tools (The Writers, The Readers and The Erasers) and their implications in cancer therapy. Eur J Pharmacol. https://dx.doi.org/10.1016/j.ejphar.2018.08.021

Boutell, C., Cuchet-Lourenco, D., Vanni, E., Orr, A., Glass, M., McFarlane, S., and Everett, R.D. (2011). A viral ubiquitin ligase has substrate preferential SUMO targeted ubiquitin ligase activity that counteracts intrinsic antiviral defence. PLoS Pathog *7*, e1002245. https://dx.doi.org/10.1371/journal.ppat.1002245

Boutell, C., and Everett, R.D. (2003). The herpes simplex virus type 1 (HSV-1) regulatory protein ICP0 interacts with and Ubiquitinates p53. J Biol Chem *278*, 36596-36602. https://dx.doi.org/10.1074/jbc.M300776200

Boutell, C., and Everett, R.D. (2013). Regulation of alphaherpesvirus infections by the ICP0 family of proteins. J Gen Virol *94*, 465-481. https://dx.doi.org/10.1099/vir.0.048900-0

Brown, J.C. (2017). Herpes Simplex Virus Latency: The DNA Repair-Centered Pathway. Adv Virol *2017*, 7028194. https://dx.doi.org/10.1155/2017/7028194

Canning, M., Boutell, C., Parkinson, J., and Everett, R.D. (2004). A RING finger ubiquitin ligase is protected from autocatalyzed ubiquitination and degradation by binding to ubiquitin-specific protease USP7. J Biol Chem *279*, 38160-38168. https://dx.doi.org/10.1074/jbc.M402885200

Carrozza, M.J., and DeLuca, N.A. (1996). Interaction of the viral activator protein ICP4 with TFIID through TAF250. Mol Cell Biol *16*, 3085-3093

Catez, F., Picard, C., Held, K., Gross, S., Rousseau, A., Theil, D., Sawtell, N., Labetoulle, M., and Lomonte, P. (2012). HSV-1 genome subnuclear positioning and associations with host-cell PML-NBs and centromeres regulate LAT locus transcription during latency in neurons. PLoS Pathog *8*, e1002852. https://dx.doi.org/10.1371/journal.ppat.1002852

Chelbi-Alix, M.K., and de The, H. (1999). Herpes virus induced proteasome-dependent degradation of the nuclear bodies-associated PML and Sp100 proteins. Oncogene *18*, 935-941. https://dx.doi.org/10.1038/sj.onc.1202366

Chong, J.A., Tapia-Ramirez, J., Kim, S., Toledo-Aral, J.J., Zheng, Y., Boutros, M.C., Altshuller, Y.M., Frohman, M.A., Kraner, S.D., and Mandel, G. (1995). REST: a mammalian silencer protein that restricts sodium channel gene expression to neurons. Cell *80*, 949-957

Cliffe, A.R., Garber, D.A., and Knipe, D.M. (2009). Transcription of the herpes simplex virus latency-associated transcript promotes the formation of facultative heterochromatin on lytic promoters. J Virol *83*, 8182-8190. https://dx.doi.org/10.1128/JVI.00712-09

Cohen, C., Corpet, A., Roubille, S., Maroui, M.A., Poccardi, N., Rousseau, A., Kleijwegt, C., Binda, O., Texier, P., Sawtell, N., *et al.* (2018). Promyelocytic leukemia (PML) nuclear bodies (NBs) induce latent/quiescent HSV-1 genomes chromatinization through a PML NB/Histone H3.3/H3.3 Chaperone Axis. PLoS Pathog *14*, e1007313. https://dx.doi.org/10.1371/journal.ppat.1007313

Cohen, E.M., Avital, N., Shamay, M., and Kobiler, O. (2020). Abortive herpes simplex virus infection of nonneuronal cells results in quiescent viral genomes that can reactivate. Proc Natl Acad Sci U S A *117*, 635-640. https://dx.doi.org/10.1073/pnas.1910537117

Cohen, E.M., and Kobiler, O. (2016). Gene Expression Correlates with the Number of Herpes Viral Genomes Initiating Infection in Single Cells. PLoS Pathog *12*, e1006082. https://dx.doi.org/10.1371/journal.ppat.1006082

Cuchet, D., Sykes, A., Nicolas, A., Orr, A., Murray, J., Sirma, H., Heeren, J., Bartelt, A., and Everett, R.D. (2011). PML isoforms I and II participate in PML-dependent restriction of HSV-1 replication. J Cell Sci *124*, 280-291. https://dx.doi.org/10.1242/jcs.075390

Cuchet-Lourenco, D., Boutell, C., Lukashchuk, V., Grant, K., Sykes, A., Murray, J., Orr, A., and Everett, R.D. (2011). SUMO pathway dependent recruitment of cellular repressors to herpes simplex virus type 1 genomes. PLoS Pathog *7*, e1002123. https://dx.doi.org/10.1371/journal.ppat.1002123

de Bruyn Kops, A., and Knipe, D.M. (1988). Formation of DNA replication structures in herpes virus-infected cells requires a viral DNA binding protein. Cell *55*, 857-868

Delboy, M.G., and Nicola, A.V. (2011). A pre-immediate-early role for tegument ICP0 in the proteasome-dependent entry of herpes simplex virus. J Virol *85*, 5910-5918. https://dx.doi.org/10.1128/JVI.00267-11

Dembowski, J.A., and DeLuca, N.A. (2015). Selective recruitment of nuclear factors to productively replicating herpes simplex virus genomes. PLoS Pathog *11*, e1004939. https://dx.doi.org/10.1371/journal.ppat.1004939

Dembowski, J.A., and DeLuca, N.A. (2018). Temporal Viral Genome-Protein Interactions Define Distinct Stages of Productive Herpesviral Infection. MBio *9*. https://dx.doi.org/10.1128/mBio.01182-18

Dembowski, J.A., Dremel, S.E., and DeLuca, N.A. (2017). Replication-Coupled Recruitment of Viral and Cellular Factors to Herpes Simplex Virus Type 1 Replication Forks for the Maintenance and Expression of Viral Genomes. PLoS Pathog *13*, e1006166. https://dx.doi.org/10.1371/journal.ppat.1006166

Deshmane, S.L., and Fraser, N.W. (1989). During latency, herpes simplex virus type 1 DNA is associated with nucleosomes in a chromatin structure. J Virol *63*, 943-947

Diner, B.A., Lum, K.K., Toettcher, J.E., and Cristea, I.M. (2016). Viral DNA Sensors IFI16 and Cyclic GMP-AMP Synthase Possess Distinct Functions in Regulating Viral Gene Expression, Immune Defenses, and Apoptotic Responses during Herpesvirus Infection. MBio *7*. https://dx.doi.org/10.1128/mBio.01553-16

Drayman, N., Karin, O., Mayo, A., Danon, T., Shapira, L., Rafael, D., Zimmer, A., Bren, A., Kobiler, O., and Alon, U. (2017). Dynamic Proteomics of Herpes Simplex Virus Infection. MBio *8*. https://dx.doi.org/10.1128/mBio.01612-17

Drayman, N., Patel, P., Vistain, L., and Tay, S. (2019). HSV-1 single-cell analysis reveals the activation of anti-viral and developmental programs in distinct sub-populations. Elife *8*. https://dx.doi.org/10.7554/eLife.46339

Dremel, S.E., and DeLuca, N.A. (2019a). Genome replication affects transcription factor binding mediating the cascade of herpes simplex virus transcription. Proc Natl Acad Sci U S A *116*, 3734-3739. https://dx.doi.org/10.1073/pnas.1818463116

Dremel, S.E., and DeLuca, N.A. (2019b). Herpes simplex viral nucleoprotein creates a competitive transcriptional environment facilitating robust viral transcription and host shut off. Elife *8*. https://dx.doi.org/10.7554/eLife.51109

Edwards, T.G., and Bloom, D.C. (2019). Lund Human Mesencephalic (LUHMES) Neuronal Cell Line Supports Herpes Simplex Virus 1 Latency In Vitro. J Virol *93*. https://dx.doi.org/10.1128/JVI.02210-18

Edwards, T.G., Bloom, D.C., and Fisher, C. (2018). The ATM and Rad3-Related (ATR) Protein Kinase Pathway Is Activated by Herpes Simplex Virus 1 and Required for Efficient Viral Replication. J Virol *92*. https://dx.doi.org/10.1128/JVI.01884-17

El Bilali, N., Duron, J., Gingras, D., and Lippe, R. (2017). Quantitative Evaluation of Protein Heterogeneity within Herpes Simplex Virus 1 Particles. J Virol *91*. https://dx.doi.org/10.1128/JVI.00320-17

Everett, R.D. (1988). Analysis of the functional domains of herpes simplex virus type 1 immediate-early polypeptide Vmw110. J Mol Biol *202*, 87-96

Everett, R.D. (2010). Depletion of CoREST does not improve the replication of ICP0 null mutant herpes simplex virus type 1. J Virol *84*, 3695-3698. https://dx.doi.org/10.1128/JVI.00021-10

Everett, R.D. (2013). The spatial organization of DNA virus genomes in the nucleus. PLoS Pathog *9*, e1003386. https://dx.doi.org/10.1371/journal.ppat.1003386

Everett, R.D., Barlow, P., Milner, A., Luisi, B., Orr, A., Hope, G., and Lyon, D. (1993). A novel arrangement of zinc-binding residues and secondary structure in the C3HC4 motif of an alpha herpes virus protein family. J Mol Biol *234*, 1038-1047. https://dx.doi.org/10.1006/jmbi.1993.1657

Everett, R.D., Boutell, C., McNair, C., Grant, L., and Orr, A. (2010). Comparison of the biological and biochemical activities of several members of the alphaherpesvirus ICP0 family of proteins. J Virol *84*, 3476-3487. https://dx.doi.org/10.1128/JVI.02544-09

Everett, R.D., and Murray, J. (2005). ND10 components relocate to sites associated with herpes simplex virus type 1 nucleoprotein complexes during virus infection. J Virol *79*, 5078-5089. https://dx.doi.org/10.1128/JVI.79.8.5078-5089.2005

Everett, R.D., Parada, C., Gripon, P., Sirma, H., and Orr, A. (2008). Replication of ICP0-null mutant herpes simplex virus type 1 is restricted by both PML and Sp100. J Virol *82*, 2661-2672. https://dx.doi.org/10.1128/JVI.02308-07

Everett, R.D., Sourvinos, G., and Orr, A. (2003). Recruitment of herpes simplex virus type 1 transcriptional regulatory protein ICP4 into foci juxtaposed to ND10 in live, infected cells. J Virol *77*, 3680-3689

Faber, S.W., and Wilcox, K.W. (1986). Association of the herpes simplex virus regulatory protein ICP4 with specific nucleotide sequences in DNA. Nucleic Acids Res *14*, 6067-6083. https://dx.doi.org/10.1093/nar/14.15.6067

Faber, S.W., and Wilcox, K.W. (1988). Association of herpes simplex virus regulatory protein ICP4 with sequences spanning the ICP4 gene transcription initiation site. Nucleic Acids Res *16*, 555-570

Frenkel, N., and Roizman, B. (1972). Separation of the herpesvirus deoxyribonucleic acid duplex into unique fragments and intact strand on sedimentation in alkaline gradients. J Virol *10*, 565-572

Gibeault, R.L., Conn, K.L., Bildersheim, M.D., and Schang, L.M. (2016). An Essential Viral Transcription Activator Modulates Chromatin Dynamics. PLoS Pathog *12*, e1005842. https://dx.doi.org/10.1371/journal.ppat.1005842

Glass, M., and Everett, R.D. (2013). Components of promyelocytic leukemia nuclear bodies (ND10) act cooperatively to repress herpesvirus infection. J Virol *87*, 2174-2185. https://dx.doi.org/10.1128/JVI.02950-12

Golding, I. (2016). Single-Cell Studies of Phage lambda: Hidden Treasures Under Occam's Rug. Annu Rev Virol *3*, 453-472. https://dx.doi.org/10.1146/annurev-virology-110615-042127

Gu, B., Kuddus, R., and DeLuca, N.A. (1995). Repression of activator-mediated transcription by herpes simplex virus ICP4 via a mechanism involving interactions with the basal transcription factors TATA-binding protein and TFIIB. Mol Cell Biol *15*, 3618-3626

Gu, H., and Roizman, B. (2007). Herpes simplex virus-infected cell protein 0 blocks the silencing of viral DNA by dissociating histone deacetylases from the CoREST-REST complex. Proc Natl Acad Sci U S A *104*, 17134-17139. https://dx.doi.org/10.1073/pnas.0707266104

Gu, H., and Roizman, B. (2009). The two functions of herpes simplex virus 1 ICP0, inhibition of silencing by the CoREST/REST/HDAC complex and degradation of PML, are executed in tandem. J Virol *83*, 181-187. https://dx.doi.org/10.1128/JVI.01940-08

Gu, H., and Zheng, Y. (2016). Role of ND10 nuclear bodies in the chromatin repression of HSV-1. Virol J *13*, 62. https://dx.doi.org/10.1186/s12985-016-0516-4

Hancock, M.H., Cliffe, A.R., Knipe, D.M., and Smiley, J.R. (2010). Herpes simplex virus VP16, but not ICP0, is required to reduce histone occupancy and enhance histone acetylation on viral genomes in U2OS osteosarcoma cells. J Virol *84*, 1366-1375. https://dx.doi.org/10.1128/JVI.01727-09

Harris, R.A., and Preston, C.M. (1991). Establishment of latency in vitro by the herpes simplex virus type 1 mutant in1814. J Gen Virol *72 (Pt 4)*, 907-913. https://dx.doi.org/10.1099/0022-1317-72-4-907

Herr, W., and Cleary, M.A. (1995). The POU domain: versatility in transcriptional regulation by a flexible two-in-one DNA-binding domain. Genes Dev *9*, 1679-1693

Herrera, F.J., and Triezenberg, S.J. (2004). VP16-dependent association of chromatin-modifying coactivators and underrepresentation of histones at immediate-early gene promoters during herpes simplex virus infection. J Virol *78*, 9689-9696. https://dx.doi.org/10.1128/JVI.78.18.9689-9696.2004

Hirai, H., Tani, T., and Kikyo, N. (2010). Structure and functions of powerful transactivators: VP16, MyoD and FoxA. Int J Dev Biol *54*, 1589-1596. https://dx.doi.org/10.1387/ijdb.103194hh

Honess, R.W., and Roizman, B. (1974). Regulation of herpesvirus macromolecular synthesis. I. Cascade regulation of the synthesis of three groups of viral proteins. J Virol *14*, 8-19

Honess, R.W., and Roizman, B. (1975). Regulation of herpesvirus macromolecular synthesis: sequential transition of polypeptide synthesis requires functional viral polypeptides. Proc Natl Acad Sci U S A *72*, 1276-1280

Hu, H.L., Srinivas, K.P., Mohr, I., Huang, T.T., and Wilson, A.C. (2020). Using Primary SCG Neuron Cultures to Study Molecular Determinants of HSV-1 Latency and Reactivation. Methods Mol Biol *2060*, 263-277. https://dx.doi.org/10.1007/978-1-4939-9814-2_14

Hu, M., Depledge, D.P., Flores Cortes, E., Breuer, J., and Schang, L.M. (2019). Chromatin dynamics and the transcriptional competence of HSV-1 genomes during lytic infections. PLoS Pathog *15*, e1008076. https://dx.doi.org/10.1371/journal.ppat.1008076

Huang, J., Kent, J.R., Placek, B., Whelan, K.A., Hollow, C.M., Zeng, P.Y., Fraser, N.W., and Berger, S.L. (2006). Trimethylation of histone H3 lysine 4 by Set1 in the lytic infection of human herpes simplex virus 1. J Virol *80*, 5740-5746. https://dx.doi.org/10.1128/JVI.00169-06

Karttunen, H., Savas, J.N., McKinney, C., Chen, Y.H., Yates, J.R., 3rd, Hukkanen, V., Huang, T.T., and Mohr, I. (2014). Co-opting the Fanconi anemia genomic stability pathway enables herpesvirus DNA synthesis and productive growth. Mol Cell *55*, 111-122. https://dx.doi.org/10.1016/j.molcel.2014.05.020

Kelly, B.J., Fraefel, C., Cunningham, A.L., and Diefenbach, R.J. (2009). Functional roles of the tegument proteins of herpes simplex virus type 1. Virus Res *145*, 173-186. https://dx.doi.org/10.1016/j.virusres.2009.07.007

Kent, J.R., Zeng, P.Y., Atanasiu, D., Gardner, J., Fraser, N.W., and Berger, S.L. (2004). During lytic infection herpes simplex virus type 1 is associated with histones bearing modifications that correlate with active transcription. J Virol *78*, 10178-10186. https://dx.doi.org/10.1128/JVI.78.18.10178-10186.2004

Kieff, E.D., Bachenheimer, S.L., and Roizman, B. (1971). Size, composition, and structure of the deoxyribonucleic acid of herpes simplex virus subtypes 1 and 2. J Virol *8*, 125-132

Knipe, D.M., and Cliffe, A. (2008). Chromatin control of herpes simplex virus lytic and latent infection. Nat Rev Microbiol *6*, 211-221. https://dx.doi.org/10.1038/nrmicro1794

Kobiler, O., Brodersen, P., Taylor, M.P., Ludmir, E.B., and Enquist, L.W. (2011). Herpesvirus replication compartments originate with single incoming viral genomes. MBio *2*. https://dx.doi.org/10.1128/mBio.00278-11

Kobiler, O., Lipman, Y., Therkelsen, K., Daubechies, I., and Enquist, L.W. (2010). Herpesviruses carrying a Brainbow cassette reveal replication and expression of limited numbers of incoming genomes. Nat Commun *1*, 146. https://dx.doi.org/10.1038/ncomms1145

Kobiler, O., and Weitzman, M.D. (2019). Herpes simplex virus replication compartments: From naked release to recombining together. PLoS Pathog *15*, e1007714. https://dx.doi.org/10.1371/journal.ppat.1007714

Komatsu, T., Nagata, K., and Wodrich, H. (2016). The Role of Nuclear Antiviral Factors against Invading DNA Viruses: The Immediate Fate of Incoming Viral Genomes. Viruses *8*. https://dx.doi.org/10.3390/v8100290

Koyuncu, O.O., MacGibeny, M.A., and Enquist, L.W. (2018). Latent versus productive infection: the alpha herpesvirus switch. Future Virol *13*, 431-443. https://dx.doi.org/10.2217/fvl-2018-0023

Kristie, T.M. (2015). Dynamic modulation of HSV chromatin drives initiation of infection and provides targets for epigenetic therapies. Virology *479-480*, 555-561. https://dx.doi.org/10.1016/j.virol.2015.01.026

Kristie, T.M., LeBowitz, J.H., and Sharp, P.A. (1989). The octamer-binding proteins form multi-protein--DNA complexes with the HSV alpha TIF regulatory protein. EMBO J *8*, 4229-4238

Kubat, N.J., Tran, R.K., McAnany, P., and Bloom, D.C. (2004). Specific histone tail modification and not DNA methylation is a determinant of herpes simplex virus type 1 latent gene expression. J Virol *78*, 1139-1149

Kulej, K., Avgousti, D.C., Sidoli, S., Herrmann, C., Della Fera, A.N., Kim, E.T., Garcia, B.A., and Weitzman, M.D. (2017). Time-resolved Global and Chromatin Proteomics during Herpes Simplex Virus Type 1 (HSV-1) Infection. Mol Cell Proteomics *16*, S92-S107. https://dx.doi.org/10.1074/mcp.M116.065987

Kwiatkowski, D.L., Thompson, H.W., and Bloom, D.C. (2009). The polycomb group protein Bmi1 binds to the herpes simplex virus 1 latent genome and maintains repressive histone marks during latency. J Virol *83*, 8173-8181. https://dx.doi.org/10.1128/JVI.00686-09

Kwong, A.D., and Frenkel, N. (1987). Herpes simplex virus-infected cells contain a function(s) that destabilizes both host and viral mRNAs. Proc Natl Acad Sci U S A *84*, 1926-1930

Kwong, A.D., Kruper, J.A., and Frenkel, N. (1988). Herpes simplex virus virion host shutoff function. J Virol *62*, 912-921

La Boissiere, S., Hughes, T., and O'Hare, P. (1999). HCF-dependent nuclear import of VP16. EMBO J *18*, 480-489. https://dx.doi.org/10.1093/emboj/18.2.480

Lacasse, J.J., and Schang, L.M. (2010). During lytic infections, herpes simplex virus type 1 DNA is in complexes with the properties of unstable nucleosomes. J Virol *84*, 1920-1933. https://dx.doi.org/10.1128/JVI.01934-09

Lacasse, J.J., and Schang, L.M. (2012). Herpes Simplex Virus 1 DNA Is in Unstable Nucleosomes throughout the Lytic Infection Cycle, and the Instability of the Nucleosomes Is Independent of DNA Replication. Journal of Virology *86*, 11287-11300. https://dx.doi.org/10.1128/Jvi.01468-12

Lakowski, B., Roelens, I., and Jacob, S. (2006). CoREST-like complexes regulate chromatin modification and neuronal gene expression. J Mol Neurosci *29*, 227-239. https://dx.doi.org/10.1385/JMN:29:3:227

Lallemand-Breitenbach, V., and de The, H. (2018). PML nuclear bodies: from architecture to function. Curr Opin Cell Biol *52*, 154-161. https://dx.doi.org/10.1016/j.ceb.2018.03.011

Lanctot, C., Cheutin, T., Cremer, M., Cavalli, G., and Cremer, T. (2007). Dynamic genome architecture in the nuclear space: regulation of gene expression in three dimensions. Nat Rev Genet *8*, 104-115. https://dx.doi.org/10.1038/nrg2041

Leary, K., Connor, J.R., and Morahan, P.S. (1985). Comparison of herpes simplex virus type 1 DNA replication and virus production in murine bone marrow-derived and resident peritoneal macrophages. J Gen Virol *66 (Pt 5)*, 1123-1129. https://dx.doi.org/10.1099/0022-1317-66-5-1123

Lee, J.S., Raja, P., and Knipe, D.M. (2016). Herpesviral ICP0 Protein Promotes Two Waves of Heterochromatin Removal on an Early Viral Promoter during

Lytic Infection. MBio *7*, e02007-02015. https://dx.doi.org/10.1128/mBio.02007-15

Lees-Miller, S.P., Long, M.C., Kilvert, M.A., Lam, V., Rice, S.A., and Spencer, C.A. (1996). Attenuation of DNA-dependent protein kinase activity and its catalytic subunit by the herpes simplex virus type 1 transactivator ICP0. J Virol *70*, 7471-7477

Lester, J.T., and DeLuca, N.A. (2011). Herpes simplex virus 1 ICP4 forms complexes with TFIID and mediator in virus-infected cells. J Virol *85*, 5733-5744. https://dx.doi.org/10.1128/JVI.00385-11

Liang, Y., Vogel, J.L., Narayanan, A., Peng, H., and Kristie, T.M. (2009). Inhibition of the histone demethylase LSD1 blocks alpha-herpesvirus lytic replication and reactivation from latency. Nat Med *15*, 1312-1317. https://dx.doi.org/10.1038/nm.2051

Lilley, C.E., Carson, C.T., Muotri, A.R., Gage, F.H., and Weitzman, M.D. (2005). DNA repair proteins affect the lifecycle of herpes simplex virus 1. Proc Natl Acad Sci U S A *102*, 5844-5849. https://dx.doi.org/10.1073/pnas.0501916102

Lilley, C.E., Chaurushiya, M.S., Boutell, C., Everett, R.D., and Weitzman, M.D. (2011). The intrinsic antiviral defense to incoming HSV-1 genomes includes specific DNA repair proteins and is counteracted by the viral protein ICP0. PLoS Pathog *7*, e1002084. https://dx.doi.org/10.1371/journal.ppat.1002084

Lilley, C.E., Chaurushiya, M.S., Boutell, C., Landry, S., Suh, J., Panier, S., Everett, R.D., Stewart, G.S., Durocher, D., and Weitzman, M.D. (2010). A viral E3 ligase targets RNF8 and RNF168 to control histone ubiquitination and DNA damage responses. EMBO J *29*, 943-955. https://dx.doi.org/10.1038/emboj.2009.400

Linnavuori, K., and Hovi, T. (1983). Restricted replication of herpes simplex virus in human monocyte cultures: role of interferon. Virology *130*, 1-9

Liu, M., Rakowski, B., Gershburg, E., Weisend, C.M., Lucas, O., Schmidt, E.E., and Halford, W.P. (2010). ICP0 antagonizes ICP4-dependent silencing of the herpes simplex virus ICP0 gene. PLoS One *5*, e8837. https://dx.doi.org/10.1371/journal.pone.0008837

Lium, E.K., Panagiotidis, C.A., Wen, X., and Silverstein, S. (1996). Repression of the alpha0 gene by ICP4 during a productive herpes simplex virus infection. J Virol *70*, 3488-3496

Lium, E.K., Panagiotidis, C.A., Wen, X., and Silverstein, S.J. (1998). The NH2 terminus of the herpes simplex virus type 1 regulatory protein ICP0 contains a promoter-specific transcription activation domain. J Virol *72*, 7785-7795

Lium, E.K., and Silverstein, S. (1997). Mutational analysis of the herpes simplex virus type 1 ICP0 C3HC4 zinc ring finger reveals a requirement for ICP0 in the expression of the essential alpha27 gene. J Virol *71*, 8602-8614

Loret, S., Guay, G., and Lippe, R. (2008). Comprehensive characterization of extracellular herpes simplex virus type 1 virions. J Virol *82*, 8605-8618. https://dx.doi.org/10.1128/JVI.00904-08

Lukashchuk, V., and Everett, R.D. (2010). Regulation of ICP0-null mutant herpes simplex virus type 1 infection by ND10 components ATRX and hDaxx. J Virol *84*, 4026-4040. https://dx.doi.org/10.1128/JVI.02597-09

Ma, J.Z., Russell, T.A., Spelman, T., Carbone, F.R., and Tscharke, D.C. (2014). Lytic gene expression is frequent in HSV-1 latent infection and correlates with the engagement of a cell-intrinsic transcriptional response. PLoS Pathog *10*, e1004237. https://dx.doi.org/10.1371/journal.ppat.1004237

Maroui, M.A., Calle, A., Cohen, C., Streichenberger, N., Texier, P., Takissian, J., Rousseau, A., Poccardi, N., Welsch, J., Corpet, A., *et al.* (2016). Latency Entry of Herpes Simplex Virus 1 Is Determined by the Interaction of Its Genome with the Nuclear Environment. PLoS Pathog *12*, e1005834. https://dx.doi.org/10.1371/journal.ppat.1005834

Marshall, K.R., Lachmann, R.H., Efstathiou, S., Rinaldi, A., and Preston, C.M. (2000). Long-term transgene expression in mice infected with a herpes simplex virus type 1 mutant severely impaired for immediate-early gene expression. J Virol *74*, 956-964

Maul, G.G., Ishov, A.M., and Everett, R.D. (1996). Nuclear domain 10 as preexisting potential replication start sites of herpes simplex virus type-1. Virology *217*, 67-75. https://dx.doi.org/10.1006/viro.1996.0094

McFarlane, S., Orr, A., Roberts, A.P.E., Conn, K.L., Iliev, V., Loney, C., da Silva Filipe, A., Smollett, K., Gu, Q., Robertson, N., *et al.* (2019). The histone chaperone HIRA promotes the induction of host innate immune defences in response to HSV-1 infection. PLoS Pathog *15*, e1007667. https://dx.doi.org/10.1371/journal.ppat.1007667

McGeoch, D.J., Rixon, F.J., and Davison, A.J. (2006). Topics in herpesvirus genomics and evolution. Virus Res *117*, 90-104. https://dx.doi.org/10.1016/j.virusres.2006.01.002

McMahon, R., and Walsh, D. (2008). Efficient quiescent infection of normal human diploid fibroblasts with wild-type herpes simplex virus type 1. J Virol *82*, 10218-10230. https://dx.doi.org/10.1128/JVI.00859-08

Merkl, P.E., Orzalli, M.H., and Knipe, D.M. (2018). Mechanisms of Host IFI16, PML, and Daxx Protein Restriction of Herpes Simplex Virus 1 Replication. J Virol *92*. https://dx.doi.org/10.1128/JVI.00057-18

Metzler, D.W., and Wilcox, K.W. (1985). Isolation of herpes simplex virus regulatory protein ICP4 as a homodimeric complex. J Virol *55*, 329-337

Misra, V., Walter, S., Yang, P., Hayes, S., and O'Hare, P. (1996). Conformational alteration of Oct-1 upon DNA binding dictates selectivity in differential interactions with related transcriptional coactivators. Mol Cell Biol *16*, 4404-4413

Mohni, K.N., Dee, A.R., Smith, S., Schumacher, A.J., and Weller, S.K. (2013). Efficient herpes simplex virus 1 replication requires cellular ATR pathway proteins. J Virol *87*, 531-542. https://dx.doi.org/10.1128/JVI.02504-12

Mohni, K.N., Livingston, C.M., Cortez, D., and Weller, S.K. (2010). ATR and ATRIP are recruited to herpes simplex virus type 1 replication compartments even though ATR signaling is disabled. J Virol *84*, 12152-12164. https://dx.doi.org/10.1128/JVI.01643-10

Mohni, K.N., Mastrocola, A.S., Bai, P., Weller, S.K., and Heinen, C.D. (2011). DNA mismatch repair proteins are required for efficient herpes simplex virus 1 replication. J Virol *85*, 12241-12253. https://dx.doi.org/10.1128/JVI.05487-11

Mou, F., Wills, E.G., Park, R., and Baines, J.D. (2008). Effects of lamin A/C, lamin B1, and viral US3 kinase activity on viral infectivity, virion egress, and the targeting of herpes simplex virus U(L)34-encoded protein to the inner nuclear membrane. J Virol *82*, 8094-8104. https://dx.doi.org/10.1128/JVI.00874-08

Muller, S., and Dejean, A. (1999). Viral immediate-early proteins abrogate the modification by SUMO-1 of PML and Sp100 proteins, correlating with nuclear body disruption. J Virol *73*, 5137-5143

Muller, S., Matunis, M.J., and Dejean, A. (1998). Conjugation with the ubiquitin-related modifier SUMO-1 regulates the partitioning of PML within the nucleus. EMBO J *17*, 61-70. https://dx.doi.org/10.1093/emboj/17.1.61

Muylaert, I., Tang, K.W., and Elias, P. (2011). Replication and recombination of herpes simplex virus DNA. J Biol Chem *286*, 15619-15624. https://dx.doi.org/10.1074/jbc.R111.233981

Narayanan, A., Ruyechan, W.T., and Kristie, T.M. (2007). The coactivator host cell factor-1 mediates Set1 and MLL1 H3K4 trimethylation at herpesvirus immediate early promoters for initiation of infection. Proc Natl Acad Sci U S A *104*, 10835-10840. https://dx.doi.org/10.1073/pnas.0704351104

Natarajan, R., Deshmane, S., Valyi-Nagy, T., Everett, R., and Fraser, N.W. (1991). A herpes simplex virus type 1 mutant lacking the ICP0 introns reactivates with normal efficiency. J Virol *65*, 5569-5573

O'Neill, F.J. (1977). Prolongation of herpes simplex virus latency in cultured human cells by temperature elevation. J Virol *24*, 41-46

Oh, H.S., Bryant, K.F., Nieland, T.J., Mazumder, A., Bagul, M., Bathe, M., Root, D.E., and Knipe, D.M. (2014). A targeted RNA interference screen reveals novel epigenetic factors that regulate herpesviral gene expression. MBio *5*, e01086-01013. https://dx.doi.org/10.1128/mBio.01086-13

Oh, J., and Fraser, N.W. (2008). Temporal association of the herpes simplex virus genome with histone proteins during a lytic infection. J Virol *82*, 3530-3537. https://dx.doi.org/10.1128/JVI.00586-07

Ojala, P.M., Sodeik, B., Ebersold, M.W., Kutay, U., and Helenius, A. (2000). Herpes simplex virus type 1 entry into host cells: reconstitution of capsid binding and uncoating at the nuclear pore complex in vitro. Mol Cell Biol *20*, 4922-4931

Ottosen, S., Herrera, F.J., Doroghazi, J.R., Hull, A., Mittal, S., Lane, W.S., and Triezenberg, S.J. (2006). Phosphorylation of the VP16 transcriptional activator protein during herpes simplex virus infection and mutational analysis of putative phosphorylation sites. Virology *345*, 468-481. https://dx.doi.org/10.1016/j.virol.2005.10.011

Parkinson, J., Lees-Miller, S.P., and Everett, R.D. (1999). Herpes simplex virus type 1 immediate-early protein vmw110 induces the proteasome-dependent degradation of the catalytic subunit of DNA-dependent protein kinase. J Virol *73*, 650-657

Pasdeloup, D., Blondel, D., Isidro, A.L., and Rixon, F.J. (2009). Herpesvirus capsid association with the nuclear pore complex and viral DNA release involve the nucleoporin CAN/Nup214 and the capsid protein pUL25. J Virol *83*, 6610-6623.JVI.02655-08 [pii] https://dx.doi.org/10.1128/JVI.02655-08

Perry, L.J., Rixon, F.J., Everett, R.D., Frame, M.C., and McGeoch, D.J. (1986). Characterization of the IE110 gene of herpes simplex virus type 1. J Gen Virol *67 (Pt 11)*, 2365-2380. https://dx.doi.org/10.1099/0022-1317-67-11-2365

Phelan, A., Dunlop, J., Patel, A.H., Stow, N.D., and Clements, J.B. (1997). Nuclear sites of herpes simplex virus type 1 DNA replication and transcription colocalize at early times postinfection and are largely distinct from RNA processing factors. J Virol *71*, 1124-1132

Pritchard, S.M., Cunha, C.W., and Nicola, A.V. (2013). Analysis of herpes simplex virion tegument ICP4 derived from infected cells and ICP4-expressing cells. PLoS One *8*, e70889. https://dx.doi.org/10.1371/journal.pone.0070889

Quinlan, M.P., Chen, L.B., and Knipe, D.M. (1984). The intranuclear location of a herpes simplex virus DNA-binding protein is determined by the status of viral DNA replication. Cell *36*, 857-868

Rai, T.S., Glass, M., Cole, J.J., Rather, M.I., Marsden, M., Neilson, M., Brock, C., Humphreys, I.R., Everett, R.D., and Adams, P.D. (2017). Histone chaperone HIRA deposits histone H3.3 onto foreign viral DNA and contributes to anti-viral intrinsic immunity. Nucleic Acids Res *45*, 11673-11683. https://dx.doi.org/10.1093/nar/gkx771

Raja, P., Lee, J.S., Pan, D., Pesola, J.M., Coen, D.M., and Knipe, D.M. (2016). A Herpesviral Lytic Protein Regulates the Structure of Latent Viral Chromatin. MBio *7*. https://dx.doi.org/10.1128/mBio.00633-16

Ralph, M., Bednarchik, M., Tomer, E., Rafael, D., Zargarian, S., Gerlic, M., and Kobiler, O. (2017). Promoting Simultaneous Onset of Viral Gene Expression Among Cells Infected with Herpes Simplex Virus-1. Front Microbiol *8*, 2152. https://dx.doi.org/10.3389/fmicb.2017.02152

Reyes, E.D., Kulej, K., Pancholi, N.J., Akhtar, L.N., Avgousti, D.C., Kim, E.T., Bricker, D.K., Spruce, L.A., Koniski, S.A., Seeholzer, S.H., *et al.* (2017). Identifying Host Factors Associated with DNA Replicated During Virus Infection. Mol Cell Proteomics *16*, 2079-2097. https://dx.doi.org/10.1074/mcp.M117.067116

Roizman, B., and Aurelian, L. (1965). Abortive Infection of Canine Cells by Herpes Simplex Virus. I. Characterization of Viral Progeny from Co-Operative Infection with Mutants Differing in Capacity to Multiply in Canine Cells. J Mol Biol *11*, 528-538

Russell, J., and Preston, C.M. (1986). An in vitro latency system for herpes simplex virus type 2. J Gen Virol *67 (Pt 2)*, 397-403. https://dx.doi.org/10.1099/0022-1317-67-2-397

Sacks, W.R., and Schaffer, P.A. (1987). Deletion mutants in the gene encoding the herpes simplex virus type 1 immediate-early protein ICP0 exhibit impaired growth in cell culture. J Virol *61*, 829-839

Samaniego, L.A., Neiderhiser, L., and DeLuca, N.A. (1998). Persistence and expression of the herpes simplex virus genome in the absence of immediate-early proteins. J Virol *72*, 3307-3320

Scheck, A.C., Wigdahl, B., De Clercq, E., and Rapp, F. (1986). Prolonged herpes simplex virus latency in vitro after treatment of infected cells with acyclovir and human leukocyte interferon. Antimicrob Agents Chemother *29*, 589-593

Sekine, E., Schmidt, N., Gaboriau, D., and O'Hare, P. (2017). Spatiotemporal dynamics of HSV genome nuclear entry and compaction state transitions using bioorthogonal chemistry and super-resolution microscopy. PLoS Pathog *13*, e1006721. https://dx.doi.org/10.1371/journal.ppat.1006721

Sexton, T., and Cavalli, G. (2015). The role of chromosome domains in shaping the functional genome. Cell *160*, 1049-1059. https://dx.doi.org/10.1016/j.cell.2015.02.040

Shapira, L., Ralph, M., Tomer, E., Cohen, S., and Kobiler, O. (2016). Histone Deacetylase Inhibitors Reduce the Number of Herpes Simplex Virus-1 Genomes Initiating Expression in Individual Cells. Front Microbiol *7*, 1970. https://dx.doi.org/10.3389/fmicb.2016.01970

Sheldrick, P., Laithier, M., Lando, D., and Ryhiner, M.L. (1973). Infectious DNA from herpes simplex virus: infectivity of double-stranded and single-stranded molecules. Proc Natl Acad Sci U S A *70*, 3621-3625

Shiraki, K., and Rapp, F. (1986). Establishment of herpes simplex virus latency in vitro with cycloheximide. J Gen Virol *67 (Pt 11)*, 2497-2500. https://dx.doi.org/10.1099/0022-1317-67-11-2497

Shivkumar, M., Lawler, C., Milho, R., and Stevenson, P.G. (2016). Herpes Simplex Virus 1 Interaction with Myeloid Cells In Vivo. J Virol *90*, 8661-8672. https://dx.doi.org/10.1128/JVI.00881-16

Silva, L., Cliffe, A., Chang, L., and Knipe, D.M. (2008). Role for A-type lamins in herpesviral DNA targeting and heterochromatin modulation. PLoS Pathog *4*, e1000071. https://dx.doi.org/10.1371/journal.ppat.1000071

Silva, L., Oh, H.S., Chang, L., Yan, Z., Triezenberg, S.J., and Knipe, D.M. (2012). Roles of the nuclear lamina in stable nuclear association and assembly of a herpesviral transactivator complex on viral immediate-early genes. MBio *3*. https://dx.doi.org/10.1128/mBio.00300-11

Smiley, J.R., and Duncan, J. (1997). Truncation of the C-terminal acidic transcriptional activation domain of herpes simplex virus VP16 produces a phenotype similar to that of the in1814 linker insertion mutation. J Virol *71*, 6191-6193

Smith, C.A., Bates, P., Rivera-Gonzalez, R., Gu, B., and DeLuca, N.A. (1993). ICP4, the major transcriptional regulatory protein of herpes simplex virus type 1, forms a tripartite complex with TATA-binding protein and TFIIB. J Virol *67*, 4676-4687

Smith, S., Reuven, N., Mohni, K.N., Schumacher, A.J., and Weller, S.K. (2014). Structure of the herpes simplex virus 1 genome: manipulation of nicks and gaps can abrogate infectivity and alter the cellular DNA damage response. J Virol *88*, 10146-10156. https://dx.doi.org/10.1128/JVI.01723-14

Smith, S., and Weller, S.K. (2015). HSV-I and the cellular DNA damage response. Future Virol *10*, 383-397. https://dx.doi.org/10.2217/fvl.15.18

Sourvinos, G., and Everett, R.D. (2002). Visualization of parental HSV-1 genomes and replication compartments in association with ND10 in live infected cells. EMBO J *21*, 4989-4997

Stegen, C., Yakova, Y., Henaff, D., Nadjar, J., Duron, J., and Lippe, R. (2013). Analysis of virion-incorporated host proteins required for herpes simplex virus type 1 infection through a RNA interference screen. PLoS One *8*, e53276. https://dx.doi.org/10.1371/journal.pone.0053276

Steiner, I., Spivack, J.G., Deshmane, S.L., Ace, C.I., Preston, C.M., and Fraser, N.W. (1990). A herpes simplex virus type 1 mutant containing a nontransinducing Vmw65 protein establishes latent infection in vivo in the absence of viral replication and reactivates efficiently from explanted trigeminal ganglia. J Virol *64*, 1630-1638

Stern, S., Tanaka, M., and Herr, W. (1989). The Oct-1 homoeodomain directs formation of a multiprotein-DNA complex with the HSV transactivator VP16. Nature *341*, 624-630. https://dx.doi.org/10.1038/341624a0

Sternsdorf, T., Jensen, K., and Will, H. (1997). Evidence for covalent modification of the nuclear dot-associated proteins PML and Sp100 by PIC1/ SUMO-1. J Cell Biol *139*, 1621-1634

Stow, N.D., and Stow, E.C. (1986). Isolation and characterization of a herpes simplex virus type 1 mutant containing a deletion within the gene encoding the immediate early polypeptide Vmw110. J Gen Virol *67 (Pt 12)*, 2571-2585. https://dx.doi.org/10.1099/0022-1317-67-12-2571

Suk, H., and Knipe, D.M. (2015). Proteomic analysis of the herpes simplex virus 1 virion protein 16 transactivator protein in infected cells. Proteomics *15*, 1957-1967. https://dx.doi.org/10.1002/pmic.201500020

Suzich, J.B., and Cliffe, A.R. (2018). Strength in diversity: Understanding the pathways to herpes simplex virus reactivation. Virology *522*, 81-91. https://dx.doi.org/10.1016/j.virol.2018.07.011

Taylor, M.P., Kobiler, O., and Enquist, L.W. (2012). Alphaherpesvirus axon-to-cell spread involves limited virion transmission. Proc Natl Acad Sci U S A *109*, 17046-17051. https://dx.doi.org/10.1073/pnas.1212926109

Taylor, T.J., and Knipe, D.M. (2004). Proteomics of herpes simplex virus replication compartments: association of cellular DNA replication, repair, recombination, and chromatin remodeling proteins with ICP8. J Virol *78*, 5856-5866. https://dx.doi.org/10.1128/JVI.78.11.5856-5866.2004

Tomer, E., Cohen, E.M., Drayman, N., Afriat, A., Weitzman, M.D., Zaritsky, A., and Kobiler, O. (2019). Coalescing replication compartments provide the opportunity for recombination between coinfecting herpesviruses. FASEB J *33*, 9388-9403. https://dx.doi.org/10.1096/fj.201900032R

Trigg, B.J., and Ferguson, B.J. (2015). Functions of DNA damage machinery in the innate immune response to DNA virus infection. Curr Opin Virol *15*, 56-62. https://dx.doi.org/10.1016/j.coviro.2015.08.001

Trigg, B.J., Lauer, K.B., Fernandes Dos Santos, P., Coleman, H., Balmus, G., Mansur, D.S., and Ferguson, B.J. (2017). The Non-Homologous End Joining Protein PAXX Acts to Restrict HSV-1 Infection. Viruses *9*. https://dx.doi.org/10.3390/v9110342

Tunnicliffe, R.B., Lockhart-Cairns, M.P., Levy, C., Mould, A.P., Jowitt, T.A., Sito, H., Baldock, C., Sandri-Goldin, R.M., and Golovanov, A.P. (2017). The herpes viral transcription factor ICP4 forms a novel DNA recognition complex. Nucleic Acids Res *45*, 8064-8078. https://dx.doi.org/10.1093/nar/gkx419

van Lint, A.L., Murawski, M.R., Goodbody, R.E., Severa, M., Fitzgerald, K.A., Finberg, R.W., Knipe, D.M., and Kurt-Jones, E.A. (2010). Herpes simplex virus immediate-early ICP0 protein inhibits Toll-like receptor 2-dependent inflammatory responses and NF-kappaB signaling. J Virol *84*, 10802-10811. https://dx.doi.org/10.1128/JVI.00063-10

Wadsworth, S., Jacob, R.J., and Roizman, B. (1975). Anatomy of herpes simplex virus DNA. II. Size, composition, and arrangement of inverted terminal repetitions. J Virol *15*, 1487-1497

Waisner, H., and Kalamvoki, M. (2019). The ICP0 Protein of Herpes Simplex Virus 1 (HSV-1) Downregulates Major Autophagy Adaptor Proteins Sequestosome 1 and Optineurin during the Early Stages of HSV-1 Infection. J Virol *93*. https://dx.doi.org/10.1128/JVI.01258-19

Wang, Q.Y., Zhou, C., Johnson, K.E., Colgrove, R.C., Coen, D.M., and Knipe, D.M. (2005). Herpesviral latency-associated transcript gene promotes assembly of heterochromatin on viral lytic-gene promoters in latent infection. Proc Natl Acad Sci U S A *102*, 16055-16059. https://dx.doi.org/10.1073/pnas.0505850102

Wang, X., Diao, C., Yang, X., Yang, Z., Liu, M., Li, X., and Tang, H. (2016). ICP4-induced miR-101 attenuates HSV-1 replication. Sci Rep *6*, 23205. https://dx.doi.org/10.1038/srep23205

Weinheimer, S.P., Boyd, B.A., Durham, S.K., Resnick, J.L., and O'Boyle, D.R., 2nd (1992). Deletion of the VP16 open reading frame of herpes simplex virus type 1. J Virol *66*, 258-269

Wilkinson, D.E., and Weller, S.K. (2003). The role of DNA recombination in herpes simplex virus DNA replication. IUBMB Life *55*, 451-458. https://dx.doi.org/10.1080/15216540310001612237

Wu, C.A., Nelson, N.J., McGeoch, D.J., and Challberg, M.D. (1988). Identification of herpes simplex virus type 1 genes required for origin-dependent DNA synthesis. J Virol *62*, 435-443

Wyler, E., Franke, V., Menegatti, J., Kocks, C., Boltengagen, A., Praktiknjo, S., Walch-Ruckheim, B., Bosse, J., Rajewsky, N., Grasser, F., *et al.* (2019). Single-cell RNA-sequencing of herpes simplex virus 1-infected cells connects NRF2 activation to an antiviral program. Nat Commun *10*, 4878. https://dx.doi.org/10.1038/s41467-019-12894-z

Wysocka, J., and Herr, W. (2003). The herpes simplex virus VP16-induced complex: the makings of a regulatory switch. Trends Biochem Sci *28*, 294-304. https://dx.doi.org/10.1016/S0968-0004(03)00088-4

Xu, P., Mallon, S., and Roizman, B. (2016). PML plays both inimical and beneficial roles in HSV-1 replication. Proc Natl Acad Sci U S A *113*, E3022-3028. https://dx.doi.org/10.1073/pnas.1605513113

Xu, P., and Roizman, B. (2017). The SP100 component of ND10 enhances accumulation of PML and suppresses replication and the assembly of HSV replication compartments. Proc Natl Acad Sci U S A *114*, E3823-E3829. https://dx.doi.org/10.1073/pnas.1703395114

Yao, F., and Courtney, R.J. (1989). A major transcriptional regulatory protein (ICP4) of herpes simplex virus type 1 is associated with purified virions. J Virol *63*, 3338-3344

Yao, F., and Courtney, R.J. (1992). Association of ICP0 but not ICP27 with purified virions of herpes simplex virus type 1. J Virol *66*, 2709-2716

You, A., Tong, J.K., Grozinger, C.M., and Schreiber, S.L. (2001). CoREST is an integral component of the CoREST- human histone deacetylase complex. Proc Natl Acad Sci U S A *98*, 1454-1458. https://dx.doi.org/10.1073/pnas.98.4.1454

Zargar, Z., and Tyagi, S. (2012). Role of host cell factor-1 in cell cycle regulation. Transcription *3*, 187-192. https://dx.doi.org/10.4161/trns.20711

Chapter 5

Invasion of the Nervous System

Orkide O. Koyuncu[1,3], Lynn W. Enquist[1,3] and Esteban A. Engel[2,3]*

[1]Department of Molecular Biology, [2]Princeton Neuroscience Institute, [3]Princeton University, Princeton, NJ 08544, USA

*eengel@princeton.edu

DOI: https://doi.org/10.21775/9781913652555.05

Abstract

In vertebrates, the nervous system (NS) is composed of a peripheral collection of neurons (the peripheral nervous system, PNS), a central set found in the brain and spinal cord (the central nervous system, CNS). The NS is protected by rather complicated multi-layer barriers that allow access to nutrients and facilitate contact with the peripheral tissues, but block entry of pathogens and toxins. Virus infections usually begin in peripheral tissues and if these barriers are weakened, they can spread into the PNS and more rarely into the CNS. Most viral infections of the NS are opportunistic or accidental pathogens that gain access via the bloodstream (e.g., HIV and various arboviruses). But a few have evolved to enter the NS efficiently by invading neurons directly and by exploiting neuronal cell biology (e.g., rhabdoviruses and alphaherpesviruses). Most NS infections are devastating and difficult to manage. Remarkably, the alphaherpesviruses (α-HVs) establish life-long quiescent infections in the PNS, with rare but often serious CNS pathology. In this chapter, we will focus on how α-HVs gain access to and spread in the NS, with particular emphasis on bidirectional transport and spread within and between neurons and neural circuits, which is regulated by complex viral-host protein interactions. Finally, we will describe the wide use of α-HVs as tools to study nerve connectivity and function in animal models.

Introduction

Alphaherpesviruses (α-HVs) are large, enveloped viruses with double-stranded, linear DNA genomes that are tightly packed in an icosahedral protein shell termed capsid (Dai and Zhou, 2018). The virions contain a characteristic layer between the envelope and the capsid, termed tegument, that is composed of several viral and host proteins as well as some mRNAs (Kramer et al., 2011; Lippe, 2014; Sciortino et al., 2001). The tegument layer is divided into the capsid-associated inner and the envelope-associated outer layers (Owen et al., 2015). α-HV infections (e.g. herpes simplex virus 1 and 2 [HSV-1 and HSV-2] and varicella zoster virus [VZV]) are among the most common viral infections in the world (Looker et al., 2015a; Looker et al., 2015b; Moffat et al., 2007). Many alphaherpesviruses are species specific, meaning that they primarily infect a single host (e.g., VZV infects humans; bovine herpesvirus type-1 [BHV-1], cattle; Marek's disease virus [MDV], chickens). However, there are also examples of α-HVs with a broader host range. Pseudorabies virus (PRV) is a swine α-HV that can cause zoonotic infections in most mammals except higher primates. HSV-1 is a human α-HV but can also infect many small laboratory animals. Both PRV and HSV-1 grow well in tissue culture yielding high titer cell-free virus particles, and their genomes have been cloned into bacterial artificial chromosomes (BAC) amenable to generate new recombinants. As a result, these two viruses have been studied widely as prototype α-HVs in cultured cells and in animal models.

The life cycle of an α-HV involves a productive infection cycle in mucosal epithelial cells followed by a life-long quiescent (often called latent) state in the peripheral nervous system (PNS) neurons that innervate the infected mucosal tissue. Generally, mucosal surfaces (e.g., nasopharyngeal cavity, genitals, and eyes) are common sites of primary infection. The productive infection proceeds in most susceptible and permissive cell types, in a well characterized cascade fashion where viral immediate early (IE) proteins interact with host cell proteins and activate transcription of early (E) viral genes (Honess and Roizman, 1974). These E proteins are required to create the optimal environment for viral DNA replication and subsequent synthesis of late (L) viral proteins, the structural components of the virions (Honess and Roizman, 1974; Roizman and Whitley, 2013). This cascade of viral gene expression results in the rapid production of large numbers of progeny virions ready to spread to other tissues and hosts. Because epithelial tissues are highly innervated by sensory neurons, progeny viral particles typically first enter the axons of these neurons. Next, viral particles move efficiently along axons and reach the neuronal cell bodies (i.e., soma), which are tightly packed within the structures called the ganglia where they are typically linked by synapses. The ganglia of the PNS contain neurons either responsible for processing sensory information (sensory neurons) or regulating the function of internal organs (autonomic neurons). Sensory neurons within dorsal root ganglia (DRG), trigeminal ganglia (TG) and sacral ganglia are the primary sites of quiescent HSV infections (Croen et al., 1988; Mitchell et al., 2003). However, neurons of autonomic sympathetic ganglia that innervate peripheral tissues such as superior cervical ganglia (SCG) also harbor quiescent viral genomes after infection of peripheral tissues (Cheung, 1989; Enquist et al., 2002; Jones, 2003; Richter et al., 2009).

Neurons are terminally differentiated and extremely polarized. The neuronal cell body contains the nucleus with all the cellular transcription machinery, but neurons also have highly specialized extensions of their cell body cytoplasm called dendrites and axons (Figure 1). Dendrites can be branched and make connections (synapses) with axons from other neurons. Single axons can extend as long as meters, which is many thousands of times the diameter of a typical cell. The maintenance and the communication of distal axons with their cell bodies require specialized signal transduction, intracellular sorting, and trafficking systems. As obligate intracellular parasites, viruses are dependent on these cellular functions that often are cell-type specific.

After the productive infection in epithelia, when the viral progeny invade axons of the peripheral neurons, long-distance transport of viral particles to the cell

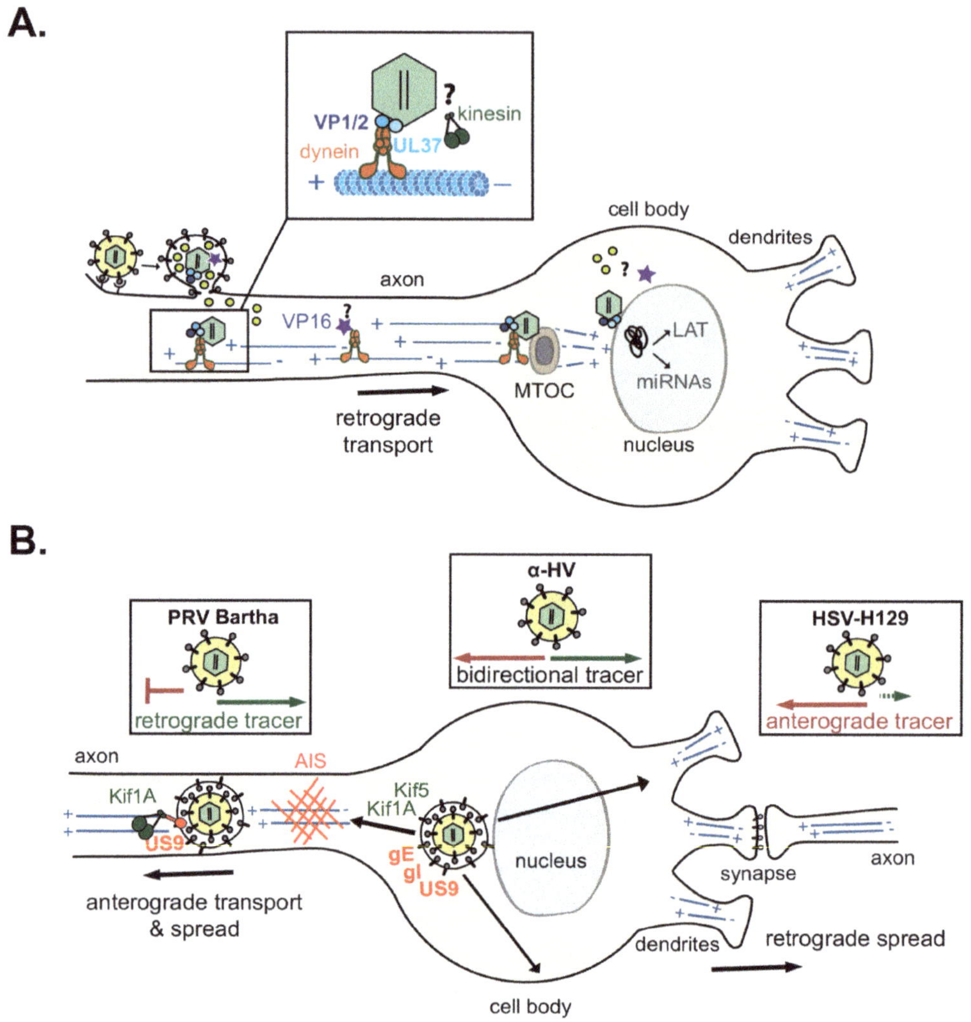

Figure 1. α-HV entry and spread in the nervous system. (A) α-HV particles enter neurons by direct fusion of their virion envelopes with the plasma membrane. Capsids with inner tegument proteins (e.g. VP1/2, UL37 and US3, blue circles) engage microtubule motors (dyneins and kinesins) for retrograde transport toward the neuronal nucleus. Microtubules in axons are unipolar; minus-ends are directed toward the microtubule organizing center, MTOC in the cell body. Somato-dendritic microtubules haves mixed polarity. The interaction of UL37 and VP1/2 may regulate the recruitment of dynein vs. kinesin motors by VP1/2 protein. The outer tegument proteins (yellow circles) including the transactivator protein VP16 (purple star) are not co-transported with the capsid-dynein complexes. This might bias the infection mode in the neuronal nucleus toward latency that is characterized by the limited expression of the viral genome (the latency associated transcript (LAT) and micro RNAs are expressed). **(B)** Kinesin motors (e.g. Kif1A and Kif5) mediate transport of α-HV virions contained within a vesicle to the cell periphery (highlighted with arrows). Interaction of viral envelope proteins (e.g. US9, gE and gI) with specific kinesin

motors is required to pass the axon initial segment (AIS) to sort the virion carrying vesicles into the axon. Kif1A is represented in the axon interacting with PRV envelope protein US9, though other kinesin motors may be involved in different stages of herpesvirus egress. α-HVs have bidirectional spread capacity in the nervous system (both anterograde and retrograde). PRV Bartha (lacking gE, gI, US9 genes) virions can only spread retrogradely between synaptically connected neurons. HSV-1 H129 strain spreads predominantly in the anterograde direction. Viral genomes do not encode genes for microtubule motor proteins. Therefore, nucleocapsids that enter axons or cell bodies must recruit the host cell dynein complex for efficient retrograde transport of the capsid complex (capsids plus inner tegument proteins) to the nucleus. The inner tegument proteins that remain associated and co-transport with the nucleocapsids are VP1/2 (aka UL36), UL37, and US3 (Granzow et al., 2005; Luxton et al., 2005). They mediate the loading of viral 'cargo' onto cellular dynein complexes both in neuronal and non-neuronal cells (Antinone and Smith, 2010; Buch et al., 2017; Coller and Smith, 2008; Luxton et al., 2005; Zaichick et al., 2013). This interaction represents a potential bottleneck in the establishment of neuronal infection because only a finite number of dynein motors and adapters are found in axons.

body nucleus - termed retrograde - occurs on microtubules and relies on the interaction of viral inner tegument proteins with the cellular motor dynein (Zaichick et al., 2013). By contrast, the outer tegument proteins are not co-transported toward the nucleus, presumably because they are not associated with capsids post-entry (Antinone et al., 2006; Luxton et al., 2005; Smith, 2012). After the capsids surrounded by the inner tegument proteins reach the nucleus, the genomes are released into the nucleus and the specialized quiescent, or latent, mode of infection initiates in the native host. During this latent phase, α-HV genomes are silenced: they are circularized and decorated with silencing nucleosomes containing histone modifications that allow transcription of only a small segment of the viral genome, particularly the latency-associated transcript (LAT) (Jarman et al., 1999; Jones, 2003; Kennedy et al., 2015).

To be sustained in a host population, the silenced viral genomes must retain the capacity to re-initiate the productive cycle (i.e., reactivate). Reactivation is usually triggered when the immunocompetent host experiences stress such as physical trauma, sunburn or fever (Cliffe and Wilson, 2016; Roizman and Whitley, 2013; Wilson and Mohr, 2012). When stress signaling pathways are activated, transcription of viral genes and replication of the viral genome occurs, and new viral particles are produced. These move back from the cell body nucleus along the axon in the anterograde direction (the opposite direction relative to that during initial infection) to the site of the original infection to ensure the spread of infection to other hosts (Smith, 2012). Surprisingly, despite the general global effect of stress on host cells, the threshold for reactivation of latently infected PNS neurons is high such that reactivation does not occur in all the neurons that harbor silenced viral genomes. Reactivation may be more likely to occur in neurons within the same ganglia that contain more copies of the silenced viral genome (Sawtell et al., 1998). Even within a single PNS nucleus, not all silenced genomes reactivate at the same time (Catez et al., 2012; Lomonte, 2017; Maroui et al., 2016; Sawtell, 1998). Currently, host and viral factors that affect the maintenance of latency and subsequent reactivation are not completely understood.

In people with healthy immune systems, reactivations are typically well controlled, and symptoms are mild, with local cold sores, genital lesions, or shingles blisters being typical manifestations. Asymptomatic reactivation and shedding of HSV-1 (Miller and Danaher, 2008) and VZV (Amlie-Lefond and Gilden, 2016; Gilden et al., 2016) are also known. However, in rare cases or in immunocompromised hosts, reactivations may result in less well contained lesions leading to oral or genital ulcerations, or spread to other locations such as eyes causing keratitis or blindness (depending on the ganglia where latency is established), and even CNS, causing encephalitis (Gnann and Whitley, 2017; Tsatsos et al., 2016). HSV-2 reactivations tend to be more frequent and do not always present as lesions. Such asymptomatic shedding of HSV-2 is common (Bertke et al., 2013; Groves, 2016). A rare but serious complication of HSV-2 (or HSV-1) shedding in the genital tract is neonatal HSV infections, which are estimated to occur in 10 of 100,000 live births globally (Looker et al., 2017). Neonatal HSV infections can spread to the CNS or other organs causing disseminated disease, which is fatal in approximately 60% of untreated cases (Corey and Wald, 2009). In rare cases, reactivating virus particles spread to the CNS, causing seizures and focal necrosis (Gilden et al., 2016; Whitley, 2015). Control of latency, reactivation, and subsequent spread of infection is affected by many cell intrinsic, tissue-specific and systemic factors that are challenging to dissect.

The mammalian nervous system, despite its vital importance in host survival, is well exploited by α-HVs. In this chapter, we will focus on how α-HV particles efficiently invade the nervous system and spread among the connected neurons, and how these properties not only lead to neuropathology, but also make α-HVs useful tools for studying the nervous system circuitry and function (Table 1).

I. Invasion of the NS: Intraneuronal Spread

Entry into axons and retrograde transport toward the neuronal nucleus

Most neurotropic virions enter neurons by endocytosis, using cellular receptors that are concentrated at nerve terminals. However, α-HV virions generally enter neurons by direct fusion of their lipid envelopes with the plasma membrane (Connolly et al., 2011; Heldwein and Krummenacher, 2008), although some examples of α-HV endocytosis have been shown (Mues et al., 2015). Nectin-1, a member of the immunoglobulin (Ig) superfamily, is a receptor for HSV and PRV virions on many cells including neurons (Kopp et al., 2009; Shukla et al., 2000; Spear et al., 2000). Binding of the conserved viral glycoprotein D (gD) on the virion surface to nectin-1 initiates the entry process (Geraghty et al., 2000; Shukla et al., 2000). The concerted action of the glycoprotein complex gH/gL and the viral fusogen gB is required for the efficient fusion of the viral envelope with the cellular membrane (Heldwein and Krummenacher, 2008). Upon membrane fusion, the capsid with inner tegument proteins is deposited into the axonal cytosol, ready for subsequent retrograde microtubule-dependent intracellular transport toward the neuronal nucleus (Figure 1A).

Microtubule networks, composed of α-tubulin and β-tubulin heterodimers, are arrays of polar cylinders that provide tracks for the long-distance transport of endocytic and exocytic vesicles, organelles, and chromosomes (Desai and Mitchison, 1997). Microtubules have a dynamic 'plus-end' that can grow and shrink rapidly while the 'minus-end' is often tethered to a microtubule-organizing center (MTOC) and is relatively stable (Luders and Stearns, 2007). Microtubules in axons have uniform polarity, and minus-ends are always directed toward the MTOC in the cell body (Conde and Caceres, 2009). On the other hand, microtubules in somato-dendritic regions of neurons have mixed polarity. Despite being nucleated by the MTOC, many of the axonal microtubules are no longer connected to the MTOC. They are distributed along the axon to form discontinuous, overlapping bundles.

Directional transport along microtubules is mediated by two classes of molecular motors (Hirokawa et al., 2010; Vale, 2003). Plus-end-directed motors are known as kinesins and they are involved in transport to the cell periphery (Kikkawa et al., 1995; Okada et al., 1995). Conventional kinesin is a heterotetramer composed of two heavy chains and two light chains. These motors are highly processive and relatively powerful, capable of moving large cargo complexes. Minus-end transport is mediated by cytoplasmic dynein that directs traffic from the periphery to the cell center (Kardon and Vale, 2009). This complex motor protein consists of two heavy chains, two intermediate chains, and multiple light intermediate and light chains. Cytoplasmic dynein function is mainly regulated by dynactin and Lis1/nudE complexes (DeSantis et al., 2017; Dohner et al., 2002; McKenney et al., 2011; Reck-Peterson et al., 2018).

Live-cell imaging experiments examining the retrograde transport of monomeric red-fluorescent protein-tagged PRV capsids and green-fluorescent protein (GFP)-tagged tegument proteins in sensory neurons revealed that outer tegument proteins, such as UL47 (VP13/14), UL48 (VP16), and UL49 (VP22), were lost during viral entry and did not move along with capsids to the nucleus (Antinone et al., 2006; Antinone and Smith, 2010; Luxton et al., 2005). In contrast, the inner tegument proteins UL36 and UL37 remained associated with capsids as they entered neurons, trafficked to the nucleus, and subsequently docked at the nuclear rim. UL36 is the largest herpesvirus protein that directly interacts with the capsid protein UL25 and tegument protein UL37 (Bottcher et al., 2008; Coller et al., 2007; Klupp et al., 2002; Lee et al., 2009). Zaichick et al., showed that a proline-rich domain near the carboxy terminus of UL36 mediates the interaction of incoming PRV capsids with dynein/dynactin (Zaichick et al., 2013). In a different set of experiments using isolated HSV-1 capsids with varying tegument protein compositions, inner tegument proteins recruited dynein motors onto incoming HSV-1 particles. Radtke et al. demonstrated that individual HSV-1 capsids with these inner tegument proteins can also recruit kinesin motors (Radtke et al., 2010). Indeed, bi-directional motility of incoming capsid in axons was detected, albeit much less frequently than the dominating retrograde motion (Scherer et al., 2016; Smith et al., 2004). This raises the question: how is the processive retrograde transport of capsids achieved early in infection if both kinesin and dynein motors are simultaneously recruited by inner tegument proteins?

Recently, UL37 has been identified as the viral protein promoting the retrograde-biased motion of both PRV and HSV-1 incoming capsids (Richards et al., 2017). Mutations in the R2 region of UL37 abrogated nervous system invasion by both PRV and HSV-1 in animal models without affecting viral replication in epithelial cells. Since UL37 has not yet been reported to directly interact with the molecular motors, the interaction of UL37 and UL36 may regulate the recruitment of dynein motors by UL36 protein (Klupp et al., 2002; Richards et al., 2017) (Figure 1A). The fact that the UL37 R2 mutation blocks neuroinvasion by two different α-HVs while not affecting viral propagation in epithelial cells, and that robust replication of this mutant in peripheral tissues allowed mice to mount an adaptive immune response against HSV-1, further prompted the idea of using this mutant as a live-attenuated vaccine candidate for HSV-1 and HSV-2 (Richards et al., 2017). Moreover, the progeny of this mutant virus is unable to invade the nervous system, reducing the possibility of latency and reactivation.

Since α-HVs are pantropic and can infect various cell types with mixed or unipolar microtubule polarity, perhaps, the inner tegument proteins evolved the ability to recruit both minus- and plus-end directed motors. These motors may engage in a tug-of-war to determine the direction of net transport. Alternatively, their action may be regulated by host or viral kinases (e.g., US3) or other viral tegument proteins (e.g., UL36 and UL37). The dual recruitment of kinesin and dynein motors and regulation of their activity might also coordinate the nuclear delivery of viral capsids from MTOC to the nuclear pore complexes in the soma (Dohner et al., 2002) (Figure 1A).

Local Control of Retrograde Infection

How do viral particles reach the neuronal cell bodies to establish productive infection after axonal invasion? The answer involves events in the initial virion-host interaction and in the way viral proteins engage cellular machineries. For example, when isolated axons in compartmented neuronal chambers are infected with a large number of PRV virions, translation of local axonal mRNAs is stimulated. This leads to local synthesis of several axonal proteins (Koyuncu et al., 2013b). When axonal protein synthesis is inhibited by cycloheximide or emetine, the efficiency of the retrograde capsid transport is reduced. Some of these newly made neuronal proteins, including a dynein regulator, Lis1, are indeed crucial for efficient virus transport toward the cell bodies (Koyuncu et al., 2013b).

Induction of new protein synthesis in neuronal axons and dendrites has been a focus of research and debate in neurobiology since the early 2000s (Jung et al., 2012). It is now known that in PNS neurons, axonal protein synthesis is essential for growth cone navigation, damage repair, and communication with the distant cell body. Some of the proteins synthesized locally engage signaling molecules transported to the cell body where they act as messengers to alarm cell bodies in case of distal axonal injury (Hanz et al., 2003; Perry et al., 2012).

Interestingly, when isolated axons are infected with a large number of PRV virions, a mechanism similar to the axonal damage response is triggered. Both injury and infection activate local protein synthesis and fast transport of signaling molecules toward neuronal nuclei. In the case of PRV axonal infection, such action leads to more efficient transport and subsequent productive infection in cultured neurons (Koyuncu et al., 2013b). If the infecting viral particle amount is below a certain threshold (below 10^3 plaque forming units [pfu] for PRV), stimulation of local protein synthesis and retrograde injury signaling is poor resulting in silenced (quiescent) infection in the neuronal nuclei (Koyuncu et al., 2015).

As there is a minimum threshold for efficient retrograde transport in axons and initiation of productive infection in the soma, there is also a limit in the number of viral particles that can be transported in axons due to the availability of host proteins. If too many particles are invading axons (e.g.10^6 pfu and above), only a fraction of the viral capsids will reach the neuronal nuclei to start infection (Koyuncu et al., 2015). When isolated axons of SCG neurons in compartmented chambers were co-infected with 10^4 pfu PRV and 10^6 pfu of UV-inactivated-PRV, the competition for the host machinery and excess amounts of inactivated particles interfered with the retrograde transport of replication-competent PRV and led to a quiescent infection in the neuronal nuclei. (Koyuncu et al., 2017; Koyuncu et al., 2015). These experiments clearly show that the number of viral particles infecting axons and reaching the neuronal soma directly affects whether the infection will be productive or quiescent.

In a natural α-HV infection, axons innervating the peripheral epithelia are exposed to the inflammatory milieu of infected epithelial cells even before the onset of neuronal invasion by the progeny virions. Because PNS neurons do not divide and are irreplaceable, they must have a high threshold for apoptosis or other modes of cell death that nonneuronal cells often use for antiviral defense. How does this complicated environment rich in cytokines affect axonal infection, particle transport, and productive infection thresholds? Song et al. investigated these questions by treating isolated axons in compartmented chambers with type I (α/β) or II (γ) interferon (IFN) before infecting them with PRV or HSV-1 (Song et al., 2016). These studies revealed that when axons were treated with either IFNβ or IFNγ, the efficiency of PRV and HSV-1 axonal transport was reduced by two different modes of local interferon responses in axons. Exposure of axons to IFNγ induced transcription of antiviral genes in the distant neuronal nuclei to further block viral replication. By contrast, treatment of axons with IFNβ did not activate transcription of antiviral genes in the cell bodies but instead activated STAT1 phosphorylation in axons by an unknown kinase (Song et al., 2016). These local antiviral responses in axons limited but did not completely block capsid transport to the neuronal nucleus. Since the productive infection of SCG neurons requires that a high number of capsids reach the cell body, the local cytokine response in axons may contribute to the establishment of a latent infection in the ganglia cell bodies by interfering with particle transport resulting in a lower number of capsids reaching the nuclei.

Besides affecting retrograde transport in axons, both type I and type II IFNs limit replication and spread of α-HVs in many cells by inducing the transcription of a multitude of interferon-stimulated genes (ISGs) (Schoggins, 2014). In the sensory ganglia of latently infected mice, interferons and other cytokines secreted by infiltrating T-cells, contribute to the establishment and maintenance of latency (Enquist and Leib, 2017; Leib et al., 1999; Mikloska and Cunningham, 2001; St Leger and Hendricks, 2011). Using porcine dissociated TG neurons cultured in a two-compartment setup, De Regge et al. demonstrated that pretreatment of neuronal cell bodies with IFNα suppressed productive infection of axonally infecting α-HV leading to a stably silenced infection similar to latency (De Regge et al., 2010). Not only do IFNβ and IFNγ promote establishment of latency, but they also block HSV-1 reactivation by interfering with an early step in the process (Linderman et al., 2017). Interference is an early event because both types of interferon fail to block reactivation when they are introduced at a later phase. Recently, type III interferons (interferon lambda; IFNλ) have been shown to restrict both HSV-1 and HSV-2 replication by inducing the JAK/STAT pathway and ISG expression (Li et al., 2011; Li et al., 2017).

Productive vs. Quiescent: Decision-making step in the neuronal nucleus

Axonal biology and long-distance transport of capsids influence the infection mode of α-HVs in neurons. Several bottlenecks (e.g. long-distance transport in axons) or thresholds (e.g., enough virus particles to induce injury signaling, protein synthesis and to engage retrograde transport machinery in axons) must be overcome to reach and establish infection in the neuronal nucleus (Koyuncu et al., 2018). This long sequence of events leading to the viral latency in the neuronal nucleus is challenging to dissect due to the difficulty of monitoring early events. The combined use of cell culture models along with *in vivo* animal models revealed that the decision to either replicate lytically or enter latency depends on outer tegument proteins, particularly on the fate of the VP16 (aka UL48) protein. VP16 interacts with host transcription factors to activate viral gene expression and thus to initiate productive infection (Campbell et al., 1984; Wilson et al., 1993; Wysocka and Herr, 2003). However, being an outer tegument protein, VP16 is not co-transported with capsids to the neuronal nucleus. Only a few retrogradely trafficking HSV-1 capsids co-localized with VP16 in axons, and no co-transport was observed during retrograde PRV capsid transport (Antinone et al., 2006; Antinone and Smith, 2010). It is still unclear whether VP16 in the tegument undergoes retrograde transport in axons, alone or with other outer tegument proteins, or not (Figure 1A). Nevertheless, if outer tegument proteins are transported separately from the viral capsids in axons, their asynchronous arrival at the cell bodies could bias the infection mode toward latency (Hafezi et al., 2012; Roizman and Sears, 1987).

By using an organotypic model established by culturing chicken trigeminal ganglia explants (TGE) in a double chamber where the two compartments are separated by a cloning cylinder ring, Hafezi et al., tested the long-standing hypothesis of separate long-distance transport of capsids and outer tegument proteins in axons leading to a quiescent infection (Hafezi et al., 2012). When the ganglionic compartment was infected with HSV-1, a productive infection was observed. However, when axons were infected with HSV-1, a quiescent infection was established in a small number of neurons. These neurons expressed the latency associate transcript (LAT) characteristic of a latent infection. These data clearly showed that axonal delivery of α-HV particles, favors establishment of a quiescent state of infection in the ganglia (Hafezi et al., 2012; Roizman and Sears, 1987).

By using SCG neurons grown in compartmented chambers that have three compartments (tri-chambers), Koyuncu et al. showed that axonally infecting low dose PRV also establishes quiescent infections that can be reactivated (Koyuncu et al., 2015). Using this model, they further uncovered two separate pathways that can trigger a shift in the mode of neuronal infection from silenced to productive: a slow, stress-mediated pathway and a fast, viral tegument-mediated pathway (Koyuncu et al., 2017). The stress-mediated pathway involves elevated cyclic AMP (cAMP), which subsequently activates protein kinase A (PKA) in the neuronal soma. If PKA was activated in the neuronal cell bodies at the same time that axons were infected with PRV at a pfu that normally results in a quiescent infection, viral genomes were not silenced. This effect was dependent on the activity of PKA and the stress signaling kinase, cJun N-terminal kinase (JNK) (Koyuncu et al., 2017). Remarkably, exposure of cell bodies to viral light particles that contain viral tegument proteins, but lack capsids, was sufficient to rapidly bypass the silencing of axonally infecting PRV genomes, which was not dependent on active PKA or JNK. This finding suggested that tegument-stimulated escape from silencing did not involve a stress activated escape pathway (Koyuncu et al., 2018; Koyuncu et al., 2017).

Anterograde axonal sorting and transport

α-HVs are unique in their capacity to spread in a bidirectional fashion between chains of synaptically connected neurons (Figure 1B). Both post-entry retrograde transport and post-replication anterograde transport, are highly regulated processes relying on different host microtubule-dependent motors rather than on free diffusion (Enquist, 2012; Koyuncu et al., 2013a; Miranda-Saksena et al., 2018; Radtke et al., 2010; Smith, 2012). While viral replication and capsid assembly occur in the nucleus, final tegument acquisition, followed by secondary envelopment at trans-Golgi-like membranes, yields mature virions in cytoplasmic vesicles (Bigalke and Heldwein, 2016; Ibáñez et al.,

2018; Smith, 2012). The direction taken by virion and virion components after neuronal infection, as well as the degree of spread within and between connected neurons, dictates the difference between a latent, reactivating PNS infection and a lethal CNS infection.

Post-replication anterograde transport occurs after episodic reactivation from latency. It is important to separate two distinct events that dictate direction of spread: sorting into axons and subsequent anterograde transport of viral cargo to axonal egress sites. For virion or virion components to be sorted into axons, they need to cross the axon initial segment (AIS). The AIS is a mesh of actin filaments that act like a selective gate, preventing physiological cargo such as transport vesicles from entering the distal axon in a non-specific fashion. Certain microtubule-dependent motors belonging to the kinesin-1 and kinesin-3 families are able to penetrate the AIS and they may provide axonal sorting specificity (Leterrier, 2018).

A controversy exists as to whether α-HVs assemble mature virions in vesicles in the cell body before axonal sorting and transport (married model), or if virion components and subassemblies are sorted independently and then transported to distant sites such as axon termini to complete assembly prior to egress (separate model) (Buch et al., 2017). For PRV, sufficient evidence supports the married model. For HSV-1, some data supports married, separated, and even a mixed model combining the elements of the first two models; however, recent research presents stronger evidence for the married model for HSV-1 as well. This debate has been detailed elsewhere (Antinone et al., 2010; Kratchmarov et al., 2012; Miranda-Saksena et al., 2000; Miranda-Saksena et al., 2009; Negatsch et al., 2010; Snyder et al., 2006; Taylor et al., 2012; Wisner et al., 2011). Regardless of the precise site of virion assembly in neurons, α-HVs display a highly regulated mechanism to sort and transport virion components in axons, facilitating the spread from presynaptic to postsynaptic neurons (Card et al., 1990; Holland et al., 1999; Salinas et al., 2010; Smith, 2012; Smith et al., 2001) (Figure 1B).

For HSV-1 and PRV, three conserved viral proteins located in the viral envelope, gI, gE, and US9, are involved in axonal sorting. Although the mechanism of axonal sorting is not fully understood, gI, gE and US9 could potentially change the balance of microtubule-dependent motor recruitment to increase the yield of viral sorting into axons. Mutant viruses lacking these proteins are unable to sort viral cargo into axons, and infection cannot spread to post-synaptically connected neurons (Awasthi and Friedman, 2016; Brideau et al., 2000; DuRaine et al., 2017; Howard et al., 2013; Husak et al., 2000; Kratchmarov et al., 2013; Polcicova et al., 2005; Snyder et al., 2008). In PRV, US9-null mutants have a strong defect in axonal sorting of virion components (Brideau et al., 2000; Lyman et al., 2007), but these mutants have no defects either in retrograde axonal transport or during infection of non-polarized cells. More recently, Daniel et el., reaffirmed that US9 is necessary for sorting viral particles into axons, but dispensable for subsequent axonal transport of particles from the proximal axon to terminals (Daniel et al., 2015). Kramer et al., showed that PRV US9 is co-transported with a microtubule-dependent motor from the kinesin-3 family, called KIF1A, which is involved in axonal sorting and transport of physiological cargo in uninfected neurons. Moreover, they showed that overexpression or disruption of KIF1A function increases or decreases capsid transport, respectively (Kramer et al., 2012). Interestingly, KIF1A was detected to be specifically targeted by the viral anterograde sorting complex for accelerated proteasomal degradation in the late phase of PRV infection (Huang et al., 2020). For HSV-1 and HSV-2, the specific roles of gI, gE and US9 in the mechanism of axonal sorting, transport and anterograde spread are still under investigation. For HSV-1, Diefenbach et al., reported that the basic domain of US9 can recruit KIF1A to facilitate egress from neurons (Diefenbach et al., 2016). Another related non-essential protein called UL56, which like US9 is a type-II membrane protein, has been shown to be a virulence factor in α-HVs (Berkowitz et al., 1994; Daniel et al., 2016). While in HSV-2, UL56 appears to associate with KIF1A (Koshizuka et al., 2005), in PRV, unlike US9, UL56 does not contribute to axonal sorting (Daniel et al., 2016). More recently, DuRaine et al., reported that motor proteins from the kinesin-1 family such as KIF5A, -5B, and -5C but not kinesin-3 proteins KIF1A and

KIF1B, promote anterograde transport of HSV-1 in axons (DuRaine et al., 2018). It is plausible to speculate that KIF1A could be essential for axonal sorting of viral components and KIF5 could play a role in axonal transport of virions. For VZV, gI and gE are critical neurovirulence factors and gI, in particular, is essential for viral spread in DRG and facilitating axonal localization of structural virion components (Christensen et al., 2013). The mechanism of axonal sorting and anterograde transport of α-HVs is not fully understood and still under ongoing investigation. It is clear that US9, gI and gE are important effector proteins but their specific role and relevance might be different among species.

II. Interneuronal spread

Neuroinvasion, neurovirulence and disease phenotype

We define a neuroinvasive virus as one with the capacity to invade the nervous system after infection in the periphery of the host. Many viruses can infect and spread within neurons after direct inoculation in the CNS, but these viruses often are not naturally neuroinvasive. By this definition, α-HVs are neuroinvasive. Although in the natural host they usually cause mild symptoms or are asymptomatic, they can cause serious disease if spread of infection is not contained by host defenses.

Neurovirulence is the degree of pathogenesis observed in the host nervous system after neuroinvasion. To be neurovirulent, a virus has to be neuroinvasive. Neurovirulence can range from extreme morbidity and mortality to more subtle effects on neuronal function and host behavior. The degree of neurovirulence depends on the interaction of viral gene products with host defensive actions. Mutant viruses that are neuroinvasive but have reduced virulence, not only help define critical viral gene products promoting pathogenesis but can also be used to study the nervous system (e.g., as neuronal circuit tracers), as we will discuss in section III. Several viral gene products define the neurovirulence range of α-HV strains in animal models, from highly attenuated to extremely neurovirulent and lethal. The identification of viral genes affecting neuroinvasion and neurovirulence properties of α-HVs has been studied primarily in various susceptible animal models using different routes of infection (Card and Enquist, 1995; Enquist et al., 1998). These models have been essential to dissect the interplay between virus replication and spread and the host's defenses.

Attenuated mutant viruses have been used as vaccines candidates, gene delivery vectors for gene therapy, and for oncolytic virotherapy. There is an extensive list of neuroinvasion and neurovirulence genes encoded by α-HVs including but not limited to RL1 (ICP34.5), UL34 (thymidine kinase), UL36 (VP1/2), UL37, UL41 (VHS), UL56, and US12 (ICP47) (Brandt et al., 2003; Chou et al., 1990; Enquist et al., 1998; Kolb et al., 2016; Lee et al., 2009; Richards et al., 2017; Rosen-Wolff et al., 1991).

Paradoxically, α-HV infections in natural hosts are normally highly neuroinvasive but with low neurovirulence. Under some conditions or in different host species, neurovirulence can be extreme. Effective intrinsic and innate immune responses are paramount to contain viral spread to the site of primary infection. If the infection is not contained quickly, it can spread to capillary beds and lymphatic organs with the risk of further viral dissemination. Cellular pattern recognition receptors (PRR) such as Toll-like receptors (TLR) and pathogen-associated molecular patterns (PAMPs), can readily detect viral components after cell attachment and entry, triggering the production of many antiviral proteins and cytokines such as IFN-α/β (Shayakhmetov et al., 2010). The action of these cytokines in cells that have appropriate receptors usually causes cell death.

If this intrinsic response in epithelial cells is insufficient to limit viral replication, the innate immune response will be amplified leading to the production of more cytokines, complement, natural killer cells (NK), neutrophils and additional granulocytes. There is a fine balance between an appropriate immune response for efficient control of viral replication and spread, and an uncontrolled innate response that eventually can cause more damage than the acute viral infection, often described as a "cytokine storm" (Enquist et al., 1998;

Koyuncu et al., 2013a). Both the virus and the natural host have co-evolved to balance offensive and defensive actions. The fact that α-HVs can establish life-long, reactivatable infections, demonstrate that they establish sophisticated antiviral responses that can be both tissue specific and time-dependent. At least five viral proteins can module the host immune response to favor a proviral rather than an antiviral state: ICP0 for example, can target interferon-inducible protein 16 (IFI16), a host viral DNA sensor for degradation. Viral proteins ICP34.5, ICP27, VHS, and US3, can directly or indirectly inhibit the host proinflammatory nuclear factor kappa-light-chain-enhancer of activated B cell (NF-κB) antiviral response (Knipe, 2015; Ma and He, 2014).

Two Modes of PRV Pathogenesis and the Role of Immune Response

α-HVs are pantropic and neuroinvasive (Goodpasture and Teague, 1923a, b). Once in the nervous system, infection spreads among synaptically connected neurons over long distances. The extent and direction of such spread determines the degree of α-HV pathogenesis. PRV has served as a model for studying the interneuronal spread of infection and pathogenic consequences of α-HV infections in animal models. Virulent PRV infection (e.g., PRV Becker) produces a severe peripheral neuropathy leading to pruritus and self-mutilation (known as the "mad itch"), as well as loss of motor coordination and ataxia. Infection of non-natural hosts with virulent PRV is lethal (Laval and Enquist, 2020; Pomeranz et al., 2005). PRV Bartha is a vaccine strain obtained by extensive passaging in the laboratory and is highly attenuated, producing minimal symptoms (Lomniczi et al., 1984; Mettenleiter et al., 1985).

Brittle, Reynolds and Enquist, employed a mouse flank infection model in which the infection is initiated in epithelial tissue, followed by spread of infection to the PNS, with the potential for further progression into the CNS. This model facilitated comparison of the spread and pathogenesis of the virulent PRV strain with the attenuated PRV strain, Bartha (Brittle et al., 2004). The PRV Bartha genome harbors a deletion in the unique short (US) region that removes the gE, gI and US9 genes- all required for sorting virions into axons (Bartha, 1961; Szpara et al., 2011). Virulent PRV infection of a single site on the mouse flank results in frantic scratching and biting of the skin at the inoculation site, as well as the corresponding dermatome (pruritus). These mice progressively became weaker and died in approximately 3 days with no detectable behavioral abnormalities indicative of CNS pathology. In contrast, mice inoculated with PRV Bartha showed virtually no symptoms (Brittle et al., 2004). These animals survived nearly three times longer than those infected with wild-type PRV. Interestingly, they neither showed any signs of pruritus, nor did they develop skin lesions, despite viral replication in the skin and corresponding ganglia. However, late in infection (6 days), PRV Bartha-infected mice developed severe neurological abnormalities such as ataxia and died approximately in 10 days with abundant virus in brain tissue. This model, which emulates a naturally occurring α-HV infection route, identified two distinct modes of pathogenesis and lethality depending on the virulence (the extent of disease) and invasiveness (the spread of infection) of the infecting virus strain. Importantly, these differences between PRV Becker and PRV Bartha reflected the functions of specific PRV-encoded genes essential for pathogenesis. However, the specific pathophysiological and immunological processes leading to the development of the characteristic neuropathic itch and the quick death of the PRV Becker infected animals remained unclear.

To elucidate the mechanism of pruritus, Granstedt et al., used an *ex-vivo* live-cell imaging of infected submandibular ganglia (SMG) and showed that PRV Becker infection, but not PRV Bartha infection, travels back-and-forth from the salivary glands and the PNS neurons innervating the gland (Granstedt et al., 2013a; Granstedt et al., 2013b). This 'round-trip' infection leads to a massive infection of peripheral neurons, which may be responsible for the severe peripheral neuropathy induced by virulent PRV but not attenuated strains (e.g. PRV Bartha). Furthermore, PRV Becker-infected neurons exhibited increased action potential (AP) rates and also became electrically coupled, which means firing synchronously (McCarthy et al., 2009). When neurons were infected with a PRV Becker mutant that could not express glycoprotein B (gB; required for membrane fusion during viral entry and cell-

cell spread), AP rates were not increased. Changes in electrical connectivity were not observed, and membrane fusion between adjacent neurons (syncytia) did not occur. These results suggest that the infection-induced elevated activity and electrical coupling result from virally induced membrane fusion and this may underlie neurological symptoms observed during infection, in particular pruritus. Granstedt et al., further showed that PRV Bartha and PRV US9 null mutants do not induce increased firing in this model.

In primary SCG neuronal cultures, as well as in differentiated PC12 cell cultures, PRV infection leads to a significant increase in intracellular Ca^{2+} accumulation during the late phase of infection (Kramer and Enquist, 2012). At least two viral membrane proteins are responsible for the elevated spontaneous activity of neurons and intracellular Ca^{2+} increase both *in vitro* and *in vivo*: gB and US9. gB mediates membrane fusion during viral entry as well as fusion of infected cells leading to syncytia, and membrane protein US9 (together with glycoproteins gE and gI) promotes axonal sorting of virus particles and glycoproteins into axons (Curanovic and Enquist, 2009; Granstedt et al., 2013b; Kramer and Enquist, 2012; Lyman et al., 2007; McCarthy et al., 2009). These findings lead to a model where viral membrane fusion proteins are transported into the axons of infected neurons, form fusion pores between adjacent non-myelinated axons, and cause the neurons to become electrically coupled. This leads to elevated and synchronous activity propagating through the neuronal network. Whether the observed increase in calcium levels or coupled electrical activity is required for efficient viral spread, is still under investigation.

The reason for the rapid death of virulent PRV infected animals has been explored by Laval et al., These authors studied the local and systemic inflammatory response during PRV infection using a mouse footpad inoculation model, where the infection spreads from PNS to CNS via the sciatic nerve, which enables a more accurate assessment of the kinetics of the immunopathological processes associated with pruritus (Laval et al., 2018). In this model, PRV Becker induced severe pruritus in the foot, and animals became moribund at 82 hours post infection. The authors found epithelial and neuronal cell necrosis, and inflammation with a massive neutrophil infiltration in the footpad and DRGs. Elevated plasma levels of circulating pro-inflammatory cytokines (IL6 and G-CSF) and chemokines (Gro-1 and MCP-1) were detected in Becker-infected mice. Specifically, IL6 and G-CSF levels were increased in several tissues, including the footpad, PNS, CNS and heart at 82 hours post infection. High plasma levels of C-reactive protein (CRP) confirmed the acute inflammatory response to PRV-Becker infection. Surprisingly, the increase in IL6 and G-CSF occurred before viral replication in DRG. Efficient replication and spread in the footpad were sufficient to induce cytokine production by DRG (Laval et al., 2019). By contrast, mice inoculated with PRV-Bartha in this model, did not show any signs of pruritus. Infection did not induce a specific and lethal systemic inflammatory response and, instead, spread further into the brain (Laval et al., 2018). These data clearly demonstrate the impact of inflammation in the clinical manifestations of PRV infection.

The importance of these observations is two-fold: The wild-type strain of PRV induces heavy inflammation in the non-native host which leads to the death of the animal even before the viral particles travel to the CNS. The attenuated strain does not induce inflammation in the non-native host which gives virus particles enough time to reach the CNS. However, the route of neuroinvasion is different from the wild-type virus since the attenuated strain cannot spread in the anterograde direction. Most probably, virus particles reach brain tissue via motor circuits in this case. In the native hosts, α-HVs do not induce a cytokine storm, and their replication, particularly in the PNS is quiet well controlled by the immune system. Even though progeny virus particles may reach the CNS after reactivation episodes, in healthy immunocompetent hosts, the innate immune system of the CNS neurons do not let productive infection of these particles.

Neonatal Herpes simplex encephalitis (HSE)

As noted above, HSV-1 and HSV-2 are highly neuroinvasive but rarely neurovirulent. Nonetheless, rare cases of herpes simplex encephalitis (HSE) occur in humans. The incidence of HSE in infants ranges from 1/500,000 to 5.9/100,000 live births, depending on the cited study (Kropp et al., 2006; Pinninti and Kimberlin, 2018; Whitley, 2006). The routes that HSV uses to invade the brain and cause HSE are not fully understood. HSE can occur either after acute primary infection or after reactivation from latency in individuals previously infected with the virus (Whitley, 2015). Neonatal patients are the most affected by HSV infections, presumably due to immature immune systems, with incidence of systemic infection ranging from 1/3,000 to 1/20,000 births (Kimberlin, 2004). When viral infection affects the CNS, mortality rates are approximately 50% (Kropp et al., 2006; Looker et al., 2017; Pinninti and Kimberlin, 2018; Whitley, 2006).

Neonatal HSV-2 infections are more common than those caused by HSV-1. Vertical infections typically occur during vaginal delivery but may also happen by transplacental transmission or during breastfeeding. Parents or other individuals with orolabial lesions can also infect newborns, and nosocomial transmission from infected hospital workers can occur as well. Strikingly, between 20% and 30% of pregnant women are seropositive for HSV-2, and 2% of neonates will show neonatal disease if mothers are shedding virus at the moment of birth. It is not well known why neonatal transmission is higher in mothers that are acutely infected during pregnancy as opposed to mothers with viral reactivation. It is possible that maternal antibodies cross the placenta and prevent, in some cases, the disease in the newborns. To further complicate the challenge of viral detection in pregnant women and prevention of neonatal transmission, more than 60% of mothers infected during pregnancy show no evident symptoms of genital infection (Kimberlin, 2004). The mortality rates of neonatal HSV infection can be as high as 80%, depending on the type of disease.

Newborns with localized disease can be classified into two groups. The group with the more severe symptoms exhibits CNS infection and HSE with or without infection of the skin, eyes, and/or mouth (SEM). The group with lesser symptoms only exhibits SEM lesions without evidence of CNS invasion, HSE or disseminated disease. Before the antiviral era, most neonates with SEM, progressed to disseminated disease of CNS and/or visceral organs with high morbidity and mortality rates. With antiviral therapy, morbidity rates were reduced from 38% to 2-12% when neonates were treated promptly during the initial SEM phase (Kimberlin, 2004). Finally, newborns with disseminated infection exhibit viral invasion of visceral organs and disseminated intravascular coagulation with or without SEM and CNS disease. Untreated disseminated infection cases have mortality rates as high as 85% (Kimberlin, 2004; Kropp et al., 2006; Poole and Kimberlin, 2018). If left untreated, 60% of patients that survive will have severe and chronic neurological damage, and even with antiviral treatment, mortality is as high as 30%. Prompt parenteral antiviral treatment can decrease mortality rates by 50% (Pinninti and Kimberlin, 2018).

Although acyclovir is the treatment of choice for neonatal HSE, multiple antiviral drugs have shown relative efficacy to treat α-HV infection. All the drugs currently used, except for one called Foscarnet, are nucleoside analogs that are incorporated into the nascent viral DNA, preventing the attachment of additional nucleosides, causing chain termination and therefore reducing viral replication. This group of antivirals includes vidarabine, trifluridine, acyclovir, valacyclovir, ganciclovir, valganciclovir, and cidofovir. Foscarnet on the other hand is a pyrophosphate that inhibits the viral DNA polymerase and can be used in cases of acyclovir resistance (Pinninti and Kimberlin, 2018; Poole and Kimberlin, 2018).

HSE due to inborn mutations affecting the intrinsic immune system

The outcome and severity of α-HV infection in the CNS depends on viral and host gene products (Kropp et al., 2006; Pinninti and Kimberlin, 2018). There is increasing evidence that some cases of human HSE are caused by inborn mutations affecting the host intrinsic immune response. One noteworthy example is the increased incidence of HSE after primary infection with HSV-1 in children with dominant-negative mutations leading to Toll-like receptor 3 (TLR3) deficiency and impaired TLR-dependent IFN-α/β production (Zhang et al., 2007). Importantly, these patients show lesions restricted only to the frontal and temporal lobes, suggesting that this specific mutation affects the capacity of neurons and oligodendrocytes in the forebrain to control and repress HSV-1 replication and spread in these otherwise healthy children (Casrouge et al., 2006; Lafaille et al., 2012). Interestingly, this particular deficiency in the intrinsic immune response seems to be specific in restricting HSV-1 from spreading further in the CNS since TLR3-deficient mice respond normally and are resistant to infection by several other viruses such as phlebovirus, West Nile virus, and influenza A virus (Casrouge et al., 2006; Guo et al., 2011; Guo et al., 2015; Herman et al., 2012; Lafaille et al., 2012; Ott et al., 2017; Perez de Diego et al., 2010; Zhang et al., 2013).

Although there is a strong correlation between TLR3-dependent IFN-α/β deficiency and HSE incidence in children and adults, HSE occurs in patients without these mutations, further complicating the etiology of this rare but exceedingly serious outcome. Signaling molecules such as ICAM-1 (an intercellular adhesion molecule) and ITGA5 (integrin alpha 5) increase in response to HSV-1 infection in otherwise healthy individuals but not in HSE patients, suggesting a TLR3-independent pathway influencing HSE (Engel et al., 2015; Perez de Diego et al., 2013; Sironi et al., 2017).

More recently, the Casanova laboratory has expanded the study of HSE cases unrelated to TLR3 deficiencies causing forebrain infections. Roughly 5% of HSE cases occur due to brainstem infections, and in 50% of these cases no other CNS region is affected by the virus. Again, inborn mutations (in this case, in RNA lariat metabolism) are strongly associated with these instances of HSE (Zhang et al., 2018). Bi-allelic mutations in DBR1 were identified in several unrelated patients with brainstem infections due to HSV-1, influenza or norovirus. DBR1 is the only identified RNA lariat debranching enzyme, which is expressed at high levels in spinal cord and brainstem. The accumulation of intron lariats could impair cell-intrinsic responses and virus recognition resulting in increased susceptibility to viral infection by cells in the brainstem. Apparently, full DBR1 activity is needed for protective antiviral immunity in the brainstem, and not all neurons are equally susceptible to a productive α-HV infection. While cortical neurons are constitutively more resistant to HSV-1 infection even in the absence of antiviral signals, trigeminal ganglia neurons are more prone to get infected because they require a preemptive activation of antiviral immunity via TLR3 or IFN-α/β receptors (Zimmer et al., 2018).

III. α-HVs as Neural Circuit Tracers in Animal Models

α-HVs, especially HSV-1, HSV-2 and PRV, have been used for roughly 50 years to define and manipulate synaptically connected circuits in the nervous system of animal models. This powerful technique relies on the fact that when α-HVs infect neurons and produce infectious progeny, this progeny spread only through chains of synaptically connected neurons (a neuronal circuit), in a process often referred to as transsynaptic spread (Figure 1B). Other viral species such as vesicular stomatitis virus (VSV), rabies virus (RABV), mouse hepatitis virus, and betanodavirus can also spread transsynaptically (Ikenaga et al., 2002; Lanciego and Wouterlood, 2020; Maturana et al., 2020; Nassi et al., 2015; Pomeranz et al., 2005). The use of recombinant viruses as transneuronal tracers of neuronal circuitry is now a widely used approach to define the synaptic organization of neural networks (Table 1). The main advantage is that virus replication and spread delivers a self-amplifying marker of neural connectivity.

The capacity of a virus infection to spread predominantly among chains of synaptically connected neurons is remarkable. Several hundred published manuscripts have demonstrated that α-HVs are unable to spread inter-axonally to physically adjacent but non-connected neural circuits. Moreover, viral spread through the nervous system will only happen via intact and synaptically connected circuits (Card and Enquist, 1995, 2014; Pomeranz et al., 2005). The infected neurons apparently do not release infectious progeny randomly, but only at or near sites of synaptic contact. It is unclear whether viral particles actually spread through a synapse. Careful analysis suggests that spread occurs between sites of close membrane apposition such as synapses or glial-neuronal contacts (Card and Enquist, 1995; Card et al., 1993; Card et al., 1990; Enquist et al., 1994; Levine et al., 1998).

Non-viral tracers have been used for more than 100 years. One of the first successful attempts was made by the Spanish neuroscientist and physician Santiago Ramon y Cajal, that stained and visualized entire neurons with their processes. Ramon y Cajal applied a staining method developed by the Italian physician Camilo Golgi that used silver nitrate and potassium dichromate to impregnate fixed brain slices. He described in great detail in more than 3000 drawings of brain structures, the shape and complexity of the nervous system (Bro and Haycock, 1977; Haycock and Bro, 1975). Ramon y Cajal suggested that the brain works by constantly changing connections between cells, a finding that was disputed by the scientific community at the time. In his memoir, he wrote: "*I expressed the surprise, which I experienced upon seeing with my own eyes the wonderful revelatory powers of the chrome-silver reaction and the absence of any excitement in the scientific world aroused by its discovery" (Katz-Sidlow, 1998).* Santiago Ramon y Cajal and Camilo Golgi received the Nobel Prize in Physiology or Medicine in 1906 for their seminal studies of the structure of the nervous system. Although Ramon y Cajal's findings with the Golgi staining method allowed important morphological advancements, it did not permit transneuronal tracing and demonstration of functional cell connections. Several decades later, scientists reported on the use of many conventional tracers including but not limited to small proteins such as horseradish peroxidase and albumin; inorganic fluorescent molecules such as fast blue, diamidino yellow and fluoro-gold; dextrans such as fluoro-ruby and biotinylated dextran amine; lectins such as wheat germ agglutinin and phaseolus vulgaris-leucoagglutinin; beads like latex microspheres; bacterial toxins such as tetanus and cholera B subunit; trophins like nerve growth factor and brain derived neurotropic factor; amino acids like tritiated-proline, tritiated-leucine and biocytin; and fluorescent long-chain dialkyl-carbocyanines such as DiI and DiO. For additional information on conventional or non-viral tracers see (Boldogkoi et al., 2004; Nassi et al., 2015; Pomeranz et al., 2005). These molecules allowed the basic understanding of nerve damage and regeneration as well as the concept of anterograde and retrograde transport in neurons (Cowan and Cuénod, 1975). Non-viral tracers are still widely used but share some important limitations. Only some of them can cross synapses, and even if they do, they will only label first-order of connected neurons (monosynaptic only, not polysynaptic). Since they do not self-propagate or amplify in the nerve circuit, they become diluted quickly as a function of time and distance from the site of injection. Consequently, they cannot inform on the polysynaptic architecture of neurons committed to a specific function or task. Fortunately for neuroanatomists and neurobiologists, nature has already evolved a great tool with the neurotropic viruses. Nearly a century ago, Goodpasture and Teague (Goodpasture, 1925; Goodpasture and Teague, 1923b) were the first to hypothesize that infectious virions are taken up by nerve endings and then retrogradely transported to the neuronal cell bodies. Fifty years later, neuroscientists were able to show that HSV-1 could spread by transneuronal transfer between and within chains of synaptically connected neurons (Cook and Stevens, 1973; Kristensson et al., 1974; Kristensson et al., 1971; Kristensson et al., 1982; Kristensson et al., 1978).

It was not until the 1980's that α-HVs become widely used for polysynaptic circuit tracing (Card and Enquist, 1995; Card et al., 1991; Dolivo et al., 1979; Hogue et al., 2015; Ugolini et al., 1987). HSV-1, HSV-2 and PRV have similar behavior as circuit tracers. They are able to infect CNS and PNS neurons, including sensory, motor, sympathetic and parasympathetic neurons, in several mammal species. They can complete their replication cycles in less than 12 hours in susceptible and permissive cells, with PRV replicating faster than HSV (Kaplan and Vatter, 1959). Both viruses infect a broad range of animal models including mouse, rat, guinea pig, hamster, gerbil, rabbit and chicken embryo. Moreover, PRV can also infect swine (the natural host), dogs, cats, cattle, sheep, and some species of birds. Unlike HSV, PRV does not infect higher primates such as apes, chimpanzees and humans (Gustafson and Kanitz, 1975; Olander et al., 1966; Saunders and Gustafson, 1964). In Table 1, we provide a comprehensive list of HSV-1 and PRV mutants that express fluorescent reporter proteins used for circuit tracing studies. The polysynaptic tracing properties of α-HVs in susceptible animals are influenced by several factors that will be briefly described below. For detailed descriptions, see (Card and Enquist, 2014; Card et al., 1999; Card et al., 1993; Card et al., 1990; Card et al., 1991; Enquist et al., 1998; Enquist et al., 1994; Wojaczynski et al., 2015).

α-HVs infections typically initiate at the periphery in mucosal epithelia from where they invade the termini of sensory neurons to undergo retrograde transport towards cell bodies. In cell bodies α-HVs can establish lifelong latency and eventually undergo reactivation and anterograde transport of newly made particles towards sites of egress, most likely back out towards the periphery (Mangold et al., 2020).

The most commonly studied laboratory strains of PRV are Becker, Bartha, Kaplan, NIA-2 and NIA-3. HSV-1 laboratory strains include MacIntyre, F, 17syn +, KOS, SC16, H129, FMC, and KJ502 while HSV-2 strains include 186 and G (Enquist et al., 1998; Norgren and Lehman, 1998; Szpara et al., 2014). Most of these α-HVs strains exhibit a wild-type phenotype and will spread in a bidirectional fashion through synaptically connected circuitry. They can also be highly neurovirulent and therefore are not optimal transsynaptic tracers.

The most effective α-HVs tracing strains are neuroinvasive mutants with reduced neurovirulence and unidirectional transsynaptic spread properties as opposed to the wild-type bidirectional spread phenotype (Table 1). Perhaps one of the better and more characterized tracing α-HVs is the attenuated vaccine strain Bartha of PRV (Bartha, 1961). Bartha was obtained after extensive passaging of a virulent PRV isolate obtained from pigs in cultured chicken cells and chicken embryos. The PRV Bartha genome was sequenced by Szpara et al. and compared to the sequence of virulent PRV strains (Szpara et al., 2011). The PRV Bartha genome contains several mutations in genes UL21, UL44 (gC), UL10 (gM), and a small 3.5 kb deletion in the US region of the viral genome resulting in the deletion of US8 (gE) and US9 and disruption of US2 and US7 (gI) that may contribute to the decreased neurovirulence. PRV

Bartha allow longer survival times than a wild-type PRV strain in the infected animal. In the laboratory, PRV Bartha grows to titers above most wild-type PRV strains and is more thermostable. Additionally, it can readily spread through the PNS and the CNS of animals. More importantly, Bartha selectively spread in the retrograde direction through neural circuits, making it an excellent transsynaptic tracing tool (Dolivo, 1980).

Another noteworthy and widely used α-HV polysynaptic tracer is HSV-1 strain H129. This is the only known virus to move preferentially in the anterograde direction trough neural circuits (Wojaczynski et al., 2015). Zemanick et al. showed that H129 was, indeed, transported preferentially in the anterograde direction after injection into the motor cortex of cebus monkeys (Zemanick et al., 1991). Although the genome of H129 was fully sequenced by Szpara et al., the precise relationship between the genotype and the mutant phenotype has not been established (Szpara et al., 2010). Genome sequence analysis determined that more than 50% of genes in H129 genome, contain missense or silent mutations, but no deletions or other null mutations. The retrograde spread defect is most likely caused by multiple genes leading to differences in somatodendritic sorting and egress, or the dendritic versus axonal partition of viral progeny (Wojaczynski et al., 2015).

Over the last twenty years, several fluorescently labelled α-HV tracers have been constructed. Most of them replicate and express gene markers in a constitutive fashion, while others can either replicate and or express transgenes in a conditional fashion, allowing for monosynaptic and polysynaptic tracing studies (Card et al., 2011b; DeFalco et al., 2001; Li et al., 2019; Lo and Anderson, 2011; Wu et al., 2014) (see Table 1 for more details).

Besides choosing the viral strain that betters suits the tracing circuit of interest, it is critical to consider the total amount of infectious virus inoculated. The quantity is often called the "titer" and is typically expressed in plaque-forming units per milliliter (PFU). The titer in PFUs is determined using standard tissue culture cells. Sufficient PFUs must be inoculated to achieve replication and transsynaptic spread in the infected animal. The optimal amounts of virus vary depending on the viral strain, animal species and age, and the circuit of interest. The quantity of virions injected will obviously affect the survival time of the infected animal. A virion is defined as an infectious particle. A parameter often ignored is the "particle to PFU" ratio, which can vary between different viral mutants and even among different stocks of the same recombinant virus. It is always a good idea to inject sufficient PFUs so that at least 50% of the animals are infected (the infectious dose 50 or ID50), assuming that the ID50 has been determined for the strain or mutant used (Card et al., 1995; Card and Enquist, 2014).

The number of synapses and extent of the circuit of study will influence the infection outcome in animal models. The density of neuron terminals at or near the site of inoculation will affect efficiency of virion uptake and possibly the spread kinetics (Card et al., 1999). Conversely different neuronal types can also show disparate permissivity and susceptibility to viral infection as shown by Zimmer et al., (Zimmer et al., 2018). The length of the axons will affect the temporal kinetics of spread and anatomical barriers such as the extracellular matrix at the tissue or organ level might also delay or prevent viral invasion.

Finally, it is important to consider that the spread of infection and subsequent labeling of synaptically connected neurons occur in a sequential fashion. Different neuronal circuits can show differences in the spread kinetics defining first-, second-, and third-order neurons. First order neurons will show signs of cytotoxicity before second or higher order neurons are infected. Furthermore, reactive nonneuronal responses are also faithful indicators of the temporal progression of viral spread through a CNS circuit (e.g. reactive glia surrounding an infected neuron are a clear indication of a late stage infection, while the absence of reactive gliosis is indicative of an earlier stage of infection) (Card and Enquist, 2014; Card et al., 1990; Enquist and Card, 2003; Levine et al., 1998; Wojaczynski et al., 2015).

IV. Discussion

Most acute and persistent viral infections often begin at epithelial or endothelial cell surfaces. Remarkably, many α-HVs efficiently enter the PNS of their natural hosts and establish a quiescent infection with little or no CNS pathogenesis. Their initial peripheral infection stimulates a well-controlled intrinsic and innate immune response, as well as a long-lived adaptive immune response. The α-HV genomes remain quiescent in PNS neurons for the life of their hosts, only occasionally reactivating to produce virions that can re-infect peripheral tissues and spread to other hosts. Perhaps the pro-survival, cytolysis-resistant nature of mature neurons, facilitates in part the establishment of such persistent and reactivatable infections.

PRV and HSV infect most if not all known primary sensory neurons, sympathetic and parasympathetic neurons, and motor neurons in a broad range of susceptible mammalian animal models. The directional spread of viral infection in the NS of the host is an essential aspect of the viral life cycle and pathogenesis. α-HVs such as HSV and PRV can spread efficiently within synaptically connected neural circuits and therefore have been used extensively to define the architecture of the mammalian nervous system (Table 1). Wild-type PRV and HSV-1 will spread in a bidirectional retrograde and anterograde fashion to both presynaptic and postsynaptic connected neurons (Figure 1B). Replicating PRV and HSV behave as self-amplifying tracers that will faithfully spread to higher-order neurons without decay or dilution of the reporter transgene expressed by the virus.

The interplay between viral and neuronal proteins, starting from the invasion of the nervous system leading to latent infections in the PNS ganglia, is complex and difficult to dissect using common cell culture techniques and currently available animal models. Therefore, many questions still remain concerning the temporal dynamics of how genome silencing in the neuronal nucleus is established in the first place, what controls reactivation cycles, and what happens to neurons after reactivation episodes. Another poorly understood issue is whether reactivating virus particles reach the CNS tissue to establish latent infections and whether subsequent reactivations in the CNS can lead to neuroinflammation predisposing the tissue for other neurodegenerative diseases such as Alzheimer`s disease. Detailed molecular investigation of α-HV infected primary PNS and CNS neurons is required to understand fundamentals of herpesvirus induced neurodegeneration.

Table 1. List of commonly used fluorescently labeled α-HV tracing recombinants.
The list includes 29 viral recombinants based on PRV Becker, PRV Bartha and HSV-1 H129. PRV Becker is a wild-type highly virulent laboratory strain that undergoes retrograde and anterograde transport in neural circuits. We only include one attenuated IE180 null and US9 null Becker recombinant that is unable to replicate and spread. PRV Bartha is an attenuated vaccine strain widely used for retrograde transneuronal tracing. Bartha infected animals live several days longer without significant symptoms compared to PRV Becker. HSV-1 strain H129 is widely used for anterograde transneuronal tracing. Transsynaptic spread of H129 is significantly more efficient in the anterograde direction than in the retrograde direction. For each viral recombinant we provide the name, strain (Becker, Bartha, H129), a brief description of the recombinant characteristics, the spread phenotype (retrograde "R", anterograde "A", anterograde mainly "A mainly", conditional spread "Cond"), and the corresponding reference.

Virus	Strain	Description	Spread	Ref
PRV-151	Becker	CMV-EGFP reporter gene cassette inserted into the gG locus	R & A	(Demmin et al., 2001)
PRV-180	Becker	mRFP-UL35 (capsid) fusion protein	R & A	(del Rio et al., 2005)
PRV-GS443	Becker	EGFP-UL35 (capsid) fusion protein	R & A	(Smith et al., 2001)
PRV-950	Becker	mTurquoise2-UL35 (capsid) fusion protein	R & A	(Hogue et al., 2014)
PRV-959	Becker	mNeonGreen-UL35 (capsid) fusion protein	R & A	(Koyuncu et al., 2015)
PRV-960	Becker	mCherry-UL35 (capsid) fusion protein	R & A	(Hogue et al., 2015)
PRV-181	Becker	mRFP-UL35 (capsid) and GFP-VP22 (tegument) fusion proteins	R & A	(del Rio et al., 2005)
PRV-823	Becker	Us3 null mutant, expressing mRFP-UL35 (capsid) fusion protein. Less virulent than Becker but not as attenuated at PRV Bartha	R & A	(Olsen et al., 2006)
PRV-833	Becker	US3 null mutant, expressing mRFP- UL35 (capsid) fusion protein and eGFP-VP22 (tegument) fusion protein. Less virulent than Becker but not as attenuated at PRV Bartha	R & A	(Olsen et al., 2006)

PRV IE180-null-hSyn-Cre	Becker	IE180 and US9 null mutant with hSyn-Cre reporter gene cassette replacing the US9 locus. There will be long-term constitutive Cre expression but no viral replication or spread will occur unless trans-complemented with IE180	Cond R	(Oyibo et al., 2014; Wu et al., 2014)
PRV-152	Bartha	CMV-EGFP reporter gene cassette inserted into the gG locus	R	(Demmin et al., 2001; Smith et al., 2000)
PRV-290	Bartha	CMV-mTurquoise2 reporter gene cassette inserted into the gG locus	R	(Hogue et al., 2018)
PRV-614	Bartha	CMV-mRFP reporter gene cassette inserted into the gG locus	R	(Banfield et al., 2003)
PRV-Ip298	Bartha	mCherry reporter gene cassette fused to UL23 (TK) ,	R	(Schneeberger et al., 2019)
PRV-263	Bartha	CMV-Brainbow 1.0L reporter gene cassette inserted into the gG locus. Default expression of tdTomato reporter unless a Cre-expressing cell is infected where there is expression EYFP and mCerulean	R	(Kobiler et al., 2010)
PRV-267	Bartha	CMV-Cre reporter gene cassette inserted into the gG locus	R	(Card et al., 2011a)
PRV-2001	Bartha	UL23 (TK) null mutant with CMV-LoxP-STOP-LoxP-Tau-EGFP-IRES-TK reporter gene cassette inserted into the gG locus. Viral replication and EGFP expression is conditional upon infecting Cre-expressing cells	Cond R	(DeFalco et al., 2001)
PRV-Introvert-GFP	Bartha	UL23 (TK) null mutant with TK1-intron-LoxN-Lox2272-intron-GFP1-2A-TK2-intron-LoxN-Lox2272-intron-GFP2 reporter gene cassette inserted into the TK locus and driven by the native TK promoter. Viral replication and EGFP expression is conditional upon infecting Cre-expressing cells	Cond R	(Pomeranz et al., 2017)

PRV-Introvert-mCherry	Bartha	UL23 (TK) mutant with TK1-intron-LoxN-Lox2272-intron-mCherry1-2A-TK2-intron-LoxN-Lox2272-intron-mCherry2 reporter gene cassette inserted into the TK locus and driven by the native TK promoter. Viral replication and mCherry expression is conditional upon infecting Cre-expressing cells	Cond R	(Pomeranz et al., 2017)
PRV-B180	Bartha	UL23 (TK) mutant with DIO-CMV-TK-GFP reporter gene cassette inserted into the gG locus. Viral replication and EGFP expression is conditional upon infecting Cre-expressing cells	Cond R	(Hanchate et al., 2020)
HSV-414	H129	CMV-EYFP reporter gene cassette inserted into the UL37/UL38 locus	A mainly	(Wojaczynski et al., 2015)
HSV-424	H129	mRFP-UL35 (capsid) fusion protein	A mainly	(Wojaczynski et al., 2015)
HSV-772	H129	CMV-EGFP reporter gene cassette inserted into the UL26/UL27 locus	A mainly	(Wojaczynski et al., 2015)
HSV-774	H129	mRFP-UL35 (capsid) fusion protein, and CMV-EGFP reporter gene cassette inserted into the UL26/UL27 locus	A mainly	(Engel)
HSV-373	H129	CMV-mCherry reporter gene cassette inserted into the UL26/UL27 locus	A mainly	(Engel)
HSV-768	H129	CMV-Brainbow 1.0L reporter gene cassette inserted into the UL26/UL27 locus. Default expression of tdTomato reporter unless a Cre-expressing cell is infected where there is expression EYFP and mCerulean	A mainly	(Engel)
HSV-floxed	H129	CMV-LoxP-EGFP-pA-LoxP-tdTomato-pA reporter gene cassette inserted into the UL26/UL27 locus. Default expression of EGFP unless a Cre-expressing cell is infected where there is expression of tdTomato	A mainly	(McGovern et al., 2015)

HSV-ΔTK-TT	H129	UL23 (TK) null mutant with CAG-LoxP-STOP-LoxP-tdTomato-2A-TK reporter gene cassette inserted into the TK locus. Replication, spread and tdTomato expression is conditioned to infecting a Cre-expressing cell	Cond A mainly	(Lo and Anderson, 2011)
H129-ΔTK-tdT	H129	UL23 (TK) null mutant with CMV-tdTomato reporter gene inserted into the TK locus. Constitutive expression of tdTomato but no viral spread unless trans-complemented with TK	Cond A mainly	(Zeng et al., 2017)

V. Web Resources

The Center for Neuroanatomy with Neurotropic Viruses (CNNV), is a national resource supported by the NIH to develop improved transneuronal tracing technology and make it available to investigators. The CNNV offers a comprehensive collection of viral recombinants that can be found in the following link: http://www.cnnv.pitt.edu/resources.htm

References

Amlie-Lefond, C., and Gilden, D. (2016). Varicella Zoster Virus: A Common Cause of Stroke in Children and Adults. J Stroke Cerebrovasc Dis *25*, 1561-1569. 10.1016/j.jstrokecerebrovasdis.2016.03.052.

Antinone, S.E., Shubeita, G.T., Coller, K.E., Lee, J.I., Haverlock-Moyns, S., Gross, S.P., and Smith, G.A. (2006). The Herpesvirus capsid surface protein, VP26, and the majority of the tegument proteins are dispensable for capsid transport toward the nucleus. Journal of virology *80*, 5494-5498. 10.1128/JVI.00026-06.

Antinone, S.E., and Smith, G.A. (2010). Retrograde axon transport of herpes simplex virus and pseudorabies virus: a live-cell comparative analysis. Journal of virology *84*, 1504-1512. JVI.02029-09 [pii]
10.1128/JVI.02029-09.

Antinone, S.E., Zaichick, S.V., and Smith, G.A. (2010). Resolving the assembly state of herpes simplex virus during axon transport by live-cell imaging. Journal of virology *84*, 13019-13030. 10.1128/JVI.01296-10.

Awasthi, S., and Friedman, H.M. (2016). Molecular association of herpes simplex virus type 1 glycoprotein E with membrane protein Us9. Archives of virology *161*, 3203-3213. 10.1007/s00705-016-3028-z.

Banfield, B.W., Kaufman, J.D., Randall, J.A., and Pickard, G.E. (2003). Development of pseudorabies virus strains expressing red fluorescent proteins: new tools for multisynaptic labeling applications. Journal of virology *77*, 10106-10112.

Bartha, A. (1961). Experiments to reduce the virulence of Aujeszky's virus. Magyar Allatorvosok Lapja *16*.

Berkowitz, C., Moyal, M., Rosen-Wolff, A., Darai, G., and Becker, Y. (1994). Herpes simplex virus type 1 (HSV-1) UL56 gene is involved in viral intraperitoneal pathogenicity to immunocompetent mice. Archives of virology *134*, 73-83.

Bertke, A.S., Ma, A., Margolis, M.S., and Margolis, T.P. (2013). Different mechanisms regulate productive herpes simplex virus 1 (HSV-1) and HSV-2 infections in adult trigeminal neurons. Journal of virology *87*, 6512-6516. 10.1128/JVI.00383-13.

Bigalke, J.M., and Heldwein, E.E. (2016). Nuclear Exodus: Herpesviruses Lead the Way. Annu Rev Virol *3*, 387-409. 10.1146/annurev-virology-110615-042215.

Boldogkoi, Z., Sik, A., Denes, A., Reichart, A., Toldi, J., Gerendai, I., Kovacs, K.J., and Palkovits, M. (2004). Novel tracing paradigms--genetically engineered herpesviruses as tools for mapping functional circuits within the CNS: present status and future prospects. Prog Neurobiol *72*, 417-445. 10.1016/j.pneurobio.2004.03.010.

Bottcher, S., Maresch, C., Granzow, H., Klupp, B.G., Teifke, J.P., and Mettenleiter, T.C. (2008). Mutagenesis of the active-site cysteine in the ubiquitin-specific protease contained in large tegument protein pUL36 of pseudorabies virus impairs viral replication in vitro and neuroinvasion in vivo. Journal of virology *82*, 6009-6016. 10.1128/JVI.00280-08.

Brandt, C.R., Kolb, A.W., Shah, D.D., Pumfery, A.M., Kintner, R.L., Jaehnig, E., and Van Gompel, J.J. (2003). Multiple determinants contribute to the virulence of HSV ocular and CNS infection and identification of serine 34 of the US1 gene as an ocular disease determinant. Invest Ophthalmol Vis Sci *44*, 2657-2668.

Brideau, A.D., Card, J.P., and Enquist, L.W. (2000). Role of pseudorabies virus Us9, a type II membrane protein, in infection of tissue culture cells and the rat nervous system. Journal of virology *74*, 834-845.

Brittle, E.E., Reynolds, A.E., and Enquist, L.W. (2004). Two modes of pseudorabies virus neuroinvasion and lethality in mice. Journal of virology *78*, 12951-12963. 10.1128/JVI.78.23.12951-12963.2004.

Bro, S., and Haycock, J.W. (1977). Visual cortex (translation of S. Ramon y Cajal) translated from L'Encorce Cerebrale Suivant les Regions L'Ecorce Visuelle, Chapter 25 in "Histologie du Systeme Nerveux de l'Homme et des Vertebres," 1911. Behav Biol *21*, 508-528.

Buch, A., Muller, O., Ivanova, L., Dohner, K., Bialy, D., Bosse, J.B., Pohlmann, A., Binz, A., Hegemann, M., Nagel, C.H., *et al.* (2017). Inner tegument proteins of Herpes Simplex Virus are sufficient for intracellular capsid motility in neurons but not for axonal targeting. PLoS pathogens *13*, e1006813. 10.1371/journal.ppat.1006813.

Campbell, M.E., Palfreyman, J.W., and Preston, C.M. (1984). Identification of herpes simplex virus DNA sequences which encode a trans-acting polypeptide responsible for stimulation of immediate early transcription. Journal of molecular biology *180*, 1-19.

Card, J.P., Dubin, J.R., Whealy, M.E., and Enquist, L.W. (1995). Influence of infectious dose upon productive replication and transynaptic passage of pseudorabies virus in rat central nervous system. J Neurovirol *1*, 349-358.

Card, J.P., and Enquist, L.W. (1995). Neurovirulence of pseudorabies virus. Crit Rev Neurobiol *9*, 137-162.

Card, J.P., and Enquist, L.W. (2014). Transneuronal circuit analysis with pseudorabies viruses. Current protocols in neuroscience / editorial board, Jacqueline N Crawley [et al] *68*, 1 5 1-1 5 39. 10.1002/0471142301.ns0105s68.

Card, J.P., Enquist, L.W., and Moore, R.Y. (1999). Neuroinvasiveness of pseudorabies virus injected intracerebrally is dependent on viral concentration and terminal field density. J Comp Neurol *407*, 438-452.

Card, J.P., Kobiler, O., Ludmir, E.B., Desai, V., Sved, A.F., and Enquist, L.W. (2011a). A dual infection pseudorabies virus conditional reporter approach to identify projections to collateralized neurons in complex neural circuits. PloS one *6*, e21141. 10.1371/journal.pone.0021141.

Card, J.P., Kobiler, O., McCambridge, J., Ebdlahad, S., Shan, Z., Raizada, M.K., Sved, A.F., and Enquist, L.W. (2011b). Microdissection of neural networks by conditional reporter expression from a Brainbow herpesvirus. Proceedings of the National Academy of Sciences of the United States of America *108*, 3377-3382. 10.1073/pnas.1015033108.

Card, J.P., Rinaman, L., Lynn, R.B., Lee, B.H., Meade, R.P., Miselis, R.R., and Enquist, L.W. (1993). Pseudorabies virus infection of the rat central nervous system: ultrastructural characterization of viral replication, transport, and pathogenesis. J Neurosci *13*, 2515-2539.

Card, J.P., Rinaman, L., Schwaber, J.S., Miselis, R.R., Whealy, M.E., Robbins, A.K., and Enquist, L.W. (1990). Neurotropic properties of pseudorabies virus: uptake and transneuronal passage in the rat central nervous system. J Neurosci *10*, 1974-1994.

Card, J.P., Whealy, M.E., Robbins, A.K., Moore, R.Y., and Enquist, L.W. (1991). Two alpha-herpesvirus strains are transported differentially in the rodent visual system. Neuron *6*, 957-969.

Casrouge, A., Zhang, S.Y., Eidenschenk, C., Jouanguy, E., Puel, A., Yang, K., Alcais, A., Picard, C., Mahfoufi, N., Nicolas, N., *et al.* (2006). Herpes simplex virus encephalitis in human UNC-93B deficiency. Science *314*, 308-312. 10.1126/science.1128346.

Catez, F., Picard, C., Held, K., Gross, S., Rousseau, A., Theil, D., Sawtell, N., Labetoulle, M., and Lomonte, P. (2012). HSV-1 genome subnuclear positioning and associations with host-cell PML-NBs and centromeres regulate LAT locus transcription during latency in neurons. PLoS pathogens *8*, e1002852. 10.1371/journal.ppat.1002852.

Cheung, A.K. (1989). Detection of pseudorabies virus transcripts in trigeminal ganglia of latently infected swine. Journal of virology *63*, 2908-2913.

Chou, J., Kern, E.R., Whitley, R.J., and Roizman, B. (1990). Mapping of herpes simplex virus-1 neurovirulence to gamma 134.5, a gene nonessential for growth in culture. Science *250*, 1262-1266.

Christensen, J., Steain, M., Slobedman, B., and Abendroth, A. (2013). Varicella-zoster virus glycoprotein I is essential for spread in dorsal root ganglia and facilitates axonal localization of structural virion components in neuronal cultures. Journal of virology *87*, 13719-13728. 10.1128/JVI. 02293-13.

Cliffe, A.R., and Wilson, A.C. (2016). Restarting Lytic Gene Transcription at the Onset of Herpes Simplex Virus Reactivation. Journal of virology10.1128/JVI. 01419-16.

Coller, K.E., Lee, J.I., Ueda, A., and Smith, G.A. (2007). The capsid and tegument of the alphaherpesviruses are linked by an interaction between the UL25 and VP1/2 proteins. Journal of virology *81*, 11790-11797. 10.1128/JVI. 01113-07.

Coller, K.E., and Smith, G.A. (2008). Two viral kinases are required for sustained long distance axon transport of a neuroinvasive herpesvirus. Traffic *9*, 1458-1470. 10.1111/j.1600-0854.2008.00782.x.

Conde, C., and Caceres, A. (2009). Microtubule assembly, organization and dynamics in axons and dendrites. Nature reviews Neuroscience *10*, 319-332. 10.1038/nrn2631.

Connolly, S.A., Jackson, J.O., Jardetzky, T.S., and Longnecker, R. (2011). Fusing structure and function: a structural view of the herpesvirus entry machinery. Nature reviews Microbiology *9*, 369-381. nrmicro2548 [pii] 10.1038/nrmicro2548.

Cook, M.L., and Stevens, J.G. (1973). Pathogenesis of herpetic neuritis and ganglionitis in mice: evidence for intra-axonal transport of infection. Infect Immun *7*, 272-288.

Corey, L., and Wald, A. (2009). Maternal and neonatal herpes simplex virus infections. N Engl J Med *361*, 1376-1385. 10.1056/NEJMra0807633.

Cowan, W.M., and Cuénod, M. (1975). The Use of axonal transport for studies of neuronal connectivity (Amsterdam ; New York: Elsevier Scientific Pub. Co.).

Croen, K.D., Ostrove, J.M., Dragovic, L.J., and Straus, S.E. (1988). Patterns of gene expression and sites of latency in human nerve ganglia are different for varicella-zoster and herpes simplex viruses. Proceedings of the National Academy of Sciences of the United States of America *85*, 9773-9777.

Curanovic, D., and Enquist, L.W. (2009). Virion-incorporated glycoprotein B mediates transneuronal spread of pseudorabies virus. Journal of virology *83*, 7796-7804. 10.1128/JVI.00745-09.

Dai, X., and Zhou, Z.H. (2018). Structure of the herpes simplex virus 1 capsid with associated tegument protein complexes. Science *360*10.1126/science.aao7298.

Daniel, G.R., Sollars, P.J., Pickard, G.E., and Smith, G.A. (2015). Pseudorabies Virus Fast Axonal Transport Occurs by a pUS9-Independent Mechanism. Journal of virology *89*, 8088-8091. 10.1128/JVI.00771-15.

Daniel, G.R., Sollars, P.J., Pickard, G.E., and Smith, G.A. (2016). The pseudorabies virus protein, pUL56, enhances virus dissemination and virulence but is dispensable for axonal transport. Virology *488*, 179-186. 10.1016/j.virol.2015.11.014.

De Regge, N., Van Opdenbosch, N., Nauwynck, H.J., Efstathiou, S., and Favoreel, H.W. (2010). Interferon alpha induces establishment of alphaherpesvirus latency in sensory neurons in vitro. PloS one *5*10.1371/journal.pone.0013076.

DeFalco, J., Tomishima, M., Liu, H., Zhao, C., Cai, X., Marth, J.D., Enquist, L., and Friedman, J.M. (2001). Virus-assisted mapping of neural inputs to a feeding center in the hypothalamus. Science *291*, 2608-2613. 10.1126/science.1056602.

del Rio, T., Ch'ng, T.H., Flood, E.A., Gross, S.P., and Enquist, L.W. (2005). Heterogeneity of a fluorescent tegument component in single pseudorabies virus virions and enveloped axonal assemblies. Journal of virology *79*, 3903-3919. 10.1128/JVI.79.7.3903-3919.2005.

Demmin, G.L., Clase, A.C., Randall, J.A., Enquist, L.W., and Banfield, B.W. (2001). Insertions in the gG gene of pseudorabies virus reduce expression of the upstream Us3 protein and inhibit cell-to-cell spread of virus infection. Journal of virology *75*, 10856-10869. 10.1128/JVI.75.22.10856-10869.2001.

Desai, A., and Mitchison, T.J. (1997). Microtubule polymerization dynamics. Annu Rev Cell Dev Biol *13*, 83-117. 10.1146/annurev.cellbio.13.1.83.

DeSantis, M.E., Cianfrocco, M.A., Htet, Z.M., Tran, P.T., Reck-Peterson, S.L., and Leschziner, A.E. (2017). Lis1 Has Two Opposing Modes of Regulating Cytoplasmic Dynein. Cell *170*, 1197-1208 e1112. 10.1016/j.cell.2017.08.037.

Diefenbach, R.J., Davis, A., Miranda-Saksena, M., Fernandez, M.A., Kelly, B.J., Jones, C.A., LaVail, J.H., Xue, J., Lai, J., and Cunningham, A.L. (2016). The Basic Domain of Herpes Simplex Virus 1 pUS9 Recruits Kinesin-1 To Facilitate Egress from Neurons. Journal of virology *90*, 2102-2111. 10.1128/JVI.03041-15.

Dohner, K., Wolfstein, A., Prank, U., Echeverri, C., Dujardin, D., Vallee, R., and Sodeik, B. (2002). Function of dynein and dynactin in herpes simplex virus capsid transport. Mol Biol Cell *13*, 2795-2809. 10.1091/mbc.01-07-0348.

Dolivo, M. (1980). A neurobiological approach to neurotropic viruses. Trends in Neurosciences *3*, 149-152. https://doi.org/10.1016/0166-2236(80)90054-5.

Dolivo, M., Honegger, P., George, C., Kiraly, M., and Bommeli, W. (1979). Enzymatic activity, ultrastructure and function in ganglia infected with a neurotropic virus. Prog Brain Res *51*, 51-57. 10.1016/S0079-6123(08)61293-5.

DuRaine, G., Wisner, T.W., Howard, P., and Johnson, D.C. (2018). Kinesin-1 Proteins KIF5A, -5B, and -5C Promote Anterograde Transport of Herpes Simplex Virus Enveloped Virions in Axons. Journal of virology *92*10.1128/JVI.01269-18.

DuRaine, G., Wisner, T.W., Howard, P., Williams, M., and Johnson, D.C. (2017). Herpes Simplex Virus gE/gI and US9 Promote both Envelopment and Sorting of Virus Particles in the Cytoplasm of Neurons, Two Processes That Precede Anterograde Transport in Axons. Journal of virology *91*10.1128/JVI.00050-17.

Engel, E.A. Personal Communicaiton

Engel, E.A., Song, R., Koyuncu, O.O., and Enquist, L.W. (2015). Investigating the biology of alpha herpesviruses with MS-based proteomics. Proteomics *15*, 1943-1956. 10.1002/pmic.201400604.

Enquist, L.W. (2012). Five questions about viral trafficking in neurons. PLoS pathogens *8*, e1002472. 10.1371/journal.ppat.1002472.

Enquist, L.W., and Card, J.P. (2003). Recent advances in the use of neurotropic viruses for circuit analysis. Curr Opin Neurobiol *13*, 603-606.

Enquist, L.W., Husak, P.J., Banfield, B.W., and Smith, G.A. (1998). Infection and spread of alphaherpesviruses in the nervous system. Adv Virus Res *51*, 237-347.

Enquist, L.W., and Leib, D.A. (2017). Intrinsic and Innate Defenses of Neurons: Detente with the Herpesviruses. Journal of virology *91*10.1128/JVI.01200-16.

Enquist, L.W., Miselis, R.R., and Card, J.P. (1994). Specific infection of rat neuronal circuits by pseudorabies virus. Gene Ther *1 Suppl 1*, S10.

Enquist, L.W., Tomishima, M.J., Gross, S., and Smith, G.A. (2002). Directional spread of an alpha-herpesvirus in the nervous system. Veterinary microbiology *86*, 5-16.

Geraghty, R.J., Jogger, C.R., and Spear, P.G. (2000). Cellular expression of alphaherpesvirus gD interferes with entry of homologous and heterologous alphaherpesviruses by blocking access to a shared gD receptor. Virology *268*, 147-158. 10.1006/viro.1999.0157
S0042-6822(99)90157-5 [pii].

Gilden, D., Grose, C., White, T., Nagae, L., Hendricks, R.L., Cohrs, R.J., and Nagel, M.A. (2016). Successful antiviral treatment after 6years of chronic progressive neurological disease attributed to VZV brain infection. J Neurol Sci *368*, 240-242. 10.1016/j.jns.2016.07.035.

Gnann, J.W., Jr., and Whitley, R.J. (2017). Herpes Simplex Encephalitis: an Update. Curr Infect Dis Rep *19*, 13. 10.1007/s11908-017-0568-7.

Goodpasture, E.W. (1925). The Axis-Cylinders of Peripheral Nerves as Portals of Entry to the Central Nervous System for the Virus of Herpes Simplex in Experimentally Infected Rabbits. Am J Pathol *1*, 11-28 15.

Goodpasture, E.W., and Teague, O. (1923a). Experimental Production of Herpetic Lesions in Organs and Tissues of the Rabbit. J Med Res *44*, 121-138 125.

Goodpasture, E.W., and Teague, O. (1923b). Transmission of the Virus of Herpes Febrilis along Nerves in experimentally infected Rabbits. J Med Res *44*, 139-184 137.

Granstedt, A.E., Bosse, J.B., Thiberge, S.Y., and Enquist, L.W. (2013a). In vivo imaging of alphaherpesvirus infection reveals synchronized activity dependent on axonal sorting of viral proteins. Proceedings of the National Academy of Sciences of the United States of America *110*, E3516-3525. 10.1073/pnas.1311062110.

Granstedt, A.E., Brunton, B.W., and Enquist, L.W. (2013b). Imaging the transport dynamics of single alphaherpesvirus particles in intact peripheral nervous system explants from infected mice. mBio *4*, e00358-00313. 10.1128/mBio.00358-13.

Granzow, H., Klupp, B.G., and Mettenleiter, T.C. (2005). Entry of pseudorabies virus: an immunogold-labeling study. Journal of virology *79*, 3200-3205. 10.1128/JVI.79.5.3200-3205.2005.

Groves, M.J. (2016). Genital Herpes: A Review. Am Fam Physician *93*, 928-934.

Guo, Y., Audry, M., Ciancanelli, M., Alsina, L., Azevedo, J., Herman, M., Anguiano, E., Sancho-Shimizu, V., Lorenzo, L., Pauwels, E., *et al.* (2011). Herpes simplex virus encephalitis in a patient with complete TLR3 deficiency: TLR3 is otherwise redundant in protective immunity. J Exp Med *208*, 2083-2098. 10.1084/jem.20101568.

Guo, Y.J., Luo, T., Wu, F., Mei, Y.W., Peng, J., Liu, H., Li, H.R., Zhang, S.L., Dong, J.H., Fang, Y., *et al.* (2015). Involvement of TLR2 and TLR9 in the anti-inflammatory effects of chlorogenic acid in HSV-1-infected microglia. Life Sci *127*, 12-18. 10.1016/j.lfs.2015.01.036.

Gustafson, D.P., and Kanitz, C.L. (1975). Pseudorabies in perspective. Proc Annu Meet U S Anim Health Assoc, 287-293.

Hafezi, W., Lorentzen, E.U., Eing, B.R., Muller, M., King, N.J., Klupp, B., Mettenleiter, T.C., and Kuhn, J.E. (2012). Entry of herpes simplex virus type 1

(HSV-1) into the distal axons of trigeminal neurons favors the onset of nonproductive, silent infection. PLoS pathogens *8*, e1002679. 10.1371/journal.ppat.1002679

PPATHOGENS-D-11-02484 [pii].

Hanchate, N.K., Lee, E.J., Ellis, A., Kondoh, K., Kuang, D., Basom, R., Trapnell, C., and Buck, L.B. (2020). Connect-seq to superimpose molecular on anatomical neural circuit maps. Proceedings of the National Academy of Sciences of the United States of America *117*, 4375-4384. 10.1073/pnas.1912176117.

Hanz, S., Perlson, E., Willis, D., Zheng, J.Q., Massarwa, R., Huerta, J.J., Koltzenburg, M., Kohler, M., van-Minnen, J., Twiss, J.L., *et al.* (2003). Axoplasmic importins enable retrograde injury signaling in lesioned nerve. Neuron *40*, 1095-1104. S0896627303007700 [pii].

Haycock, J.W., and Bro, S. (1975). Corpus striatum (Translation of S. Ramon y Cajal). translated from Corps Strie, chapter 23, in "Histologie du systeme nerveux de l'homme et des vertebres" 1911. Behav Biol *14*, 387-402.

Heldwein, E.E., and Krummenacher, C. (2008). Entry of herpesviruses into mammalian cells. Cell Mol Life Sci *65*, 1653-1668. 10.1007/s00018-008-7570-z.

Herman, M., Ciancanelli, M., Ou, Y.H., Lorenzo, L., Klaudel-Dreszler, M., Pauwels, E., Sancho-Shimizu, V., Perez de Diego, R., Abhyankar, A., Israelsson, E., *et al.* (2012). Heterozygous TBK1 mutations impair TLR3 immunity and underlie herpes simplex encephalitis of childhood. J Exp Med *209*, 1567-1582. 10.1084/jem.20111316.

Hirokawa, N., Niwa, S., and Tanaka, Y. (2010). Molecular motors in neurons: transport mechanisms and roles in brain function, development, and disease. Neuron *68*, 610-638. 10.1016/j.neuron.2010.09.039.

Hogue, I.B., Bosse, J.B., Engel, E.A., Scherer, J., Hu, J.R., Del Rio, T., and Enquist, L.W. (2015). Fluorescent Protein Approaches in Alpha Herpesvirus Research. Viruses *7*, 5933-5961. 10.3390/v7112915.

Hogue, I.B., Bosse, J.B., Hu, J.R., Thiberge, S.Y., and Enquist, L.W. (2014). Cellular mechanisms of alpha herpesvirus egress: live cell fluorescence microscopy of pseudorabies virus exocytosis. PLoS pathogens *10*, e1004535. 10.1371/journal.ppat.1004535.

Hogue, I.B., Card, J.P., Rinaman, L., Staniszewska Goraczniak, H., and Enquist, L.W. (2018). Characterization of the neuroinvasive profile of a pseudorabies virus recombinant expressing the mTurquoise2 reporter in single and multiple injection experiments. J Neurosci Methods *308*, 228-239. 10.1016/j.jneumeth.2018.08.004.

Holland, D.J., Miranda-Saksena, M., Boadle, R.A., Armati, P., and Cunningham, A.L. (1999). Anterograde transport of herpes simplex virus proteins in axons of peripheral human fetal neurons: an immunoelectron microscopy study. Journal of virology *73*, 8503-8511.

Honess, R.W., and Roizman, B. (1974). Regulation of herpesvirus macromolecular synthesis. I. Cascade regulation of the synthesis of three groups of viral proteins. Journal of virology *14*, 8-19.

Howard, P.W., Howard, T.L., and Johnson, D.C. (2013). Herpes simplex virus membrane proteins gE/gI and US9 act cooperatively to promote transport of capsids and glycoproteins from neuron cell bodies into initial axon segments. Journal of virology *87*, 403-414. 10.1128/JVI.02465-12.

Huang, H., Koyuncu, O.O., and Enquist, L.W. (2020). Pseudorabies Virus Infection Accelerates Degradation of the Kinesin-3 Motor KIF1A. Journal of virology *94*10.1128/JVI.01934-19.

Husak, P.J., Kuo, T., and Enquist, L.W. (2000). Pseudorabies virus membrane proteins gI and gE facilitate anterograde spread of infection in projection-specific neurons in the rat. Journal of virology *74*, 10975-10983.

Ibáñez, F.J., Farías, M.A., Gonzalez-Troncoso, M.P., Corrales, N., Duarte, L.F., Retamal-Díaz, A., and González, P.A. (2018). Experimental Dissection of the Lytic Replication Cycles of Herpes Simplex Viruses in vitro. Frontiers in microbiology *9*10.3389/fmicb.2018.02406.

Ikenaga, T., Tatecho, Y., Nakai, T., and Uematsu, K. (2002). Betanodavirus as a novel transneuronal tracer for fish. Neurosci Lett *331*, 55-59.

Jarman, R.G., Wagner, E.K., and Bloom, D.C. (1999). LAT expression during an acute HSV infection in the mouse. Virology *262*, 384-397. 10.1006/viro.1999.9861.

Jones, C. (2003). Herpes simplex virus type 1 and bovine herpesvirus 1 latency. Clinical microbiology reviews *16*, 79-95.

Jung, H., Yoon, B.C., and Holt, C.E. (2012). Axonal mRNA localization and local protein synthesis in nervous system assembly, maintenance and repair. Nature reviews Neuroscience *13*, 308-324. 10.1038/nrn3210.

Kaplan, A.S., and Vatter, A.E. (1959). A comparison of herpes simplex and pseudorabies viruses. Virology *7*, 394-407.

Kardon, J.R., and Vale, R.D. (2009). Regulators of the cytoplasmic dynein motor. Nat Rev Mol Cell Biol *10*, 854-865. 10.1038/nrm2804.

Katz-Sidlow, R.J. (1998). The formulation of the neuron doctrine: the Island of Cajal. Arch Neurol *55*, 237-240.

Kennedy, P.G., Rovnak, J., Badani, H., and Cohrs, R.J. (2015). A comparison of herpes simplex virus type 1 and varicella-zoster virus latency and reactivation. The Journal of general virology *96*, 1581-1602. 10.1099/vir.0.000128.

Kikkawa, M., Ishikawa, T., Wakabayashi, T., and Hirokawa, N. (1995). Three-dimensional structure of the kinesin head-microtubule complex. Nature *376*, 274-277. 10.1038/376274a0.

Kimberlin, D.W. (2004). Neonatal herpes simplex infection. Clinical microbiology reviews *17*, 1-13.

Klupp, B.G., Fuchs, W., Granzow, H., Nixdorf, R., and Mettenleiter, T.C. (2002). Pseudorabies virus UL36 tegument protein physically interacts with the UL37 protein. Journal of virology *76*, 3065-3071.

Knipe, D.M. (2015). Nuclear sensing of viral DNA, epigenetic regulation of herpes simplex virus infection, and innate immunity. Virology *479-480*, 153-159. 10.1016/j.virol.2015.02.009.

Kobiler, O., Lipman, Y., Therkelsen, K., Daubechies, I., and Enquist, L.W. (2010). Herpesviruses carrying a Brainbow cassette reveal replication and expression of limited numbers of incoming genomes. Nat Commun *1*, 146. 10.1038/ncomms1145.

Kolb, A.W., Lee, K., Larsen, I., Craven, M., and Brandt, C.R. (2016). Quantitative Trait Locus Based Virulence Determinant Mapping of the HSV-1 Genome in Murine Ocular Infection: Genes Involved in Viral Regulatory and

Innate Immune Networks Contribute to Virulence. PLoS pathogens *12*, e1005499. 10.1371/journal.ppat.1005499.

Kopp, S.J., Banisadr, G., Glajch, K., Maurer, U.E., Grunewald, K., Miller, R.J., Osten, P., and Spear, P.G. (2009). Infection of neurons and encephalitis after intracranial inoculation of herpes simplex virus requires the entry receptor nectin-1. Proceedings of the National Academy of Sciences of the United States of America *106*, 17916-17920. 0908892106 [pii]

10.1073/pnas.0908892106.

Koshizuka, T., Kawaguchi, Y., and Nishiyama, Y. (2005). Herpes simplex virus type 2 membrane protein UL56 associates with the kinesin motor protein KIF1A. The Journal of general virology *86*, 527-533. 10.1099/vir.0.80633-0.

Koyuncu, O.O., Hogue, I.B., and Enquist, L.W. (2013a). Virus infections in the nervous system. Cell host and microbe *13*, 379-393. 10.1016/j.chom.2013.03.010.

Koyuncu, O.O., MacGibeny, M.A., and Enquist, L.W. (2018). Latent versus productive infection: the alpha herpesvirus switch. Future Virol *13*, 431-443. 10.2217/fvl-2018-0023.

Koyuncu, O.O., MacGibeny, M.A., Hogue, I.B., and Enquist, L.W. (2017). Compartmented neuronal cultures reveal two distinct mechanisms for alpha herpesvirus escape from genome silencing. PLoS pathogens *13*, e1006608. 10.1371/journal.ppat.1006608.

Koyuncu, O.O., Perlman, D.H., and Enquist, L.W. (2013b). Efficient retrograde transport of pseudorabies virus within neurons requires local protein synthesis in axons. Cell host and microbe *13*, 54-66. 10.1016/j.chom.2012.10.021.

Koyuncu, O.O., Song, R., Greco, T.M., Cristea, I.M., and Enquist, L.W. (2015). The number of alphaherpesvirus particles infecting axons and the axonal protein repertoire determines the outcome of neuronal infection. mBio *6*10.1128/mBio.00276-15.

Kramer, T., and Enquist, L.W. (2012). Alphaherpesvirus infection disrupts mitochondrial transport in neurons. Cell host and microbe *11*, 504-514. 10.1016/j.chom.2012.03.005.

Kramer, T., Greco, T.M., Enquist, L.W., and Cristea, I.M. (2011). Proteomic characterization of pseudorabies virus extracellular virions. Journal of virology *85*, 6427-6441. 10.1128/JVI.02253-10.

Kramer, T., Greco, T.M., Taylor, M.P., Ambrosini, A.E., Cristea, I.M., and Enquist, L.W. (2012). Kinesin-3 mediates axonal sorting and directional transport of alphaherpesvirus particles in neurons. Cell host and microbe *12*, 806-814. 10.1016/j.chom.2012.10.013.

Kratchmarov, R., Kramer, T., Greco, T.M., Taylor, M.P., Ch'ng, T.H., Cristea, I.M., and Enquist, L.W. (2013). Glycoproteins gE and gI are required for efficient KIF1A-dependent anterograde axonal transport of alphaherpesvirus particles in neurons. Journal of virology *87*, 9431-9440. 10.1128/JVI.01317-13.

Kratchmarov, R., Taylor, M.P., and Enquist, L.W. (2012). Making the case: Married versus Separate models of alphaherpes virus anterograde transport in axons. Rev Med Virol *22*, 378-391. 10.1002/rmv.1724.

Kristensson, K., Ghetti, B., and Wisniewski, H.M. (1974). Study on the propagation of Herpes simplex virus (type 2) into the brain after intraocular injection. Brain Res *69*, 189-201.

Kristensson, K., Lycke, E., and Sjostrand, J. (1971). Spread of herpes simplex virus in peripheral nerves. Acta Neuropathol *17*, 44-53.

Kristensson, K., Nennesmo, L., Persson, L., and Lycke, E. (1982). Neuron to neuron transmission of herpes simplex virus. Transport of virus from skin to brainstem nuclei. J Neurol Sci *54*, 149-156.

Kristensson, K., Vahlne, A., Persson, L.A., and Lycke, E. (1978). Neural spread of herpes simplex virus types 1 and 2 in mice after corneal or subcutaneous (footpad) inoculation. J Neurol Sci *35*, 331-340.

Kropp, R.Y., Wong, T., Cormier, L., Ringrose, A., Burton, S., Embree, J.E., and Steben, M. (2006). Neonatal herpes simplex virus infections in Canada: results of a 3-year national prospective study. Pediatrics *117*, 1955-1962. 10.1542/peds.2005-1778.

Lafaille, F.G., Pessach, I.M., Zhang, S.Y., Ciancanelli, M.J., Herman, M., Abhyankar, A., Ying, S.W., Keros, S., Goldstein, P.A., Mostoslavsky, G., *et al.*

(2012). Impaired intrinsic immunity to HSV-1 in human iPSC-derived TLR3-deficient CNS cells. Nature *491*, 769-773. 10.1038/nature11583.

Lanciego, J.L., and Wouterlood, F.G. (2020). Neuroanatomical tract-tracing techniques that did go viral. Brain structure and function10.1007/s00429-020-02041-6.

Laval, K., and Enquist, L.W. (2020). The Neuropathic Itch Caused by Pseudorabies Virus. Pathogens *9*10.3390/pathogens9040254.

Laval, K., Van Cleemput, J., Vernejoul, J.B., and Enquist, L.W. (2019). Alphaherpesvirus infection of mice primes PNS neurons to an inflammatory state regulated by TLR2 and type I IFN signaling. PLoS pathogens *15*, e1008087. 10.1371/journal.ppat.1008087.

Laval, K., Vernejoul, J.B., Van Cleemput, J., Koyuncu, O.O., and Enquist, L.W. (2018). Virulent PRV infection induces a specific and lethal systemic inflammatory response in mice. Journal of virology10.1128/JVI.01614-18.

Lee, J.I., Sollars, P.J., Baver, S.B., Pickard, G.E., Leelawong, M., and Smith, G.A. (2009). A herpesvirus encoded deubiquitinase is a novel neuroinvasive determinant. PLoS pathogens *5*, e1000387. 10.1371/journal.ppat.1000387.

Leib, D.A., Harrison, T.E., Laslo, K.M., Machalek, M.A., Moorman, N.J., and Virgin, H.W. (1999). Interferons regulate the phenotype of wild-type and mutant herpes simplex viruses in vivo. J Exp Med *189*, 663-672.

Leterrier, C. (2018). The Axon Initial Segment: An Updated Viewpoint. J Neurosci *38*, 2135-2145. 10.1523/JNEUROSCI.1922-17.2018.

Levine, J.M., Enquist, L.W., and Card, J.P. (1998). Reactions of oligodendrocyte precursor cells to alpha herpesvirus infection of the central nervous system. Glia *23*, 316-328.

Li, J., Hu, S., Zhou, L., Ye, L., Wang, X., Ho, J., and Ho, W. (2011). Interferon lambda inhibits herpes simplex virus type I infection of human astrocytes and neurons. Glia *59*, 58-67. 10.1002/glia.21076.

Li, J., Liu, T., Dong, Y., Kondoh, K., and Lu, Z. (2019). Trans-synaptic Neural Circuit-Tracing with Neurotropic Viruses. Neurosci Bull *35*, 909-920. 10.1007/s12264-019-00374-9.

Li, Z., Lu, X., Zhu, Y., Cheng, P., Liu, S., Zhang, Y., Tang, J., Yang, S., and Zhou, L. (2017). Lambda-Interferons Inhibit Herpes Simplex Virus Type 2

Replication in Human Cervical Epithelial Cells by Activating the JAK/STAT Pathway. Jpn J Infect Dis *70*, 416-422. 10.7883/yoken.JJID.2016.465.

Linderman, J.A., Kobayashi, M., Rayannavar, V., Fak, J.J., Darnell, R.B., Chao, M.V., Wilson, A.C., and Mohr, I. (2017). Immune Escape via a Transient Gene Expression Program Enables Productive Replication of a Latent Pathogen. Cell reports *18*, 1312-1323. 10.1016/j.celrep.2017.01.017.

Lippe, R. (2014). Characterization of extracellular HSV-1 virions by proteomics. Methods in molecular biology *1144*, 181-190. 10.1007/978-1-4939-0428-0_12.

Lo, L., and Anderson, D.J. (2011). A Cre-dependent, anterograde transsynaptic viral tracer for mapping output pathways of genetically marked neurons. Neuron *72*, 938-950. 10.1016/j.neuron.2011.12.002.

Lomniczi, B., Blankenship, M.L., and Ben-Porat, T. (1984). Deletions in the genomes of pseudorabies virus vaccine strains and existence of four isomers of the genomes. Journal of virology *49*, 970-979.

Lomonte, P. (2017). Herpesvirus Latency: On the Importance of Positioning Oneself. Adv Anat Embryol Cell Biol *223*, 95-117. 10.1007/978-3-319-53168-7_5.

Looker, K.J., Magaret, A.S., May, M.T., Turner, K.M., Vickerman, P., Gottlieb, S.L., and Newman, L.M. (2015a). Global and Regional Estimates of Prevalent and Incident Herpes Simplex Virus Type 1 Infections in 2012. PloS one *10*, e0140765. 10.1371/journal.pone.0140765.

Looker, K.J., Magaret, A.S., May, M.T., Turner, K.M.E., Vickerman, P., Newman, L.M., and Gottlieb, S.L. (2017). First estimates of the global and regional incidence of neonatal herpes infection. Lancet Glob Health *5*, e300-e309. 10.1016/S2214-109X(16)30362-X.

Looker, K.J., Magaret, A.S., Turner, K.M., Vickerman, P., Gottlieb, S.L., and Newman, L.M. (2015b). Global estimates of prevalent and incident herpes simplex virus type 2 infections in 2012. PloS one *10*, e114989. 10.1371/journal.pone.0114989.

Luders, J., and Stearns, T. (2007). Microtubule-organizing centres: a re-evaluation. Nat Rev Mol Cell Biol *8*, 161-167. 10.1038/nrm2100.

Luxton, G.W., Haverlock, S., Coller, K.E., Antinone, S.E., Pincetic, A., and Smith, G.A. (2005). Targeting of herpesvirus capsid transport in axons is coupled to association with specific sets of tegument proteins. Proceedings of the National Academy of Sciences of the United States of America *102*, 5832-5837. 10.1073/pnas.0500803102.

Lyman, M.G., Feierbach, B., Curanovic, D., Bisher, M., and Enquist, L.W. (2007). Pseudorabies virus Us9 directs axonal sorting of viral capsids. Journal of virology *81*, 11363-11371. 10.1128/JVI.01281-07.

Ma, Y., and He, B. (2014). Recognition of herpes simplex viruses: toll-like receptors and beyond. Journal of molecular biology *426*, 1133-1147. 10.1016/j.jmb.2013.11.012.

Mangold, C.A., Engel, E.A., Ostler, J.B., Jr., Wilson, A.C., and Cohrs, R.J. (2020). 2019 Colorado Alphaherpesvirus Latency Society symposium. J Neurovirol *26*, 297-309. 10.1007/s13365-019-00798-z.

Maroui, M.A., Calle, A., Cohen, C., Streichenberger, N., Texier, P., Takissian, J., Rousseau, A., Poccardi, N., Welsch, J., Corpet, A., *et al.* (2016). Latency Entry of Herpes Simplex Virus 1 Is Determined by the Interaction of Its Genome with the Nuclear Environment. PLoS pathogens *12*, e1005834. 10.1371/journal.ppat.1005834.

Maturana, C.J., Verpeut, J.L., Pisano, T.J., Dhanerawala, Z.M., Esteves, A., Enquist, L.W., and Engel, E.A. (2020). Small Alphaherpesvirus Latency-Associated Promoters Drive Efficient and Long-Term Transgene Expression in the CNS. Mol Ther Methods Clin Dev *17*, 843-857. 10.1016/j.omtm.2020.04.004.

McCarthy, K.M., Tank, D.W., and Enquist, L.W. (2009). Pseudorabies virus infection alters neuronal activity and connectivity in vitro. PLoS pathogens *5*, e1000640. 10.1371/journal.ppat.1000640.

McGovern, A.E., Driessen, A.K., Simmons, D.G., Powell, J., Davis-Poynter, N., Farrell, M.J., and Mazzone, S.B. (2015). Distinct brainstem and forebrain circuits receiving tracheal sensory neuron inputs revealed using a novel conditional anterograde transsynaptic viral tracing system. J Neurosci *35*, 7041-7055. 10.1523/JNEUROSCI.5128-14.2015.

McKenney, R.J., Weil, S.J., Scherer, J., and Vallee, R.B. (2011). Mutually exclusive cytoplasmic dynein regulation by NudE-Lis1 and dynactin. The Journal of biological chemistry *286*, 39615-39622. 10.1074/jbc.M111.289017.

Mettenleiter, T.C., Lukacs, N., and Rziha, H.J. (1985). Pseudorabies virus avirulent strains fail to express a major glycoprotein. Journal of virology *56*, 307-311.

Mikloska, Z., and Cunningham, A.L. (2001). Alpha and gamma interferons inhibit herpes simplex virus type 1 infection and spread in epidermal cells after axonal transmission. Journal of virology *75*, 11821-11826. 10.1128/JVI.75.23.11821-11826.2001.

Miller, C.S., and Danaher, R.J. (2008). Asymptomatic shedding of herpes simplex virus (HSV) in the oral cavity. Oral Surg Oral Med Oral Pathol Oral Radiol Endod *105*, 43-50. 10.1016/j.tripleo.2007.06.011.

Miranda-Saksena, M., Armati, P., Boadle, R.A., Holland, D.J., and Cunningham, A.L. (2000). Anterograde transport of herpes simplex virus type 1 in cultured, dissociated human and rat dorsal root ganglion neurons. Journal of virology *74*, 1827-1839.

Miranda-Saksena, M., Boadle, R.A., Aggarwal, A., Tijono, B., Rixon, F.J., Diefenbach, R.J., and Cunningham, A.L. (2009). Herpes simplex virus utilizes the large secretory vesicle pathway for anterograde transport of tegument and envelope proteins and for viral exocytosis from growth cones of human fetal axons. Journal of virology *83*, 3187-3199. 10.1128/JVI.01579-08.

Miranda-Saksena, M., Denes, C.E., Diefenbach, R.J., and Cunningham, A.L. (2018). Infection and Transport of Herpes Simplex Virus Type 1 in Neurons: Role of the Cytoskeleton. Viruses *10*10.3390/v10020092.

Mitchell, B.M., Bloom, D.C., Cohrs, R.J., Gilden, D.H., and Kennedy, P.G. (2003). Herpes simplex virus-1 and varicella-zoster virus latency in ganglia. J Neurovirol *9*, 194-204. 10.1080/13550280390194000.

Moffat, J., Ku, C.C., Zerboni, L., Sommer, M., and Arvin, A. (2007). VZV: pathogenesis and the disease consequences of primary infection. In Human Herpesviruses: Biology, Therapy, and Immunoprophylaxis, A. Arvin, G. Campadelli-Fiume, E. Mocarski, P.S. Moore, B. Roizman, R. Whitley, and K. Yamanishi, eds. (Cambridge).

Mues, M.B., Cheshenko, N., Wilson, D.W., Gunther-Cummins, L., and Herold, B.C. (2015). Dynasore disrupts trafficking of herpes simplex virus proteins. Journal of virology *89*, 6673-6684. 10.1128/JVI.00636-15.

Nassi, J.J., Cepko, C.L., Born, R.T., and Beier, K.T. (2015). Neuroanatomy goes viral! Front Neuroanat *9*, 80. 10.3389/fnana.2015.00080.

Negatsch, A., Granzow, H., Maresch, C., Klupp, B.G., Fuchs, W., Teifke, J.P., and Mettenleiter, T.C. (2010). Ultrastructural analysis of virion formation and intraaxonal transport of herpes simplex virus type 1 in primary rat neurons. Journal of virology *84*, 13031-13035. 10.1128/JVI.01784-10.

Norgren, R.B., Jr., and Lehman, M.N. (1998). Herpes simplex virus as a transneuronal tracer. Neurosci Biobehav Rev *22*, 695-708.

Okada, Y., Yamazaki, H., Sekine-Aizawa, Y., and Hirokawa, N. (1995). The neuron-specific kinesin superfamily protein KIF1A is a unique monomeric motor for anterograde axonal transport of synaptic vesicle precursors. Cell *81*, 769-780.

Olander, H.J., Saunders, J.R., Gustafson, D.P., and Jones, R.K. (1966). Pathologic findings in swine affected with a virulent strain of Aujeszky's virus. Pathol Vet *3*, 64-82.

Olsen, L.M., Ch'ng, T.H., Card, J.P., and Enquist, L.W. (2006). Role of pseudorabies virus Us3 protein kinase during neuronal infection. Journal of virology *80*, 6387-6398. 10.1128/JVI.00352-06.

Ott, M., Jing, L., Lorenzo, L., Casanova, J.L., Zhang, S.Y., and Koelle, D.M. (2017). T-cell Responses to HSV-1 in Persons Who Have Survived Childhood Herpes Simplex Encephalitis. Pediatr Infect Dis J *36*, 741-744. 10.1097/INF.0000000000001631.

Owen, D.J., Crump, C.M., and Graham, S.C. (2015). Tegument Assembly and Secondary Envelopment of Alphaherpesviruses. Viruses *7*, 5084-5114. 10.3390/v7092861.

Oyibo, H.K., Znamenskiy, P., Oviedo, H.V., Enquist, L.W., and Zador, A.M. (2014). Long-term Cre-mediated retrograde tagging of neurons using a novel recombinant pseudorabies virus. Front Neuroanat *8*, 86. 10.3389/fnana.2014.00086.

Perez de Diego, R., Mulvey, C., Crawford, M., Trotter, M.W., Lorenzo, L., Sancho-Shimizu, V., Abel, L., Zhang, S.Y., Casanova, J.L., and Godovac-Zimmermann, J. (2013). The proteome of Toll-like receptor 3-stimulated human immortalized fibroblasts: implications for susceptibility to herpes simplex virus encephalitis. The Journal of allergy and clinical immunology *131*, 1157-1166. 10.1016/j.jaci.2013.01.008.

Perez de Diego, R., Sancho-Shimizu, V., Lorenzo, L., Puel, A., Plancoulaine, S., Picard, C., Herman, M., Cardon, A., Durandy, A., Bustamante, J., *et al.* (2010). Human TRAF3 adaptor molecule deficiency leads to impaired Toll-like receptor 3 response and susceptibility to herpes simplex encephalitis. Immunity *33*, 400-411. 10.1016/j.immuni.2010.08.014.

Perry, R.B., Doron-Mandel, E., Iavnilovitch, E., Rishal, I., Dagan, S.Y., Tsoory, M., Coppola, G., McDonald, M.K., Gomes, C., Geschwind, D.H., *et al.* (2012). Subcellular knockout of importin beta1 perturbs axonal retrograde signaling. Neuron *75*, 294-305. 10.1016/j.neuron.2012.05.033.

Pinninti, S.G., and Kimberlin, D.W. (2018). Neonatal herpes simplex virus infections. Semin Perinatol *42*, 168-175. 10.1053/j.semperi.2018.02.004.

Polcicova, K., Biswas, P.S., Banerjee, K., Wisner, T.W., Rouse, B.T., and Johnson, D.C. (2005). Herpes keratitis in the absence of anterograde transport of virus from sensory ganglia to the cornea. Proceedings of the National Academy of Sciences of the United States of America *102*, 11462-11467. 10.1073/pnas.0503230102.

Pomeranz, L.E., Ekstrand, M.I., Latcha, K.N., Smith, G.A., Enquist, L.W., and Friedman, J.M. (2017). Gene Expression Profiling with Cre-Conditional Pseudorabies Virus Reveals a Subset of Midbrain Neurons That Participate in Reward Circuitry. J Neurosci *37*, 4128-4144. 10.1523/JNEUROSCI.3193-16.2017.

Pomeranz, L.E., Reynolds, A.E., and Hengartner, C.J. (2005). Molecular biology of pseudorabies virus: impact on neurovirology and veterinary medicine. Microbiology and molecular biology reviews : MMBR *69*, 462-500. 10.1128/MMBR.69.3.462-500.2005.

Poole, C.L., and Kimberlin, D.W. (2018). Antiviral Approaches for the Treatment of Herpes Simplex Virus Infections in Newborn Infants. Annu Rev Virol *5*, 407-425. 10.1146/annurev-virology-092917-043457.

Radtke, K., Kieneke, D., Wolfstein, A., Michael, K., Steffen, W., Scholz, T., Karger, A., and Sodeik, B. (2010). Plus- and minus-end directed microtubule motors bind simultaneously to herpes simplex virus capsids using different inner tegument structures. PLoS pathogens *6*, e1000991. 10.1371/journal.ppat.1000991.

Reck-Peterson, S.L., Redwine, W.B., Vale, R.D., and Carter, A.P. (2018). The cytoplasmic dynein transport machinery and its many cargoes. Nat Rev Mol Cell Biol *19*, 382-398. 10.1038/s41580-018-0004-3.

Richards, A.L., Sollars, P.J., Pitts, J.D., Stults, A.M., Heldwein, E.E., Pickard, G.E., and Smith, G.A. (2017). The pUL37 tegument protein guides alpha-herpesvirus retrograde axonal transport to promote neuroinvasion. PLoS pathogens *13*, e1006741. 10.1371/journal.ppat.1006741.

Richter, E.R., Dias, J.K., Gilbert, J.E., 2nd, and Atherton, S.S. (2009). Distribution of herpes simplex virus type 1 and varicella zoster virus in ganglia of the human head and neck. J Infect Dis *200*, 1901-1906. 10.1086/648474.

Roizman, B., and Sears, A.E. (1987). An inquiry into the mechanisms of herpes simplex virus latency. Annual review of microbiology *41*, 543-571. 10.1146/annurev.mi.41.100187.002551.

Roizman, B., and Whitley, R.J. (2013). An inquiry into the molecular basis of HSV latency and reactivation. Annual review of microbiology *67*, 355-374. 10.1146/annurev-micro-092412-155654.

Rosen-Wolff, A., Lamade, W., Berkowitz, C., Becker, Y., and Darai, G. (1991). Elimination of UL56 gene by insertion of LacZ cassette between nucleotide position 116030 to 121753 of the herpes simplex virus type 1 genome abrogates intraperitoneal pathogenicity in tree shrews and mice. Virus Res *20*, 205-221.

Salinas, S., Schiavo, G., and Kremer, E.J. (2010). A hitchhiker's guide to the nervous system: the complex journey of viruses and toxins. Nature reviews Microbiology *8*, 645-655. 10.1038/nrmicro2395.

Saunders, J.R., and Gustafson, D.P. (1964). Serological and experimental studies of pseudorabies in swine. Proc Annu Meet U S Anim Health Assoc *68*, 256-265.

Sawtell, N.M. (1998). The probability of in vivo reactivation of herpes simplex virus type 1 increases with the number of latently infected neurons in the ganglia. Journal of virology *72*, 6888-6892.

Sawtell, N.M., Poon, D.K., Tansky, C.S., and Thompson, R.L. (1998). The latent herpes simplex virus type 1 genome copy number in individual neurons is virus strain specific and correlates with reactivation. Journal of virology *72*, 5343-5350.

Scherer, J., Yaffe, Z.A., Vershinin, M., and Enquist, L.W. (2016). Dual-color Herpesvirus Capsids Discriminate Inoculum from Progeny and Reveal Axonal Transport Dynamics. Journal of virology10.1128/JVI.01122-16.

Schneeberger, M., Parolari, L., Das Banerjee, T., Bhave, V., Wang, P., Patel, B., Topilko, T., Wu, Z., Choi, C.H.J., Yu, X., *et al.* (2019). Regulation of Energy Expenditure by Brainstem GABA Neurons. Cell *178*, 672-685 e612. 10.1016/j.cell.2019.05.048.

Schoggins, J.W. (2014). Interferon-stimulated genes: roles in viral pathogenesis. Curr Opin Virol *6*, 40-46. 10.1016/j.coviro.2014.03.006.

Sciortino, M.T., Suzuki, M., Taddeo, B., and Roizman, B. (2001). RNAs extracted from herpes simplex virus 1 virions: apparent selectivity of viral but not cellular RNAs packaged in virions. Journal of virology *75*, 8105-8116.

Shayakhmetov, D.M., Di Paolo, N.C., and Mossman, K.L. (2010). Recognition of virus infection and innate host responses to viral gene therapy vectors. Mol Ther *18*, 1422-1429. 10.1038/mt.2010.124.

Shukla, D., Dal Canto, M.C., Rowe, C.L., and Spear, P.G. (2000). Striking similarity of murine nectin-1alpha to human nectin-1alpha (HveC) in sequence and activity as a glycoprotein D receptor for alphaherpesvirus entry. Journal of virology *74*, 11773-11781.

Sironi, M., Peri, A.M., Cagliani, R., Forni, D., Riva, S., Biasin, M., Clerici, M., and Gori, A. (2017). TLR3 Mutations in Adult Patients With Herpes Simplex Virus and Varicella-Zoster Virus Encephalitis. J Infect Dis *215*, 1430-1434. 10.1093/infdis/jix166.

Smith, B.N., Banfield, B.W., Smeraski, C.A., Wilcox, C.L., Dudek, F.E., Enquist, L.W., and Pickard, G.E. (2000). Pseudorabies virus expressing enhanced green fluorescent protein: A tool for in vitro electrophysiological analysis of transsynaptically labeled neurons in identified central nervous system circuits. Proceedings of the National Academy of Sciences of the United States of America *97*, 9264-9269.

Smith, G. (2012). Herpesvirus transport to the nervous system and back again. Annual review of microbiology *66*, 153-176. 10.1146/annurev-micro-092611-150051.

Smith, G.A., Gross, S.P., and Enquist, L.W. (2001). Herpesviruses use bidirectional fast-axonal transport to spread in sensory neurons. Proceedings of the National Academy of Sciences of the United States of America *98*, 3466-3470. 10.1073/pnas.061029798.

Smith, G.A., Pomeranz, L., Gross, S.P., and Enquist, L.W. (2004). Local modulation of plus-end transport targets herpesvirus entry and egress in sensory axons. Proceedings of the National Academy of Sciences of the United States of America *101*, 16034-16039. 10.1073/pnas.0404686101.

Snyder, A., Polcicova, K., and Johnson, D.C. (2008). Herpes simplex virus gE/gI and US9 proteins promote transport of both capsids and virion glycoproteins in neuronal axons. Journal of virology *82*, 10613-10624. 10.1128/JVI.01241-08.

Snyder, A., Wisner, T.W., and Johnson, D.C. (2006). Herpes simplex virus capsids are transported in neuronal axons without an envelope containing the viral glycoproteins. Journal of virology *80*, 11165-11177. 10.1128/JVI.01107-06.

Song, R., Koyuncu, O.O., Greco, T.M., Diner, B.A., Cristea, I.M., and Enquist, L.W. (2016). Two Modes of the Axonal Interferon Response Limit Alphaherpesvirus Neuroinvasion. mBio *7*, e02145-02115. 10.1128/mBio.02145-15.

Spear, P.G., Eisenberg, R.J., and Cohen, G.H. (2000). Three classes of cell surface receptors for alphaherpesvirus entry. Virology *275*, 1-8. 10.1006/viro.2000.0529

S0042-6822(00)90529-4 [pii].

St Leger, A.J., and Hendricks, R.L. (2011). CD8+ T cells patrol HSV-1-infected trigeminal ganglia and prevent viral reactivation. J Neurovirol *17*, 528-534. 10.1007/s13365-011-0062-1.

Szpara, M.L., Parsons, L., and Enquist, L.W. (2010). Sequence variability in clinical and laboratory isolates of herpes simplex virus 1 reveals new mutations. Journal of virology *84*, 5303-5313. 10.1128/JVI.00312-10.

Szpara, M.L., Tafuri, Y.R., Parsons, L., Shamim, S.R., Verstrepen, K.J., Legendre, M., and Enquist, L.W. (2011). A wide extent of inter-strain diversity in virulent and vaccine strains of alphaherpesviruses. PLoS pathogens *7*, e1002282. 10.1371/journal.ppat.1002282.

Szpara, M.L., Tafuri, Y.R., Parsons, L., Shreve, J.T., Engel, E.A., and Enquist, L.W. (2014). Genome Sequence of the Anterograde-Spread-Defective Herpes Simplex Virus 1 Strain MacIntyre. Genome Announc *2*10.1128/genomeA.01161-14.

Taylor, M.P., Kramer, T., Lyman, M.G., Kratchmarov, R., and Enquist, L.W. (2012). Visualization of an alphaherpesvirus membrane protein that is essential for anterograde axonal spread of infection in neurons. mBio *3*10.1128/mBio.00063-12.

Tsatsos, M., MacGregor, C., Athanasiadis, I., Moschos, M.M., Hossain, P., and Anderson, D. (2016). Herpes simplex virus keratitis: an update of the pathogenesis and current treatment with oral and topical antiviral agents. Clin Exp Ophthalmol *44*, 824-837. 10.1111/ceo.12785.

Ugolini, G., Kuypers, H.G., and Simmons, A. (1987). Retrograde transneuronal transfer of herpes simplex virus type 1 (HSV 1) from motoneurones. Brain Res *422*, 242-256.

Vale, R.D. (2003). The molecular motor toolbox for intracellular transport. Cell *112*, 467-480.

Whitley, R.J. (2006). Herpes simplex encephalitis: adolescents and adults. Antiviral Res *71*, 141-148. 10.1016/j.antiviral.2006.04.002.

Whitley, R.J. (2015). Herpes Simplex Virus Infections of the Central Nervous System. Continuum (Minneap Minn) *21*, 1704-1713. 10.1212/CON.0000000000000243.

Wilson, A.C., Cleary, M.A., Lai, J.S., LaMarco, K., Peterson, M.G., and Herr, W. (1993). Combinatorial control of transcription: the herpes simplex virus VP16-induced complex. Cold Spring Harb Symp Quant Biol *58*, 167-178.

Wilson, A.C., and Mohr, I. (2012). A cultured affair: HSV latency and reactivation in neurons. Trends Microbiol *20*, 604-611. S0966-842X(12)00148-5 [pii] 10.1016/j.tim.2012.08.005.

Wisner, T.W., Sugimoto, K., Howard, P.W., Kawaguchi, Y., and Johnson, D.C. (2011). Anterograde transport of herpes simplex virus capsids in neurons by both separate and married mechanisms. Journal of virology *85*, 5919-5928. 10.1128/JVI.00116-11.

Wojaczynski, G.J., Engel, E.A., Steren, K.E., Enquist, L.W., and Patrick Card, J. (2015). The neuroinvasive profiles of H129 (herpes simplex virus type 1) recombinants with putative anterograde-only transneuronal spread properties. Brain structure and function *220*, 1395-1420. 10.1007/s00429-014-0733-9.

Wu, B.W., Engel, E.A., and Enquist, L.W. (2014). Characterization of a Replication-Incompetent Pseudorabies Virus Mutant Lacking the Sole Immediate Early Gene IE180. mBio *5*10.1128/mBio.01850-14.

Wysocka, J., and Herr, W. (2003). The herpes simplex virus VP16-induced complex: the makings of a regulatory switch. Trends in biochemical sciences *28*, 294-304. 10.1016/S0968-0004(03)00088-4.

Zaichick, S.V., Bohannon, K.P., Hughes, A., Sollars, P.J., Pickard, G.E., and Smith, G.A. (2013). The herpesvirus VP1/2 protein is an effector of dynein-mediated capsid transport and neuroinvasion. Cell host and microbe *13*, 193-203. 10.1016/j.chom.2013.01.009.

Zemanick, M.C., Strick, P.L., and Dix, R.D. (1991). Direction of transneuronal transport of herpes simplex virus 1 in the primate motor system is strain-dependent. Proceedings of the National Academy of Sciences of the United States of America *88*, 8048-8051.

Zeng, W.B., Jiang, H.F., Gang, Y.D., Song, Y.G., Shen, Z.Z., Yang, H., Dong, X., Tian, Y.L., Ni, R.J., Liu, Y., *et al.* (2017). Anterograde monosynaptic transneuronal tracers derived from herpes simplex virus 1 strain H129. Mol Neurodegener *12*, 38. 10.1186/s13024-017-0179-7.

Zhang, S.Y., Abel, L., and Casanova, J.L. (2013). Mendelian predisposition to herpes simplex encephalitis. Handb Clin Neurol *112*, 1091-1097. 10.1016/B978-0-444-52910-7.00027-1.

Zhang, S.Y., Clark, N.E., Freije, C.A., Pauwels, E., Taggart, A.J., Okada, S., Mandel, H., Garcia, P., Ciancanelli, M.J., Biran, A., *et al.* (2018). Inborn Errors of RNA Lariat Metabolism in Humans with Brainstem Viral Infection. Cell *172*, 952-965 e918. 10.1016/j.cell.2018.02.019.

Zhang, S.Y., Jouanguy, E., Ugolini, S., Smahi, A., Elain, G., Romero, P., Segal, D., Sancho-Shimizu, V., Lorenzo, L., Puel, A., *et al.* (2007). TLR3 deficiency in patients with herpes simplex encephalitis. Science *317*, 1522-1527. 10.1126/science.1139522.

Zimmer, B., Ewaleifoh, O., Harschnitz, O., Lee, Y.S., Peneau, C., McAlpine, J.L., Liu, B., Tchieu, J., Steinbeck, J.A., Lafaille, F., *et al.* (2018). Human iPSC-derived trigeminal neurons lack constitutive TLR3-dependent immunity that protects cortical neurons from HSV-1 infection. Proc Natl Acad Sci U S A10.1073/pnas.1809853115.

Chapter 6

Alphaherpesvirus Latency and Reactivation with a Focus on Herpes Simplex Virus

Nancy M. Sawtell[1]* and Richard L. Thompson[2]*

[1]Infectious Diseases, Cincinnati Children's Hospital Medical Center, 3333 Burnet Ave, Cincinnati Ohio, 45229-3039, USA
[2]Molecular Genetics, Microbiology, and Biochemistry, University of Cincinnati College of Medicine, 231 Albert Sabin Way, Cincinnati, Ohio, 45267-0524, USA

*nancy.sawtell@cchmc.org, richard.thompson@uc.edu

DOI: https://doi.org/10.21775/9781913652555.06

Abstract

We are at an interesting time in the understanding of alpha herpesvirus latency and reactivation and their implications to human disease. Conceptual advances have come from both animal and neuronal culture models. This chapter focuses on the concept that the tegument protein and viral transactivator VP16 plays a major role in the transition from latency to the lytic cycle. During acute infection, regulation of VP16 transactivation balances spread in the nervous system, establishment of latent infections and virulence. Reactivation is dependent on this transactivator to drive entry into the lytic cycle. In vivo de novo expression of VP16 protein is mediated by sequences conferring pre-immediate early transcription embedded in the normally leaky late promoter. In vitro, alternate mechanisms regulating VP16 expression in the context of latency have come from the SCG neuron culture model and include the concepts that (i) generalized transcriptional derepression of the viral genome and sequestration of VP16 in the cytoplasm for ~48 hours (Phase I) precedes and is required for VP16-dependent reactivation (Phase II); and (ii) a histone methyl/phospho switch during Phase I is required for Phase II reactivation. The challenge to the field is reconciling these data into a unified model of virus reactivation.

*The greatest enemy of **knowledge** is not ignorance, it is the illusion of **knowledge**.-- Daniel J. Boorstin.*

The task of compiling this chapter was uncomfortably humbling, as if cataloging the stars in the universe. While not completely dark, our night sky is missing a multitude of studies which are among the many points of light contributing to our field. This chapter is a focused review in which we discuss from the vantage point of our expertise, just a handful of concepts that have or are emerging. A lookback at some of the pioneering work that grounds our field is also included.

Introduction

The mechanisms of latency and reactivation of the human alpha herpesviruses, herpes simplex virus 1 and 2 (HSV1, HSV2) and varicella zoster virus (VZV), remain major unsolved mysteries. Each year, recurrent HSV disease contributes to ~200 million new HSV infections (worldwide, extrapolated from US infection rates) (Azwa and Barton, 2009; Pepose et al., 2006), two-thirds of new sexually acquired HIV infections (Glynn et al., 2009), ~10 million cases of eye disease (Pepose et al., 2006), ~150,000 cases of encephalitis (Stone and Hawkins, 2007), and devastating neonatal disease (Looker et al., 2017; Roberts, 2009). Herpesvirus infections are also linked to type II diabetes (Sun et al., 2005), cardiovascular disease (Mukamal et al., 2004; Visser and Vercellotti, 1993), and now strongly implicated in Alzheimer's disease (Fulop et al., 2018; Itzhaki, 2018; Itzhaki and Wozniak, 2008; Lathe et al., 2019). The ubiquity of HSV results from the ability of the virus to establish latent infections that periodically reactivate and transmit the infection to new hosts (Ahmed and Stevens, 1990; Knipe, 2007; Roizman et al.).

The availability of animal models that support HSV latency and reactivation, and the work of many laboratories over many years, have contributed to a collective understanding of the HSV life cycle. Infection occurs at the body surface, usually a mucosal epithelium, where the virus replicates, spreads, and gains access to the innervating neuron axonal endings. At this point, nucleocapsids deposited into the axonal cytosol use retrograde axonal transport to reach neuronal cell bodies that are housed within peripheral ganglia (a collection of sensory or autonomic nerve cell bodies innervating specific tissues). The trigeminal ganglia (TG), which innervate the oral mucosa and other regions of the face, is the typical site of HSV1 latent infection, whereas the sacral ganglia (SG) innervates the genital mucosa and serves as the predominant site of latent HSV2 infection.

Overview

Our current understanding of HSV latency and reactivation arises out of a blend of human studies and findings from various animal and neuronal culture models. It is important to recognize the limitations of these models when summing up what we know. Importantly, new technologies and approaches with increased sensitivity have driven new waves of data acquisition which requires assimilation into what we think we know about an extremely complex process. There is increasing evidence for the concept that HSV1 latency/ reactivation can be a risk factor in the development of neurodegenerative disorders. Also arising from the human model, the concept of "continuous" ganglionic reactivation has emerged as an explanation for the unexpectedly high frequency of viral DNA in human genital swabs in some patients in the absence of detectable infectious virus.

This chapter focuses on an unexpected observation that the tegument protein and viral transactivator, VP16, plays a major role in the transition from latent into the lytic cycle and is required for balancing the latent/lytic transition (Figure 1). During acute infection, regulation of VP16 transactivation balances spread into the nervous system, establishment of latency, and virulence. In

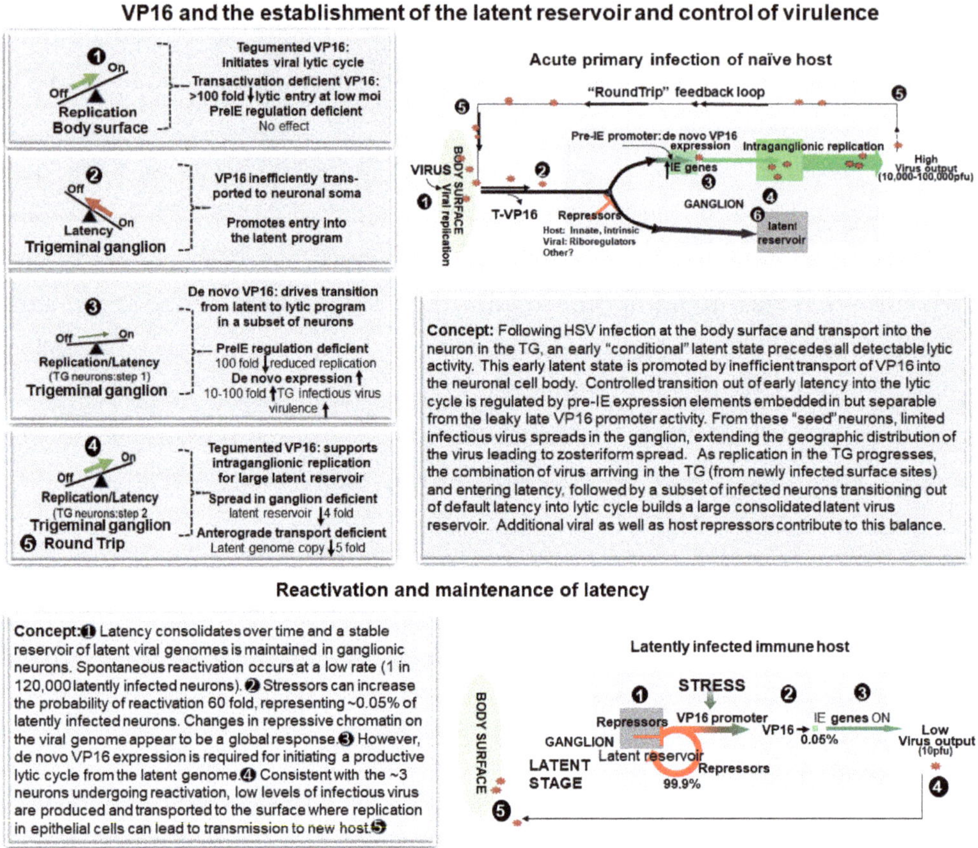

Figure 1. Regulation of VP16 function balances latent/lytic programs. ***Entering the neuron:*** The VP16 tegument protein, which is a potent transactivator of viral immediate-early gene expression, is not available to perform this function in neurons infected via axonal endings because of inefficient transport of VP16 via axon to the cell body during acute infection. The viral DNA enters the nucleus where it forms a circular episome and becomes associated with nucleosomes, which promotes the latent transcriptional program.

turn, reactivation is dependent on this transactivator to drive entry into the lytic cycle. A potential mechanism for regulating VP16 expression in the context of *in vitro* latency has come from the superior cervical ganglion (SCG) neuron culture model (Cliffe et al., 2015; Kim et al., 2012). The concepts that have emerged from this work are as follows: (i) generalized derepression of the viral genome (Phase I) precedes VP16-dependent reactivation and is required for it (Phase II), and (ii) a histone methyl/phospho switch during Phase I is required for Phase II reactivation. Notably, in vitro neuronal models of latency vary in this regard and some do not display this dual phase reactivation mechanism (Edwards and Bloom, 2019). It will be important to reconcile the findings in diverse in vitro and in vivo models of alpha herpesvirus latency and reactivation to develop testable hypotheses and determine relevant regulatory molecular mechanisms. To frame the present, we start by discussing the early seminal findings.

Creating the latent reservoir: The latent reservoir (which is essential for viral reactivation) is established upon HSV1 reaching neuronal nuclei. The number of latently infected neurons and the number viral genomes residing within each neuron are important parameters that contribute to the potential of future reactivation events. Although latency is the default outcome of neuronal infection in some neurons, HSV1 enters the lytic cycle during acute infection initiated via preIE expression of VP16. This regulated entry into the lytic cycle in the ganglionic neuron modulates virulence while supporting the viral replication that ultimately increases the size of the latent reservoir in the ganglia which occurs during this acute stage of infection. Virus produced in the ganglion is transported back to the body surface, resulting in a positive feedback loop and zosteriform spread. Exit from this "default latency" pathway is balanced by negative factors such as virus and host riboregulators including microRNAs that target viral genes, other repressors such as intrinsic immune functions (promyelocytic leukemia bodies (PML)) and positive factors (*de novo* VP16, multiple viral genome copies).

Preserving the latent reservoir: During latent infection, modification in the chromatin associated with the viral genome evolves into an increasingly repressive state. Despite this repression, there is a low level of transcription from the genome related to lytic genes. The significance of this transcription is not known but could reflect a repressive mechanism working in tandem with the LATs and microRNAs. Replicating virus and viral proteins are not detected during latency in the TG. Long-term studies in the mouse and rabbit reveal the latent reservoir appears to be stable. Expression from the LAT locus is important to maintain reactivation competent latent infections.

Reactivation. Stress results in *de novo* expression of VP16 which coordinates activation of the viral immediate-early (IE) genes. A productive lytic cycle ensues in 0.05% of latently infected neurons per event although more generalized changes in the chromatin associated with the latent genomes occurs. Thus, HSV has devised a complex regulatory strategy that maintains the vast majority of its latent genomes in latency, while allowing the release into a productive lytic cycle in an extremely rare number of neurons. During reactivation, virus does not spread to neighboring neurons in the ganglia but is transported back the body surface where subsequent replication in epithelial cells amplifies virus output and facilitates transmission to new hosts. Spontaneous reactivation in sensory ganglia has been documented in mouse models. Adapted from (Thompson et al., 2009).

How 150 years of research have shaped our ideas about herpesvirus pathogenesis

The earliest scientific investigations on alpha herpesvirus latency were primarily clinical observations. Herpes, an ancient Greek word meaning "to creep", was used by Hippocrates to describe diverse skin diseases. The association between fever and blisters around the mouth and nose was recognized nearly 2500 years ago by the Roman physician Herodotus who described herpes febrilis (Wildy, 1973). The terms "fever blisters" and "cold sores" are still in use today. Through the ensuing millennia the recurrent nature of these types of diseases was appreciated.

Why do herpetic lesions recur? Are they from an infectious agent? What is the nature of the agent?

1863-1930s. Available Toolbox:

- Knowledge of transmission of infection and infectious agents including bacteria and "filterable agents" (the latter the earliest indication of viruses)
- Koch's famous postulates delivered in 1890 (Koch, 1890)
- Identification of a "neutralizing" substance in the blood that countered infections (Behring and Kitasato, 1890)
- Generation of these neutralizing substances, including those to herpes simplex
- Methods to store or serially passage herpes in animal brains (Perdrau, 1925)
- Light microscopy combined with histological staining to study tissue. Neurons distinguished from other cells used in conjunction with agents transmitted through nerves contributed to knowledge of brain structure and innervation from the periphery (Doer, 1920; Friedenwald, 1923; Goodpasture and Teague, 1923)

The association of herpetic lesions with the nervous system was appreciated as early as the mid-19th century (Von Barensprung, 1863). However, it was not until the beginning of the 20th century that clinicians and scientists began to appreciate a relationship between herpetic lesions on the body surface and the sensory nerve endings innervating the skin. In 1892 von Bokay observed that children often developed varicella after exposure to an adult suffering from zoster (von Bokay, 1909). In 1900 Head and Campbell deduced from the pattern of lesions on their patients that herpes zoster must be related to sensory ganglia. Their tour-de-force study (Head and Campbell, 1900), reprinted in part in 1997 (Head et al., 1997), was the first to show how herpetic viral diseases can be employed to help understand the anatomy of the nervous system, a practice that continues today (Sarno and Robison, 2018). Relying heavily on their work, Howard noted that human herpetic lesions of the face were associated with trigeminal ganglionitis and pneumonitis early in the 20th century (Howard, 1903). At about the same time Cushing reported that some individuals treated for trigeminal neuralgia by surgical resection of the trigeminal ganglion and nerve roots developed herpetic lesions in areas innervated by the contralateral (opposite side) nerve but not on the ipsilateral (resected) side (Cushing, 1905), supporting the hypothesis that herpetic lesions were associated with stress or damage to the peripheral nervous system.

Vidal first demonstrated the infectious nature of herpes (Vidal, 1873), but of greater significance to those interested in latency were later studies involving transmission of herpetic stromal keratitis. At the end of World War I the transmission of human herpetic stromal keratitis to rabbit corneas by Gruter (Gruter, 1920; Kraupa, 1920) and Loewenstein (Loewenstein, 1919, 1920) and subsequently transmission back to a human (reviewed in (Holden, 1932)) cemented the idea that herpes was an infectious agent, and provided an animal model for study. During this same time, varicella was transmitted to naive children using vesicle fluid from children with varicella lesions (Kundratitz, 1925); however, attempts to infect laboratory animals failed (Rivers and Tillett, 1924). The lack of an animal model is a difficult challenge for the study of VZV pathogenesis and latency that persists to the present day (Mahalingam et al., 2019; Ouwendijk and Verjans, 2015). A creative approach to overcome the species specificity of VZV is the use of human tissue xenografts in mice with severe combined immunodeficiency (SCID). This model allows the analysis of VZV infection in differentiated human cells in the context appropriate tissue microenvironments, providing many insights into VZV pathogenesis (Moffat et al., 1995; Zerboni and Arvin, 2015; Zerboni et al., 2014). More recent studies using dissociated human TG have provided insight into VZV reactivation (Cohrs et al., 2017).

Several groups noted that herpes infection on the rabbit eye led to infection of the central nervous system (CNS) and changes in rabbit behavior (e.g. turning the head to a particular side), both suggesting a neural route of transmission (Doer, 1920; Friedenwald, 1923; Goodpasture and Teague, 1923). Goodpasture began a series of histological studies delineating herpes spread within rabbits (Goodpasture, 1925a, b; Goodpasture and Teague, 1923) that culminated in the astounding and prescient conclusion that: "Following a primary infection, it seems quite probable that the virus remains in a latent state within the ganglia after the local lesion has healed. A second cutaneous eruption may occur as a result of injury…or the disturbed physiological states... which sets in activity a latent virus" (Goodpasture, 1929). Despite the seemingly overwhelming evidence of nerve involvement in herpes infection this idea was not universally accepted.

1930s-1980s. Toolbox additions include:

- The nature of viruses (Stanely, 1935)
- Tissue and virus culture as well as plaque assays for animal viruses (Dulbecco and Vogt, 1953)
- Ultra-low freezers for virus storage
- Electron microscopy
- Refined histological approaches, fixation methods and immunohistochemistry;
- Conditionally lethal viral mutants
- The "phage Church" and Lambda phage latency, DNA structure and code (Watson and Crick, 1953)
- The central dogma of DNA to RNA to protein
- Liquid DNA/RNA hybridizations
- Bacterial restriction-modification system (S.E. and M.L., 1952) and commercial restriction enzymes
- *In situ* hybridization for DNA and RNA (Gall and Pardue, 1969)
- Radiolabeling of proteins and nucleic acids and various separation methods including chromatography, electrophoretic gel systems, and blotting and probing (Southern, 1975)
- Cloning and cloning vectors (Bolivar et al., 1977)
- Engineered viral mutations
- Genetically modified and engineered animal models, especially mice
- Maxim and Gilbert, then Sanger sequencing

Good and Campbell employed Perdrau's rabbit model of herpetic encephalitis (Perdrau, 1925, 1938) to demonstrate that anaphylactic shock could "precipitate" herpetic encephalitis in previously infected rabbits pre-sensitized to egg albumin (Good and Campbell, 1948). Rabbits were infected with HSV and one to three months after they appeared to be disease free, anaphylactic shock was induced by exposure to egg albumin. Encephalitis occurred in 19/44 tests and virus was recovered from the brains of all rabbits that died and many that recovered. Virus was not detected in the brains of rabbits prior to anaphylaxis. Arguably this was the first well controlled evidence of latency and reactivation of HSV in the nervous system *in vivo*. We now know that intramuscular injection of HSV most likely resulted in virus replication in the skin at the site of injection and subsequent spread to the innervating sensory ganglia. Anaphylactic shock presumably caused virus reactivation within sensory neurons with subsequent spread to the brain resulting in fatal encephalitis but virus reactivation within brain neurons was not ruled out.

Schmidt and Rasmussen explored alternate methods to "precipitate" herpes leading them to favor HSV latency occurs in the skin. Of many methods tried, only intramuscular injections of adrenalin "precipitated" encephalitis (Schmidt and Rasmussen, 1960). Encephalomyelitis was "precipitated" in 60% of the rabbits given intramuscular adrenalin injections and, importantly, herpesvirus was detected in all six of these rabbit brains. A few years earlier at the Wisconsin meeting on Latency and Masking in Viral and Rickettsial Infections (Andrewes, 1957), six types of "latent" infections were described. Herpes was thought to be a unique agent because it could not be cultured from skin between episodes. Herpes latency was defined as the period of time in which skin was negative between outbreaks. Rasmussen speculated "that temporary vasoconstriction resulting from increased adrenalin output, could produce a local anoxia {e.g. local reducing conditions} in the skin and consequent 'reactivation' of residual herpesvirus" (Schmidt and Rasmussen, 1960). This hypothesis harkened back to the early observations of Perdrau. Early virologists stored their virus stocks as bits of infected rabbit brains in glycerin in ice boxes. Exclusion of air greatly increased the time such a stock remained infectious. Perdrau discovered that oxidation destroyed herpesvirus activity, but the virus stocks could be "reactivated" by subsequent reduction (Perdrau, 1931).

Rasmussen replaced the commonly employed term "precipitate" with "reactivate" directing thought toward the concept that an active agent has been muted to a latent agent, pre-existing *in toto*, that is somehow reactivated (presumably by reducing conditions in vasoconstricted skin) to become infectious. Latent herpes was thought by most to be present in the skin and this was an early example of a "skin trigger" hypothesis: an idea that has recently regained some attention and is further discussed below.

The early neural work by Goodpasture was supported by the formation of cutaneous herpes lesions following surgery in the trigeminal nerve tract. For example Carton and Kilbourne noted that within a few days after section (axotomy) of the fifth cranial nerve, oral or facial lesions occurred in 90% of patients if the ganglion and nerve were not destroyed (Carton and Kilbourne, 1952). The existence of a dormant form of herpes in the TG was supported by the failure of several groups to isolate the infectious agent from human TG (Burnet and Williams, 1939; Carton and Kilbourne, 1952). However, observation of HSV lesions in skin following "blowout" fractures that destroyed the nerve tract innervating the site of recurrence suggested that the virus must have already been in the skin (Hoyt and Billson, 1976). We now know the complexity of innervation of the facial skin provides a plethora of alternate routes for virus transport to the skin.

HSV-2 is discovered and VZV is proposed to be a virus that goes latent in sensory ganglia

Plummer determined that there were two serotypes of herpes simplex virus (Plummer, 1964), with different biological properties (Plummer et al., 1968), and was the first to show reactivation of HSV in the peripheral nervous system (albeit in the absence of a critical control as detailed below) (Plummer et al., 1967). But, in the absence of a tractable animal model for varicella zoster virus

(VZV) infection, the idea that zoster was the result of reactivation of latent VZV arose instead from a long-term epidemiological study (Hope-Simpson, 1965). Formal proof that the same strain of VZV could cause both chickenpox and, subsequently, shingles in the same patient was obtained 20 years later (Straus et al., 1984)

While HSV had been isolated from skin, saliva, tears and mucosal surfaces (especially genital mucosa) by many different groups in the absence of frank lesions, the source of this virus was unknown (reviewed in (Stevens, 1975a, b)). Transplantation of facial skin from sites where lesions occurred to other parts of the body did not result in virus recurrence at those sites, and most efforts to isolate virus from such skin either directly or following explant into culture failed (reviewed in (Finlay and MacCallim, 1940; Hill, 1985)). Notably, in the absence of lesions, very low titers are usually found (Agyemang et al., 2018). And yet, viral DNA is detectable by PCR in genital swabs much more frequently than is infectious virus as discussed below. A potential role for extra-neuronal sites of HSV latency in these phenomena should not be discounted.

During the 1950s to late 1960s laboratory animal models of herpetic disease were further developed, notably in rabbits, guinea pigs, and mice. The use of these models has led to several seminal observations, including the discovery that the two serotypes of HSV (HSV1 and HSV2) caused different pathologies in mice (e.g. HSV-2 is much more virulent than HSV-1 and causes larger pox (plaques) on liver following intraperitoneal injection) (Plummer, 1964; Plummer et al., 1974; Plummer et al., 1968). In 1970 the first experimental evidence that HSV2 remained in neural tissues for many months was presented. HSV2 was isolated *in vivo* after adrenalin injection, and also by cultivation of trypsinized rabbit neural tissue with indicator cells that presumably resulted from reactivation *in vitro* (Plummer et al., 1970).

What is the cellular site of HSV latency and reactivation?

Cook and Stevens confirmed and extended the observation that HSV remained latent in neural tissues by developing a method to reactivate HSV-1 from dorsal root ganglia (DRG) of mice infected via the rear footpad. Following infection, virus could be isolated from feet, the PNS, and CNS with replication ceasing about 8 days post infection (dpi). Thirty dpi DRG were either excised and homogenized immediately, or explanted and co-cultivated on monolayers of susceptible cells. Infectious virus was not detected in DRG that were directly homogenized, but virus was produced by the great majority of explanted ganglia 7 to 14 days post explant. This report is often considered the first definitive proof that HSV resides within sensory ganglia in a non-infectious or latent state and is reactivated following the stress of axotomy and explant into culture (Stevens and Cook, 1971). Numerous groups subsequently applied this approach to diverse human tissues and detected reactivated HSV1 in sensory and autonomic ganglia (Baringer and Swoveland, 1973; Bastian et al., 1972; Plummer, 1973).

Stevens and colleagues went on to show that latent HSV1 could be reactivated only from central and peripheral neural tissues, including the adrenal medulla, following intravenous injection of virus to induce an artificial viremia. This solidified the idea that nervous tissues are the predominant sites of HSV1 latency (Cook and Stevens, 1976). This group also sought to demonstrate that the sensory neuron was the site of viral reactivation that led to skin lesions through the use of immunohistochemical and electron microscopic techniques (Cook et al., 1974), but their data cannot be unambiguously interpreted since tissue was examined at 48 hours and later following axotomy, a time now known to be confounded by secondary spread of reactivated virus within the explanted ganglia. The virus could have reactivated in cells other than neurons and then spread to them. The earliest solid evidence for virus reactivation in neurons was provided by the studies McLennan and Darby who employed temperature sensitive (ts) mutants in mice. At a core body temperature of 38.5 degrees, replication of the ts mutants was restricted and viral proteins in mice in which nerves were resected to induce reactivation *in vivo* were confined to neurons (McLennan and Darby, 1980). These neurons "appeared" to die.

Are any essential viral gene products required for the establishment or maintenance of latency?

Watson et al. and Lofgren et al. explored whether viral proteins essential for the viral lytic cycle were required for the establishment of latency by intra-cranially inoculating mice with ts viral mutants and asking if the mutants could later be reactivated from brain tissue at the permissive temperature *in vitro* (Lofgren et al., 1977; Watson et al., 1980). The major viral immediate early transactivator ICP4 was initially thought to be essential for latency establishment, a hypothesis disproved much later by Stevens (Sedarati et al., 1993). Therefore, even viral proteins essential for viral replication like ICP4 are not absolutely required for the establishment or maintenance of latency.

Is the latent HSV genome integrated into the host genome?

Alpha herpesvirus genomes are linear double-stranded DNA that circularize upon entry into the nucleus (Sheldrick and Berthelot, 1975). The physical state of the viral genome during latency was the subject of much speculation. Fraser and colleagues first demonstrated "endless" HSV DNA in latently infected mouse brain tissue by restriction endonuclease digestion and Southern blotting. Later, this and other groups confirmed that circularized genomes are also the predominant form found in latently-infected mouse TG (Efstathiou et al., 1986; Fraser et al., 1981; Rock and Fraser, 1983). We now know that the great majority of HSV and other human alpha herpesvirus latent genomes exist as extra-chromosomal circular episomes (Azarkh et al., 2010). While integration events cannot be ruled out, they are not thought to be biologically relevant. Curiously, pseudorabies virus (PRV), an alpha herpesvirus of swine, is reported to persist largely in a linear form along with some circularized genomes (Rziha et al., 1986).

Spontaneous reactivation from latency in mice

HSV is commonly stated to not spontaneously reactivate in the mouse. This misconception is argued against by the consistent levels of spontaneous reactivation that have been reported for over three decades. In the late 70s and early 80s, Hill and Blyth developed an informative model of *in vivo* reactivation, recurrence, and recurrent disease (recrudescence) in mice. Infection of the mouse ear pinna resulted in latent infections in superior cervical ganglia (SCG) that could be reactivated by diverse stimuli including UV light, DMSO on the pinna, or stripping of the pinna with cellophane tape (Hill et al., 1975). In this model, virus could be routinely isolated from the pinna in about 10% of the animals on any given day, consistent with later estimates of spontaneous productive reactivation seen within mouse TG (Blyth et al., 1984; Blyth et al., 1981). They showed that some procedures to induce *in vivo* reactivation resulted in detectable virus within ganglia and in skin with no evidence of recurrent disease, whereas other procedures induced virus reactivation within ganglia and recurrent lesions that contained infectious virus. This led them to postulate that there were "ganglionic triggers" and "skin triggers", the latter leading to recurrent lesions with the possible involvement of prostaglandins (known to enhance the replication of HSV in cultured cells), which were found to be elevated in the skin selectively by "skin triggers" and not "ganglionic triggers" (Harbour et al., 1978; Harbour et al., 1977), for review see (Hill, 1985). This hypothesis may still be relevant today as it might explain why infectious virus, and more frequently viral DNA, can be detected in the absence of a frank lesion.

The development of a sensitive procedure to detect viral protein expression in the whole ganglion allowed extensive analysis of latently infected ganglia for the expression of viral proteins at a single cell level (Sawtell, 2003). Whole ganglion IHC (WGIHC) allowed for detection of single neurons positive for viral proteins. A comprehensive long-term study revealed that in mice, as in humans (Agyemang et al., 2018), the frequency of spontaneous HSV reactivation (based on detection of TG neurons positive for infectious virus and viral proteins) declined during the time period between 15 and 40 days and then remained stable over the 240 days examined (Sawtell, 2003). The frequency of spontaneous reactivation in the TG was similar to the frequency observed in SCG, which also corresponded to the frequency of spontaneously positive mouse ear pinnas as reported by Hill and colleagues (Blyth et al., 1981) (reviewed in (Hill, 1985)). Analyzing sectioned ganglia for HSV proteins,

Feldman et al. similarly found that 10% of mouse TG pairs contained neurons positive for HSV proteins and called this "molecular reactivation" (Feldman et al., 2002). Later, the same group determined that 10% of unstressed mice were positive for infectious virus in TG (Margolis et al., 2007) demonstrating that molecular reactivation is consistent with the well-recognized spontaneous reactivation rate of mice as previously reported by others. This rate is likely dependent upon the viral strain, latent reservoir, and mouse strain. The animal models noted for spontaneous reactivation, rabbit and guinea pig, also have a time-dependent reduction in the frequency of reactivation, with the highest rate of spontaneous reactivation observed being 20-35 dpi. In rabbits, reactivation is quantified by virus shedding in tear films. The source of this virus is not known but thought to be neurons in TG. In guinea pigs, vaginal lesions are scored as reactivation events. Viral DNA can be recovered from these lesions (Roizman et al., 2013). Thus, spontaneous reactivation of HSV occurs in diverse animal models including mice, rabbits and guinea pigs.

The timing of viral reactivation in neurons in vitro or in vivo following stress.

The kinetics of induced reactivation, both *in vivo* and in explanted ganglia, remained a major gap in our understanding of alpha herpesvirus infections. Although knowing when a triggering event occurred that resulted in spontaneous reactivation is not possible, an increased frequency of ocular shedding occurred in rabbits over a period of two weeks following a three-day procedure of iontophoresis of epinephrine into corneas to induce reactivation (presumably initiated in the innervating neurons) (Toma et al., 2008). Likewise reactivation in explanted ganglia was thought to take several days or even weeks (Stevens, 1975a). However, the rate of reactivation was faster in subsequent studies. While a variety of stimuli caused virus reactivation and spread to the mouse ear pinna in 3 to 5 days, Harbour et al. found that virus could be isolated from a few SCG as early as one day post treatment (Harbour et al., 1983). Furthermore, using a mouse model of *in vivo* HSV1 reactivation induced by a 10 minute hyperthermic stress (42.5 C), virus was detected in TGs of mice as early as 14 hrs with 70% of mice positive by 22 hrs (Sawtell and Thompson, 1992a) (reviewed in (Webre et al., 2012)).

The same timing of primary reactivation events is evident in axotomized and explanted TG with virus being detectable by 14 hours post explant. However, unlike the *in vivo* situation where virus does not spread from the original reactivating neurons, virus rapidly spreads within the explanted ganglia (Doll and Sawtell, 2017; Pesola et al., 2005; Sawtell and Thompson, 2004). The potential for reactivation to occur in explants within 14 hrs and the rapid spread of virus within the ganglion is inconsistent with conclusions drawn from earlier studies that reactivation requires several days to occur in explanted sensory ganglia. However, it should be noted that many factors could affect the levels of latency established in animal models and in turn the subsequent frequency and perhaps timing of viral reactivation in vitro and in vivo. These parameters include the method of inoculation, the virus inoculation titer, the strain of virus employed, and strain of animals employed.

What is the role of the immune system in maintaining latency?

As is the case today, the role of the host immune system in alpha herpesvirus latency and reactivation was of great interest to investigators. However, immunosuppression of latently infected mice failed to induce reactivation or produce recurrent disease (Blyth et al., 1981; Hurd and Robinson, 1977; Stevens and Cook, 1973), with only a low incidence of recurrence in hairless mice treated with the immunosuppressant drug, prednisone, being observed (Underwood and Weed, 1974). Conflicting with these findings is the more recent hypothesis that CD8 T-cells directed against a specific epitope on glycoprotein B may play a role in the maintenance of HSV1 latency (Bourne et al., 2018; Held and Derfuss, 2011; Knickelbein et al., 2008; Lahmidi et al., 2017; Liu et al., 2000; Treat et al., 2017). In contrast to HSV, depletion of CD4 cells induced reactivation of simian varicella virus in Rhesus Macaques (Traina-Dorge et al., 2019). The role of the host immune system in HSV and VZV latency and reactivation (Gershon et al., 2015) is a topic worthy of a its own chapter and is not covered in further depth here.

In vitro primary neuronal quiescence/latency and the NGF depletion reactivation model

The application of cultured primary neuron models of HSV latency led to several new concepts in reactivation. Establishment of a quiescent/latent like state in cells including neurons requires the use of antiviral compounds, or the use of replication deficient viral mutants (Harris and Preston, 1991; O'Neill et al., 1972; Wilcox and Johnson, 1987). During the late 1980s Wilcox and colleagues developed an *in vitro* model of HSV latency in cultured primary neurons. Quiescent/latent infections were established by infecting the neurons at very low MOI and maintaining them in the presence of human IgG (which contains significant amounts of anti-HSV antibodies). The antibodies were removed after fourteen days and the cultures remained latently infected for at least five weeks. These autonomic neurons derived from SCG were dependent on nerve growth factor (NGF) for their survival and deprivation of NGF (by anti-NGF antibody depletion) resulted in reactivation of latent virus (Wilcox and Johnson, 1987; Wilcox et al., 1990). These cultures were interesting in that they mimicked some of what was known about latency *in vivo*, including expression of the LATs (stable LAT introns) in a subset of neurons (Doerig et al., 1991). However, they did not parallel the *in vivo* situation in all cases. For example, thymidine kinase negative mutants established latency and reactivated within these cultures (Wilcox et al., 1992), which is not the case for either *in vivo* reactivation in mice, or reactivation in explanted latently-infected mouse ganglia (Field et al., 1982; Field et al., 1979; Field and Wildy, 1978; Pyles and Thompson, 1994b). Reactivation in these cultures could be induced by cyclic AMP repressors and was dependent on caspase three (Hunsperger and Wilcox, 2003). This model was eventually abandoned in part because latent infections (identified by LATs expression) and reactivation was rare at the cellular level and it was not thought likely that widespread death of neurons caused by NGF deprivation would be relevant biologically except with a few exceptions such as nerve resection caused by "blowout" fractures or surgery (Hunsperger and Wilcox, 2003).

A resurgence of similar in vitro models of HSV latency are the topic of recent reviews (Bloom, 2016; Thellman and Triezenberg, 2017) and new in vitro models of VZV latency are under development (Baird et al., 2019; Depledge et al., 2018b; Laemmle et al., 2019). The ability to differentiate pluripotent human embryonic stem cells into neurons is being exploited to investigate latency and reactivation of HSV and VZV (for example see (D'Aiuto et al., 2019; Kurapati et al., 2017; Markus et al., 2015; Pourchet et al., 2017; Sadaoka et al., 2018)). Rodent fetal and adult neuronal culture models are also being used and are discussed further below.

Using molecular approaches to address fundamental questions

Discovery of the latency-associated transcripts (LATs) or latency related RNAs (LRRs)

There was great interest in determining whether any viral gene expression was required to maintain latency. The first report of abundant RNA transcription during HSV latency in neurons was published by Tenser and colleagues. However, the entire viral genome was employed as a probe and so the genomic location of this transcription was not known (Tenser et al., 1982). In 1986 two abstracts describing significant RNA transcription from the terminal repeat of HSV-1 during latency were submitted to the 11th annual International Herpesvirus Workshop. Stevens and colleagues described the transcription of ICP0 during latency but employed a double stranded probe. Rock and Nesburn also detected transcription but from the strand opposite of that of ICP0. Stevens et al. became aware of Rock's strand specific finding, confirmed that the transcription they detected was from the strand opposite of that of ICP0 and first described the latency associated transcripts (LATs) (Stevens et al., 1987). Arguably, credit for this discovery might be shared.

The discovery of the LATs led to speculation that they might encode a protein or a function important for the establishment, maintenance and/or reactivation of latent infections. However, the earliest reports suggested that the LATs played no role in the establishment or reactivation of HSV latency (Sedarati et al., 1989) (reviewed in (Fraser et al., 1992)). Subsequent early studies led to confusion and controversy, largely due to the complex nature of transcription from this region that was not appreciated at the time. We now know that long, short, and micro non-coding RNAs from both DNA strands as well as unusually stable introns are generated from the LAT locus, and that most of these can be detected during both latency and during acute viral replication. Presumably expression of the LATs introns and miRNAs during the lytic cycle are the result of read through transcription late in the infection cycle. Activity from the LAT promoter is restricted during acute infection (Batchelor and O'Hare, 1990). Additionally, there are reports noting potential open reading frames (ORFs) in some of these RNAs (Bloom et al., 1996; Thomas et al., 1999; Thomas et al., 2002), but no latency related protein has yet been convincingly described. In contrast, the recently discovered VZV LAT RNA does encode a protein (Depledge et al., 2018a). The role of the LAT locus will be discussed in greater detail below.

Quantification of the number of genomes in the latently infected ganglion by the polymerase chain reaction (PCR)

Katz et al. first employed PCR to detect latent viral genomes, which laid the groundwork for the use of qPCR to address questions in HSV latency (Katz et al., 1990). These investigators confirmed the early finding of Stevens and colleagues that latent infections in sensory neurons could be established even with HSV mutants that did not replicate efficiently, including TK- and ICP4-deficient mutants, and added to this list mutants lacking ribonucleotide reductase, ICP27, and DNA polymerase (Katz et al., 1990; Lofgren et al., 1977; Watson et al., 1980). More startling was the high numbers of viral genomes Katz et al. detected in latently infected ganglia (Katz et al., 1990). While many groups had observed only a few to several hundred neurons expressing LATs, PCR detected hundreds of thousands to millions of viral genomes (reviewed in (Phelan et al., 2017)). This suggests that either neurons contain multiple copies of the viral genome or that many neurons harboring latent genomes do not express detectable LATs, or both. We now know that there are 1,000s of latently infected neurons in the TGs of mice and humans, each containing 10-100s of viral genomes and most of these do not express detectable LAT stable introns. (Sawtell, 1997, 1998; Sawtell et al., 2001; Wang et al., 2005b).

How many genomes are in individual latently infected neurons? Latency at the single neuron level

While the qPCR method provided unprecedented sensitivity for detecting and quantifying viral genomes in latently infected TG, methods to examine viral DNA in single cells were needed to address outstanding questions. For example, how many neurons were latently infected, were other cells also harboring viral genomes, how many viral genomes were within individual latently infected neurons, and was this genome number uniform or variable? Towards that goal, an approach termed contextual analysis (CXA) was develop. TGs were perfusion-fixed to stop all metabolic processes and to preserve DNA, RNA and protein, and the cells were dissociated and purified on Percoll gradients. Following treatment with DNase linked to beads to remove DNA from the outside of the cells, the intracellular viral genome content of individual neurons and other cells was analyzed by qPCR. Viral genomes were detected in about 1/4th of the neurons, only in neurons and with copy numbers ranging from ~1 to >10,000 (Sawtell, 1997). Similar percentages of positive neurons and ranges of viral genome copies were later detected in human TG neurons using laser capture technologies combined with qPCR (Wang et al., 2005b). Other single-cell approaches addressing these questions were applied to VZV in human TG (LaGuardia et al., 1999; Levin et al., 2003).

The viral genome copy number in individual latently-infected neurons varies across HSV1 strains (Sawtell et al., 1998). Neurons latently infected with the virulent HSV1 strains McKrae or 17Syn+ contained an average of 81 ± 36 or 50 ± 12 genome copies, respectively. However, neurons latently infected with the avirulent strain KOS contained only 7 ± 2 viral genome copies, which may explain why it reactivates less well than the other strains (Sawtell et al., 1998). Yet, the percentage of latently infected neurons was similar with all three strains, varying between 26 and 32%. The reason for the low genome copy number in KOS latently-infected TG neurons is not known but may be a reflection of the mutations in the KOS strain that disrupt the US9 gene TATA box, truncate US9 at aa58, and eliminate the native stop codon of pUS8A (Negatsch et al., 2011). pUS9 is required for efficient anterograde axonal transport (Howard et al., 2013), and pUS8A is a neurovirulence factor (Kato et al., 2016). Thus, the multiple "round trips" of virus from the body surface to nerve cells and back again, which contributes to the higher viral genome copy number latency (Thompson and Sawtell, 2000), may not occur with the HSV1 KOS strain and its derivatives.

Do any viral proteins enhance the establishment of latency?

A number of viral genes contribute to the efficiency with which latency is established. However, many of these are required for efficient lytic replication at body surfaces or in the nervous system. Surface replication efficiency is correlated with the number of latent sites established and, therefore, viral proteins important for lytic replication such as thymidine kinase appear to increase latent infections indirectly (Katz et al., 1990; Thompson and Sawtell, 2000). Other virally-encoded enzymatic functions such as dUTPase and uracil glycosylase enhanced establishment of latency and reactivation frequency likely by promoting replication within the trigeminal ganglia (TG) (Pyles et al., 1992; Pyles and Thompson, 1994a). The lack of these enzymes may also have led to a higher mutation frequency in the latent viral genomes that could exert a deleterious effect on reactivation from latency (Pyles and Thompson, 1994b). In aggregate these and similar studies on diverse viral proteins (i.e. TK, ribonucleotide reductase, virion host shut off, and various glycoproteins) demonstrate that no viral proteins are absolutely essential for the establishment of latency (Aggarwal et al., 2012; Diefenbach et al., 2008; Field et al., 1982; Field and Wildy, 1978; Izumi and Stevens, 1990; Jacobson et al., 1989; Johnson et al., 1986; Lam et al., 1996; Meignier et al., 1988; Smith, 2012; Strelow and Leib, 1995; Wang et al., 2005a; Wang et al., 2010). However, all viral proteins likely contribute to the efficiency with which latency is established indirectly by enhancing virus replication at body surfaces and in the nervous system.

Viral lytic gene transcripts are present in latently infected ganglia in the absence of detectable reactivation

In a landmark study, Kramer and Coen found transcripts for ICP4 and TK in latently infected mouse TG. Quantification of these transcripts revealed that they were present at extremely low levels. These transcripts did not necessarily initiate at recognized promoters and may have been the products of random transcriptional activity (Kramer and Coen, 1995). Thompson and Sawtell also reported RNAs related to ICP4, ICP0, ICP22, ICP27 and ICP47 in TG latently infected with HSV1. Transcription was found upstream of normal mRNA initiation sites. Furthermore, the ICP0 transcripts were not spliced as is the case during productive infection, again suggesting these might not be properly processed mRNAs, but rather random transcriptional events (Thompson and Sawtell, 2006). The presence of HSV related RNAs during latency has complicated the interpretation of results that utilize viral transcriptional activity as a marker of reactivation. This is especially true in vivo where reactivation occurs in only one or a very few neurons per ganglion and the neuronal distribution of the RNAs detected is not known. Sensitive whole ganglion approaches to detect viral transcription at the single neuron level during latency and in vivo reactivation are needed.

In depth considerations of latency models and mechanisms

The role of VP16 transactivation function in the switch between latent and lytic infection

Virion protein 16 (VP16, α-TIF, pUL48) is produced as a leaky late protein and is packaged into the viral tegument. Upon viral fusion into cultured cells and concomitant loss of the envelope, VP16 is released from the nucleocapsid/ tegument and is thought to complex with the host cell factor 1 (HCF-1) in the cytosol. The HCF-1/VP16 dimer is transported into the nucleus where it forms the trimeric VP16-induced complex (VIC) with Octamer-binding protein-1 (Oct-1). The VIC binds to the consensus sequences, TAATGARAT, which are present in the promoters of the five immediate early (IE) genes of HSV. The acidic carboxy terminal domain of VP16 is a strong transcriptional activator, which then initiates the virus lytic gene transcription program by transactivating the IE genes (Ace et al., 1988; Campbell et al., 1984; Kristie and Roizman, 1987; Mackem and Roizman, 1982; O'Hare, 1993), reviewed in (Wysocka and Herr, 2003). The importance of VP16 in initiating the viral lytic cycle is revealed at low moi. In its absence, entry into the lytic cycle (plaquing efficiency) is reduced 100 to 1,000-fold (Ace et al., 1989; Smiley and Duncan, 1997).

Based on what is known about the critical role played by VP16 in initiation of the viral lytic cycle at low moi, its absence or presence would also be expected to play central roles in the establishment of, or reactivation from latency. Indeed, in cultured neurons, VP16 deposited into axons as part of the tegument complex is not transported retrograde to the neural soma with the nucleocapsid, and in the absence of its delivery to neuronal nuclei latency establishment may be favored (Aggarwal et al., 2012; Antinone and Smith, 2010). Importantly, cultured neurons infected via axons favors the entry of virus into a quiescent/latent state (Hafezi et al., 2012) and the addition of pseudorabies virus tegument proteins to the neuronal soma can shift the outcome to the lytic cycle (Koyuncu et al., 2017), a result similar to early studies of HSV light particles containing tegument proteins including VP16 (Dargan et al., 1995). As discussed further below, VP16 and its unique regulation as a pre immediate early gene in neurons orchestrates the choice between lytic and latent viral programs.

The generation of a transactivation-deficient mutant (VP16TF), in1814, provided the critical tool needed to test the role of VP16TF during latency and reactivation (Ace et al., 1989). This mutant contains a 12 bp insertion at aa379 in VP16 that retains the protein's essential contribution to virion structure but selectively disrupts the interaction of VP16 with Oct-1, thus preventing the formation of the VIC (Ace et al., 1988; Ace et al., 1989; Campbell et al., 1984; Wysocka and Herr, 2003). This and other VP16TF mutants are severely deficient in replication at low moi (McFarlane et al., 1992; Preston and McFarlane, 1998; Smiley and Duncan, 1997). Despite this, in1814 establishes latent infections in mice and subsequently reactivates with wild type kinetics from latently-infected ganglia explanted into culture (Steiner et al., 1990). Moreover, attempts to artificially-induce the expression of VP16 *in vivo* did not disrupt the balance between latent and lytic infection (Sears et al., 1991). These two influential reports seemingly disproved the hypothesis that VP16 played an important role in regulating HSV latency and reactivation.

However, whether the VP16TF studies truly ruled out an initiating role of VP16 in reactivation requires a nuanced consideration of experimental design and interpretation. First, the method used to evaluate reactivation is a central factor. The most widely used approach to evaluate the ability of latent virus within a ganglion to produce infectious virus is to dissect the infected animal, sever the ganglionic neurons by axotomy, and explant the ganglion into culture (Stevens and Cook, 1971), sometimes also dissociating or mincing the ganglion prior to explantation (Blyth et al., 1981; Leib et al., 1989a; Nicholls and Blyth, 1989). Using this method, virus was detected five to ten days post explant, which was not different than the five to six days required for wildtype and rescued virus reactivation (Steiner et al., 1990). At the time, these studies and others led to the conclusion that VP16 transactivation was not involved in reactivation from latency, and as a consequence the IE protein ICP0 superseded VP16 as the major contender for this important role. However, whether explant-based reactivation accurately models *in vivo* reactivation is an important consideration (Sawtell and Thompson, 2004; Thompson et al., 2009).

Although there is general consensus that ICP0 is required for efficient reactivation from latency. *In vivo*, ICP0 is required for efficient viral replication and establishment of latency (Cai et al., 1993; Cai and Schaffer, 1992; Everett, 2000; Halford and Schaffer, 2001; Leib et al., 1989b), and ectopic expression of ICP0 can disrupt latency in explanted ganglion neurons (Halford et al., 2001). The intersection of latency establishment levels and reactivation efficiency have clouded the issue but it is clear that at equivalent levels of latency (number of neurons and copy number profile), ICP0 null mutants enter the lytic cycle (based on viral protein expression) but do not generate detectable levels of infectious virus (Thompson and Sawtell, 2006). Thus, while ICP0 is essential for amplification of infectious virus production during reactivation *in vivo* this protein is not required, and appears to play no major direct role, in the initiation of reactivation *in vivo*.

Whether the contribution of ICP0 is to the initiation or progression of reactivation may seem to be a hairsplitting issue, but as discussed above the goal of these studies is to identify the interfaces between the critical host cell factors and their mechanisms of regulating essential viral functions. An early effort to achieve this used a viral mutant, ΔTFI, that was designed to gain insight into the role of cis-acting sequences in the ICP0 promoter that potentially regulate ICP0 expression during acute infection, establishment of latency, and reactivation (Davido and Leib, 1996). A deletion removed a number of sites in the proximal ICP0 promoter including Sp1, NF-kB, C/EBP, F2, and four TAATGARAT motifs. Strikingly, although there were reduced levels of ICP0 protein expression at early times post-infection in Vero cells and *in vivo*, and reduced levels of blepharitis in CD1 mice and death in SCID mice, neither establishment of latency nor explant reactivation were different from the wild type or the marker rescue viruses (Davido and Leib, 1996).

Our group was also focused on ICP0 and its role in reactivation, but we were utilizing an *in vivo* reactivation model triggered by hyperthermic stress to the intact animal (Sawtell and Thompson, 1992b). Whether reactivation *in vivo* was different than explant reactivation was of great interest. Remarkably, in1814 failed to reactivate and reactivation of ΔTFI was greatly reduced compared to wild type or marker rescue viruses *in vivo* (Thompson and Sawtell, 2006) (Thompson et al., 2009). Importantly, viral genome copy number and the number of latently infected neurons were not different, ruling out the possibility that the establishment of latency by in1814 and ΔTFI were reduced in this model. These studies revealed for the first time that reactivation in the *in vivo* context may have distinct constraints, which are relaxed in the explant setting.

These studies also suggested that the TAATGARAT motifs in the ICP0 promoter might be important in regulating reactivation *in vivo*. Specific restoration of a single TATA box proximal TAATGARAT element to the ICP0 promoter in ΔTFI restored wild type *in vivo* reactivation frequency, emphasizing the importance of VP16 interaction at TAATGARAT elements for the regulation of ICP0 expression during *in vivo* reactivation (Thompson and Sawtell, 2006). It was not yet clear if reactivation of the VP16TF mutant was actually blocked, or whether in1814 did reactivate and was not detected because of the ~100-fold reduced plaquing efficiency of the mutant (Ace et al., 1989). Even with wild type virus the levels of virus recovered from a reactivation event *in vivo* are low, and virus does not spread within the ganglion during in vivo reactivation as it does during explant induced reactivation (Doll and Sawtell, 2017). Thus, reactivation of in1814 in vivo may have occurred without being detected. This was tested by the inclusion of hexamethylene bisacetamide (HMBA), a cyto-differentiating agent used to complement the plaquing deficiency of in1814 (Ace et al., 1989), in the assays, reactivation in vivo (infectious virus production) was not detected in assays in the presence of HMBA.

We hypothesized that some yet unknown regulatory mechanism altered the normally leaky late expression of VP16 to that of a pre-immediate early gene in a neuron destined to reactivate, thereby permitting it to coordinate the expression of the five important IE genes and initiate the lytic cascade (Thompson et al., 2009). If this were the case, then in contrast to ΔTFI, in which viral proteins were expressed in neurons "starting" to reactivate, neurons expressing viral proteins in in1814 infected TG would be absent following stress. Using a whole ganglion approach that allows quantification of viral protein expression at the individual neuron level, the lack of neurons expressing viral proteins was confirmed (Thompson et al., 2009). This extended our understanding of the reactivation process now to include VP16 transactivation of ICP0 (and presumably the other IE genes as well but this has not been directly tested) as a key and absolutely required step in the earliest stages of viral protein expression from the latent genome. Of great importance, mutations in the VP16 promoter, including a four base substitution in a G+C rich region scoring as a putative overlapping Egr-1/SP1 site ablated the ability of the virus to exit the latent state (Sawtell and Thompson, 2016; Thompson et al., 2009; Thompson and Sawtell, 2019). These findings are consistent with the hypotheses that both a functional ICP0 and VP16 transactivation were required for efficient virus production, but de novo pre-immediate early expression of VP16 is required to initiate viral gene expression during reactivation.

Additional supporting data comes from a series of refined viral mutants containing single or double amino acid changes in the core domain critical for VIC formation (Sawtell and Thompson, 2016; Sawtell et al., 2011), including the first viral mutant to specifically disrupt the HCF-1 interaction. The *in vivo* reactivation phenotypes observed for these mutants align with predictions based on in vitro biochemical analyses of amino acids critical for VIC formation (Stern et al., 1989; Wysocka and Herr, 2003). Together, there is now strong evidence that VP16 interactions with Oct-1 and HCF are critical for the earliest stages for reactivation *in vivo* (Figure 1). That the core domain mutant phenotype was a result of a deficit in VP16 transactivation was supported further by the *in vivo* reactivation impairment of a mutant in which the core domain was intact, but the acidic transactivation domain (TAD) was lacking (Thompson et al., 2009). Importantly, all of these mutants reactivate following ganglion explantation, emphasizing the importance of the reactivation model used and the associated neuronal metabolic state. Efforts to develop and characterize additional *in vivo* reactivation triggers to ask whether hyperthermic stress is representative of other stressors (i.e., do diverse stressors flow into a common interface on the viral genome) is in progress. In vivo reactivation by additional triggers, for example, local skin trauma (see below), is dependent upon VP16 transactivation.

Modeling reactivation *in vitro*

The development of an *in vitro* latency model that provides a platform for dissecting host cell signaling pathways involved in viral genome entry into latency, its maintenance, and its reactivation would be a valuable resource. Ideally such models could yield hypotheses testable in the more complex *in vivo* setting. Defining the limitations to these models and the approaches used to probe outcomes is critical to interpreting results obtained from them.

There is a growing number of *in vitro* latency models for both HSV and VZV that are the topic of several recent reviews (Bloom, 2016; Depledge et al., 2018b; Thellman and Triezenberg, 2017). *In vitro* models have been exploited to study both the establishment of latency and subsequent reactivation of VZV. Major efforts in the use of human embryonic stem cells for the study of VZV have led to several critical improvements, including axon-restricted infection protocols and methods to obtain enriched populations of specific types of neurons (Kurapati et al., 2017; Markus et al., 2015; Pourchet et al., 2017; Sadaoka et al., 2016; Sadaoka et al., 2018). These models are being explored for HSV latency and reactivation as well (Pourchet et al., 2017) and have brought forward the concept of a two-phase paradigm of transcription leading to reactivation. It should be noted that the establishment of latency in most of these models requires the use of antiviral compounds such as acyclovir, and latency is considered to be established after just a few days. It is not yet clear how this "forced" latency parallels latency in vivo. A primary limitation of these models is the disruption of the intricate context of the neuron and the very different mode of entry of the virus into the neuron. Roizman noted that satellite cells which surround neurons and communicate constantly with them may play roles in the establishment and maintenance of latency. Also noted is the fact that the acyclovir employed to suppress virus replication does not prevent the initial stages of reactivation and in effect might serve to eliminate those neurons in which spontaneous reactivation occurs (Roizman and Whitley, 2013). Immune functions also are thought to play important roles in these processes and are absent (for review see (St Leger and Hendricks, 2011)).

Reactivation in the superior cervical ganglion neuron culture model

As discussed above a primary SCG neuronal culture model of HSV quiescence/latency and reactivation was reported to require nerve growth factor (NGF) to maintain HSV latency/quiescence (Wilcox and Johnson, 1987, 1988; Wilcox et al., 1990). One current hypothesis is that the sustained signaling program mediated by NGF via the TrkA receptor, PI3 kinase p110a isoform, PDK1 and Akt represses HSV replication, maintaining the quiescent state (Camarena et al., 2010; Kim et al., 2012). A mutant containing a GFP-US11 fusion gene was employed so that fluorescence would mark lytic gene expression. The PI3 kinase inhibitor LY294002 was found to parallel NGF withdrawal in that the number of GFP positive wells (6 days post treatment) increased 3.5-fold from the baseline of 20% (time 0) to 70% (48-72 hrs post LY294002 treatment). Switching of wells from GFP negative to positive (a surrogate for reactivation) did not occur in 30% of wells and the number of neurons expressing GFP (the marker of lytic cycle employed) averaged 5% (48/1000) (Cliffe et al., 2015). The number of neurons containing latent viral genomes was estimated at 25-50% (stated in the discussion of (Kim et al., 2012)); thus, reactivation occurs in the range of 10-20% of infected neurons. PI3 kinase inhibition does not induce reactivation in most of the neurons in the well that contain the viral genome.

Presumably all of these embryonic SCG express the NGF TrkA receptor, indicating that additional factors are preventing viral genome activation following inhibition of PI3 kinase. Considering reactivation at the level of the individual latently infected neuron treated with the inhibitor is actually rare, sustained signaling through PI3 kinase and Akt would not seem to be universally required to maintain latency in this culture model (Camarena et al., 2010). This model system allowed detailed analysis of the role of individual growth factors; a study not practical in the *in vivo* context. Interestingly, these authors demonstrated that other growth factors expressed in SCG neurons, including epidermal growth factor (EGF) and glial cell-line derived neurotropic factor (GDNF), both of which act through the PI3-kinase pathway, differed in their ability to maintain quiescence/latency in these cultures. This was based on the abilities of these growth factors to provide sustained signaling through PI3-K and Akt. More recently, the importance of neurotropic factors in adult sympathetic and sensory neurons was examined (Yanez et al., 2017). While NGF and GDNF withdrawal induced HSV1 reactivation in adult sympathetic neurons, in adult sensory neurons NGF deprivation had no effect. Neurturin (NTN) and GDNF withdrawal induced HSV1 and HSV2 reactivation, respectively. Thus, the nature of receptor tyrosine kinase (RTK) signaling appears to be a key host parameter that regulates the HSV1 quiescent to lytic activation in a subset of neurons even in this "homogeneous neuronal" model system. While the importance of neurotrophic factors during development is well established (Indo, 2018), in the adult nervous system a role for continuous NGF or other neurotropic support is less clear as is the relevance of neurotropic factor withdrawal as a trigger of reactivation (Skaper, 2017).

A two phase VP16TF dependent HSV reactivation model

In a second study using the SCG neuron culture model, the expression of 4-5 viral genes from IE, E, Lg_1, Lg_2 kinetic classes were profiled following LY294002 treatment (Kim et al., 2012). Two distinct phases of transcriptional activity were observed. During the first at 20-24hr (Phase I), extremely low levels of UL30 (658 RNA copies/sample*) and UL48 (3,470 RNA copies/ sample*) transcripts were detected when protein synthesis was inhibited (cycloheximide, which blocks all protein synthesis, was added 10 hrs after LY294002). In the absence of the protein synthesis inhibitor, neither infectious virus nor increases in viral DNA were observed during the 25hr Phase I. Phase II encompassed 48hr and onward, during which time modest replication of the viral genome (2.5 rise over baseline) and infectious virus production (250 pfu at 72 hrs) were detected. Thus, Phase II has the hallmarks of a lytic infection. Examination of the SCG neurons for viral protein expression during both phases included three proteins: VP16, ICP0, and ICP27. Curiously, VP16 was detected in 5% of the neurons/well at the time of LY294002 treatment, representing 10-20% of the infected neurons. Following LY294002 treatment and during Phase I, VP16-positive neurons increased 5-fold by 20hrs to 25% of cells which is 50-100% of the infected neurons/well. ICP0 and ICP27 expressing neurons were detected during Phase I but the frequency was not indicated. These proteins appear in clusters of neurons suggestive of cell to cell spread. Characterization of the spectrum of viral proteins expressed during Phase I will be important to determine. The assumption is made that the Phase I transcriptional burst is the source of the viral proteins and, furthermore, that these proteins drive the productive Phase II event. 10%-20% of infected neurons/well express VP16 *before* the transcriptional burst is induced by LY294002 treatment. Whether other proteins are present at this time is not reported but is quite important, as this may represent persistent infection.

This model includes a forced form of latency requiring suppression of viral replication by the anti-viral compound acyclovir (ACV) for a 1-week period, which is quite different than the events and time frame that occur *in vivo*. While it appears that the resulting cultures conform to latency as broadly defined, analysis of latent viral genomes in neurons has revealed that within a given nucleus their association with promyelocytic leukemia protein (PML) containing nuclear bodies and histone variants are features of latency. These are under investigation and likely to be important (Cohen et al., 2018; Lomonte, 2016; Maroui et al., 2016). The potential that this model is characterized by a low level of persistent infection is likely. *In vivo* the presence of viral protein has been examined extensively using approaches at least as sensitive as those employed in the in vitro studies. In vivo viral protein is rarely detected and, in extensive studies using wild type virus, its presence correlates with the detection of infectious virus in what are likely spontaneously reactivating neurons (Sawtell, 1998, 2003; Sawtell et al., 1999; Sawtell et al., 2001). From this perspective, neuronal cultures do not appear to model *in vivo* latency in TG. Even if only 0.1% of latently infected TG neurons expressed cytoplasmic VP16, these ~6 neurons would be detectable assuming similar expression levels of protein *in vivo* as in the cultured neurons. It is possible that host immune functions effectively "remove" these viral protein positive neurons. This is observed in vivo during the final stages of a reactivation event in reactivating neurons and late during the lytic cycle in ganglia (Doll et al., 2020;

Goodpasture, 1929). However, if this were the case and such neurons were removed through time, a reduction in the size of the in vivo latent reservoir would be expected over time, which is not observed (Thompson and Sawtell, 2011). These differences could arise from differences between SCG and sensory neurons, and examination of latent infection and reactivation at the neuronal level in SCG *in vivo* may be informative.

An additional feature of the SCG neuron culture model is the hypothesis is that VP16 remains sequestered in the cytoplasm with HCF-1 until HCF-1 is transported to the nucleus, which is triggered by LY294002 treatment. Whether these HCF-1/ LY294002 studies were done in infected or uninfected cultures is difficult to interpret (Kim et al., 2012). While the HCF-1 correlation is of interest, further studies are needed to test this hypothesis. Simple co-localization studies would provide strong support for the hypothesis that VP16 is sequestered in the cytoplasm by HCF-1 and is then translocated to the nucleus by HCF following LY294002 treatment. These cultures systems are ideal for addressing this type of question. The GFP-expressing neurons would be expected to result from the nuclear translocation of VP16 by HCF-1. Drawing these relationships is especially important since infection is expected to promote cytoplasmic to nuclear translocation of HCF-1 six days prior to LY294002 treatment. During acute infection of TG *in vivo*, HCF-1 relocates to the nucleus in many neurons. Importantly, this nuclear translocation of HCF-1 during acute infection does not return to the pre-infection cytoplasmic localization. Thus, in a latently infected ganglia, there is a persistent high percentage of neurons in which HCF-1 is nuclear.

A histone methyl/phospho switch is required for HSV reactivation

A third study using the SCG neuronal culture model revealed novel concepts regarding the mechanism of HSV reactivation with respect to Phase I and its requirement for Phase II. The conclusions drawn from this study are that (i) a methyl/phospho switch at H3S10 during Phase I is required for HSV reactivation and that (ii) the DLK/JIP-3 JNK pathway is directly linked to this switch. While this study has the potential to be quite important in providing insight into the molecular signaling underlying HSV reactivation, aspects of the experimental design raise questions and emphasize the challenges faced. First, interpretation of the study relies heavily on the specificity of two inhibitors of JNK, SP 600125 and AS 601245. In an extensive evaluation of the selectivity of protein kinase inhibitors, SP 600125 was tested and 19 of the 69 protein kinases tested were inhibited with similar or greater potency than JNK isoforms (Bain et al., 2003; Bain et al., 2007; Fabian et al., 2005; Tanemura et al., 2010; Uitdehaag et al., 2012). AS 601245 was also determined not to be selective for JNK. Importantly, AS 601245 was found to be an exceptionally potent inhibitor of PIM1, PIM3 and GSK3 with IC_{50} values in the nanomolar range, 50-100 fold lower than the IC_{50} values for JNK1 and JNK2 (Bain et al., 2003). This does not mean that JNK was not inhibited in this study, but it does mean that JNK was not the only kinase inhibited. Thus, it is not possible to attribute the phenotypes observed to JNK inhibition without further studies. It is also possible that the observed effects of the inhibitor over time (Phase I 18hr) and (Phase II 72hr) are related to the inhibition of different kinases and the phenotypes observed in phase I are unrelated to phase II making interpretation of the results difficult.

A second issue arises from additional limitations of the experimental design. The viral genome related phenotypes, i.e., either transcription (ICP27 related RNA) or the histone modifications related to specific viral promoters were measured from the entire population of cells in a given well. There is no information on the distribution of these phenotypes among the many cells in the well. This is critically important because these averaged phenotypes are being connected to the very minor population of reactivating neurons in the well. Thus, while an event occurs in Phase II at 72 hrs post treatment (GFP expression in a small number of cells), whether the RNA detected in Phase I at 18hrs is or is not expressed in these particular cells (at 18 hrs) is not known. Whether the Phase I transcriptional activity has any effect on reactivation remains an open question. VP16 expression in this study was not reported but could have been informative. This becomes even more complex given that the limited number of modified histones detected cannot be ascribed to the transcriptional activity from any given viral promoter. The fact that GSK-J4 treated neurons during phase I showed no difference in the relative levels of ICP27 RNA in the total cell population does not mean that the distribution of the RNA did not change among individual neurons.

As mentioned, the use of inhibitors or depletion strategies to eliminate or reduce the function of key host proteins without the required supporting studies is problematic. While this type of approach can indicate the importance of that function to the reactivation process (for example GSK-J4 in explant reactivation) whether this is a direct effect (i.e. methylase function on the viral genome) or more likely, a result of effects on host cell functions that then influence reactivation is often difficult to parse out. Additional studies are required before drawing direct links between the host function and the viral genome. For reference, a more detailed comparison between the in vivo reactivation model and the SCG model with respect to latency and reactivation is shown in Figures 2 and 3.

HCF-1: A potential VP16 independent inducer of reactivation

HCF-1 is an essential transcriptional co-regulator that is important for several aspects of cell proliferation based on cell culture genetic studies (reviewed in (Minocha et al., 2016; Wysocka and Herr, 2003; Zargar and Tyagi, 2012)). HCF-1 protein is located at the transcriptional start sites of many genes (over 5400 genes in Hela cells), associating with both sequence-specific DNA-binding proteins, chromatin-modifying enzymes (Dejosez et al., 2010; Michaud et al., 2013), transcription initiation factors, and elongation complexes (Alfonso-Dunn et al., 2017). As such, HCF-1 is a scaffold protein central to normal cell function, bringing together regulatory effector proteins. As discussed above, HCF-1 was first identified as a component of the VP16 induced complex (VIC) and a similar complex Oct-1/ORF10/HCF-1 was identified for IE gene expression regulation in VZV (Narayanan et al., 2005).

The potential ability of HCF-1 to be recruited to the viral genome through binding partners other than VP16 raises the possibility that HCF-1 could play a role in initiating expression of the viral IE genes apart from its interaction with VP16. This hypothesis was proposed more than a decade ago (Kristie et al., 2010; Whitlow and Kristie, 2009) and is a viable possibility during high multiplicity infection (cultured cells) or explant induced reactivation. These are both *ex vivo* contexts in which viral IE gene expression and viral lytic cycle entry are not dependent upon VP16 transactivation function (Ace et al., 1989; Steiner et al., 1990). However, *in vivo* replication in the absence of VP16 is severely impaired at peripheral sites of infection and in the nervous system (Sawtell and Thompson, 2016; Sawtell et al., 2011; Thompson et al., 2009), indicating that if HCF-1 is facilitating IE gene expression in the absence of VP16 it is quite limited in the *in vivo* context. Consistent with this, hyperthermic stress induced reactivation is VP16 dependent. Although the possibility that distinct triggers could interface with the viral genome *in vivo* in diverse ways has not been fully explored, a VP16-independent role for HCF-1 in driving IE gene expression *in vivo* does not currently have support. Identification of potential HCF-1 binding partner(s) that recruit HCF-1 to the latent viral genome in the absence of VP16 would support the hypothesis. However, VP16 transactivation function mutants do not express any detectable viral proteins in latently infected neurons in vivo following reactivation inducing stresses (Sawtell and Thompson, 2016; Sawtell et al., 2011; Thompson et al., 2009). This strongly suggests any such VP16 independent recruitment of HCF-1 to the latent genome does not initiate viral reactivation and detectable viral protein expression.

JQ1 and reactivation: manipulation of an epigenetic reader perturbs viral homeostasis

Epigenetic features of herpesvirus genomes, including the chromatin associated with latency and changes observed during reactivation of the alpha herpesviruses are the subject of multiple reviews (Balakrishnan and Milavetz, 2017; Lieberman, 2016). Newer findings include the role played by CTCF binding sites on the HSV genome and their cognate binding proteins to the establishment and maintenance of latent infections (Washington et al., 2018a; Washington et al., 2019).

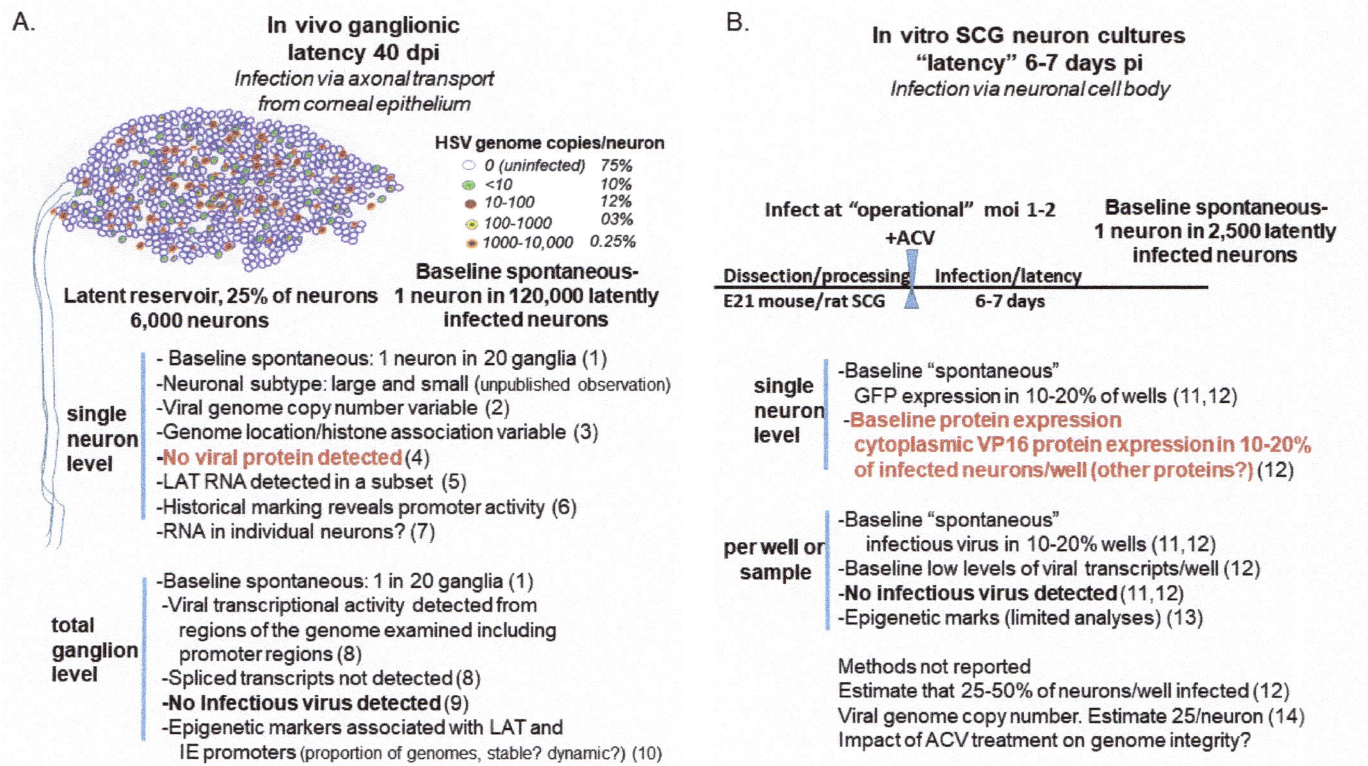

Figure 2. Comparison of latency mouse TG in vivo and in vitro in SCG rodent neuron model. The SCG neuron model appears to have hallmarks of a persistent infection including viral protein expression in 10-20% of latently infected neurons/well and a "spontaneous reactivation rate 50-fold greater than observed in vivo. References (1)= for example (Sawtell, 1998); (2) = for example (Sawtell et al., 1998); (3)= for example (Cohen et al., 2018); (4)= for example (Sawtell and Thompson, 2016); (5)= for example (Stevens et al., 1987); (6)= for example (Proenca et al., 2008; Proenca et al., 2015; Russell and Tscharke, 2016); (7)= (Russell and Tscharke, 2016); (8)= (Thompson and Sawtell, 2006); (10) reviewed in (Knipe, 2015); (11+12)= (Camarena et al., 2010; Kim et al., 2012); (13)= (Cliffe et al., 2015); (14)= (Kobayashi et al., 2012).

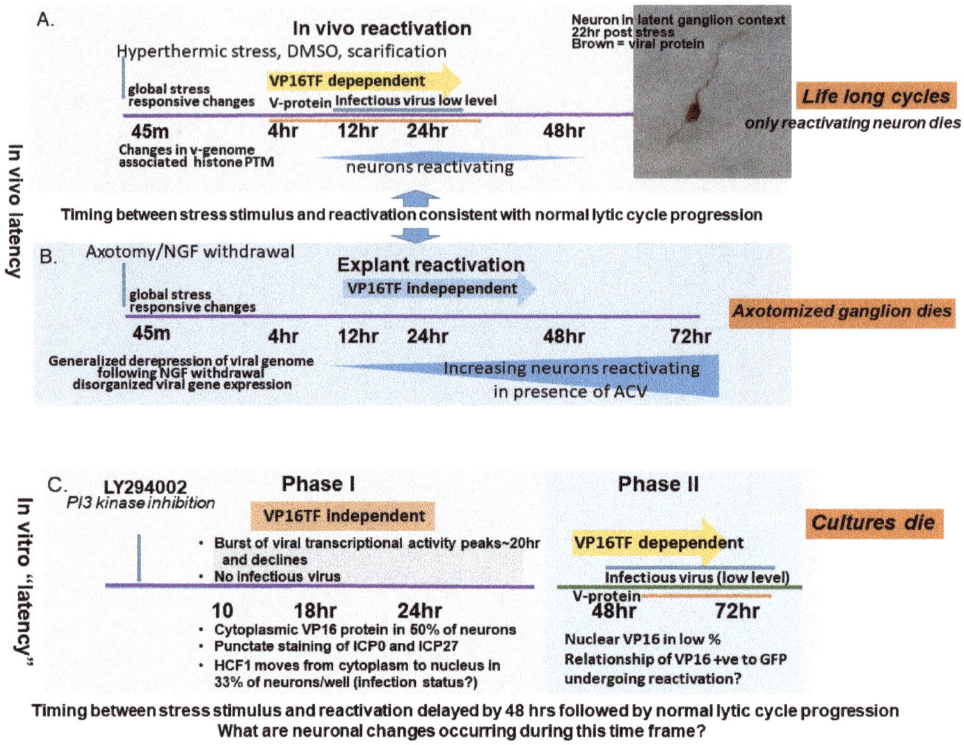

Figure 3. Comparison of the timing of reactivation in latently infected ganglia and SCG cultured neurons. Infectious virus can be detected in latently ganglia following stress in vivo (A) or axotomy (B) within 14 hrs post stimulus. In conrast, infectious virus is not detected until 48 hrs after PI3 kinase inhibition(C).

The availability of small molecule inhibitors of epigenetic readers and writers designed as potential cancer therapeutics have provided probes to test for effects on viral transcription and replication. An epigenetic compound library screen to test for effects on HSV replication in Vero cells revealed that BET (Bromodomain and Extraterminal Domain) protein inhibitor JQ1 increased HSV1 and 2 lytic infection by as much as 100 fold (Ren et al., 2016). This effect was specifically dependent upon the BET protein, Brd4. Paradoxically, knock down of Brd4 inhibited viral replication. The requirement for Brd4 for transcription, replication and other functions is reported for a number of viruses (McKinney et al., 2016; Wang et al., 2013).

Alfonso-Dunn, et al, tested the ability of JQ1 to modify explant reactivation, finding that consistent with the effect of the cyto-differentiating agent HMBA (that also inhibits Bet bromodomain proteins), JQ1 modestly increased the number of neurons expressing viral protein at 48 hrs post explantation from 2 to 7 (Alfonso-Dunn et al., 2017). The shared function in the release of positive transcription elongation factor, P-TEFb, from the inhibitory 7SK small nuclear ribonucleoprotein (7SK snRNP) complex of the two compounds (HMBA and JQ1) and their well characterized effect on HIV latency, suggest that transient release of P-TEFb facilitated reactivation, although further studies are required to demonstrate this (Abner et al., 2018; Bensussen et al., 2018; Khoury et al., 2018). The effect of JQ1 was also tested *in vivo* with latently infected mice being injected with JQ1 twice at 24-hour intervals. Increased levels of ICP27 and gC transcripts were detected in TG and 2-3-fold enhancement of viral DNA in some eyes were detected at 48hrs. These findings are intriguing but in the absence of infectious virus or viral protein detection, whether these changes represent bone fide reactivation remains an open question. Nonetheless, the ability of JQ1 to perturb viral latency *in vivo* is significant. JQ1 is a broad BET domain inhibitor and not selective for Brd4. Thus, in the absence of additional studies, the mechanism underlying the observed phenotypes is yet to be illuminated.

Frequency of HSV2 genital shedding implies HSV2 is continuously "reactivating" in the ganglion. Corollary: the reactivating neuron must survive the production and release of infectious virus

Multiple studies have now solidified not only the shedding frequency but also the peak amounts of viral DNA and rate of decay with which HSV2 DNA is detected in human genital swabs collected from a specified regional map (Agyemang et al., 2018; Ramchandani et al., 2017; Sacks et al., 2004). These studies have substantially altered what had been established concepts of the frequency of viral presence at the genital mucosal surface and raise important questions regarding the transmission potential of these frequent PCR detectable shedding events. The results of these studies were best fit by a mathematical model that assumed a nearly constant release of small numbers of virions from ganglionic neurons (Schiffer et al., 2009). While a legitimate hypothesis, it is worth noting that the source of viral DNA detected by PCR in the genital swab is not known but rather assumed to be the innervating sensory ganglia. Thus, the concept is based on a mathematical model which assumes that the only source of viral DNA is (or could be) the sensory ganglion. More recent mathematical modelling from this group is focused on immune control in the mucosa (Gottlieb et al., 2017; Schiffer et al., 2018). Based on the shedding frequency, neurons presumably survive reactivation and are able to undergo repeated reactivation events; otherwise, frequent shedding cannot easily be explained. However, the ability of HSV2 to persist in other cell types is well described in the older literature, including HSV in hair follicles, skin, T cells, B cells, and myeloid cells (Al-Saadi et al., 1988; Claoue et al., 1987; Easty et al., 1987; Nicholls et al., 1996; Scriba, 1977, 1981; Shimeld et al., 1986; Shimeld et al., 1982; Tullo et al., 1985). Additionally, the idea that the reactivating neuron must survive based on this level of shedding does not appear to consider the number of ganglia that could be potentially alternate sources for the viral DNA in genital skin or mucosa. Progress in developing animal models to gain insight into the contribution of other ganglia to vaginal reactivation (Bertke et al., 2007; Lee et al., 2015; Pieknik et al., 2019; Yanez et al., 2017) as well as developing the tools in HSV2 to test hypotheses *in vivo* (Kawamura et al., 2018; Pieknik et al., 2018) will undoubtedly deepen current understanding of HSV1 and 2 reactivation and the distinctions between them.

A recent report detailed the resolution of reactivation of HSV1, characterizing the surrounding cellular context and morphological changes in individual neurons undergoing reactivation in vivo (Doll et al., 2020). Infectious virus was detected in a high percentage (60-80%) of the trigeminal ganglia at 20 hours post-reactivation stimulus but declined by 48 hours post-stimulus (0-13%). Importantly, in addition to intact viral protein positive neurons, fragmented viral protein positive neurons morphologically consistent with apoptotic bodies and containing cleaved caspase-3 were detected. The frequency of this fragmented phenotype increased through time post-reactivation. These fragmented neurons were surrounded by Iba1+ cells, consistent with phagocytic removal of dead neurons. Treatment with anti-CD4/CD8 depleting antibodies did not alter this outcome. In contrast, blocking viral DNA replication during reactivation prevented neuronal fragmentation within the 48-hour time frame, even though viral protein was expressed. These findings suggest that reactivation in vivo of HSV1 is resolved through the destruction of the neurons which appears to be independent of antigen-mediated T cell cytotoxicity but does require viral replication. Similar studies with HSV2 are needed.

The role of the latency associated transcription unit (LATU) in latency and reactivation: a reexamination of what we think we know

Functions of products from the LATU locus

As more knowledge concerning the rather subtle effects that non-coding RNAs exert on many different processes including transcription, translation, post translational modification, and chromatin structure have been described, the appearance that the non-coding RNAs generated in the LAT locus had no or minor effects on latency/reactivation can be attributed at least in part to the general lack of assays sensitive enough to measure their influences individually and in aggregate (Chen and Aravin, 2015; Dhanoa et al., 2018; He et al., 2014; Zheng et al., 2017). Space restrictions prevent presentation in detail of all of these earlier studies on the LAT locus and its functions here, and readers are directed to a recently published comprehensive review on this subject (Phelan et al., 2017). In brief, the latency associated transcript (LAT) is a ~8.5 kb mRNA that maps largely to the long terminal repeat of the virus and is antisense to the ICP0 and ICP34.5 genes (Wechsler et al., 1988). It is the only mRNA abundantly transcribed during latent infection of neurons and it expression is strongly repressed during acute infection by an ICP4 binding site that includes the transcriptional start site of LAT (Batchelor et al., 1994). The LAT mRNA is spliced, which produces two co-linear introns of 2.0 and 1.5 kb that share the same splice donner and acceptor sites, differing by an internal splice in the 5' end of smaller intron (Farrell et al., 1991; Phelan et al., 2017). These introns are stable due to an unusual lariat structure that is not resolve efficiently and are partially antisense to the 3' end of the ICP0 mRNA and together are known as the LATs (Mukerjee et al., 2004). The LAT mRNA does not encode any known proteins and has a short half-life. It is processed by cellular mechanisms into several microRNAs and stable small RNAs, the functions of which are the subjects of current research efforts. The LAT locus is more complex, having multiple transcripts originating from both strands, none of which encode known proteins. These mRNAs are also processed into microRNAs and are expressed during lytic and latent infections at low levels. The patterns of microRNA expression and sequences from the LAT transcriptional unit locus (LATU) are not the same for HSV-1 and HSV-2. Curiously, deletion of a small ~200 bp region of the basal LAT promoter including the TATA box and mRNA start site perturbs the expression of most if not all of the transcripts originating in the LATU (Phelan et al., 2017). Thus, many of the phenotypes originally attributed to "LAT", may actually map to other nearby regions or on the strand opposite of the LAT mRNA (Phelan et al., 2017; Roizman et al., 2013).

Leib et al. provided evidence that a LAT null mutant that replicated normally *in vivo* displayed a reduced reactivation phenotype in a dissociated ganglia explant model. One caveat was that the slot blot analysis of latent viral genome DNA may not have been sufficiently sensitive to determine a defect in the establishment of latency by the mutant (Leib et al., 1989a). Sawtell and Thompson first determined that LAT locus mutants establish latent infection less efficiently than wild type. While the mechanism(s) involved still remain controversial, this phenotype is one that has stood the test of time and is shared among the various mouse rabbit and guinea pig models (Phelan et al., 2017; Roizman et al., 2013) (Krause et al., 1995; Perng et al., 2000b; Sawtell and Thompson, 1992a; Thompson and Sawtell, 1997, 2001).

Two important questions remained. Is the LAT locus required solely for the efficient establishment and/or maintenance of latency, or does it also exert a positive effect on reactivation from latency? What is the mechanism underlying the reduced establishment of latent infections and/or the increase in reactivation from latency? Answering these questions required more sensitive quantitative assays. Using a quantitative single neuron approach, it was found that fewer latent infections were established in mice infected with LAT-null mutants. When methods were employed to reach equivalent latency between wild type and mutant strains, no evidence for a specific defect in viral reactivation from latency was detected (Thompson and Sawtell, 1997). Counting the total number of neurons present in uninfected or latently infected mouse ganglia latently infected with wild type or LAT null mutants in strain 17syn+ revealed twice the number of neurons were killed in mouse TG infected with the LAT-null mutants compared to the rescuants and wild type. Very low levels and similar numbers of neurons undergoing apoptosis were seen in both groups (Thompson and Sawtell, 2001). Whether transcription from the LATU locus exerts a negative effect on virus entry into the lytic cycle in neurons and in its absence viral lytic infection kills more neurons, or something made from the LATU protects neurons from cytopathic effects associated with HSV replication and thereby increases successful establishment and/or reactivation frequencies, is unknown. Whatever the mechanism, LATU expression reduces lytic cycle gene expression during acute infection of mouse TG neurons (Garber et al., 1997). Increased neuro-invasiveness was displayed by mutants lacking specific regions of the LAT locus (Jiang et al., 2016; Jiang et al., 2015; Thompson and Sawtell, 1997). More recent findings demonstrate that the LAT transcription program is expressed early and first in TG neurons during acute *in vivo* infection. Thus, LAT is expressed at a time when it could negatively influence entry into the lytic cycle in neurons. (Sawtell and Thompson, 2016).

HSV LATs and neuronal apoptosis

The hypothesis that the LAT locus protects newly infected sensory neurons from apoptosis immediately after the lytic cycle has ended in the ganglia (hypothesized to be induced as the result of toxic viral virion proteins) is steeped in the lore of HSV1 latency (Perng et al., 2000a). As recently reviewed this may not be true (Roizman et al., 2013). Significantly, Perng et al. reported extensive apoptosis (>25% of neurons per section) in 66% of TG sections infected with a LAT mutant and in 4% of sections from TG infected with the rescuant or from TG of uninfected rabbits on day 7 postinfection (Perng et al., 2000a). Unfortunately, this experiment has never been repeated. A manuscript that is often cited to support the idea that the LAT locus blocks apoptosis in the mouse model actually reported very low numbers of neurons undergoing apoptosis (<0.01 to <1%) 30 days postinfection with LAT mutants (Branco and Fraser, 2005), and the relationship of this minor amount of TUNEL staining at 30 day postinfection to the "extensive apoptosis" detected by (Henderson et al., 2004; Jin et al., 2005) and colleagues in rabbit TG only at day 7 postinfection, but not at earlier or later times is not obvious. Further studies with mutants in which anti-apoptotic proteins were inserted in place of LAT are considered to support this hypothesis (Jin et al., 2005; Perng et al., 2002). However, only the downstream phenotypes of ocular shedding (rabbit) or explant reactivation (mouse) were examined, and not the apoptosis of neurons per se. That Wechsler and colleagues were unable to demonstrate more extensive apoptosis in mouse TG infected with LAT null mutants but did find suppression of the reactivation deficient phenotype in mice suggests other mechanisms may have been responsible. The pleomorphic functions of the inserted proteins and the absence of assessment of apoptosis precludes attributing the observed effect on the inhibition of apoptosis. Further experimentation is warranted to test this hypothesis, and experimentation might be productively extended to studies in human neuronal cultures as well.

Ectopic expression of portions of the LAT locus interferes with apoptosis induced with drugs like etoposide in cultured cells (Jin et al., 2003; Peng et al., 2003). Two small non-coding RNAs from this locus can inhibit viral replication and can also cooperate to reduce apoptosis in cultured cells (Shen et al., 2009). These functions may be important during productive infection by the virus, which encodes many functions that inhibit apoptosis including ICP0, ICP34.5, gJ, gD, US3, and ribonucleotide reductase (for review see (Yu and He, 2016)).

The LAT locus serves to maintain latent infections

Regardless of the mechanisms involved, it is clear that the latency associated transcript locus of HSV-1 is required for long-term maintenance of reactivation competent latent infections (Thompson and Sawtell, 2011). Mice were infected with a LAT null mutant and a rescued variant maintained for 30 days post inoculation at which time were each divided into two groups. One half of the latently infected mice were maintained an additional 294 days. Additional mice from these same groups were subjected to the hyperthemic stress procedure an average of 2.5 times per week during 280 days after the initial 30-day period. After week 40 the mice receiving the multiple stress treatments were rested for two weeks, and then all of the mice were subjected to a final stress procedure to induce reactivation in vivo. Two measures were employed to quantify the outcome in TG at 22 hours post HS, (i) the quantification of infectious virus and (ii) quantification of the number of neurons exiting latency using whole ganglion immunohistochemistry (WGIHC) for viral proteins.

Does the LAT locus specifically enhance reactivation from latency in the rabbit model?

As discussed above, when the level of latency established by LAT null mutants in mouse TG is equivalent to that of wild type virus, stress induced reactivation from latency is also equivalent, and even more efficient on a per latently infected neuron basis (Thompson and Sawtell, 1997). This locus is also required for the long-term maintenance of latent infections (Thompson and Sawtell, 2011), emphasizing that in the mouse model it serves to reduce lytic infection. However in the rabbit a number of studies support the idea of a direct role for the LAT locus in promoting viral reactivation from latency (Bloom et al., 1996; Hill et al., 1990) as reviewed in (Toma et al., 2008). This may indeed be a difference between the rabbit and murine models as in the rabbit LAT locus transcription is associated with an increase in lytic gene transcripts such as ICP4, TK and gC, (Giordani et al., 2008) whereas the exact opposite, a decrease in these transcripts, was seen in the presence of an intact LAT locus (Garber et al., 1997).

A recent report suggests LAT null mutants may have a specific reactivation deficit. Using an interesting new approach that circumvents this problem, Bloom and colleagues expressed hammerhead ribozymes directed at the 5' exon of the primary LAT (or a control ribozyme) in TG neurons of rabbits latently infected with wild type HSV1. A significant reduction of induced reactivation was found in rabbits receiving the anti-LAT ribozyme (Watson et al., 2018). If this approach and result stand the test of time, a pro reactivation function of the LAT locus will have strong support. Similar approaches in the mouse model could reveal such a pro-reactivation function of LAT as well.

HSV is also well documented to establish latent infections that can be reactivated in a variety of non-neural tissues including denervated mouse footpads or footpads following long-term acyclovir treatment, various non-neural tissues from rabbit, mouse, rat, and guinea pig including cornea, conjunctiva, and hair follicles in humans and LAT expression has been detected in many of these ((Al-Saadi et al., 1988; Claoue et al., 1987; Easty et al., 1987; Nicholls et al., 1996; Scriba, 1977, 1981; Shimeld et al., 1986; Shimeld et al., 1982; Tullo et al., 1985) and see (Kennedy et al., 2011) for a relatively recent review). These extra-neural sites of latency are usually dismissed but may be biologically relevant, especially in recurrent ocular shedding and eye disease in the rabbit and humans. Whether the LAT locus plays a direct role in promoting reactivation from such sites deserves further study.

How does the virus go latent in the first place?

The ubiquity of HSV1 in the human population is the result of the efficient establishment of latent infections that subsequently reactivate throughout the host's lifetime and transmit to new hosts. A long-lived question is how the virus evolved to interact with sensory neurons to promote the efficient establishment of latency, which requires viral replication within the sensory ganglia but also a self-limiting mechanism to prevent neurovirulence and transmission into the CNS. The discovery of the LAT (Stevens et al., 1987) and mapping of its promoter (Dobson et al., 1989) provided a powerful tool to examine latent phase transcription in TG neurons by *in situ* hybridization for the LATs or by using LAT promoter/beta-galactosidase reporter mutants (Margolis et al., 1992; Sawtell and Thompson, 1992a; Simmons et al., 1992; Speck and Simmons, 1992).

A somewhat unexpected result common to all these studies was that the acute stage of infection in the TG (examined at 48 hours postinfection and later) was characterized by neurons expressing viral proteins and others expressing only the LATs or the LAT promoter. A few neurons were detected that expressed both, but these were rare. These data were interpreted to suggest that latent and lytic pathways in neurons were district from each other (Lachmann and Efstathiou, 1997; Margolis et al., 1992; Sawtell and Thompson, 1992b; Simmons et al., 1992; Speck and Simmons, 1992), and that the lytic and latent viral transcriptional programs in neurons were regulated in ways that were not understood. Hypotheses were proposed to explain this transcriptional programmatic duality including differential lytic/latent expression in neuronal subtypes, (Cabrera et al., 2018; Yang et al., 2000), transition from some acute gene expression into the latent program (Proenca et al., 2008), failure of VP16 to be transported to neuronal nuclei (Sears et al., 1991; Steiner et al., 1990), alternative regulation of gene expression (Kosz-Vnenchak et al., 1993), strict nonnuclear compartmentalization of essential co-activators such as HCF-1 specifically in neurons (Kristie et al., 1999), or that its interaction with alternate neuronally expressed cofactors like HCF-2, Oct-2 or Brn3 might inhibit its function (Liang et al., 2009; Lillycrop et al., 1993; Nogueira et al., 2004). Any or all of these mechanisms may act to constrain virus entry into the lytic phase in neurons, although replication in TG neurons is required to promote the efficient establishment of latency (Field et al., 1982; Field and Wildy, 1978; Katz et al., 1990; Kramer and Coen, 1995; Thompson and Sawtell, 2000), for review see (Hill, 1985; Roizman et al., 2013). To avoid injury or death of the host, this entry into the lytic cycle in neurons must be tightly controlled.

Several lines of investigation support the concept that VP16 and its unusual regulation in neurons plays a central role in the choice between latent and lytic infection (Sawtell and Thompson, 2016; Sawtell et al., 2011; Thompson et al., 2009). In the absence of the VP16 transactivation function latency is favored over acute viral replication in TG neurons *in vivo* (Sawtell and Thompson, 2016; Sawtell et al., 2011; Steiner et al., 1990; Thompson et al., 2009). During retrograde transport of HSV nucleocapsid/tegument through axons in cultured neurons loosely associated tegument proteins including VP16 are left behind and this presumably favors quiescent/latent infection in cultured neurons infected via axons (Aggarwal et al., 2012; Diefenbach et al., 2008; Hafezi et al., 2012; Koyuncu et al., 2017). Live-cell analysis by fluorescence time-lapse recording of capsid/tegument complexes transporting retrograde in axons of cultured sensory neurons has revealed that while the majority of VP16 is lost from capsids prior to transport to the neural soma, small amounts of the protein co-transport in this experimental setting that could promote a small amount of lytic neuronal infection (Antinone and Smith, 2010; Smith, 2012). This mechanism likely promotes latency in neurons (Hafezi et al., 2012; Koyuncu et al., 2017). Importantly, in all *in vitro* and *in vivo* models, significant viral replication occurs in cultured and sensory ganglion neurons demonstrating that these neurons can support lytic viral replication if the lytic cycle is initiated.

Replacing the VP16 gene promoter with that of another leaky late gene, VP5, resulted in normal viral replication on the mouse eye, but >100-fold reduced virus replication in TG (Thompson et al., 2009). Likewise, mutations in putative factor binding sites near and downstream of the VP16 TATA box also resulted in wild type replication on mouse eyes, and greatly reduced viral replication in TG (Sawtell and Thompson, 2016). In the latter study promoters of various kinetic classes (LAT, immediate early ICP0, and leaky late VP16) were employed to drive expression of E. coli beta-galactosidase (B-gal) in otherwise wild type HSV1 strain 17syn+. Whole ganglion analysis of viral protein expression by immunohistochemistry and B-gal expression from the diverse promoters was employed to examine the earliest events that occur within TG neurons infected via axons from the cornea. The expected asynchrony of TG neuron infection was evident in the analysis of the time postinfection and the

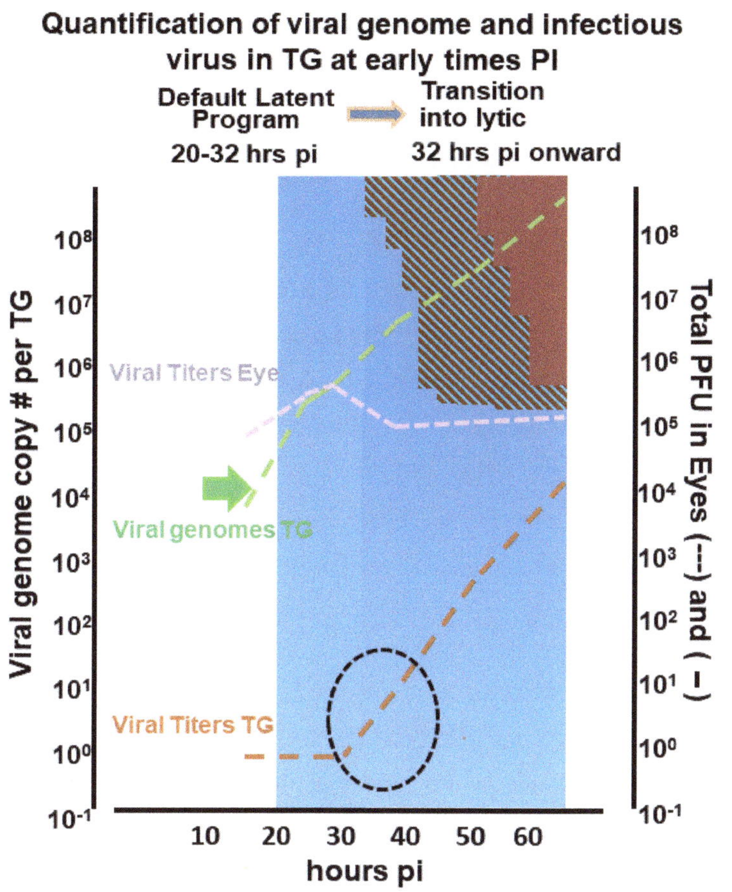

Figure 4. The colored field depicts the percent of positive neurons expressing the LAT promoter (Blue), only viral proteins (brown) or both (striped). Also shown are virus titers in the eye and TG and viral genomes in the TG (dashed lines).

number of neurons expressing viral proteins and promoter-driven B-gal prior to viral spread within the TG (up to 32 hours postinfection, Figure 4). At the earliest times examined (18 to 30 hpi), only LAT promoter-driven B-gal was detected in TG neurons.

Thus, the latent transcriptional program was engaged first in neurons infected via axons, consistent with the idea that VP16 did not arrive with the viral genome to initiate the lytic cascade. Of importance, B-gal activity and cognate viral protein expression in neurons was detected simultaneously in neurons (e.g. viral proteins including VP16 were detected in all neurons expressing B-gal from the VP16 promoter). The same was true of neurons marked by the ICP0 promoter, indicating simultaneous expression of the native promoters and the transgenes.

Viral proteins were first detected about 14 hours after LAT promoter driven B-gal, and only in neurons previously marked by LAT promoter expression. Since the LAT promoter is strongly repressed by ICP4 in cells including neurons during acute productive infection (Batchelor and O'Hare, 1990; Farrell et al., 1994; Margolis et al., 1992; Sawtell and Thompson, 1992a), we concluded that the viral protein expression observed at these early times postinfection occurred in neurons transitioning out of the acute stage latent program into the lytic cycle.

Mutants expressing a second copy of wild type VP16, the VP16TF deficient mutant, or a third copy of ICP0 driven by the LAT promoter were examined to determine whether the *de novo* expression of either ICP0 or VP16 was sufficient to push neurons out of the default latent transcription program into the lytic program. If so, viral protein expression and virus production in TG would be expected to occur about 14 hours earlier than that seen with wild type HSV1 infection. Early viral protein expression and infectious virus production was seen in TG of mice infected with the mutant expressing VP16 from the LAT promoter, but not in those infected with the LAT driven VP16TF mutant or ICP0 gene. Therefore, *de novo* expression of VP16, but not of ICP0, precipitated entry into lytic infection in TG neurons and the transactivation function of VP16 was required for this.

However, the requirement for the VP16 transactivation function is not absolute in so far as VP16TF viruses replicate to some extent in TG albeit at a ~100 fold reduction (Sawtell et al., 2011; Tal-Singer et al., 1999; Thompson et al., 2009). High genome copy number could potentially override a requirement for VP16TF and this is supported by the finding that strategies increasing surface replication also increase replication in the TG (Sawtell et al., 2011; Thompson et al., 2009).

A specific region of the VP16 promoter regulated the transition into the lytic transcription program

The region of the VP16 promoter near and downstream of the VP16 TATA box is dispensable for normal leaky late kinetics (Lieu and Wagner, 2000). Thompson and Sawtell found that 3 bp mutations in two putative factor binding sites in this region resulted in wild type virus replication in cultured cells and on the mouse eye, but nearly completely ablated the ability of the virus to transition out of the default latent pathway (2% transition vs. 30% at 44 hours postinfection (Sawtell and Thompson, 2016)). Recently a targeted replacement the HSV1 VP16 promoter sequences with those of HSV2 was generated and tested in mice. This demonstrated that the type 2 VP16 promoter also directs pre-immediate early VP16 expression and does so with greater efficiency during acute infection and following hyperthermic stress induced reactivation from latency (Thompson and Sawtell, 2019).

Combined, the studies offer strong support for the hypotheses that:

(i) VP16 is generally not transported retrograde in axons, favoring latency in neurons.

(ii) The VP16 promoter contains elements required for the *de novo* pre-immediate early production of VP16, which drives exit from a default latent pathway into lytic viral replication in neurons.

(iii) Production of this VP16 is not the result of random "phase 1" type transcription seen in persistently infected cultures of primary neurons treated with AKT or protease inhibitors (Cliffe et al., 2015; Kim et al., 2012).

(iv) Latency is established by default in sensory neurons because VP16 does not arrive with the viral genome. Subsequent reactivation of HSV from the default, or consolidated latent state requires *de novo* VP16 production, and at least for exit from the default latent state this is mediated specifically by the VP16 promoter (Sawtell and Thompson, 2016; Thompson et al., 2009).

Events during the consolidation of HSV latency

Assembly of chromatin structures on latent viral genomes is thought to be an important aspect of silencing the genome. In cultured cells viral DNA is rapidly associated with histones and chromatin is rapidly assembled, occurring in the first thirty minutes post infection (Roizman et al., 2005). This appears to be a multi-step and curiously slow process in TG neurons *in vivo* with initial events detectable at 7 days postinfection and modifications associated with silencing of genes such as heterochromatin formation occurring on lytic promoters after two weeks post infection, whereas the LAT promoter is associated with euchromatin at this time (for review see (Conn and Schang, 2013; Knipe, 2015; Kristie, 2015; Roizman et al., 2013)). Since viral gene expression ceases well before this time, this process may be more important for maintaining latency rather than the initial formation of latent infections. In support of this idea, recent studies suggest that chromatin insulator sequences and their binding proteins may maintain boundaries between regions of the latent HSV genome that are actively transcribed (e.g. the LAT locus) from the lytic phase genes that are silenced (Lee et al., 2018; Washington et al., 2018b; Watson et al., 2018).

Part of the difficulty in understanding the effects of chromatin modifications and boundaries is that latent viral genomes are diverse in these properties. Approaches employed on whole tissues can only detect changes averaged over these millions of latent genomes. Lomonte and co-workers have developed methods to take such analyses to the next level in single neuronal nuclei. As expected, they detected multiple genomes in many neurons. What was not expected is that some of these are intimately associated with PML containing nuclear bodies, whereas others in the same nucleus are not. Also, of interest, some genomes express LAT locus RNAs, whereas others in the same nucleus do not. The silencing properties of these PML-NBs are thought to provide an intrinsic anti-viral defense mechanism that serves to promote latency (Cohen et al., 2018; Lomonte, 2016; Maroui et al., 2016).

Efstathiou and colleagues developed a method to historically interrogate viral promoter expression in individual neurons that survive the acute stage of infection. Employing the Cre reporter mouse strain ROSA26R in combination with viral mutants that express Cre from various promoters they discovered that promoter expression in neurons that survive is far more frequent than previously thought (Proenca et al., 2008; Proenca et al., 2011; Proenca et al., 2015). However, the extent of production of the relevant viral proteins is not yet known. Another group employed this assay to examine the potential expression of viral promoters for genes encoding proteins that are important targets for immune surveillance. Evidence for the expression of these promoters suggest periodic expression of these promoters might be one mechanism whereby long-term immunity is maintained (Russell and Tscharke, 2016).

A role for HSV 1 latency/reactivation in the development of neurodegenerative disorders

The neurotropic nature of HSV and VZV, the extensive data identifying these viruses in the human CNS at autopsy or identification of virus in the cerebrospinal fluid together with the concept of viral latency and reactivation in the trigeminal ganglion led to the hypothesis "that reactivation of the same dormant viral material travelling centripetally instead might be the cause of the "degenerative" lesions typical both of AD and of the normal aged brain." -MJ Ball (Ball, 1982). In 1997 Itzhaki et al. provided evidence that the combination of HSV1 in brain and carriage of an APOE-e4 allele is a strong risk factor for AD (Itzhaki et al., 1997; Lin et al., 1997). Evidence is now emerging that directly supports the concept that HSV1 infection is a major risk factor in the development of Alzheimer's Disease (reviewed in (Itzhaki, 2018; Lathe et al., 2019)). This concept has been met with much skepticism over the years for reasons including the high prevalence of HSV infection relative to AD and the correlative nature of the studies. Clinical trials are now ongoing to test the efficacy of valacyclovir in individuals with mild AD (ClinicalTrials.gov Identifier: NCT03282916). Animal models are under development, and in human-ApoE4-targeted replacement mouse, long-term HSV1 infected but not mock-infected mice exhibited spatial memory deficits and CNS pathology consistent with AD. HSV protein was detected in the hippocampus but was extremely rarely (work in progress). Such models will allow efficacy of antiviral treatments to be tested and investigation into the mechanisms of disease initiation and progression.

Conclusions

Increasingly sensitive methods to detect viral transcriptional activity from the latent viral genome have modified the notion of a strictly "silent" vs "active" genome (Collins-McMillen and Goodrum, 2017). However, the basic concepts of latency and reactivation based on detection of infectious virus remain solid fundamental concepts. From our vantage point, the field has three major challenges. The first is recognizing that all models have limitations. Recognizing the limitations and advantages of diverse models is necessary for integration of information from all models. Second, harnessing the advancing single-cell approaches for the investigation of HSV latency and reactivation in complex as well as simpler culture contexts. Third, recognizing that *both* complexity and reductionism have a role to play in future progress in understanding the intricacies of latency and reactivation in the nervous system. An excellent example of the combined power of these approaches was recently published (Cabrera et al., 2018). All models have advantages and limitations and the goal remains to relate what is learned to what is known about human/virus interactions, expanding our overall understanding of HSV pathogenesis.

Reactivation is highly restricted at the cellular level both *in vivo* and *in vitro*, consistent with the fact that maintaining latency is essential for the long-term survival of the virus in the host. Importantly, the mere presence of RNA related to viral lytic genes in the context of latency *in vivo* does not necessarily correlate with either viral protein expression (exit from latency) or reactivation (infectious virus). More startling is the revelation that even in homogenous cultures of neurons collectively stimulated by a potent inhibitor, reactivation occurs in only a small percentage of neurons. This feature of reactivation makes it extremely difficult to interpret analyses on the whole population.

Distinguishing viral pro-reactivation from anti-reactivation transcriptional activity and how this activity is distributed on the viral genome as well as among the individual latently infected neurons is required before sufficient understanding of the significance of global transcriptional activation profiles can be interpreted.

We do not yet completely understand the significance of RNA transcription detected in latently infected ganglia, or under phase 1 conditions in cultured neurons. For example, Krause and colleagues demonstrated that viral transcription during latency that occurs in the absence of ICP27 protein results in splicing at sites that are normally suppressed, disrupting viral ORFs. Thus, they hypothesize HSV-1 may take advantage of host splicing machinery to restrict expression of randomly activated antigenic viral genes to achieve immune evasion (Tang et al., 2019). As we search for and query models in which viral reactivation is more robust, it is important to remember that the alpha herpesviruses evolved to remain latent most of the time and that productive reactivation from latency is a rare event at the cellular and viral genome level. Major remaining challenges include the characterization of the composition of the viral transcriptional activity within individual latently infected cells following a reactivation stimulus, the functional relevance of this activity in promoting or suppressing viral protein production, and whether this activity proceeds to reactivation. In the great majority of latently infected neurons the 100s of thousands of latent genomes distributed in them remain latent. From this perspective, the methodologies are rapidly evolving that will support the analysis of transcriptomes at the single neuron level isolated from the relevant tissues (Cao et al., 2017; Hu et al., 2018; Rosenberg et al., 2018).

While the animal models of HSV latency/reactivation are not perfect, they have proven to support the complex viral life cycle with reasonable fidelity and these models approximate the level of complexity encountered in the human host. The simple act of axotomizing and explanting the ganglia results in changes that yield outcomes that do not align with the requirements for productive reactivation in the *in vivo* context such as the requirement of the VP16 transactivation function to initiate detectable viral protein synthesis (compare (Steiner et al., 1990) to (Thompson et al., 2009). It is becoming increasingly apparent that properties of a complex system cannot be reduced to the components of a system because they depend on many interactions. Emergent properties of a complex system are lost when a system is stripped down to eliminate "extraneous" interactions or "complicating" interactions, with surrounding cells including satellite cells, resident glial and microglial cells, and infiltrating immune cells. In the pursuit of understanding HSV reactivation, a revival of reductionist approaches and models has arisen. The development of an *in vitro* latency model that provides a platform for dissecting host cell signaling pathways involved in viral genome entry into latency, its maintenance, and its activation would be an important tool. Ideally such a model could yield hypotheses testable in the more complex *in vivo* setting. However, defining the limitations to all models and approaches used to probe outcomes in these models is a critical component of the scientific process.

References

Abner, E., Stoszko, M., Zeng, L., Chen, H.C., Izquierdo-Bouldstridge, A., Konuma, T., Zorita, E., Fanunza, E., Zhang, Q., Mahmoudi, T., *et al.* (2018). A New Quinoline BRD4 Inhibitor Targets a Distinct Latent HIV-1 Reservoir for Reactivation from Other "Shock" Drugs. Journal of virology *92*.

Ace, C.I., Dalrymple, M.A., Ramsay, F.H., Preston, V.G., and Preston, C.M. (1988). Mutational analysis of the herpes simplex virus type 1 trans-inducing factor Vmw65. The Journal of general virology *69*, 2595-2605.

Ace, C.I., McKee, T.A., Ryan, J.M., Cameron, J.M., and Preston, C.M. (1989). Construction and characterization of a herpes simplex virus type 1 mutant unable to transinduce immediate-early gene expression. Journal of virology *63*, 2260-2269.

Aggarwal, A., Miranda-Saksena, M., Boadle, R.A., Kelly, B.J., Diefenbach, R.J., Alam, W., and Cunningham, A.L. (2012). Ultrastructural Visualization of Individual Tegument Protein Dissociation during Entry of Herpes Simplex Virus 1 into Human and Rat Dorsal Root Ganglion Neurons. Journal of virology *86*, 6123-6137.

Agyemang, E., Magaret, A.S., Selke, S., Johnston, C., Corey, L., and Wald, A. (2018). Herpes Simplex Virus Shedding Rate: Surrogate Outcome for Genital Herpes Recurrence Frequency and Lesion Rates, and Phase 2 Clinical Trials End Point for Evaluating Efficacy of Antivirals. The Journal of infectious diseases.

Ahmed, R., and Stevens, J.G. (1990). Viral persistence. In B N Fields Virology, Raven Press, New York, NY.

Al-Saadi, S.A., Gross, P., and Wildy, P. (1988). Herpes simplex virus type 2 latency in the footpad of mice: effect of acycloguanosine on the recovery of virus. The Journal of general virology *69 (Pt 2)*, 433-438.

Alfonso-Dunn, R., Turner, A.W., Jean Beltran, P.M., Arbuckle, J.H., Budayeva, H.G., Cristea, I.M., and Kristie, T.M. (2017). Transcriptional Elongation of HSV Immediate Early Genes by the Super Elongation Complex Drives Lytic Infection and Reactivation from Latency. Cell host and microbe *21*, 507-517 e505.

Andrewes, C.H. (1957). Latent Virus Infections. Nature *180*, 788.

Antinone, S.E., and Smith, G.A. (2010). Retrograde axon transport of herpes simplex virus and pseudorabies virus: a live-cell comparative analysis. Journal of virology *84*, 1504-1512.

Azarkh, Y., Gilden, D., and Cohrs, R.J. (2010). Molecular characterization of varicella zoster virus in latently infected human ganglia: physical state and abundance of VZV DNA, Quantitation of viral transcripts and detection of VZV-specific proteins. Current topics in microbiology and immunology *342*, 229-241.

Azwa, A., and Barton, S.E. (2009). Aspects of herpes simplex virus: a clinical review. J Fam Plann Reprod Health Care *35*, 237-242.

Bain, J., McLauchlan, H., Elliott, M., and Cohen, P. (2003). The specificities of protein kinase inhibitors: an update. The Biochemical journal *371*, 199-204.

Bain, J., Plater, L., Elliott, M., Shpiro, N., Hastie, C.J., McLauchlan, H., Klevernic, I., Arthur, J.S., Alessi, D.R., and Cohen, P. (2007). The selectivity of protein kinase inhibitors: a further update. The Biochemical journal *408*, 297-315.

Baird, N.L., Zhu, S., Pearce, C.M., and Viejo-Borbolla, A. (2019). Current In Vitro Models to Study Varicella Zoster Virus Latency and Reactivation. Viruses *11*.

Balakrishnan, L., and Milavetz, B. (2017). Epigenetic Regulation of Viral Biological Processes. Viruses *9*.

Ball, M.J. (1982). "Limbic predilection in Alzheimer dementia: is reactivated herpesvirus involved?". Can J Neurol Sci *9*, 303-306.

Baringer, J.R., and Swoveland, P. (1973). Herpesvirus hominis: Isolation from Human Trigeminal Ganglion. New England Journal of Medicine *288*, 648-650.

Bastian, F.O., Rabson, A.S., Yee, C.L., and Tralka, T.S. (1972). Herpesvirus hominis: Isolation from Human Trigeminal Ganglion. Science *178*, 306-307.

Batchelor, A.H., and O'Hare, P. (1990). Regulation and cell-type-specific activity of a promoter located upstream of the latency-associated transcript of herpes simplex virus type 1. Journal of virology *64*, 3269-3279.

Batchelor, A.H., Wilcox, K.W., and O'Hare, P. (1994). Binding and repression of the latency-associated promoter of herpes simplex virus by the immediate early 175K protein. The Journal of general virology *75*, 753-767.

Behring, E.A., and Kitasato, S. (1890). Uber das Zustandekommen der Diphtherie-Immunität und der Tetanus, Immunität bei Thieren. Deutsch Med Wochenschrift *49*, 1130-1145.

Bensussen, A., Torres-Sosa, C., Gonzalez, R.A., and Diaz, J. (2018). Dynamics of the Gene Regulatory Network of HIV-1 and the Role of Viral Non-coding RNAs on Latency Reversion. Frontiers in physiology *9*, 1364.

Bertke, A.S., Patel, A., and Krause, P.R. (2007). Herpes simplex virus latency-associated transcript sequence downstream of the promoter influences type-specific reactivation and viral neurotropism. Journal of virology *81*, 6605-6613.

Bloom, D.C. (2016). Alphaherpesvirus Latency: A Dynamic State of Transcription and Reactivation. Advances in virus research *94*, 53-80.

Bloom, D.C., Hill, J.M., Devi-Rao, G., Wagner, E.K., Feldman, L.T., and Stevens, J.G. (1996). A 348-base-pair region in the latency-associated transcript facilitates herpes simplex virus type 1 reactivation. Journal of virology *70*, 2449-2459.

Blyth, W.A., Harbour, D.A., and Hill, T.J. (1984). Pathogenesis of zosteriform spread of herpes simplex virus in the mouse. The Journal of general virology *65*, 1477-1486.

Blyth, W.A., Hill, T.J., and Harbour, D.A. (1981). Mechanisms of control of latent herpes simplex virus infection in the skin, and eye, . in: Herpetic Eye Disease(Blyth, W A, Hill, T J, and Harbour, D A, ed), *Springer-Verlag (Bergmann)*, 43-47.

Bolivar, F., Rodriguez, R.L., Betlach, M.C., and Boyer, H.W. (1977). Construction and characterization of new cloning vehicles. I. Ampicillin-resistant derivatives of the plasmid pMB9. Gene analysis techniques *2*, 75-93.

Bourne, N., Perry, C.L., Banasik, B.N., Miller, A.L., White, M., Pyles, R.B., Schafer, H., and Milligan, G.N. (2018). Increased frequency of virus shedding by herpes simplex virus type 2-infected guinea pigs in the absence of CD4+ T lymphocytes. Journal of virology.

Branco, F.J., and Fraser, N.W. (2005). Herpes simplex virus type 1 latency-associated transcript expression protects trigeminal ganglion neurons from apoptosis. Journal of virology *79*, 9019-9025.

Burnet, F.M., and Williams, S.W. (1939). Herpes Simplex: A point of view. Med J Aust *1*, 637-642.

Cabrera, J.R., Charron, A.J., and Leib, D.A. (2018). Neuronal subtype determines HSV-1 Latency-Associated-Transcript (LAT) promoter activity during latency. Journal of virology.

Cai, W., Astor, T.L., Liptak, L.M., Cho, C., Coen, D.M., and Schaffer, P.A. (1993). The herpes simplex virus type 1 regulatory protein ICP0 enhances virus replication during acute infection and reactivation from latency. Journal of virology *67*, 7501-7512.

Cai, W., and Schaffer, P.A. (1992). Herpes simplex virus type 1 ICP0 regulates expression of immediate- early, early, and late genes in productively infected cells. Journal of virology *66*, 2904-2915.

Camarena, V., Kobayashi, M., Kim, J.Y., Roehm, P., Perez, R., Gardner, J., Wilson, A.C., Mohr, I., and Chao, M.V. (2010). Nature and duration of growth factor signaling through receptor tyrosine kinases regulates HSV-1 latency in neurons. Cell host and microbe *8*, 320-330.

Campbell, M.E., Palfreyman, J.W., and Preston, C.M. (1984). Identification of herpes simplex virus DNA sequences which encode a trans-acting polypeptide responsible for stimulation of immediate early transcription. Journal of molecular biology *180*, 1-19.

Cao, J., Packer, J.S., Ramani, V., Cusanovich, D.A., Huynh, C., Daza, R., Qiu, X., Lee, C., Furlan, S.N., Steemers, F.J., *et al.* (2017). Comprehensive single-cell transcriptional profiling of a multicellular organism. Science *357*, 661-667.

Carton, C.A., and Kilbourne, E.D. (1952). Activation of latent herpes simplex by trigeminal sensory-root section. New Engl J Med *246*, 172-176.

Chen, Y.A., and Aravin, A.A. (2015). Non-Coding RNAs in Transcriptional Regulation: The review for Current Molecular Biology Reports. Current molecular biology reports *1*, 10-18.

Claoue, C.M., Blyth, W.A., Hill, T.J., and Easty, D.L. (1987). Does herpes simplex virus establish latency in the eye of the mouse? Eye *1 (Pt 4)*, 525-528.

Cliffe, A.R., Arbuckle, J.H., Vogel, J.L., Geden, M.J., Rothbart, S.B., Cusack, C.L., Strahl, B.D., Kristie, T.M., and Deshmukh, M. (2015). Neuronal Stress Pathway Mediating a Histone Methyl/Phospho Switch Is Required for Herpes Simplex Virus Reactivation. Cell host and microbe *18*, 649-658.

Cohen, C., Corpet, A., Roubille, S., Maroui, M.A., Poccardi, N., Rousseau, A., Kleijwegt, C., Binda, O., Texier, P., Sawtell, N., *et al.* (2018). Promyelocytic leukemia (PML) nuclear bodies (NBs) induce latent/quiescent HSV-1 genomes chromatinization through a PML NB/Histone H3.3/H3.3 Chaperone Axis. PLoS pathogens *14*, e1007313.

Cohrs, R.J., Badani, H., Baird, N.L., White, T.M., Sanford, B., and Gilden, D. (2017). Induction of varicella zoster virus DNA replication in dissociated human trigeminal ganglia. Journal of neurovirology *23*, 152-157.

Collins-McMillen, D., and Goodrum, F.D. (2017). The loss of binary: Pushing the herpesvirus latency paradigm. Current clinical microbiology reports *4*, 124-131.

Conn, K.L., and Schang, L.M. (2013). Chromatin dynamics during lytic infection with herpes simplex virus 1. Viruses *5*, 1758-1786.

Cook, M.L., Bastone, V.B., and Stevens, F.G. (1974). Evidence that neurons harbor latent herpes simplex virus. Infect Immun *9*, 946-951.

Cook, M.L., and Stevens, J.G. (1976). Latent herpetic infections following experimental viraemia. The Journal of general virology *31*, 75-80.

Cushing, H. (1905). The surgical aspects of major neuralgia of the trigeminal nerve.a report of twenty cases of operation on the gasserian ganglion, with anatomic and physiologic notes on the consequences of its removal. Journal of the American Medical Association *XLIV*, 860-865.

D'Aiuto, L., Bloom, D.C., Naciri, J.N., Smith, A., Edwards, T.G., McClain, L., Callio, J.A., Jessup, M., Wood, J., Chowdari, K., *et al.* (2019). Modeling Herpes Simplex Virus 1 Infections in Human Central Nervous System Neuronal Cells Using Two- and Three-Dimensional Cultures Derived from Induced Pluripotent Stem Cells. Journal of virology *93*.

Dargan, D.J., Patel, A.H., and Subak-Sharpe, J.H. (1995). PREPs: herpes simplex virus type 1-specific particles produced by infected cells when viral DNA replication is blocked. Journal of virology *69*, 4924-4932.

Davido, D.J., and Leib, D.A. (1996). Role of cis-acting sequences of the ICPO promoter of herpes simplex virus type 1 in viral pathogenesis, latency and reactivation. The Journal of general virology *77*, 1853-1863.

Dejosez, M., Levine, S.S., Frampton, G.M., Whyte, W.A., Stratton, S.A., Barton, M.C., Gunaratne, P.H., Young, R.A., and Zwaka, T.P. (2010). Ronin/Hcf-1 binds to a hyperconserved enhancer element and regulates genes involved in the growth of embryonic stem cells. Genes and development *24*, 1479-1484.

Depledge, D., Ouwendijk, W.J.D., Sadaoka, T., E. Braspenning, S., Mori, Y.M., Cohrs, R., Verjans, G., and Breuer, J. (2018a). A spliced latency-associated VZV transcript maps antisense to the viral transactivator gene 61, Vol 9.

Depledge, D.P., Sadaoka, T., and Ouwendijk, W.J.D. (2018b). Molecular Aspects of Varicella-Zoster Virus Latency. Viruses *10*.

Dhanoa, J.K., Sethi, R.S., Verma, R., Arora, J.S., and Mukhopadhyay, C.S. (2018). Long non-coding RNA: its evolutionary relics and biological implications in mammals: a review. Journal of animal science and technology *60*, 25.

Diefenbach, R.J., Miranda-Saksena, M., Douglas, M.W., and Cunningham, A.L. (2008). Transport and egress of herpes simplex virus in neurons. Reviews in medical virology *18*, 35-51.

Dobson, A.T., Sederati, F., Devi-Rao, G., Flanagan, W.M., Farrell, M.J., Stevens, J.G., Wagner, E.K., and Feldman, L.T. (1989). Identification of the latency-associated transcript promoter by expression of rabbit beta-globin mRNA in mouse sensory nerve ganglia latently infected with a recombinant herpes simplex virus. Journal of virology *63*, 3844-3851.

Doer, R. (1920). Etudes sur le virus de l'herpes febrile. Rev Gen Ophthalmol Paris *34*, 409-421.

Doerig, C., Pizer, L.I., and Wilcox, C.L. (1991). Detection of the latency-associated transcript in neuronal cultures during the latent infection with herpes simplex virus type 1. Virology *183*, 423-426.

Doll, J.R., Hoebe, K., Thompson, R.L., and Sawtell, N.M. (2020). Resolution of herpes simplex virus reactivation in vivo results in neuronal destruction. PLoS pathogens *16*, e1008296.

Doll, J.R., and Sawtell, N.M. (2017). Analysis of Herpes Simplex Virus Reactivation in Explant Reveals a Method-Dependent Difference in Measured Timing of Reactivation. Journal of virology *91*.

Dulbecco, R., and Vogt, M. (1953). Some problems of animal virology as studied by the plaque technique. . Cold Spring Harbor Symp Quant Biol *18*, 273-279.

Easty, D.L., Shimeld, C., Claoue, C.M., and Menage, M. (1987). Herpes simplex virus isolation in chronic stromal keratitis: human and laboratory studies. Current eye research *6*, 69-74.

Edwards, T.G., and Bloom, D.C. (2019). Lund Human Mesencephalic (LUHMES) Neuronal Cell Line Supports Herpes Simplex Virus 1 Latency In Vitro. Journal of virology *93*.

Efstathiou, S., Minson, A.C., Field, H.J., Anderson, J.R., and Wildy, P. (1986). Detection of herpes simplex virus-specific DNA sequences in latently infected mice and in humans. Journal of virology *57*, 446-455.

Everett, R.D. (2000). ICP0, a regulator of herpes simplex virus during lytic and latent infection. BioEssays : news and reviews in molecular, cellular and developmental biology *22*, 761-770.

Fabian, M.A., Biggs, W.H., 3rd, Treiber, D.K., Atteridge, C.E., Azimioara, M.D., Benedetti, M.G., Carter, T.A., Ciceri, P., Edeen, P.T., Floyd, M., *et al.* (2005). A small molecule-kinase interaction map for clinical kinase inhibitors. Nature biotechnology *23*, 329-336.

Farrell, M.J., Dobson, A.T., and Feldman, L.T. (1991). Herpes simplex virus latency-associated transcript is a stable intron. Proceedings of the National Academy of Sciences of the United States of America *88*, 790-794.

Farrell, M.J., Margolis, T.P., Gomes, W.A., and Feldman, L.T. (1994). Effect of the transcription start region of the herpes simplex virus type 1 latency-associated transcript promoter on expression of productively infected neurons in vivo. Journal of virology *68*, 5337-5343.

Feldman, L.T., Ellison, A.R., Voytek, C.C., Yang, L., Krause, P., and Margolis, T.P. (2002). Spontaneous molecular reactivation of herpes simplex virus type 1 latency in mice. Proceedings of the National Academy of Sciences of the United States of America *99*, 978-983.

Field, H.J., Anderson, J.R., and Wildy, P. (1982). Atypical patterns of neural infection produced in mice by drug- resistant strains of herpes simplex virus. The Journal of general virology *59*, 91-99.

Field, H.J., Bell, S.E., Elion, G.B., Nash, A.A., and Wildy, P. (1979). Effect of acycloguanosine treatment of acute and latent herpes simplex infections in mice. Antimicrobial agents and chemotherapy *15*, 554-561.

Field, H.J., and Wildy, P. (1978). The pathogenicity of thymidine kinase-deficient mutants of herpes simplex virus in mice. The Journal of hygiene *81*, 267-277.

Finlay, G.M., and MacCallim, F.O. (1940). Lancet, 259-261.

Fraser, N.W., Block, T.M., and Spivack, J.G. (1992). The latency-associated transcripts of herpes simplex virus: RNA in search of function. Virology *191*, 1-8.

Fraser, N.W., Lawrence, W.C., Wroblewska, Z., Gilden, D.H., and Koprowski, H. (1981). Herpes simplex type 1 DNA in human brain tissue. Proc Natl Acad Sci USA *78*, 6461-6465.

Friedenwald, J.S. (1923). Studies in the virus of herpes simplex. Arch Ophthalmol *52*, 105-131.

Fulop, T., Itzhaki, R.F., Balin, B.J., Miklossy, J., and Barron, A.E. (2018). Role of Microbes in the Development of Alzheimer's Disease: State of the Art - An International Symposium Presented at the 2017 IAGG Congress in San Francisco. Frontiers in genetics *9*, 362.

Gall, J.G., and Pardue, M.L. (1969). Formation and detection of RNA-DNA hybrid molecules in cytological preparations. Proc Natl Acad Sci USA *63*, 378-383.

Garber, D.A., Schaffer, P.A., and Knipe, D.M. (1997). A LAT-associated function reduces productive-cycle gene expression during acute infection of murine sensory neurons with herpes simplex virus type 1. Journal of virology *71*, 5885-5893.

Gershon, A.A., Breuer, J., Cohen, J.I., Cohrs, R.J., Gershon, M.D., Gilden, D., Grose, C., Hambleton, S., Kennedy, P.G., Oxman, M.N., *et al.* (2015). Varicella zoster virus infection. Nature reviews Disease primers *1*, 15016.

Giordani, N.V., Neumann, D.M., Kwiatkowski, D.L., Bhattacharjee, P.S., McAnany, P.K., Hill, J.M., and Bloom, D.C. (2008). During herpes simplex virus type 1 infection of rabbits, the ability to express the latency-associated transcript increases latent-phase transcription of lytic genes. Journal of virology *82*, 6056-6060.

Glynn, J.R., Biraro, S., and Weiss, H.A. (2009). Herpes simplex virus type 2: a key role in HIV incidence. Aids *23*, 1595-1598.

Good, R.A., and Campbell, B. (1948). The Precipitation of Latent Herpes simplex Encephalitis by Anaphylaetie Shock. . Proc Soc Expt Bio Med *68*, 82-87.

Goodpasture, E.W. (1925a). The axis cylinders of peripheral nerves as portals of entry to the central nervous system for the virus of herpes simplex in experimentally infected rabbits. . Amer J Path *1*, 11-28.

Goodpasture, E.W. (1925b). The pathways of infection of the central nervous system in herpetic encephalitis of rabbits contracted by contact, with a comparative comment on medullary lesions in a case of human poliomyelitis. Amer J Path *1*, 29-46.

Goodpasture, E.W. (1929). Herpetic infections with special reference to involvement of the nervous system. Medicine *8*, 223-235.

Goodpasture, E.W., and Teague, O. (1923). Transmission of the virus of herpes fibrilis along nerves in experimentally infected rabbits. J med Res *44*, 139-184.

Gottlieb, S.L., Giersing, B., Boily, M.C., Chesson, H., Looker, K.J., Schiffer, J., Spicknall, I., Hutubessy, R., Broutet, N., and Group, W.H.V.I.M.M.W. (2017). Modelling efforts needed to advance herpes simplex virus (HSV) vaccine development: Key findings from the World Health Organization Consultation on HSV Vaccine Impact Modelling. Vaccine.

Gruter, W. (1920). Experimental and clinical studies on herpes cornea. Klin Monatsbl Augenh *65*, 398-399.

Hafezi, W., Lorentzen, E.U., Eing, B.R., Muller, M., King, N.J., Klupp, B., Mettenleiter, T.C., and Kuhn, J.E. (2012). Entry of herpes simplex virus type 1 (HSV-1) into the distal axons of trigeminal neurons favors the onset of nonproductive, silent infection. PLoS pathogens *8*, e1002679.

Halford, W.P., Kemp, C.D., Isler, J.A., Davido, D.J., and Schaffer, P.A. (2001). ICP0, ICP4, or VP16 expressed from adenovirus vectors induces reactivation of latent herpes simplex virus type 1 in primary cultures of latently infected trigeminal ganglion cells. Journal of virology *75*, 6143-6153.

Halford, W.P., and Schaffer, P.A. (2001). ICP0 is required for efficient reactivation of herpes simplex virus type 1 from neuronal latency. Journal of virology *75*, 3240-3249.

Harbour, D.A., Blyth, W.A., and Hill, T.J. (1978). Prostaglandins enhance spread of herpes simplex virus in cell cultures. . J Gen Virol *41*, 87-95.

Harbour, D.A., Hill, T.J., and Blyth, W.A. (1983). Recurrent herpes simplex in the mouse: inflammation in the skin and activation of virus in the ganglia following peripheral stimulation. The Journal of general virology *64 (Pt 7)*, 1491-1498.

Harbour, D.A., Hill, T.J., and Blyth, W.A., 1977, (1977). The effect of ultraviolet light on primary herpes virus infection in the mouse,. Arch Virol *54*, 367-372.

Harris, R.A., and Preston, C.M. (1991). Establishment of latency in vitro by the herpes simplex virus type 1 mutant in1814. The Journal of general virology *72*, 907-913.

He, J.H., Han, Z.P., and Li, Y.G. (2014). Association between long non-coding RNA and human rare diseases (Review). Biomedical reports *2*, 19-23.

Head, H., and Campbell, A.W. (1900). The pathology of herpes zoster and its bearing on sensory localization. Brain *23*, 353-523.

Head, H., Campbell, A.W., and Kennedy, P.G.E. (1997). The pathology of Herpes Zoster and its bearing on sensory localisation. Reviews in Medical Virology *7*, 131-143.

Held, K., and Derfuss, T. (2011). Control of HSV-1 latency in human trigeminal ganglia--current overview. Journal of neurovirology *17*, 518-527.

Henderson, G., Perng, G.C., Nesburn, A.B., Wechsler, S.L., and Jones, C. (2004). The latency-related gene encoded by bovine herpesvirus 1 can suppress caspase 3 and caspase 9 cleavage during productive infection. Journal of neurovirology *10*, 64-70.

Hill, J.M., Sedarati, F., Javier, R.T., Wagner, E.K., and Stevens, J.G. (1990). Herpes simplex virus latent phase transcription facilitates in vivo reactivation. Virology *174*, 117-125.

Hill, T. (1985). The herpesviruses Vol 3, Vol 3 (Plenum Press, New York).

Hill, T.J., Field, H.J., and Blyth, W.A. (1975). Acute and recurrent infection with herpes simplex virus in the mouse: a model for studying latency and recurrent disease. The Journal of general virology *28*, 341-353.

Holden, M. (1932). The Nature and Properties of the Virus of Herpes. The Journal of infectious diseases *50*, 218-236.

Hope-Simpson, R.E. (1965). The nature of herpes zoster: a long-term study and a new hypothesis. Proc R Soc Med *58*, 9-20.

Howard, P.W., Howard, T.L., and Johnson, D.C. (2013). Herpes simplex virus membrane proteins gE/gI and US9 act cooperatively to promote transport of capsids and glycoproteins from neuron cell bodies into initial axon segments. Journal of virology *87*, 403-414.

Howard, W.T. (1903). The pathology of labial and nasal herpes and of herpes of the body occurring in acute croupous pneumonia and their relation to so-called herpes zoster. . Amer J med Sci *125*, 256-272.

Hoyt, C.S., and Billson, F.A. (1976). , Herpes simplex infection after blowout fractures. Lancet *2*, 1364-1365.

Hu, P., Liu, J., Zhao, J., Wilkins, B.J., Lupino, K., Wu, H., and Pei, L. (2018). Single-nucleus transcriptomic survey of cell diversity and functional maturation in postnatal mammalian hearts. Genes and development *32*, 1344-1357.

Hunsperger, E.A., and Wilcox, C.L. (2003). Caspase-3-dependent reactivation of latent herpes simplex virus type 1 in sensory neuronal cultures. Journal of neurovirology *9*, 390-398.

Hurd, J., and Robinson, T.W.E. (1977). Herpes simplex: Aspects of reactivation in a mouse model. J Antimicrob Chemother *3*, 99-106.

Indo, Y. (2018). NGF-dependent neurons and neurobiology of emotions and feelings: Lessons from congenital insensitivity to pain with anhidrosis. Neuroscience and biobehavioral reviews *87*, 1-16.

Itzhaki, R.F. (2018). Corroboration of a Major Role for Herpes Simplex Virus Type 1 in Alzheimer's Disease. Frontiers in aging neuroscience *10*, 324.

Itzhaki, R.F., Lin, W.R., Shang, D., Wilcock, G.K., Faragher, B., and Jamieson, G.A. (1997). Herpes simplex virus type 1 in brain and risk of Alzheimer's disease. Lancet *349*, 241-244.

Itzhaki, R.F., and Wozniak, M.A. (2008). Herpes simplex virus type 1 in Alzheimer's disease: the enemy within. Journal of Alzheimer's disease : JAD *13*, 393-405.

Izumi, K.M., and Stevens, J.G. (1990). Molecular and biological characterization of a herpes simplex virus type 1 (HSV-1) neuroinvasiveness gene. The Journal of experimental medicine *172*, 487-496.

Jacobson, J.G., Leib, D.A., Goldstein, D.J., Bogard, C.L., Schaffer, P.A., Weller, S.K., and Coen, D.M. (1989). A herpes simplex virus ribonucleotide reductase deletion mutant is defective for productive acute and reactivatable latent infections of mice and for replication in mouse cells. Virology *173*, 276-283.

Jiang, X., Brown, D., Osorio, N., Hsiang, C., BenMohamed, L., and Wechsler, S.L. (2016). Increased neurovirulence and reactivation of the herpes simplex virus type 1 latency-associated transcript (LAT)-negative mutant dLAT2903 with a disrupted LAT miR-H2. Journal of neurovirology *22*, 38-49.

Jiang, X., Brown, D., Osorio, N., Hsiang, C., Li, L., Chan, L., BenMohamed, L., and Wechsler, S.L. (2015). A herpes simplex virus type 1 mutant disrupted for microRNA H2 with increased neurovirulence and rate of reactivation. Journal of neurovirology *21*, 199-209.

Jin, L., Peng, W., Perng, G.C., Brick, D.J., Nesburn, A.B., Jones, C., and Wechsler, S.L. (2003). Identification of herpes simplex virus type 1 latency-associated transcript sequences that both inhibit apoptosis and enhance the spontaneous reactivation phenotype. Journal of virology *77*, 6556-6561.

Jin, L., Perng, G.C., Mott, K.R., Osorio, N., Naito, J., Brick, D.J., Carpenter, D., Jones, C., and Wechsler, S.L. (2005). A herpes simplex virus type 1 mutant expressing a baculovirus inhibitor of apoptosis gene in place of latency-associated transcript has a wild-type reactivation phenotype in the mouse. Journal of virology *79*, 12286-12295.

Johnson, D.C., McDermott, M.R., Chrisp, C., and Glorioso, J.C. (1986). Pathogenicity in mice of herpes simplex virus type 2 mutants unable to express glycoprotein C. Journal of virology *58*, 36-42.

Kato, A., Ando, T., Oda, S., Watanabe, M., Koyanagi, N., Arii, J., and Kawaguchi, Y. (2016). Roles of Us8A and Its Phosphorylation Mediated by Us3 in Herpes Simplex Virus 1 Pathogenesis. Journal of virology *90*, 5622-5635.

Katz, J.P., Bodin, E.T., and Coen, D.M. (1990). Quantitative polymerase chain reaction analysis of herpes simplex virus DNA in ganglia of mice infected with replication-incompetent mutants. Journal of virology *64*, 4288-4295.

Kawamura, Y., Bosch-Marce, M., Tang, S., Patel, A., and Krause, P.R. (2018). Herpes Simplex Virus 2 Latency-Associated Transcript (LAT) Region Mutations Do Not Identify a Role for LAT-Associated MicroRNAs in Viral Reactivation in Guinea Pig Genital Models. Journal of virology *92*.

Kennedy, D.P., Clement, C., Arceneaux, R.L., Bhattacharjee, P.S., Huq, T.S., and Hill, J.M. (2011). Ocular herpes simplex virus type 1: is the cornea a reservoir for viral latency or a fast pit stop? Cornea *30*, 251-259.

Khoury, G., Mota, T.M., Li, S., Tumpach, C., Lee, M.Y., Jacobson, J., Harty, L., Anderson, J.L., Lewin, S.R., and Purcell, D.F.J. (2018). HIV latency reversing agents act through Tat post translational modifications. Retrovirology *15*, 36.

Kim, J.Y., Mandarino, A., Chao, M.V., Mohr, I., and Wilson, A.C. (2012). Transient reversal of episome silencing precedes VP16-dependent transcription during reactivation of latent HSV-1 in neurons. PLoS pathogens *8*, e1002540.

Knickelbein, J.E., Khanna, K.M., Yee, M.B., Baty, C.J., Kinchington, P.R., and Hendricks, R.L. (2008). Noncytotoxic lytic granule-mediated CD8+ T cell inhibition of HSV-1 reactivation from neuronal latency. Science *322*, 268-271.

Knipe, D. (2007). Fields Virology, Vol 2, 5 edn (Philadelphia: Lippincott Williams and Wilkins).

Knipe, D.M. (2015). Nuclear sensing of viral DNA, epigenetic regulation of herpes simplex virus infection, and innate immunity. Virology *479-480*, 153-159.

Kobayashi, M., Kim, J.Y., Camarena, V., Roehm, P.C., Chao, M.V., Wilson, A.C., and Mohr, I. (2012). A primary neuron culture system for the study of herpes simplex virus latency and reactivation. Journal of visualized experiments : JoVE.

Koch, R. (1890). Report of address at 10th International Medical Congress, Berlin, August 4, 1890. Deut Med Wochenschr *16*, 756-757.

Kosz-Vnenchak, M., Jacobson, J., Coen, D.M., and Knipe, D.M. (1993). Evidence for a novel regulatory pathway for herpes simplex virus gene expression in trigeminal ganglion neurons. Journal of virology *67*, 5383-5393.

Koyuncu, O.O., MacGibeny, M.A., Hogue, I.B., and Enquist, L.W. (2017). Compartmented neuronal cultures reveal two distinct mechanisms for alpha herpesvirus escape from genome silencing. PLoS pathogens *13*, e1006608.

Kramer, M.F., and Coen, D.M. (1995). Quantification of transcripts from the ICP4 and thymidine kinase genes in mouse ganglia latently infected with herpes simplex virus. Journal of virology *69*, 1389-1399.

Kraupa, E. (1920). Zu Grüters ätiologischen Untersuchungen über den fieberhaften Herpes. Munch Med Wochenschr *67*.

Krause, P.R., Stanberry, L.R., Bourne, N., Connelly, B., Kurawadwala, J.F., Patel, A., and Straus, S.E. (1995). Expression of the herpes simplex virus type 2 latency-associated transcript enhances spontaneous reactivation of genital herpes in latently infected guinea pigs. The Journal of experimental medicine *181*, 297-306.

Kristie, T.M. (2015). Dynamic modulation of HSV chromatin drives initiation of infection and provides targets for epigenetic therapies. Virology *479-480*, 555-561.

Kristie, T.M., Liang, Y., and Vogel, J.L. (2010). Control of alpha-herpesvirus IE gene expression by HCF-1 coupled chromatin modification activities. Biochimica et biophysica acta *1799*, 257-265.

Kristie, T.M., and Roizman, B. (1987). Host cell proteins bind to the cis-acting site required for virion-mediated induction of herpes simplex virus 1 alpha genes. Proceedings of the National Academy of Sciences of the United States of America *84*, 71-75.

Kristie, T.M., Vogel, J.L., and Sears, A.E. (1999). Nuclear localization of the C1 factor (host cell factor) in sensory neurons correlates with reactivation of herpes simplex virus from latency. Proceedings of the National Academy of Sciences of the United States of America *96*, 1229-1233.

Kundratitz, K. (1925). U¨ ber die A¨ tiologie des Zoster and u¨ber seine Beziehungen zu Varizellen. Wien Klin Wochenschr *38*, 502-503.

Kurapati, S., Sadaoka, T., Rajbhandari, L., Jagdish, B., Shukla, P., Ali, M.A., Kim, Y.J., Lee, G., Cohen, J.I., and Venkatesan, A. (2017). Role of the JNK Pathway in Varicella-Zoster Virus Lytic Infection and Reactivation. Journal of virology *91*.

Lachmann, R.H., and Efstathiou, S. (1997). Utilization of the herpes simplex virus type 1 latency-associated regulatory region to drive stable reporter gene expression in the nervous system. Journal of virology *71*, 3197-3207.

Laemmle, L., Goldstein, R.S., and Kinchington, P.R. (2019). Modeling Varicella Zoster Virus Persistence and Reactivation - Closer to Resolving a Perplexing Persistent State. Frontiers in microbiology *10*, 1634.

LaGuardia, J.J., Cohrs, R.J., and Gilden, D.H. (1999). Prevalence of varicella-zoster virus DNA in dissociated human trigeminal ganglion neurons and nonneuronal cells. Journal of virology *73*, 8571-8577.

Lahmidi, S., Yousefi, M., Dridi, S., Duplay, P., and Pearson, A. (2017). Dok-1 and Dok-2 Are Required To Maintain Herpes Simplex Virus 1-Specific CD8(+) T Cells in a Murine Model of Ocular Infection. Journal of virology *91*.

Lam, Q., Smibert, C.A., Koop, K.E., Lavery, C., Capone, J.P., Weinheimer, S.P., and Smiley, J.R. (1996). Herpes simplex virus VP16 rescues viral mRNA from destruction by the virion host shutoff function. The EMBO journal *15*, 2575-2581.

Lathe, R., Tzeng, N.S., and Itzhaki, R. (2019). Herpes Infections and Dementia: Rebutting Alternative Fact. Neurotherapeutics : the journal of the American Society for Experimental NeuroTherapeutics *16*, 176-179.

Lee, J.S., Raja, P., Pan, D., Pesola, J.M., Coen, D.M., and Knipe, D.M. (2018). CCCTC-Binding Factor Acts as a Heterochromatin Barrier on Herpes Simplex Viral Latent Chromatin and Contributes to Poised Latent Infection. mBio *9*.

Lee, S., Ives, A.M., and Bertke, A.S. (2015). Herpes Simplex Virus 1 Reactivates from Autonomic Ciliary Ganglia Independently from Sensory Trigeminal Ganglia To Cause Recurrent Ocular Disease. Journal of virology *89*, 8383-8391.

Leib, D.A., Bogard, C.L., Kosz-Vnenchak, M., Hicks, K.A., Coen, D.M., Knipe, D.M., and Schaffer, P.A. (1989a). A deletion mutant of the latency-associated transcript of herpes simplex virus type 1 reactivates from the latent state with reduced frequency. Journal of virology *63*, 2893-2900.

Leib, D.A., Coen, D.M., Bogard, C.L., Hicks, K.A., Yager, D.R., Knipe, D.M., Tyler, K.L., and Schaffer, P.A. (1989b). Immediate-early regulatory gene mutants define different stages in the establishment and reactivation of herpes simplex virus latency. Journal of virology *63*, 759-768.

Levin, M.J., Cai, G.Y., Manchak, M.D., and Pizer, L.I. (2003). Varicella-zoster virus DNA in cells isolated from human trigeminal ganglia. Journal of virology *77*, 6979-6987.

Liang, Y., Vogel, J.L., Narayanan, A., Peng, H., and Kristie, T.M. (2009). Inhibition of the histone demethylase LSD1 blocks alpha-herpesvirus lytic replication and reactivation from latency. Nature medicine *15*, 1312-1317.

Lieberman, P.M. (2016). Epigenetics and Genetics of Viral Latency. Cell host and microbe *19*, 619-628.

Lieu, P.T., and Wagner, E.K. (2000). Two leaky-late HSV-1 promoters differ significantly in structural architecture. Virology *272*, 191-203.

Lillycrop, K.A., Estridge, J.K., and Latchman, D.S. (1993). The octamer binding protein Oct-2 inhibits transactivation of the herpes simplex virus immediate-early genes by the virion protein Vmw65. Virology *196*, 888-891.

Lin, W.R., Casas, I., Wilcock, G.K., and Itzhaki, R.F. (1997). Neurotropic viruses and Alzheimer's disease: a search for varicella zoster virus DNA by the polymerase chain reaction. Journal of neurology, neurosurgery, and psychiatry *62*, 586-589.

Liu, T., Khanna, K.M., Chen, X., Fink, D.J., and Hendricks, R.L. (2000). CD8(+) T cells can block herpes simplex virus type 1 (HSV-1) reactivation from latency in sensory neurons. The Journal of experimental medicine *191*, 1459-1466.

Loewenstein, A. (1919). Experimental transmission of HSV-1 from human lesions to rabbit corneas. Munch Med Wochenschr *66*, 769-770.

Loewenstein, A. (1920). Etiology of febrile herpes. Munch med Wochenschr *67*, 1236-1237.

Lofgren, K.W., Stevens, J.G., Marsden, H.S., and Subak-Sharpe, J.H. (1977). Temperature-sensitive mutants of herpes simplex virus differ in the capacity to establish latent infections in mice. Virology *76*, 440-443.

Lomonte, P. (2016). The interaction between herpes simplex virus 1 genome and promyelocytic leukemia nuclear bodies (PML-NBs) as a hallmark of the entry in latency. Microb Cell *3*, 569-572.

Looker, K.J., Magaret, A.S., May, M.T., Turner, K.M.E., Vickerman, P., Newman, L.M., and Gottlieb, S.L. (2017). First estimates of the global and regional incidence of neonatal herpes infection. The Lancet Global health *5*, e300-e309.

Mackem, S., and Roizman, B. (1982). Structural features of the herpes simplex virus alpha gene 4, 0, and 27 promoter-regulatory sequences which confer alpha regulation on chimeric thymidine kinase genes. Journal of virology *44*, 939-949.

Mahalingam, R., Gershon, A., Gershon, M., Cohen, J.I., Arvin, A., Zerboni, L., Zhu, H., Gray, W., Messaoudi, I., and Traina-Dorge, V. (2019). Current In Vivo Models of Varicella-Zoster Virus Neurotropism. Viruses *11*.

Margolis, T.P., Elfman, F.L., Leib, D., Pakpour, N., Apakupakul, K., Imai, Y., and Voytek, C. (2007). Spontaneous reactivation of herpes simplex virus type 1 in latently infected murine sensory ganglia. Journal of virology *81*, 11069-11074.

Margolis, T.P., Sedarati, F., Dobson, A.T., Feldman, L.T., and Stevens, J.G. (1992). Pathways of viral gene expression during acute neuronal infection with HSV-1. Virology *189*, 150-160.

Markus, A., Lebenthal-Loinger, I., Yang, I.H., Kinchington, P.R., and Goldstein, R.S. (2015). An in vitro model of latency and reactivation of varicella zoster virus in human stem cell-derived neurons. PLoS pathogens *11*, e1004885.

Maroui, M.A., Calle, A., Cohen, C., Streichenberger, N., Texier, P., Takissian, J., Rousseau, A., Poccardi, N., Welsch, J., Corpet, A.*, et al.* (2016). Latency Entry of Herpes Simplex Virus 1 Is Determined by the Interaction of Its Genome with the Nuclear Environment. PLoS pathogens *12*, e1005834.

McFarlane, M., Daksis, J.I., and Preston, C.M. (1992). Hexamethylene bisacetamide stimulates herpes simplex virus immediate early gene expression in the absence of trans-induction by Vmw65. The Journal of general virology *73*, 285-292.

McKinney, C.C., Kim, M.J., Chen, D., and McBride, A.A. (2016). Brd4 Activates Early Viral Transcription upon Human Papillomavirus 18 Infection of Primary Keratinocytes. mBio *7*.

McLennan, J.L., and Darby, G. (1980). Herpes simplex virus latency: the cellular location of virus in dorsal root ganglia and the fate of the infected cell following virus activation. The Journal of general virology *51*, 233-243.

Meignier, B., Longnecker, R., and Roizman, B. (1988). In vivo behavior of genetically engineered herpes simplex viruses R7017 and R7020: construction and evaluation in rodents. The Journal of infectious diseases *158*, 602-614.

Michaud, J., Praz, V., James Faresse, N., Jnbaptiste, C.K., Tyagi, S., Schutz, F., and Herr, W. (2013). HCFC1 is a common component of active human CpG-island promoters and coincides with ZNF143, THAP11, YY1, and GABP transcription factor occupancy. Genome research *23*, 907-916.

Minocha, S., Sung, T.L., Villeneuve, D., Lammers, F., and Herr, W. (2016). Compensatory embryonic response to allele-specific inactivation of the murine X-linked gene Hcfc1. Developmental biology *412*, 1-17.

Moffat, J.F., Stein, M.D., Kaneshima, H., and Arvin, A.M. (1995). Tropism of varicella-zoster virus for human CD4+ and CD8+ T lymphocytes and epidermal cells in SCID-hu mice. Journal of virology *69*, 5236-5242.

Mukamal, K.J., Kronmal, R.A., Tracy, R.P., Cushman, M., and Siscovick, D.S. (2004). Traditional and novel risk factors in older adults: cardiovascular risk assessment late in life. Am J Geriatr Cardiol *13*, 69-80.

Mukerjee, R., Kang, W., Suri, V., and Fraser, N.W. (2004). A non-consensus branch point plays an important role in determining the stability of the 2-kb LAT intron during acute and latent infections of herpes simplex virus type-1. Virology *324*, 340-349.

Narayanan, A., Nogueira, M.L., Ruyechan, W.T., and Kristie, T.M. (2005). Combinatorial transcription of herpes simplex virus and varicella zoster virus immediate early genes is strictly determined by the cellular coactivator HCF-1. The Journal of biological chemistry *280*, 1369-1375.

Negatsch, A., Mettenleiter, T.C., and Fuchs, W. (2011). Herpes simplex virus type 1 strain KOS carries a defective US9 and a mutated US8A gene. The Journal of general virology *92*, 167-172.

Nicholls, S.M., and Blyth, W.A. (1989). Quantification of herpes simplex virus infection in cervical ganglia of mice. The Journal of general virology *70*, 1779-1788.

Nicholls, S.M., Shimeld, C., Easty, D.L., and Hill, T.J. (1996). Recurrent herpes simplex after corneal transplantation in rats. Investigative ophthalmology and visual science *37*, 425-435.

Nogueira, M.L., Wang, V.E., Tantin, D., Sharp, P.A., and Kristie, T.M. (2004). Herpes simplex virus infections are arrested in Oct-1-deficient cells. Proceedings of the National Academy of Sciences of the United States of America *101*, 1473-1478.

O'Hare, P. (1993). The virion transactivator of herpes simplex virus. Seminars in Virology *4*, 145-155.

O'Neill, F.J., Goldberg, R.J., and Rapp, F. (1972). Herpes simplex virus latency in cultured human cells following treatment with cytosine arabinoside. The Journal of general virology *14*, 189-197.

Ouwendijk, W.J., and Verjans, G.M. (2015). Pathogenesis of varicelloviruses in primates. The Journal of pathology *235*, 298-311.

Peng, W., Henderson, G., Perng, G.C., Nesburn, A.B., Wechsler, S.L., and Jones, C. (2003). The gene that encodes the herpes simplex virus type 1 latency-associated transcript influences the accumulation of transcripts (Bcl-x(L) and Bcl-x(S)) that encode apoptotic regulatory proteins. Journal of virology *77*, 10714-10718.

Pepose, J.S., Keadle, T.L., and Morrison, L.A. (2006). Ocular herpes simplex: changing epidemiology, emerging disease patterns, and the potential of vaccine prevention and therapy. American journal of ophthalmology *141*, 547-557.

Perdrau, J.R. (1925). The virus of herpes: Its immune reactions and its relation to that of encephalitis lethargica. Brit J Exper Path *6*, 41-52.

Perdrau, J.R. (1931). Inactivation and Reactivation of the Virus of Herpes. Proceedings of the Royal Society of London *109*, 304-308.

Perdrau, J.R. (1938). Persistence of the virus of herpes in rabbits immunised with living virus. The Journal of pathology and bacteriology *47*, 447-455.

Perng, G.C., Jones, C., Ciacci-Zanella, J., Stone, M., Henderson, G., Yukht, A., Slanina, S.M., Hofman, F.M., Ghiasi, H., Nesburn, A.B., *et al.* (2000a). Virus-induced neuronal apoptosis blocked by the herpes simplex virus latency-associated transcript. Science *287*, 1500-1503.

Perng, G.C., Maguen, B., Jin, L., Mott, K.R., Osorio, N., Slanina, S.M., Yukht, A., Ghiasi, H., Nesburn, A.B., Inman, M., *et al.* (2002). A gene capable of blocking apoptosis can substitute for the herpes simplex virus type 1 latency-associated transcript gene and restore wild-type reactivation levels. Journal of virology *76*, 1224-1235.

Perng, G.C., Slanina, S.M., Yukht, A., Ghiasi, H., Nesburn, A.B., and Wechsler, S.L. (2000b). The latency-associated transcript gene enhances establishment of herpes simplex virus type 1 latency in rabbits. Journal of virology *74*, 1885-1891.

Pesola, J.M., Zhu, J., Knipe, D.M., and Coen, D.M. (2005). Herpes simplex virus 1 immediate-early and early gene expression during reactivation from latency under conditions that prevent infectious virus production. Journal of virology *79*, 14516-14525.

Phelan, D., Barrozo, E.R., and Bloom, D.C. (2017). HSV1 latent transcription and non-coding RNA: A critical retrospective. Journal of neuroimmunology *308*, 65-101.

Pieknik, J.R., Bertke, A.S., and Krause, P.R. (2019). Herpes Simplex Virus 2 in Autonomic Ganglia: Evidence for Spontaneous Reactivation. Journal of virology *93*.

Pieknik, J.R., Bertke, A.S., Tang, S., and Krause, P.R. (2018). A VP26-mNeonGreen Capsid Fusion HSV-2 Mutant Reactivates from Viral Latency in the Guinea Pig Genital Model with Normal Kinetics. Viruses *10*.

Plummer, G. (1964). Serological Comparison of the herpes simple viruses. Br J Exp Pathol *45*, 135-141.

Plummer, G. (1973). Isolation of herpesviruses from the trigeminal ganglia of man, monkeys and cats. J Infect Dis *128*, 345-348.

Plummer, G., Cleveland, P.H., and Stevens, C. (1967). Herpes simplex virus and paralysis of rabbits. Activation of the paralysis by adrenalin. British journal of experimental pathology *48*, 390-394.

Plummer, G., Goodheart, C.R., Miyagi, M., Skinner, G.R., Thouless, M.E., and Wildy, P. (1974). Herpes simplex viruses: discrimination of types and correlation between different characteristics. Virology *60*, 206-216.

Plummer, G., Hollingsworth, D.C., Phuangsab, A., and Bowling, C.P. (1970). Chronic infections by herpes simplex viruses and by the horse and cat herpesviruses. Infect Immun *1*, 351-355.

Plummer, G., Waner, J.L., and Bowling, C.P. (1968). Comparative studies of type 1 and type 2 and 'herpes simplex' viruses. British journal of experimental pathology *49*, 202-208.

Pourchet, A., Modrek, A.S., Placantonakis, D.G., Mohr, I., and Wilson, A.C. (2017). Modeling HSV-1 Latency in Human Embryonic Stem Cell-Derived Neurons. Pathogens *6*.

Preston, C.M., and McFarlane, M. (1998). Cytodifferentiating agents affect the replication of herpes simplex virus type 1 in the absence of functional VP16. Virology *249*, 418-426.

Proenca, J.T., Coleman, H.M., Connor, V., Winton, D.J., and Efstathiou, S. (2008). A historical analysis of herpes simplex virus promoter activation in vivo reveals distinct populations of latently infected neurones. The Journal of general virology *89*, 2965-2974.

Proenca, J.T., Coleman, H.M., Nicoll, M.P., Connor, V., Preston, C.M., Arthur, J., and Efstathiou, S. (2011). An investigation of HSV promoter activity compatible with latency establishment reveals VP16 independent activation of HSV immediate early promoters in sensory neurones. The Journal of general virology.

Proenca, J.T., Nelson, D., Nicoll, M.P., Connor, V., and Efstathiou, S. (2015). Analyses of HSV-1 latency and reactivation at the single cell level using fluorescent reporter mice. The Journal of general virology.

Pyles, R.B., Sawtell, N.M., and Thompson, R.L. (1992). Herpes simplex virus type 1 dUTPase mutants are attenuated for neurovirulence, neuroinvasiveness, and reactivation from latency. Journal of virology *66*, 6706-6713.

Pyles, R.B., and Thompson, R.L. (1994a). Evidence that the herpes simplex virus type 1 uracil DNA glycosylase is required for efficient viral replication and latency in the murine nervous system. Journal of virology *68*, 4963-4972.

Pyles, R.B., and Thompson, R.L. (1994b). Mutations in accessory DNA replicating functions alter the relative mutation frequency of herpes simplex virus type 1 strains in cultured murine cells. Journal of virology *68*, 4514-4524.

Ramchandani, M., Selke, S., Magaret, A., Barnum, G., Huang, M.W., Corey, L., and Wald, A. (2017). Prospective cohort study showing persistent HSV-2 shedding in women with genital herpes 2 years after acquisition. Sexually transmitted infections.

Ren, K., Zhang, W., Chen, X., Ma, Y., Dai, Y., Fan, Y., Hou, Y., Tan, R.X., and Li, E. (2016). An Epigenetic Compound Library Screen Identifies BET Inhibitors That Promote HSV-1 and -2 Replication by Bridging P-TEFb to Viral Gene Promoters through BRD4. PLoS pathogens *12*, e1005950.

Rivers, T., and Tillett, W.S. (1924). Further observations on th ephenomena encountered in attepting to transmit varecella to rabbits. J Exp Med *39*, 777-803.

Roberts, S. (2009). Herpes simplex virus: incidence of neonatal herpes simplex virus, maternal screening, management during pregnancy, and HIV. Curr Opin Obstet Gynecol *21*, 124-130.

Rock, D.L., and Fraser, N.W. (1983). Detection of HSV-1 genome in central nervous system of latently infected mice. Nature *302*, 523-525.

Roizman, B., Gu, H., and Mandel, G. (2005). The first 30 minutes in the life of a virus: unREST in the nucleus. Cell cycle *4*, 1019-1021.

Roizman, B., Knipe, D.M., and Whitley, R.J. (2013). Fields Virology,, Vol 2 (Lippincott Williams and Wilkins, Philadelphia, PA,).

Roizman, B., and Whitley, R.J. (2013). An Inquiry into the Molecular Basis of HSV Latency and Reactivation. Annual review of microbiology *67*, 355-374.

Rosenberg, A.B., Roco, C.M., Muscat, R.A., Kuchina, A., Sample, P., Yao, Z., Graybuck, L.T., Peeler, D.J., Mukherjee, S., Chen, W., *et al.* (2018). Single-cell profiling of the developing mouse brain and spinal cord with split-pool barcoding. Science *360*, 176-182.

Russell, T.A., and Tscharke, D.C. (2016). Lytic Promoters Express Protein during Herpes Simplex Virus Latency. PLoS pathogens *12*, e1005729.

Rziha, H.J., Mettenleiter, T.C., Ohlinger, V., and Wittmann, G. (1986). Herpesvirus (pseudorabies virus) latency in swine: occurrence and physical state of viral DNA in neural tissues. Virology *155*, 600-613.

S.E., L., and M.L., H. (1952). A nonhereditary, host-induced variation of bacterial viruses. J Bacteriol *64*, 557-569.

Sacks, S.L., Griffiths, P.D., Corey, L., Cohen, C., Cunningham, A., Dusheiko, G.M., Self, S., Spruance, S., Stanberry, L.R., Wald, A., *et al.* (2004). HSV shedding. Antiviral research *63 Suppl 1*, S19-26.

Sadaoka, T., Depledge, D.P., Rajbhandari, L., Venkatesan, A., Breuer, J., and Cohen, J.I. (2016). In vitro system using human neurons demonstrates that varicella-zoster vaccine virus is impaired for reactivation, but not latency. Proceedings of the National Academy of Sciences of the United States of America *113*, E2403-2412.

Sadaoka, T., Schwartz, C.L., Rajbhandari, L., Venkatesan, A., and Cohen, J.I. (2018). Human Embryonic Stem Cell-Derived Neurons Are Highly Permissive for Varicella-Zoster Virus Lytic Infection. Journal of virology *92*.

Sarno, E., and Robison, A.J. (2018). Emerging role of viral vectors for circuit-specific gene interrogation and manipulation in rodent brain. Pharmacology, biochemistry, and behavior *174*, 2-8.

Sawtell, N.M. (1997). Comprehensive quantification of herpes simplex virus latency at the single-cell level. Journal of virology *71*, 5423-5431.

Sawtell, N.M. (1998). The probability of in vivo reactivation of herpes simplex virus type 1 increases with the number of latently infected neurons in the ganglia. Journal of virology *72*, 6888-6892.

Sawtell, N.M. (2003). Quantitative Analysis of Herpes Simplex Virus Reactivation In Vivo Demonstrates that Reactivation in the Nervous System Is Not Inhibited at Early Times Postinoculation. Journal of virology *77*, 4127-4138.

Sawtell, N.M., Bernstein, D.I., and Stanberry, L.R. (1999). A temporal analysis of acyclovir inhibition of induced herpes simplex virus type 1 In vivo reactivation in the mouse trigeminal ganglia [In Process Citation]. The Journal of infectious diseases *180*, 821-823.

Sawtell, N.M., Poon, D.K., Tansky, C.S., and Thompson, R.L. (1998). The latent herpes simplex virus type 1 genome copy number in individual neurons is virus strain specific and correlates with reactivation. Journal of virology *72*, 5343-5350.

Sawtell, N.M., and Thompson, R.L. (1992a). Herpes simplex virus type 1 latency-associated transcription unit promotes anatomical site-dependent establishment and reactivation from latency. Journal of virology *66*, 2157-2169.

Sawtell, N.M., and Thompson, R.L. (1992b). Rapid in vivo reactivation of herpes simplex virus in latently infected murine ganglionic neurons after transient hyperthermia. Journal of virology *66*, 2150-2156.

Sawtell, N.M., and Thompson, R.L. (2004). Comparison of herpes simplex virus reactivation in ganglia in vivo and in explants demonstrates quantitative and qualitative differences. Journal of virology *78*, 7784-7794.

Sawtell, N.M., and Thompson, R.L. (2016). De Novo Herpes Simplex Virus VP16 Expression Gates a Dynamic Programmatic Transition and Sets the Latent/Lytic Balance during Acute Infection in Trigeminal Ganglia. PLoS pathogens *12*, e1005877.

Sawtell, N.M., Thompson, R.L., Stanberry, L.R., and Bernstein, D.I. (2001). Early intervention with high-dose acyclovir treatment during primary herpes simplex virus infection reduces latency and subsequent reactivation in the nervous system in vivo. The Journal of infectious diseases *184*, 964-971.

Sawtell, N.M., Triezenberg, S.J., and Thompson, R.L. (2011). VP16 serine 375 is a critical determinant of herpes simplex virus exit from latency in vivo. Journal of neurovirology *17*, 546-551.

Schiffer, J.T., Abu-Raddad, L., Mark, K.E., Zhu, J., Selke, S., Magaret, A., Wald, A., and Corey, L. (2009). Frequent release of low amounts of herpes simplex virus from neurons: results of a mathematical model. Science translational medicine *1*, 7ra16.

Schiffer, J.T., Swan, D.A., Prlic, M., and Lund, J.M. (2018). Herpes simplex virus-2 dynamics as a probe to measure the extremely rapid and spatially localized tissue-resident T-cell response. Immunological reviews *285*, 113-133.

Schmidt, J.R., and Rasmussen, A.F. (1960). Activation of Latent Herpes Simplex Encephalitis by Chemical Means. The Journal of infectious diseases *106*, 154-158.

Scriba, M. (1977). Extraneural localisation of herpes simplex virus in latently infected guinea pigs. Nature *267*, 529-531.

Scriba, M. (1981). Persistence of herpes simplex virus (HSV) infection in ganglia and peripheral tissues of guinea pigs. Medical microbiology and immunology *169*, 91-96.

Sears, A.E., Hukkanen, V., Labow, M.A., Levine, A.J., and Roizman, B. (1991). Expression of the herpes simplex virus 1 alpha transinducing factor (VP16) does not induce reactivation of latent virus or prevent the establishment of latency in mice. Journal of virology *65*, 2929-2935.

Sedarati, F., Izumi, K.M., Wagner, E.K., and Stevens, J.G. (1989). Herpes simplex virus type 1 latency-associated transcription plays no role in establishment or maintenance of a latent infection in murine sensory neurons. Journal of virology *63*, 4455-4458.

Sedarati, F., Margolis, T.P., and Stevens, J.G. (1993). Latent infection can be established with drastically restricted transcription and replication of the HSV-1 genome. Virology *192*, 687-691.

Sheldrick, P., and Berthelot, N. (1975). Inverted repetitions in the chromosome of herpes simplex virus. Cold Spring Harbor symposia on quantitative biology *39 Pt 2*, 667-678.

Shen, W., Sa e Silva, M., Jaber, T., Vitvitskaia, O., Li, S., Henderson, G., and Jones, C. (2009). Two small RNAs encoded within the first 1.5 kilobases of the herpes simplex virus type 1 latency-associated transcript can inhibit productive infection and cooperate to inhibit apoptosis. Journal of virology *83*, 9131-9139.

Shimeld, C., Lewkowicz-Moss, S.J., Lipworth, K.M., Hill, T.J., Blyth, W.A., and Easty, D.L. (1986). Antigens of herpes simplex virus in whole corneal epithelial sheets from mice. Archives of ophthalmology *104*, 1830-1834.

Shimeld, C., Tullo, A.B., Easty, D.L., and Thomsitt, J. (1982). Isolation of herpes simplex virus from the cornea in chronic stromal keratitis. The British journal of ophthalmology *66*, 643-647.

Simmons, A., Slobedman, B., Speck, P., Arthur, J., and Efstathiou, S. (1992). Two patterns of persistence of herpes simplex virus DNA sequences in the nervous systems of latently infected mice. The Journal of general virology *73*, 1287-1291.

Skaper, S.D. (2017). Nerve growth factor: a neuroimmune crosstalk mediator for all seasons. Immunology *151*, 1-15.

Smiley, J.R., and Duncan, J. (1997). Truncation of the C-terminal acidic transcriptional activation domain of herpes simplex virus VP16 produces a phenotype similar to that of the in1814 linker insertion mutation. Journal of virology *71*, 6191-6193.

Smith, G. (2012). Herpesvirus transport to the nervous system and back again. Annual review of microbiology *66*, 153-176.

Southern, E.M. (1975). Detection of specific sequences among DNA fragments separated by gel electrophoresis. J Mol Biol *98*, 503-517.

Speck, P.G., and Simmons, A. (1992). Synchronous appearance of antigen-positive and latently infected neurons in spinal ganglia of mice infected with a virulent strain of herpes simplex virus. The Journal of general virology *73*, 1281-1285.

St Leger, A.J., and Hendricks, R.L. (2011). CD8+ T cells patrol HSV-1-infected trigeminal ganglia and prevent viral reactivation. Journal of neurovirology *17*, 528-534.

Stanely, W.M. (1935). Isolation of a crystalline protein possing the properties of tobaccoo -mosaic virus. Science *81*, 644-645.

Steiner, I., Spivack, J.G., Deshmane, S.L., Ace, C.I., Preston, C.M., and Fraser, N.W. (1990). A herpes simplex virus type 1 mutant containing a nontransinducing Vmw65 protein establishes latent infection in vivo in the absence of viral replication and reactivates efficiently from explanted trigeminal ganglia. Journal of virology *64*, 1630-1638.

Stern, S., Tanaka, M., and Herr, W. (1989). The Oct-1 homoeodomain directs formation of a multiprotein-DNA complex with the HSV transactivator VP16. Nature *341*, 624-630.

Stevens, J.G. (1975a). Herpes simplex viral latency - a review. IARC scientific publications, 67-72.

Stevens, J.G. (1975b). Latent herpes simplex virus and the nervous system. Current topics in microbiology and immunology *70*, 31-50.

Stevens, J.G., and Cook, M.L. (1971). Latent herpes simplex virus in spinal ganglia of mice. Science *173*, 843-845.

Stevens, J.G., and Cook, M.L. (1973). Latent herpes simplex infection. in: Virus Research (C F Fox and W S Robinson, eds), *Academic Press, New York*, 437.

Stevens, J.G., Wagner, E.K., Devi-Rao, G.B., Cook, M.L., and Feldman, L.T. (1987). RNA complementary to a herpesvirus alpha gene mRNA is prominent in latently infected neurons. Science *235*, 1056-1059.

Stone, M.J., and Hawkins, C.P. (2007). A medical overview of encephalitis. Neuropsychol Rehabil *17*, 429-449.

Straus, S.E., Reinhold, W., Smith, H.A., Ruyechan, W.T., Henderson, D.K., Blaese, R.M., and Hay, J. (1984). Endonuclease analysis of viral DNA from varicella and subsequent zoster infections in the same patient. The New England journal of medicine *311*, 1362-1364.

Strelow, L.I., and Leib, D.A. (1995). Role of the virion host shutoff (vhs) of herpes simplex virus type 1 in latency and pathogenesis. Journal of virology *69*, 6779-6786.

Sun, Y., Pei, W., Wu, Y., and Yang, Y. (2005). An association of herpes simplex virus type 1 infection with type 2 diabetes. Diabetes Care *28*, 435-436.

Tal-Singer, R., Pichyangkura, R., Chung, E., Lasner, T.M., Randazzo, B.P., Trojanowski, J.Q., Fraser, N.W., and Triezenberg, S.J. (1999). The transcriptional activation domain of VP16 is required for efficient infection and establishment of latency by HSV-1 in the murine peripheral and central nervous systems. Virology *259*, 20-33.

Tanemura, S., Yamasaki, T., Katada, K., Hiroshi, and Nishina, H. (2010). Utility and limitations of sp600125 an inhibitor of stress-responsive c-jun n-terminal kinase. Current Enzyme Inhibition *6*, 26-33.

Tang, S., Patel, A., and Krause, P.R. (2019). Hidden regulation of herpes simplex virus 1 pre-mRNA splicing and polyadenylation by virally encoded immediate early gene ICP27. PLoS pathogens *15*, e1007884.

Tenser, R.B., Dawson, M., Ressel, S.J., and Dunstan, M.E. (1982). Detection of herpes simplex virus mRNA in latently infected trigeminal ganglion neurons by in situ hybridization. Annals of neurology *11*, 285-291.

Thellman, N.M., and Triezenberg, S.J. (2017). Herpes Simplex Virus Establishment, Maintenance, and Reactivation: In Vitro Modeling of Latency. Pathogens *6*.

Thomas, S.K., Gough, G., Latchman, D.S., and Coffin, R.S. (1999). Herpes simplex virus latency-associated transcript encodes a protein which greatly enhances virus growth, can compensate for deficiencies in immediate-early gene expression, and is likely to function during reactivation from virus latency. Journal of virology *73*, 6618-6625.

Thomas, S.K., Lilley, C.E., Latchman, D.S., and Coffin, R.S. (2002). A protein encoded by the herpes simplex virus (HSV) type 1 2-kilobase latency-associated transcript is phosphorylated, localized to the nucleus, and overcomes the repression of expression from exogenous promoters when inserted into the quiescent HSV genome. Journal of virology *76*, 4056-4067.

Thompson, R.L., Preston, C.M., and Sawtell, N.M. (2009). De novo synthesis of VP16 coordinates the exit from HSV latency in vivo. PLoS pathogens *5*, e1000352.

Thompson, R.L., and Sawtell, N.M. (1997). The herpes simplex virus type 1 latency-associated transcript gene regulates the establishment of latency. Journal of virology *71*, 5432-5440.

Thompson, R.L., and Sawtell, N.M. (2000). Replication of herpes simplex virus type 1 within trigeminal ganglia is required for high frequency but not high viral genome copy number latency. Journal of virology *74*, 965-974.

Thompson, R.L., and Sawtell, N.M. (2001). Herpes simplex virus type 1 latency-associated transcript gene promotes neuronal survival. Journal of virology *75*, 6660-6675.

Thompson, R.L., and Sawtell, N.M. (2006). Evidence that the herpes simplex virus type 1 ICP0 protein does not initiate reactivation from latency in vivo. Journal of virology *80*, 10919-10930.

Thompson, R.L., and Sawtell, N.M. (2011). The herpes simplex virus type 1 latency associated transcript locus is required for the maintenance of reactivation competent latent infections. Journal of neurovirology *17*, 552-558.

Thompson, R.L., and Sawtell, N.M. (2019). Targeted Promoter Replacement Reveals That Herpes Simplex Virus Type-1 and 2 Specific VP16 Promoters Direct Distinct Rates of Entry Into the Lytic Program in Sensory Neurons in vivo. Frontiers in microbiology *10*.

Toma, H.S., Murina, A.T., Areaux, R.G., Jr., Neumann, D.M., Bhattacharjee, P.S., Foster, T.P., Kaufman, H.E., and Hill, J.M. (2008). Ocular HSV-1 latency, reactivation and recurrent disease. Seminars in ophthalmology *23*, 249-273.

Traina-Dorge, V., Palmer, B.E., Coleman, C., Hunter, M., Frieman, A., Gilmore, A., Altrock, K., Doyle-Meyers, L., Nagel, M.A., and Mahalingam, R. (2019). Reactivation of Simian Varicella Virus in Rhesus Macaques after CD4 T Cell Depletion. Journal of virology *93*.

Treat, B.R., Bidula, S.M., Ramachandran, S., St Leger, A.J., Hendricks, R.L., and Kinchington, P.R. (2017). Influence of an immunodominant herpes simplex virus type 1 CD8+ T cell epitope on the target hierarchy and function of subdominant CD8+ T cells. PLoS pathogens *13*, e1006732.

Tullo, A.B., Easty, D.L., Shimeld, C., Stirling, P.E., and Darville, J.M. (1985). Isolation of herpes simplex virus from corneal discs of patients with chronic stromal keratitis. Transactions of the ophthalmological societies of the United Kingdom *104 (Pt 2)*, 159-165.

Uitdehaag, J.C., Verkaar, F., Alwan, H., de Man, J., Buijsman, R.C., and Zaman, G.J. (2012). A guide to picking the most selective kinase inhibitor tool compounds for pharmacological validation of drug targets. British journal of pharmacology *166*, 858-876.

Underwood, G.E., and Weed, S.D. (1974). Recurrent cutaneous herpes simplex in hairless mice. Infect Immun *10*, 471-474.

Vidal, E. (1873). Inoculabilite des pustules d'ecthyma. Ann Derm Syphiligr *4*, 350-358.

Visser, M.R., and Vercellotti, G.M. (1993). Herpes simplex virus and atherosclerosis. Eur Heart J *14 Suppl K*, 39-42.

Von Barensprung, F.G.F. (1863). Beitrage zur Kenntnis des Zoster. Annalen Charite-Krunkenhauses zu Berlin *11*, 96-104.

von Bokay, J. (1909). Uber den a¨tiologischen Zusammenhang der Varizellen mit gewissen Fa¨llen von Herpes Zoster. Wien Klin Wochenschr *22*, 1323-1326.

Wang, F., Tang, W., McGraw, H.M., Bennett, J., Enquist, L.W., and Friedman, H.M. (2005a). Herpes simplex virus type 1 glycoprotein e is required for axonal localization of capsid, tegument, and membrane glycoproteins. Journal of virology *79*, 13362-13372.

Wang, F., Zumbrun, E.E., Huang, J., Si, H., Makaroun, L., and Friedman, H.M. (2010). Herpes simplex virus type 2 glycoprotein E is required for efficient virus spread from epithelial cells to neurons and for targeting viral proteins from the neuron cell body into axons. Virology *405*, 269-279.

Wang, K., Lau, T.Y., Morales, M., Mont, E.K., and Straus, S.E. (2005b). Laser-capture microdissection: refining estimates of the quantity and distribution of latent herpes simplex virus 1 and varicella-zoster virus DNA in human trigeminal Ganglia at the single-cell level. Journal of virology *79*, 14079-14087.

Wang, X., Helfer, C.M., Pancholi, N., Bradner, J.E., and You, J. (2013). Recruitment of Brd4 to the human papillomavirus type 16 DNA replication complex is essential for replication of viral DNA. Journal of virology *87*, 3871-3884.

Washington, S.D., Edenfield, S.I., Lieux, C., Watson, Z.L., Taasan, S.M., Dhummakupt, A., Bloom, D.C., and Neumann, D.M. (2018a). Depletion of the Insulator Protein CTCF Results in Herpes Simplex Virus 1 Reactivation In Vivo. Journal of virology *92*.

Washington, S.D., Edenfield, S.I., Lieux, C., Watson, Z.L., Taasan, S.M., Dhummakupt, A., Bloom, D.C., and Neumann, D.M. (2018b). Depletion of the insulator protein CTCF results in HSV-1 reactivation in vivo. Journal of virology.

Washington, S.D., Singh, P., Johns, R.N., Edwards, T.G., Mariani, M., Frietze, S., Bloom, D.C., and Neumann, D.M. (2019). The CCCTC Binding Factor, CTRL2, Modulates Heterochromatin Deposition and the Establishment of Herpes Simplex Virus 1 Latency In Vivo. Journal of virology *93*.

Watson, J.D., and Crick, F.H. (1953). The structure of DNA. Cold Spring Harbor symposia on quantitative biology *18*, 123-131.

Watson, K., Stevens, J.G., Cook, M.L., and Subak-Sharpe, J.H. (1980). Latency competence of thirteen HSV-1 temperature-sensitive mutants. The Journal of general virology *49*, 149-159.

Watson, Z.L., Washington, S.D., Phelan, D.M., Lewin, A.S., Tuli, S.S., Schultz, G.S., Neumann, D.M., and Bloom, D.C. (2018). In Vivo Knockdown of the Herpes Simplex Virus 1 Latency-Associated Transcript Reduces Reactivation from Latency. Journal of virology *92*, e00812-00818.

Webre, J.M., Hill, J.M., Nolan, N.M., Clement, C., McFerrin, H.E., Bhattacharjee, P.S., Hsia, V., Neumann, D.M., Foster, T.P., Lukiw, W.J., *et al.* (2012). Rabbit and mouse models of HSV-1 latency, reactivation, and recurrent eye diseases. Journal of biomedicine and biotechnology *2012*, 612316.

Wechsler, S.L., Nesburn, A.B., Watson, R., Slanina, S.M., and Ghiasi, H. (1988). Fine mapping of the latency-related gene of herpes simplex virus type 1: alternative splicing produces distinct latency-related RNAs containing open reading frames. Journal of virology *62*, 4051-4058.

Whitlow, Z., and Kristie, T.M. (2009). Recruitment of the transcriptional coactivator HCF-1 to viral immediate-early promoters during initiation of reactivation from latency of herpes simplex virus type 1. Journal of virology *83*, 9591-9595.

Wilcox, C.L., Crnic, L.S., and Pizer, L.I. (1992). Replication, latent infection, and reactivation in neuronal culture with a herpes simplex virus thymidine kinase-negative mutant. Virology *187*, 348-352.

Wilcox, C.L., and Johnson, E.M., Jr. (1987). Nerve growth factor deprivation results in the reactivation of latent herpes simplex virus in vitro. Journal of virology *61*, 2311-2315.

Wilcox, C.L., and Johnson, E.M., Jr. (1988). Characterization of nerve growth factor-dependent herpes simplex virus latency in neurons in vitro. Journal of virology *62*, 393-399.

Wilcox, C.L., Smith, R.L., Freed, C.R., and Johnson, E.M., Jr. (1990). Nerve growth factor-dependence of herpes simplex virus latency in peripheral sympathetic and sensory neurons in vitro. The Journal of neuroscience : the official journal of the Society for Neuroscience *10*, 1268-1275.

Wildy, P. (1973). Herpes: history and classification. In: Kaplan, AS ed, The Herpesviruses, Academic Press NY, NY.

Wysocka, J., and Herr, W. (2003). The herpes simplex virus VP16-induced complex: the makings of a regulatory switch. Trends in biochemical sciences *28*, 294-304.

Yanez, A.A., Harrell, T., Sriranganathan, H.J., Ives, A.M., and Bertke, A.S. (2017). Neurotrophic Factors NGF, GDNF and NTN Selectively Modulate HSV1 and HSV2 Lytic Infection and Reactivation in Primary Adult Sensory and Autonomic Neurons. Pathogens *6*.

Yang, L., Voytek, C.C., and Margolis, T.P. (2000). Immunohistochemical analysis of primary sensory neurons latently infected with herpes simplex virus type 1. Journal of virology *74*, 209-217.

Yu, X., and He, S. (2016). The interplay between human herpes simplex virus infection and the apoptosis and necroptosis cell death pathways. Virology journal *13*, 77.

Zargar, Z., and Tyagi, S. (2012). Role of host cell factor-1 in cell cycle regulation. Transcription *3*, 187-192.

Zerboni, L., and Arvin, A. (2015). Neuronal Subtype and Satellite Cell Tropism Are Determinants of Varicella-Zoster Virus Virulence in Human Dorsal Root Ganglia Xenografts In Vivo. PLoS pathogens *11*, e1004989.

Zerboni, L., Sen, N., Oliver, S.L., and Arvin, A.M. (2014). Molecular mechanisms of varicella zoster virus pathogenesis. Nature reviews Microbiology *12*, 197-210.

Zheng, Y., Liu, L., and Shukla, G.C. (2017). A comprehensive review of web-based non-coding RNA resources for cancer research. Cancer letters *407*, 1-8.

Chapter 7

Host Innate Immune Response and Viral Immune Evasion During Alphaherpesvirus Infection

Krystal K. Lum and Ileana M. Cristea*

Department of Molecular Biology, Princeton University, Lewis Thomas Laboratory, Washington Road, Princeton, NJ 08544, USA

*icristea@princeton.edu

DOI: https://doi.org/10.21775/9781913652555.07

Abstract

Both the development of the mammalian innate immune system and the antagonistic strategies acquired by alphaherpesviruses to dismantle it have been shaped by co-evolving virus-host interactions over millions of years. Here, we review mechanisms employed by mammalian cells to detect pathogen molecules, such as viral glycoproteins and nucleic acids, and induce innate immune signaling upon infection with alphaherpesviruses. We further explore strategies acquired by these viruses to bypass immune detection and activation, thereby supporting virus replication and spread. Finally, we discuss the contributions of advanced "omics" and microscopy methods to these discoveries in immune signaling and highlight emerging technologies that can help to further our understanding of the dynamic interplay between host innate immune responses and virus immune evasion.

Introduction

Mammalian innate immunity has been shaped by ancestral and continuous challenges from viruses for millions of years. Phylogenetic analyses have predicted that the alphaherpesvirinae subfamily of herpesviruses arose approximately 180 to 220 million years ago, and that mammalian herpesviruses have been cospeciating with hosts as far back as placental mammals have existed (McGeoch and Cook, 1994; McGeoch et al., 1995). The capacity of mammalian cells to induce immune responses against human alphaherpesviruses, including herpes simplex virus types 1 and 2 (HSV-1, HSV-2) and Varicella-Zoster virus (VZV), is a critical arm of host defense. HSV initiates primary infection via local inoculation of mucocutaneous tissue or oral mucosa, causing localized lesions, while VZV infects upper respiratory mucosa that results in varicella (Arvin, 1996; Karasneh and Shukla, 2011). Upon mucosal cell entry, VZV interacts locally with immune cells, which traffic to the skin and clinically manifest as vesicular skin lesions. Exemplifying their success as pathogens, a defining attribute of alphaherpesviruses is their ability to latently persist in sensory neurons of the nervous system. Thus, not only adaptive immune responses, but also innate immune responses within differentiated cells of the skin and epithelium, are critical for controlling both primary and reactivated alphaherpesvirus infections.

The last decade has provided significant advances in the identification and characterization of host restriction factors that drive cellular innate immunity during infection. Indispensable components that initiate immune signal transduction are specialized pattern recognition receptors (PRR) that survey intracellular spaces for pathogen-associated molecular patterns (PAMP). Alphaherpesvirus glycoproteins and nucleic acids are two PAMP classes that elicit immune signaling from the cell surface or intracellularly, respectively. Viral double-stranded DNA (dsDNA) or dsRNA intermediates are recognized by PRRs in multiple subcellular compartments, including endosomes, the cytosol, and the nucleus. These recognition events signal through adaptor molecules in a PRR-dependent manner and lead to the transcription of antiviral interferons (IFN) and other cytokines. IFN expression further amplifies the transcription of additional IFN molecules and interferon-stimulated genes (ISG) in both the host and neighboring cells (Platanias, 2005). The activated ISGs represent several hundred genes that regulate a diverse set of immune modulatory functions, prime cells for pathogen detection, arrest the cell cycle, or inhibit stages of the viral life cycle (Schneider et al., 2014).

Concurrently, the establishment of a productive viral infection relies on the ability of the virus to evade these IFN- and cytokine-stimulatory programs. Indeed, despite the sophisticated innate immune arsenal, human alphaherpesviruses maintain a close relationship with humans throughout the lifetime of the host via latent persistence in the central nervous system. To accomplish this, alphapherpesviruses have undergone successive cycles of adaptation with mammalian cells, acquiring effective mechanisms that use virus-host interactions to dismantle the host defense repertoire.

Here, we review the current understanding of the host innate immune response to alphaherpesvirus infection. Specifically, we discuss the central host PRRs that detect HSV and VZV components, and the mechanisms underlying immune signaling cascades. We further explore the strategies evolved by these viruses to inhibit innate immune responses by either limiting the production or the downstream signal transduction of ISGs. In closing, we examine how technological advances in omics and microscopy have contributed to these findings and highlight the value of emerging technologies that promise to help further uncover the complex relationship between alphaherpesviruses and mammalian cells.

Innate immunity against alphaherpesviruses

Thus far, we understand that the mammalian innate immune system has established capabilities to detect two classes of alphaherpesvirus PAMPs - glycoproteins embedded in the lipid envelopes of virion particles and intracellular virus-derived nucleic acids (dsDNA and RNA). Both PAMP classes present challenges to host cells because, unlike bacterial and fungal pathogens that contain pathogen-specific biochemical signatures in their cell walls, viruses lack conserved physical characteristics across virus types (Kawai and Akira, 2006; Medzhitov, 2007; Barbalat et al., 2009). Sensing viral glycoproteins may appear inefficient when accounting for both the absence of consistent protein homology across viruses and the potential of viruses to evade immune detection via mutations that maintain protein function(s). Despite this, several PRRs have been identified that demonstrate viral glycoprotein recognition and innate immune activation. Similarly, considering the essentiality of the viral DNA and RNA intermediates for viral gene expression and replication, the ability of the immune system to target pathogenic nucleic acids seems strategically effective. However, the likeness of viral and host DNA molecules led to the initial thinking that viral DNA can only be sensed in subcellular compartments devoid of human DNA.

Indeed, PRRs that sense alphaherpesvirus DNA have been identified in endolysosomes and the cytoplasm. One conundrum of this DNA sensing mechanism is that the virus DNA is protected by the capsid following virus entry into the cell and its trafficking towards the nucleus. A possible answer to this puzzle was given by the observation that the integrity of the capsid can be compromised during infection by proteasomal degradation in certain cell types, such as macrophages, thereby resulting in viral genome leakage and sensing of viral DNA within the cytoplasm (Horan et al., 2013a). Additionally, in recent years it was further recognized that mammalian cells have the ability to sense viral DNA in the nuclei of infected cells. In this section, we review the broad classes of innate immune sensing of alphaherpesviruses by the aforementioned PAMPs and place this knowledge in the context of the subcellular locations for PAMP detection.

Cell surface glycoprotein detection

At the plasma membrane, mammalian cells are equipped with Toll-like receptors (TLR), transmembrane PRRs that have been shown to recognize a variety of microbial membrane components, including lipids, proteins (e.g., lipoproteins, glycoproteins, flagellin), polysaccharides, and other glycans (Lester and Li, 2014). Specific to alphaherpesviruses, TLR detection of viral glycoproteins has been shown for HSV-1, VZV, and certain HSV-2 strains (Kurt-Jones et al., 2004; Wang et al., 2005; Sato et al., 2006). TLR2 is known to form heterodimers *in vivo* with either TLR1 or TLR6, enabling extracellular binding of ligands (Figure 1, step 1a) (Lien et al., 1999; Hajjar et al., 2001). Ligand-bound TLR2 heterodimers initiate immune signal transduction via signaling domains that project into the cytosol and recruit the adaptor protein MyD88. MyD88 activates the cytoplasmic-to-nuclear translocation of several transcription factors, including NF-κB, IRF3, and IRF7. In non-immune stimulated cells, NF-κB remains in an inhibited state via association with IκB. To initiate NF-κB translocation, MyD88 facilitates the activation of the IκB kinase (IKK) complex, which mediates the phosphorylation and subsequent proteasomal degradation of the NF-κB inhibitor, IκB. Via its death domain, MyD88 further recruits a set of interleukin (IL)-1 receptor-associated kinases (IRAK)-1, IRAK-2, and IRAK-4, IRF7, and the E3 ubiquitin ligase TRAF6 (Takeda et al., 2003). The recruited complex both phosphorylates IRF7, thereby activating its translocation to the nucleus, as well as activates the TANK-binding kinase 1 (TBK-1), which subsequently also phosphorylates IRF3. These transcription factors are responsible for inducing the expression of type I IFNs (IFNα, IFNβ) and pro-inflammatory cytokines (IL-6, IL-8, IL-12) (West et al., 2012) (Figure 1).

TLR sensing of alphaherpesviral glycoprotein was first demonstrated during HSV-1 infection (Kurt-Jones et al., 2004). In HEK 293 cells transfected with TLR2, IL-6 expression increased significantly upon HSV-1 infection in a NF-κB-dependent manner, whereas infected TLR2$^{-/-}$ mice demonstrated attenuated IL-6 expression compared to wild-type (WT) or TLR4$^{-/-}$ mice. Surprisingly, the authors additionally observed that TLR2 activity may not necessarily confer a global protective state *in vivo*. Specifically, infected TLR2$^{-/-}$ mice exhibited lower severity of viral encephalitis, suggesting that TLR2 expression contributes to lethal HSV-1-induced encephalitis. Supporting a mechanism of TLR2 hetero-dimerization in immune signaling upon alphaherpesviral infection, TLR2 and TLR6, and to a lesser extent TLR2 and TLR1, were shown to contribute to cytokine induction in HEK 293T cells infected with clinical and laboratory isolates of TLR2-activating HSV-1 (Sato et al., 2006).

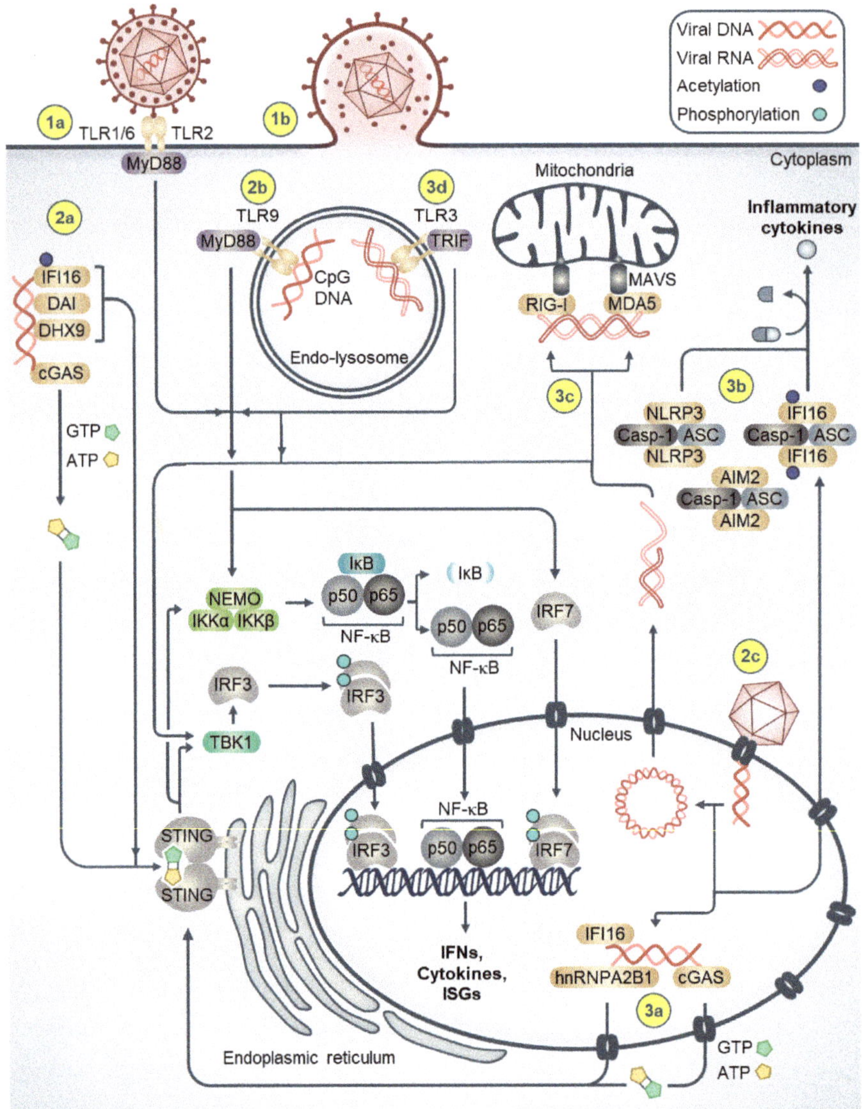

Figure 1. Mechanisms of innate immune signaling during alphaherpesvirus infection. (1a) TLR2 forms heterodimers with TLR1/6, binds extracellular viral glycoproteins, and intracellularly recruits the adaptor protein MyD88. MyD88 activates the transcription factors NF-κB, IRF3, IRF7, which upregulate type I IFNs, cytokines, and ISGs. (1b) Virion fusion with the plasma membrane facilitates entry of the viral nucleocapsid and tegument proteins. (2a) Cytosolic DNA sensors bind viral DNA genomes and signal to STING at the ER. Activated cGAS generates cGAMP, which binds STING. Activated STING induces immune signaling via TBK-1 and the IKK complex, resulting in cytoplasmic to nuclear translocation of IRF3 and NF-κB and induction of type I IFNs, cytokines, and ISGs. (2b) Endosomal detection of CpG-rich viral DNA occurs via TLR9, which also recruits MyD88 to induce immune signaling. (2c) Viral nucleocapsids docking at the nuclear pore eject dsDNA genomes into the nucleus, (3a) facilitating DNA sensing by nuclear IFI16, hnRNPA2B1, and cGAS. (3b) Viral DNA-bound IFI16, NLRP3, and AIM2 inflammasomes in the cytoplasm generate pro-inflammatory cytokines. (3c) RNA intermediates generated during viral replication are sensed by RIG-I and MDA5 in the cytoplasm. (3d) Endosomal RNA is sensed by TLR3, which recruits the adaptor TRIF to activate downstream transcription factors.

It was later recognized that virions from WT or ultraviolet (UV)-inactivated HSV-2 similarly induced cytokines (IL-6 and IL12p40) and IFNα/β in mouse-derived dendritic cells via a TLR2-dependent mechanism (Sorensen et al., 2008). This finding solidified that TLR2 immune signaling did not require active virus replication or gene expression. Yet until more recently, the alphaherpesviral components recognized by TLR2 remained unknown. HSV-1 glycoproteins gH/gL and gB, which are conserved across all human herpesviruses and are essential components for virus attachment and entry into permissive cells (Reske et al., 2007), were identified as independent TLR2 ligands (Leoni et al., 2012; Cai et al., 2013). While gH/gL was found sufficient for activating TLR2-mediated immunity via NF-κB, the activation by gB remains less certain and warrants future *in vivo* investigation. Thus far the reports have been divided, as Leoni *et al.* did not observe expression of NF-κB-responsive genes in HEK 293T cells transfected with TLR2 and a NF-κB luciferase reporter plasmid. In opposition, Cai *et al.* observed an effect when using a similar transfection method and cells. Additionally, gB was found to co-immunoprecipitate not only with TLR2, but also with TLR1 and TLR6 (Cai et al., 2013).

Similar to sensing HSV-1 and HSV-2, TLR2 also recognizes VZV virion components and induces immune signaling. VZV infection of human monocyte-derived macrophages induced IL-6 expression in a TLR2- and NF-κB-dependent manner (Wang et al., 2005). Wang *et al.* determined that CD14, an accessory molecule and common co-receptor of TLR2, was also required to elicit cytokine expression. It was subsequently determined by Cai, *et al.* that TLR2 sensing during HSV-1 infection did not necessitate CD14, indicating differences between immune activation by alphaherpesviruses (Cai et al., 2013). Further emphasizing the divergence of TLR2 immune responses between HSV and VZV infections, no NF-κB activation was observed in mouse cell lines infected with VZV (Wang et al., 2005). However, human cells expressing murine TLR2 were capable of activating NF-κB, suggesting that mouse cells either lack TLR2-specific adaptor molecule(s) upstream of NF-κB, or VZV has evolved a species-specific strategy to subvert immune signaling in mouse cells.

Endolysosomal DNA sensing

Aside from cell surface TLRs that sense alphaherpesviral glycoproteins, other PRRs function intracellularly to sense viral nucleic acids once virion particles have entered into cells. The first PRR discovered to detect herpesviral DNA was TLR9, which is localized to endolysosomes, and contains ligand-binding motifs that face the endolysosomal lumen (Figure 1, step 2b) (Lund et al., 2003; Roers et al., 2016). Based on experiments using cells derived from $TLR9^{-/-}$ mice, as well as treatments with UV-inactivated virus, inhibitory CpG oligonucleotides, and inhibitors of endosome acidification, TLR9 was proposed to upregulate IFNα expression during HSV-2 infection in a MyD88-dependent fashion. This suggested that TLR9 senses CpG-rich herpesviral DNA from virion particles that have entered mammalian cells via endocytosis and traversed the cytoplasm through the endocytic pathway. Consistent with these findings, in comparison to beta- and gamma-herpesviruses and adenoviruses, alphaherpesviruses (HSV-1, HSV-2, VZV, and bovine herpesvirus-1) were found to contain the highest number of CpG hexamer motifs, when normalized to the respective genome sizes (Lundberg et al., 2003). It was later clarified that TLR9 induced IFN expression by recruiting MyD88 and subsequently activating NF-κB and IRF7 (Kawai and Akira, 2011). In plasmacytoid dendritic cells infected with VZV, introduction of inhibitory CpG oligodeoxynucleotides suppressed IFNα levels, indicating that TLR9 contributed to VZV-induced IFNα expression (Yu et al., 2011). Yet, Yu *et al.* also showed that cells treated with UV-inactivated VZV induced lower IFNα levels than cells treated with WT VZV, further suggesting that additional TLR9-independent innate immune signaling exists during VZV infection. Indeed, the concurrent characterization of nucleic acid sensing pathways localized outside of the endolysosomal compartment have started to build a view of the complex immune response mechanisms against alphaherpesviruses.

Cytosolic DNA sensing

Over the past decade, seminal discoveries have been made in clarifying the immune signaling pathways underlying cytosolic detection of alphaherpesviral DNA genomes. We now understand that several DNA sensors exist in the cytosol (Figure 1, step 2a), including cyclic GMP-AMP synthase (cGAS, also known as C6ORF150 and Mab-21 Domain Containing 1 (MB21D1)), interferon-inducible protein 16 (IFI16), DEAD-box helicase 41 (DDX41), RNA polymerase III (RNAP III), absent in melanoma 2 (AIM2), DHX9, DHX36, leucine-rich repeat flightless-interacting protein 1 (LRRFIP1), DNA-dependent activator of IFN-regulatory factors (DAI), and DNA-dependent protein kinase (DNA-PK). As some of these sensors have since been identified to function redundantly, emphasis will be placed on describing recently characterized sensors and pathways that strongly induce innate immune responses during alphaherpesviral infection.

Historically, the central immune signaling axis that is activated by viral DNA detection was characterized prior to the discovery of the DNA sensors. This axis is comprised of the endoplasmic reticulum adapter protein STING (also known as ERIS, MITA, and TMEM173), TBK-1, and IRF3 (Figure 1). Initially, several human and murine cDNA screening studies identified STING as a potent inducer of IFNs and ISGs (Ishikawa and Barber, 2008; Zhong et al., 2008; Sun et al., 2009). STING was subsequently implicated in immune signaling when STING$^{-/-}$ mice were shown to exhibit greater sensitivity to HSV-1 infection (Ishikawa et al., 2009; Parker et al., 2015). It has since been demonstrated that, upon its activation, STING dimerizes and traffics from the ER to both an intermediate ER-Golgi compartment and the Golgi apparatus (Saitoh et al., 2009; Ishikawa et al., 2009; Dobbs et al., 2015). Activated STING thereafter recruits TBK-1, facilitating TBK-1 phosphorylation and subsequent phosphorylation and mobilization of IRF3 to the nucleus, where it binds to consensus interferon-stimulated response elements (Pomerantz and Baltimore, 1999; Tanaka and Chen, 2012). Additionally, STING activates IKK, relieving NF-κB inhibition and facilitating its nuclear translocation for transcriptional activity (Ishikawa and Barber, 2008).

Characterization of the STING-TBK-1-IRF3 immune signaling axis fueled efforts to identify the cytosolic DNA sensor(s) that activate this pathway during viral infection. Specific to sensing alphaherpesvirus DNA, not long after the elucidation of this axis, IFI16 was shown to bind to cytosolic viral DNA, associate with STING, and contribute to IFN production in monocyte-derived THP-1 macrophages (Unterholzner et al., 2010). Upon HSV-1 infection, knockdown of IFI16 or the mouse ortholog p204 in macrophage cells resulted in loss of IRF3 and NF-κB activation, as well as the accompanying IFN and ISG induction (Unterholzner et al., 2010). The localization of IFI16 to the cytoplasm or nucleus was found to be impacted by acetylation within a multipartite nuclear localization signal (NLS) between its oligomerizing pyrin and DNA-binding HIN200 domains, whereupon acetylation retained cytoplasmic IFI16 by inhibiting its nuclear import (Li et al., 2012). As further evidence of IFI16 immune activity during infection, in THP-1 macrophages, STING and IFI16 were observed to co-localize in the cytoplasm during HSV-1 infection (Horan et al., 2013b). In myeloid dendritic cells, DDX41 was characterized as another DNA sensor that required STING to promote IFN expression upon HSV-1 infection (Zhang et al., 2011). Horan, *et al.* proceeded to show that knockdown of DDX41 additionally repressed IFNβ levels upon HSV-1 infection in THP-1 cells (Horan et al., 2013b). Despite these findings, it remains unclear if and how cytosolic IFI16 and DDX41 directly activate STING during infection, and whether these DNA sensors function cooperatively.

Soon after, cGAS was discovered as another DNA sensor and the uncovered signaling mechanism provided a compelling explanation for how STING can be activated. cGAS was found to generate a cyclic dinucleotide that binds directly to STING, thereby robustly activating IFN and ISG expression via the STING-TBK-1-IRF3 pathway (Figure 1, step 2a) (Sun et al., 2013; Wu et al., 2013). Specifically, upon binding to viral DNA, cGAS catalyzes the production of the cyclic second messenger 2',3' cGAMP from intracellular pools of ATP and GTP. This molecule, containing unique mixed 2'-5' and 3'-5' phosphodiester bonds, binds and activates STING to induce the production of type I IFNs. The discovery of cGAS and 2'3' cGAMP cemented a fundamental function for cyclic dinucleotides (CDN) as second messengers in anti-pathogen innate immune signaling of mammalian cells. This was at first unexpected, as prior to this finding, CDNs were believed to function predominantly in prokaryotes, which possess exclusive 3'-5' linked CDNs (Danilchanka and Mekalanos, 2013).

Following these studies, several critical observations were made that solidified a role for cGAS in sensing HSV-1 via the STING-TBK-1-IRF3 pathway. 2',3' cGAMP was detected by mass spectrometry in HSV-1-infected human fibroblasts (Sun et al., 2013). cGAS$^{-/-}$ mice and lung fibroblasts from these mice exhibited attenuated IFN responses to HSV-1 infection (Li et al., 2013b). Lung fibroblasts containing a functionally-inactive form of STING, STING$^{gt/gt}$, were similarly incapable of mounting a type I IFN response to HSV-1. Although cGAS has demonstrated immune signaling capacity during HSV-1 infection, its possible role during VZV infection has yet to be fully clarified. Recently, it was shown that knockdown of STING in human dermal fibroblasts and immortalized HaCaT keratinocytes depressed levels of phosphorylated IRF3 and IFNβ upon VZV infection (Kim et al., 2017). IFNλ levels were also attenuated in VZV-infected cells lacking STING, suggesting that STING promotes an antiviral state in VZV-infected cells via the expression of both Type I and III IFNs (Kim et al., 2017). Considering the apparent parallels in the virus replication cycles and structure between HSV and VZV, assessing the possibility of shared cGAS sensing amongst alphaherpesviruses warrants future investigation.

A DNA sensor that was found to detect VZV DNA was the DNA-dependent RNAP III. RNAP III was shown to bind cytosolic AT-rich DNA, converting it into 5'-triphosphate RNA, which served as a ligand for the type I IFN-inducing RNA sensor, retinoic acid-induced gene I (RIG-I) (Chiu et al., 2009). Several children exhibiting severe VZV infections were found to have heterozygous missense mutations in the RNAP III subunits, POLR3A and POLR3C (Ogunjimi et al., 2017). During VZV infection, leukocytes from these patients were defective in producing IFN, which was restored upon transduction with respective WT alleles of the subunits. Consistent with these findings, mutations in POLR3A and POLR3E were identified in adults with reactivated VZV in the central nervous system (Carter-Timofte et al., 2018). Cells from these patients similarly exhibited impaired IFN expression when challenged with poly(dA:dT) and higher viral gene expression. Considering that the VZV genome contains several spans of AT-rich regions only otherwise partially found in HSV (Paludan et al., 2011; Khalil et al., 2015), these findings support the notion that RNAP III senses VZV genomes via its AT-rich regions. Indeed, another report showed that cGAS-deficient lung fibroblasts derived from mice did not have impaired IFN responses to transfection with AT-rich DNA in the form of poly (dA:dT) (Li et al., 2013b). The role of RNAP III in sensing HSV-1 remains unclear and warrants future research, as several reports uncover conflicting results in mouse and human cells lines (Chiu et al., 2009; Choi et al., 2009; Melchjorsen et al., 2010; Unterholzner et al., 2010).

In addition to the induction of type I IFNs, DNA sensors can also mediate the production of pro-inflammatory cytokines. Within the cytosol and the nucleus, this is accomplished in the presence of pathogen-derived DNA by the formation of distinct multimeric protein assemblies known as inflammasomes (Gram et al., 2012). These structures were evident during alphaherpesvirus infection via the PRRs NLRP3, AIM2, and IFI16 (Figure 1, step 3b), although the roles of IFI16 and AIM2 remain contentious. Upon recognizing viral PAMPs, these sensors recruit the adaptor protein ASC, which cleaves procaspase-1 into its mature and active form, caspase-1. Caspase-1 subsequently promotes the maturation of cytokines IL-18 and IL-1β. NLRP3 was found to generate cytosolic inflammasomes during VZV infection in THP-1 cells and during HSV-1 infection in primary human fibroblasts (Nour et al., 2011; Johnson et al., 2013). Johnson, *et al.* additionally identified a role for IFI16 inflammasomes during HSV-1 infection, in which IFI16-DNA puncta were first seen in the nucleus and later translocated with ASC to the cytosol (Johnson et al., 2013). These observations of nuclear IFI16 puncta during infection were consistent with a growing field of research on DNA sensing within the nucleus.

Nuclear DNA sensing

While the actions of most DNA sensors have been predominantly identified and studied in the cytosol and endolysosomes, more recent evidence has provided an additional focus on sensing pathogen-derived DNA in the host nucleus. This is particularly relevant to alphaherpesviruses, which deposit their DNA genomes into the nucleus for replication (Figure 1, step 2c). Although IFI16 was initially characterized as a cytoplasmic DNA sensor (Unterholzner et al., 2010), several observations pointed to an additional nuclear function. IFI16 was reported to localize predominantly to the nucleus in several cell types, including epithelial cells, endothelial cells, primary fibroblast cells, and immune cells (Choubey and Lengyel, 1992; Aglipay et al., 2003; Cristea et al., 2010; Unterholzner et al., 2010; Kerur et al., 2011; Li et al., 2012). Additionally, the ability of IFI16 to bind to the viral genome in the nucleus was reported even prior to the knowledge of its viral DNA sensing function (Cristea et al., 2010). Yet, this was observed during infection with the betaherpesvirus human cytomegalovirus (HCMV) and was connected to a transcriptional regulatory function. IFI16 was subsequently demonstrated to sense nuclear viral DNA during infection with the gamma herpesvirus Kaposi's sarcoma-associated herpesvirus (KSHV) (Kerur et al., 2011). Kerur, *et al.* reported that nuclear IFI16 formed inflammasomes upon co-localization with KSHV DNA.

Next, the finding of the IFI16 multi-partite NLS and its acetylation enabled interrogation of localization-dependent IFI16 DNA sensing, this time upon infection with the alphaherpesvirus HSV-1 (Li et al., 2012). Li, *et al.* showed that NLS mutants of IFI16 that could not localize to the nucleus were significantly impaired in their ability to bind nuclear HSV-1 DNA and induce IFN response upon infection. Orzalli, *et al.* subsequently demonstrated the necessity of IFI16 and STING for effective IFN and ISG expression in primary human fibroblasts during infection with a mutant HSV-1 strain known to promote strong cytokine expression relative to WT HSV-1, *d109* (Orzalli et al., 2012). Prevention of HSV-1 genome release during *d109* HSV-1 infection via proteasome inhibitor treatment with tolylsulfonylphenylalanyl chloromethyl ketone, impaired STING- and IFI16-dependent innate immune response.

Subsequently, IFI16 was shown to form nuclear aggregates via its pyrin domain (Li et al., 2013a). Consistent with this phenotype, Morrone, *et al.* established that IFI16 forms filaments and cooperatively assembles onto DNA (Morrone et al., 2014). More recently, during HSV-1 infection in primary fibroblasts, IFI16 was shown to form these oligomeric puncta at the nuclear periphery upon viral nucleocapsid docking (Cuchet-Lourenco et al., 2013; Everett, 2016; Diner et al., 2016; Lum et al., 2019). Characterization of IFI16 pyrin oligomerization via predicted solvent-exposed charged residues further suggested that IFI16 rapidly binds to incoming viral DNA at nuclear pores in an oligomerization-dependent manner (Lum et al., 2019). In these studies, CRISPR-mediated knockout of IFI16 or expression of oligomerization-incapable forms of IFI16 significantly impaired both cytokine expression and suppression of viral genes during infection.

Another member of the protein family containing pyrin and HIN200 domains, the interferon-inducible protein X (IFIX), was also demonstrated to have the capacity to bind to dsDNA in a sequence-independent manner and to induce IFN response (Diner et al., 2015a). Similar to IFI16, IFIX was reported to be enriched at the nuclear periphery during early HSV-1 infection and to further promote nucleus-derived immune signaling in fibroblasts (Crow and Cristea, 2017). Altogether, these findings suggested a role for nuclear DNA sensors in directly signaling to the STING-TBK-1-IRF3 pathway for cytokine induction. In support of this possibility, Dunphy *et al.* demonstrated the capacity of nuclear IFI16 to translocate from the nucleus to STING-containing membranous fractions upon DNA damage (Dunphy et al., 2018). Most recently, the heterogeneous nuclear ribonucleoprotein A2B1 (hnRNPA2B1) was identified as a nuclear-to-cytoplasmic translocating DNA sensor against HSV-1 (Wang et al., 2019). Upon HSV-1 infection in murine macrophages and human THP-1 cells, nuclear hnRNPA2B1 was found to associate with HSV-1 DNA, engaging its homodimerization, demethylation, and translocation from the nucleus to the cytoplasm, where it associated with STING and TBK-1.

Several of these observations of STING activation by DNA sensors during HSV-1 infection incited efforts to explore the possibility of synergistic, cooperative, or sequential activities among the sensors (e.g., nuclear and cytosolic IFI16 pools signaling to STING without known IFI16-specific ligands; the identification of a cGAS-specific ligand that directly activates STING; characterization of nuclear-to-cytoplasmic translocating hnRNPA2B1 that binds STING and TBK-1). Yet, the precise mechanisms governing many of these observations are still being uncovered. For instance, how do IFI16 immune signals reach the ER- and Golgi apparatus-associated STING, and are there pools of nuclear cGAS that impact the antiviral state of the cell during alphaherpesvirus infection? Indeed, initial explorations into the localization of endogenous cGAS using antibodies and immunoblotting suggested that a subset of cGAS resides within the nucleus of primary human fibroblasts and normal oral keratinocytes (Orzalli et al., 2015). cGAS was also found to interact with nuclear IFI16, promoting IFI16 protein-level stability during HSV-1 infection. Such an interaction implicates cGAS in maintaining the presence of IFI16-derived immune signals, or sequentially coordinating with IFI16 to activate STING upon infection. Several studies have since identified roles for nuclear cGAS during viral infection and DNA damage (Lahaye et al., 2018; Liu et al., 2018). In further support of DNA sensor-mediated maintenance of immune responses, hnRNPA2B1 was found to promote N6-methyladenosine modification and nuclear-to-cytoplasmic translocation of IFI16, cGAS, and STING mRNAs (Wang et al., 2019). This served to maintain and amplify STING-activated immune responses. Taken together, these findings further highlight how the regulation of nuclear DNA sensing during viral infection is only just beginning to be clarified.

In addition to directly regulating immune signaling pathways, nuclear innate immune proteins can also modulate alphaherpesviral transcription. For example, IFI16 was found to epigenetically promote nucleosome loading and addition of heterochromatin-associated markers onto HSV-1 genomes (Orzalli et al., 2013). These observations suggest that IFI16 functions may be in-part associated with those of a set of sub-nuclear multi-protein complexes, Nuclear

Domain 10 (ND10) bodies (also known as PML nuclear bodies). Components of ND10 bodies are IFN-inducible and are known to rapidly associate with viral DNA genomes being extruded into the nucleus from viral capsids docked at the nuclear pore. This association functions to suppress viral gene expression (Everett et al., 2006; Cuchet et al., 2011). In primary human fibroblasts, stable knockdowns of ND10 complex components, including PML, hDaxx, ATRX, or Sp100, increased titers of an HSV-1 strain lacking ICP0, which is an immediate-early viral protein that degrades several ND10 body components (Chelbi-Alix and de The, 1999). Consistent with the concerted action of innate immune factors, interactions between the DNA sensors IFI16 and IFIX with ND10 body components and mediators of chromatin assembly and transcription have been identified (Diner et al., 2015a; Diner et al., 2015b; Diner et al., 2016; Lum et al., 2019).

RNA sensing

During the replication of DNA viruses, RNA intermediates have long been known to accumulate within the cytoplasm (Jacquemont and Roizman, 1975; Weber et al., 2006). This occurrence is thought to be facilitated in-part by symmetrical, overlapping transcription that generates RNA-RNA annealing. As mammalian cells are armed with an ensemble of RNA sensing components, these pathways have been found to contribute to the induction of an innate immune response in cells during alphaherpesvirus infection. Upon HSV-1 and HSV-2 infections, dsRNA intermediates have been shown to serve as agonists of the endolysosome-localized TLR3 sensor and adaptor protein, TIR-domain-containing adapter-inducing interferon-β (TRIF) (Figure 1, step 3d) (Zhang et al., 2007; Peri et al., 2008; Davey et al., 2010; Willmann et al., 2010). Notably, two children with HSV-1 encephalitis were found to be homozygous for a dominant-negative form of TLR3, which could not be identified in over 3,000 chromosomes examined in healthy control individuals (Zhang et al., 2007). Yet how viral RNA is localized to endolysosomes is still unclear. Most recently, TLR3-mediated IFN expression was shown to additionally rely upon coordination with the receptor, TRAF3 (Sato et al., 2018).

While initially induced by AT-rich DNA sensing by RNAP III, 5' triphosphate RNA generated by RNAP III is sensed by RIG-I, which signals to the adaptor protein, mitochondrial antiviral signaling protein (MAVS), to activate TBK-1 and IKK for inducing IFNs (Chiu et al., 2009; Choi et al., 2009). An additional RIG-I-independent mechanism for sensing RNA during HSV-1 infection has since been characterized via melanoma differentiation-associated protein 5 (MDA5) and MAVS (Melchjorsen et al., 2010) (Figure 1, step 3c). In primary macrophages, knockdown of MDA5 resulted in reduced levels of IFNβ and IFNλ1 upon HSV-1 infection, whereas knockdown of RIG-I had no effect. The authors further observed an increase in TNFα levels during infection that was independent of either RIG-I or MDA5, indicating the potential for alternative pathways, such as via TLR3, in promoting a cytokine-rich antiviral state in the cell.

Alphaherpesvirus innate immune evasion strategies

The ability of alphaherpesviruses to rapidly engage host factors and subvert innate immune responses is a fundamental determinant of the outcome of infection. Highlighting the multifunctionality of viral proteins, several have known capabilities to inhibit immune response via diverse mechanisms, from avoiding PRR detection of PAMPs to preventing downstream cytokine and effector ISG expression (Table 1, Figure 2). This functional versatility enables alphaherpesviruses to effectively constrain host defenses by simultaneously impeding multiple steps of immune response pathways. To ensure a productive viral infection, the implementation of these immune evasion strategies has to match the rapidity of host immune response and, hence, to take place early in infection. This is illustrated by the fact that a considerable number of known immune modulatory viral proteins are either already present as high-copy number components within virion particles that invade a permissive cell (i.e., tegument proteins), or are expressed quickly during productive infection (i.e., immediate-early proteins). In this section, we explore the array of mechanisms employed by alphaherpesviruses to circumvent host immune activation, including the degradation of immune signaling components, as well as their inactivation via binding, post-translational modifications, and sequestration.

Table 1. Alphapherpesvirus innate immune evasion strategies.

Immune evasion strategy	Virus	Viral Protein	Functional consequence	Reference(s)
Degradation	HSV-1	vhs	Degradation of ZAP, viperin, IFI16, cGAS mRNA	(Shen et al., 2014; Su et al., 2015; Orzalli et al., 2016) (Su and Zheng, 2017)
		ICP0	Degradation of MyD88, Mal	(van Lint et al., 2010)
	HSV-1, HSV-2	ICP0	Degradation of IFI16, ND10 components, DNA-PK_{cs}, NF-κB p50	(Parkinson et al., 1999; Boutell et al., 2003; Orzalli et al., 2012; Cuchet-Lourenco et al., 2013; Zhang et al., 2013; Diner et al., 2015b)
	BHV-1	ICP0	Degradation of IRF3	(Saira et al., 2007)
	VZV	ORF61p	Degradation of IRF3	(Zhu et al., 2011)
	PRV	pUL50	Degradation of IFNAR1	(Zhang et al., 2017)
PTM	HSV-1	pUL37	Deamidation of cGAS	(Zhang et al., 2018)
		pUL36	Deubiquitination of TRAF3, IKKα	(Wang et al., 2013a) (Ye et al., 2017)
		pUS3	Atypical IRF3, NF-κB p65 phosphorylation	(Wang et al., 2013b; Wang et al., 2014)
			Hyperphosphorylation, deacetylation prevention of HDAC1/2	(Poon et al., 2006; Walters et al., 2010; Shapira et al., 2016)
	VZV	ORF47p	Atypical IRF3 phosphorylation	(Vandevenne et al., 2011)
		ORF66p	Hyperphosphorylation of HDAC1/2	(Walters et al., 2009)
Molecular sink, sequestration	FHV-1	pUS3	Kinase-independent prevention of IRF3 dimerization	(Tian et al., 2018)
	HSV-1	pUS11	Binding of PKR, RIG-I, and MDA5	(Peters et al., 2002) (Xing et al., 2012)
		VP22	Binding of cGAS, prevention of AIM2 inflammasome formation	(Huang et al., 2018) (Maruzuru et al., 2018)
		pUL46	Binding of STING, TBK-1	(Deschamps and Kalamvoki, 2017)
Molecular sink, sequestration	HSV-1	ICP27	Binding of TBK-1	Christensen et al., 2016
		$\gamma_1$34.5	Binding of TBK-1	Verpooten et al., 2009)
		ICP0	Sequestration of IRF3, CBP/p300 in subnuclear puncta	(Melroe et al., 2007)
		VP24	Binding of IRF3	(Zhang et al., 2016)
		VP16	Binding of IRF3, CBP, NF-κB p65	(Xing et al., 2013)
		pUL36	Binding of IFNAR2	(Yuan et al., 2018)
Unknown		pUL13	Attenuation of CXCL9 expression	(Koyanagi et al., 2017)
Unknown		pUS3	Inhibition of TLR3 expression	(Peri et al., 2008)
Unknown	VZV	IE62	Prevention of IRF3 phosphorylation, independent of binding TBK-1, IRF3	(Sen et al., 2010)
Unknown	PRV	gE, gI	Attenuate IFNα expression	(Lamote et al., 2017)

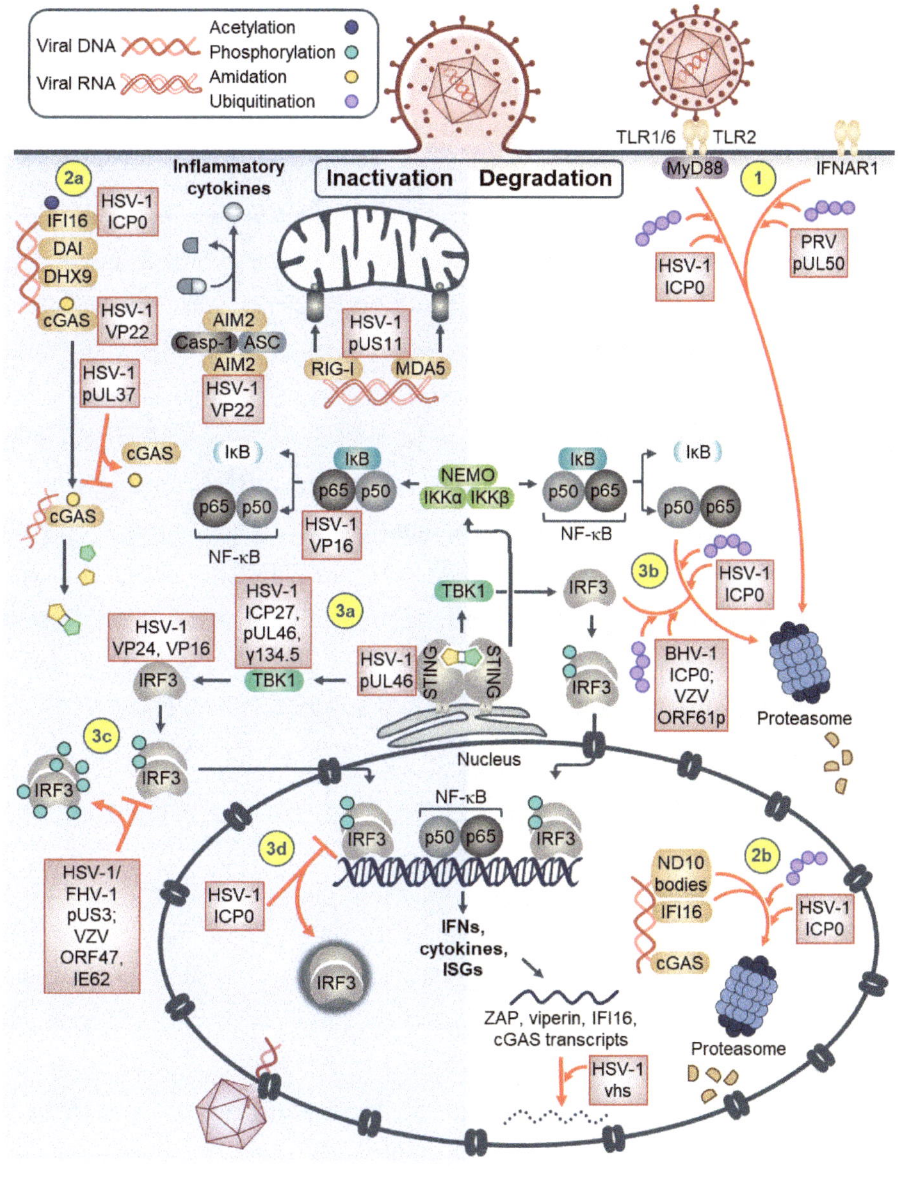

Figure 2. Mechanisms of alphaherpesvirus-mediated innate immune evasion. Signal transduction can be inhibited via inactivation (left), or degradation (right) of immune response components. (1) At the plasma membrane, IFNAR1 and TLR2-recruited MyD88 can be targeted for proteasomal degradation. (2a) Cytosolic DNA sensors IFI16 and cGAS are inhibited by direct binding. (2b) In the nucleus, ND10 body components and IFI16 are targeted for proteasomal degradation via HSV-1 ICP0. Upon activation of the STING-TBK-1-IRF3 pathway, (3a) STING, TBK-1, IRF3, and NF-κB p65 are inhibited by alphaherpesvirus proteins via binding. (3b) Additionally, IRF3 and NF-κB p50 can be targeted for degradation. (3c) IRF3 can also be atypically phosphorylated, preventing proper dimerization and nuclear translocation. (3d) HSV-1 ICP0 can also sequester IRF3 into nuclear puncta, preventing its transcriptional activity. (4) Degradation can occur at the RNA level (e.g., ISG transcripts ZAP, viperin, IFI16, cGAS). (5) Sensors AIM2, RIG-I, and MDA5 can be inhibited by binding of HSV-1 proteins.

Degradation of innate immune factor mRNA and protein

Perhaps one of the most understood and straight-forward mechanisms of virus immune evasion is the ability of alphaherpesviruses to target host innate immune factors for degradation at either the transcriptional or protein level (Figure 2). The major transcriptional strategy employed by HSV-1 is host shutoff via global mRNA destabilization. Once virion particles enter the cell, the abundant HSV-1 tegument protein, virion host shutoff (vhs), functions as an mRNA-specific endonuclease that rapidly degrades host and viral mRNA in the cytoplasm prior to *de novo* viral gene expression (Elgadi et al., 1999). This allows the virus to co-opt cellular transcriptional and translational machinery, as well as engage in the temporal cascade of viral gene expression, which relies in-part on the prior degradation of host and viral mRNA as signaling cues (Read and Frenkel, 1983; Oroskar and Read, 1987, 1989) (Table 1). Notably included as degradation targets are ISGs, such as ZAP and viperin. During WT HSV-1 infection, ZAP and viperin mRNA levels were attenuated relative to those in uninfected cells. In cells infected with HSV-1 strains lacking vhs, both ZAP and viperin mRNA destabilization was prevented (Shen et al., 2014; Su et al., 2015). In line with this function, vhs was more recently demonstrated to additionally target cGAS mRNA for degradation (Su and Zheng, 2017). Interestingly, vhs was found to specifically target adenosine and uridine (AU)-rich elements in the 3' untranslated portions of mRNA, and was shown to be recruited to AU-rich elements via the cellular protein, tristetraprolin (von Roretz et al., 2011; Shu et al., 2015; Rivas et al., 2016). These findings are consistent with the observation that mRNA encoded by many immune response genes are enriched in AU-rich elements (Schott and Stoecklin, 2010; Schwerk and Sayan, 2015), suggesting a sequence-specific mechanism for vhs-dependent destabilization of immune-related mRNA targets.

Alphaherpesvirus-induced degradation at the protein level relies on the exploitation of the cellular proteasome pathway. To maintain a balanced intracellular state, the proteasome pathway is essential for mediating the homeostatic turnover of proteins. Relevant to active innate immune signal transduction, cells require the proteasomal degradation of IκB to release the inhibition of NF-κB and promote its translocation into the nucleus (Karin and Ben-Neriah, 2000). Upon recognition and binding of proteins covalently modified with poly-ubiquitin chains, the proteasomal 26S unfolds and catabolizes proteins into oligopeptides (Lecker et al., 2006).

A prime example of virus-mediated degradation of cellular immune proteins involves the HSV-1 immediate-early protein ICP0. ICP0 contains an E3 ubiquitin ligase activity via a RING finger domain and promotes the proteasome-dependent degradation of several cellular and viral proteins via ubiquitination. ICP0 is known to target a growing number of host defense receptors for degradation, including a glycoprotein sensor, DNA sensor, chromatin modulators, and transcription factors. During infection, van Lint *et al.* determined that TLR2-mediated immune signaling, via IL-6 production and NF-κB-directed transcription, was suppressed due to the ICP0 E3 ubiquitin ligase activity on the TLR2 adaptor proteins, MyD88 and Mal (van Lint et al., 2010).

Critical to nuclear sensing of HSV-1, ICP0 was found to promote the degradation of IFI16 in the nucleus, at least in-part via its E3 ubiquitin ligase function (Orzalli et al., 2012; Cuchet-Lourenco et al., 2013; Diner et al., 2015b). Specifically, ICP0 was found to target the oligomerizing pyrin domain of IFI16 (Diner et al., 2016). Cells infected with an HSV-1 strain containing an E3 ligase-null form of ICP0 (*ICP0-RF* HSV-1) showed a rescue in IFI16 protein levels. Yet, these initial reports conducted in HFFs and HEK 293Ts remain divided over whether ICP0 is sufficient to induce IFI16 degradation, or whether additional viral or hijacked cellular cofactors are also involved in this process. Consistent with this notion, the HSV-1-mediated degradation of the DNA sensor IFIX was not relieved upon infection with *ICP0-RF* HSV-1 (Crow and Cristea, 2017). Furthermore, the mode of IFI16 degradation by HSV-1 was observed to be cell type-dependent, as HSV-1 infection resulted in reduced IFI16 protein levels in HFFs, normal oral keratinocytes, and HeLa cells, but not in U2OS cells. In HeLa cells, the HSV-1 tegument protein vhs additionally facilitated IFI16 mRNA turnover, consistent with its known roles in engaging ribonuclease-induced mRNA destabilization (Orzalli et al., 2016). Noteworthy,

ICP0 was shown to be necessary and sufficient to target several other host defense proteins for degradation, including ND10 body components (i.e., PML, ATRX, hDaxx, Sp100, SUMO1), the DNA sensor DNA-PK_{cs}, and the NF-κB subunit p50 (Parkinson et al., 1999; Boutell et al., 2003; Cuchet-Lourenco et al., 2013; Zhang et al., 2013). Highlighting an interesting conservation of this ICP0 immune-inhibiting function across strains and species, ICP0 encoded by bovine herpesvirus 1 was shown to target IRF3 for degradation in HEK 293s (Saira et al., 2007), yet the capacity of HSV-1 ICP0 to accomplish this has yet to be observed.

Given the uncontested prominence of the STING-TBK-1-IRF3 pathway in promoting an antiviral state in the cell, one might expect several more examples of alphaherpesviral proteins that target these immune components for degradation. Indeed, the VZV immediate-early protein ORF61p co-immunoprecipitated with IRF3, leading to IRF3 ubiquitination and proteasome-dependent degradation. IRF3 degradation was subsequently rescued by treatment with the proteasome inhibitor, MG132 (Zhu et al., 2011). Furthermore, the porcine alphaherpesvirus pseudorabies virus (PRV) encodes pUL50, which abrogates IFNα expression by degrading interferon α/β receptor subunit 1 (IFNAR1) via lysosomes (Zhang et al., 2017).

Inactivation by virus-induced post-translational modifications

In light of the dynamic and temporal role of protein post-translation modifications (PTM) in signal transduction, it is not surprising that alphaherpesviruses encode proteins that catalyze both the addition and removal of PTMs from cellular immune signaling components. At the PRR level, the HSV-1 inner tegument protein pUL37 was shown to deamidate human and mouse cGAS (Zhang et al., 2018). A critical asparagine residue, N210, was identified in the human and mouse cGAS activation loop, which impinges upon the conformational rearrangement of the cGAMP-producing catalytic site (Zhang et al., 2018). In both human THP-1 macrophages and murine L929 fibroblasts, this site was deamidated by pUL37 during HSV-1 infection, suppressing cGAS capacity to generate 2,3'-cGAMP and induce cytokines. pUL37 deamidation did not affect cGAS abilities to bind DNA and dimerize. Infection with a pUL37 deamidation-deficient HSV-1 strain in human macrophages and in BL6 mice resulted in higher cGAS- and STING-dependent cytokine expression relative to infection with WT HSV-1. Unexpectedly, pUL37 had no effect on non-human primate cGAS, suggesting that HSV-1 has evolved to exploit the natural species sequence variation of cGAS. Consistent with this finding, the authors identified that the N210 residue was not conserved in non-human primate species. Further supporting a crucial role for deamidation in subverting host innate immunity during HSV-1 infection, pUL37 was also shown to deamidate the RNA sensor, RIG-I, thereby attenuating IFN expression (Zhao et al., 2016). As pUL37 is evolutionarily conserved across all three herpesvirus subfamilies (alpha, beta, and gamma), and both cGAS and RIG-I have been shown to promote IFN immune signaling upon infection with these herpesviruses, it is possible that pUL37 has a similar PRR deamidation function as these other types of herpesvirus infection.

Downstream of PRR activation, virus-induced deubiquitination and phosphorylation, via both cellular- and virus-encoded enzymes, have been shown to inactivate components of the TLR and STING-TBK-1-IRF3 signaling pathways. Sun *et al.* showed that the cellular protein USP13 catalyzed the deconjugation of K27-linked polyubiquitin chains from STING in mouse embryonic fibroblasts, thereby preventing TBK-1 activation and expression of antiviral cytokines during HSV-1 infection (Sun et al., 2017). This finding suggests that HSV-1 may hijack USP13 to inactivate STING-dependent IFN and ISG expression.

For those immune-evasive enzymes encoded by viruses, the capacity to recognize and modify several host receptors confers an efficient and adaptive advantage against the host defense. This broad host substrate specificity has been identified for several alphaherpesviral proteins. The HSV-1 ubiquitin-specific protease, pUL36, was demonstrated to suppress IFN and cytokine production by modifying both the receptor TRAF3 and NF-KB inhibitor, IKKα. Wang, *et al*. showed that these pUL36 functions relied upon the activity of the deubiquitinase (DUB) motif that was previously identified by Schlieker, *et al.* (Schlieker et al., 2005; Wang et al., 2013a). Via a series of ectopic expression tests in HEK 293Ts and HSV-1 infection in C57BL/6 mice, the authors observed that pUL36-mediated deubiquitination of TRAF3 prevented the formation of a TRAF3-TBK-1 signaling complex, which interfered with IRF3 dimerization. It was later shown that pUL36 was further capable of cleaving K48-linked polyubiquitin chains from IKKα in HEK 293Ts and HFFs, protecting it from proteasomal degradation, and thus maintaining NF-κB in an inactive state (Ye et al., 2017).

Indicating the potential for conserved immune-evasion activity of viral enzymes, the HSV-1 kinase pUS3 and the predicted VZV homolog ORF47p, are known to atypically phosphorylate several cellular factors that promote IFN and ISG induction. Canonically, activated TBK-1 phosphorylates IRF3 at Ser396. However, VZV ORF47p expression in cells was found to interact with IRF3, preventing Ser396 phosphorylation and subsequent transcription of target genes (Vandevenne et al., 2011). HSV-1 pUS3 was observed to hyperphosphorylate IRF3 at Ser175, inhibiting its homodimerization, and the NF-κB subunit p65 at Ser75, preventing its translocation into the nucleus (Wang et al., 2013b; Wang et al., 2014). To identify the specific residues phosphorylated by pUS3, the authors conducted either IRF3 or p65 immunoaffinity isolations in cells co-expressing epitope-tagged US3 and either IRF3 or p65, and analyzed the eluates by mass spectrometry. Consistent with the immune-activating role of Ser396 phosphorylation, this modification was not detected upon pUS3 expression, further indicating that atypical modification of IRF3 impedes the ability of TBK-1 to modify and activate it. An astute demonstration of virus strain-specific divergence in pUS3 inhibition of IRF3 was given in feline kidney cells infected with the alphaherpesvirus, feline herpesvirus 1 (FHV-1) (Tian et al., 2018). FHV-1 pUS3 similarly prevented IRF3 homodimerization and subsequent IFNβ expression yet accomplished this via a kinase-independent mechanism by competitively binding the IRF association domain. As this study identified several other FHV-1 ORFs that inhibit IFN expression (i.e., UL1, UL4, UL3.5, UL11, UL27, UL45, UL48, and

UL55), future explorations may reveal additional layers of virus-host interactions and cooperative or distinct immune suppression mechanisms.

Host factor inactivation and sequestration

To obstruct signal transduction, alphaherpesviruses encode proteins that effectively act as molecular sinks for host innate immune factors by associating with and saturating these receptors. At the level of PAMP detection, the dsRNA sensor protein kinase R (PKR), was found to be antagonized by the HSV-1 tegument protein, pUS11. In uninfected cells, PKR binds RNA, facilitating its autophosphorylation (Romano et al., 1998). Subsequently, PKR activates the translation initiation factor eIF-2α, which engages in translational inhibition and thus blocks viral protein synthesis. As strong expression of pUS11 was previously found to rescue premature translational stasis during infection of cells with a mutant HSV-1 strain (Mohr and Gluzman, 1996; Mulvey et al., 1999), Peters, *et al.* hypothesized that pUS11 could relieve the translational repression imparted by PKR (Peters et al., 2002). Indeed, *in vitro* affinity purification assays revealed that the HSV-1 tegument protein pUS11 was capable of binding directly to PKR, preventing the function of a PKR activator, PACT, and thus inhibiting PKR-mediated translational shutdown in cells.

In addition to PKR inhibition, the HSV-1 tegument protein VP22 was found to bind to the DNA sensors, cGAS and AIM2, suppressing their immune activities (Huang et al., 2018; Maruzuru et al., 2018). With estimates of well over 1,000 copies per virion particle (Heine et al., 1974; Leslie et al., 1996), it may not be surprising that VP22 is simultaneously capable of binding and nearly saturating endogenous levels of these sensors. In HFFs, knocking down cGAS promoted the replication of an HSV-1 strain lacking VP22, but not WT HSV-1, further supporting the cGAS-specific antagonism of this viral protein. The identified VP22-AIM2 association prevented AIM2 oligomerization, formation of inflammasome complexes, and expression of the pro-inflammatory IL-1β and IL-18 in human macrophages (Maruzuru et al., 2018). In a similar fashion, the HSV-1 tegument protein pUS11 was found to bind RIG-I and MDA5. These associations prevented their oligomeric aggregation with the MAVS adaptor for inducing downstream cytokine expression in HEK 293Ts (Xing et al., 2012).

Inactivation of the STING-TBK-1-IRF3 pathway via physical and stable virus-host protein interaction has been demonstrated for each pathway component during HSV-1 infection. Deschamps and Kalamvoki established that the abundant tegument protein pUL46 binds STING to prevent IFN expression in immortalized HEL cells (Deschamps and Kalamvoki, 2017). When treated with 2',3'-cGAMP, cells exogenously expressing pUL46 were incapable of rescuing IFN production. This finding suggested that the pUL46-STING association prevents 2'3'-cGAMP and/or IFI16 binding to STING, STING dimerization, or STING recruitment and activation of TBK-1. The authors also found that pUL46 associated with TBK-1, indicating that pUL46 may physically inhibit the STING-TBK-1 complex. Additional examples of HSV-1 inhibition of TBK-1 are provided by the associations with the immediate-early protein ICP27 and leaky-late protein $\gamma_1$34.5 (Verpooten et al., 2009; Christensen et al., 2016). As expected, these associations prevented IRF3 phosphorylation and activation in each of the studies.

At the level of antagonizing transcription factors downstream of STING activation, the HSV-1 immediate-early protein ICP0 has been shown to re-localize activated IRF3 and its transcriptional co-activating binding partners, CBP and p300, into subnuclear puncta (Melroe et al., 2007). This sequestration strategy prevents IRF3 from docking at ISRE regions to promote the expression of target host genes. Subsequently, Zhang, *et al.* demonstrated that HSV-1 encodes the serine protease, VP24, which directly bound IRF3, inhibiting its activation by TBK-1, its phosphorylation, and homodimerization (Zhang et al., 2016). In a similar manner, the HSV-1 tegument protein VP16 was found to bind IRF3 and its coactivator CBP, preventing the formation of active IRF3-CBP transcriptional complexes (Xing et al., 2013). In support of this finding, the authors did not see an effect of VP16 on IRF3 homodimerization or nuclear translocation. VP16 was additionally observed to inhibit the activation of NF-κB via a VP16-p65 subunit association.

Furthermore, as IFNs are secreted from infected cells to signal to neighboring cells, it would stand to reason that alphaherpesviruses have evolved strategies to prevent this immune activity. Indeed, aside from its deubiquitinase capacity, HSV-1 pUL36 was shown to competitively bind the IFN-α/β receptor, IFNAR2. This association prevented binding of the kinase-signal transducer and activator of transcription, JAK1 (Yuan et al., 2018).

Contributions of advances in omic and microscopy techniques

Concurrent improvements in technologies and methodology have facilitated many of the above-discussed biological discoveries in immune response and virus immune subversion strategies. In this section, we highlight how transcriptomics, proteomics, and microscopy approaches have contributed to our understanding of innate immunity during alphaherpesvirus infection.

Gene profiling reveals immune modulating factors

The application of transcriptomic approaches to exploring virus-host dynamics has promoted the identification of cellular immune factors and associated alphaherpesviral antagonists. In these studies, genes were assessed for their capacity to modulate IFN and ISG expression. Prior to characterizing the DNA sensing activities of cGAS, Schoggins, *et al.* used an ISG screening approach to discover that cGAS was one of the genes that possessed broad IFN-based antiviral function against several viruses, including yellow fever virus, chikungunya virus, Venezuelan equine encephalitis virus, and Sindbis virus (Schoggins et al., 2011). Cells were treated with a lentiviral vector bicistronically bearing one of several hundred ISGs and a red fluorescent protein (RFP) tag. Upon infection with different viruses expressing a distinct green fluorescent protein (GFP) tag, FACS was used to quantify virus replication in RFP-positive cells. Similarly, the ability of FHV-1 pUS3 to antagonize two downstream effector proteins of cGAS, STING and IRF3, was identified from screens of cellular and viral genes, respectively, which affected IFN and ISG levels (Ishikawa and Barber, 2008; Tian et al., 2018). Furthermore, the HSV-1 pUL37 was initially considered as a candidate for suppressing cGAS-mediated IFNβ based on RNA sequencing of host defense genes in samples infected with HSV-1 strains containing mutant pUL37 (Zhang et al., 2018).

In addition to the use of transcriptomic strategies, concurrent explorations into protein interactions uncovered the DNA sensors and associated immune-modulatory protein complexes. These protein interaction studies complemented the findings from gene profiling and enabled investigations of associations between immune proteins with either DNA, other proteins, or small molecules.

Protein interaction studies identify DNA sensors and immune complexes

Over the past decade, the use of immunoaffinity purification coupled to mass spectrometry (IP-MS) has contributed to several landmark discoveries, including the identification of several viral DNA sensors (i.e., AIM2, DHX9, DHX36, IFI16, DNA-PK$_{cs}$, IFIX, and cGAS) and alphaherpesviral antagonists. The first of these proteins to have been identified and characterized as a DNA sensor was AIM2 in 2009 (Burckstummer et al., 2009). In this study, the authors indeed sought to discover a cytoplasmic DNA sensor that mediated inflammasome formation. In three different mouse cell lines, the authors incubated cell extracts with biotinylated 45-base pair interferon-stimulatory DNA (ISD) immobilized to streptavidin resin and subjected eluates to mass spectrometry analysis. As an orthogonal approach, a microarray analysis of genes induced by IFNβ was conducted in parallel. From these assays, AIM2 was uncovered as one of three DNA-binding proteins belonging to the interferon-inducible PYHIN protein family that was additionally transcriptionally regulated by IFNβ. Further experimentation demonstrated that AIM2 was cytoplasmic and capable of forming inflammasomes in the presence of DNA. As would be expected based on the DNA sensors that are known today, one of the other PYHIN proteins identified, but not further explored in the study, was the mouse homolog of IFI16, IFI204. In fact, not long after this initial study, IFI16 was similarly identified as a bona fide cytoplasmic DNA sensor (Unterholzner et al., 2010). Using biotinylated oligomers of 70 base pairs derived from Vaccinia virus, IP-MS was conducted from the cytosolic extracts of THP-1 cells.

The versatility of IP-MS to identify both proteins and small molecules facilitated the identifications of cGAS and cGAMP, respectively. cGAS was discovered as a robust DNA sensor using a modified mass spectrometry approach on lysates from cells pre-stimulated to produce cGAMP (Wu et al., 2013). Via a series of three separate biochemical fractionation strategies conducted in parallel, in which IP-MS using biotinylated ISD was conducted following one of the strategies, cGAS was found to co-purify with cGAMP. cGAMP was also characterized by IP-MS, upon its co-isolation with Flag-tagged STING (Sun et al., 2013). Sun, *et al.* subsequently employed a sensitive targeted mass spectrometry approach, selected reaction monitoring, to quantify cGAMP levels and demonstrate that fractions containing cGAMP were those associated with activated IRF3.

The application of IP-MS to studying DNA sensors has since further expanded to exploring their protein-protein interactions. An interaction network for all PYHIN proteins (AIM2, IFI16, IFIX and MNDA) was generated using cells inducibly expressing each GFP-tagged PYHIN protein (Diner et al., 2015a). The study revealed shared associations involved in functions such as chromatin remodeling and antiviral response, as well as IFIX interactions with PML and other ND10 components. From this study, IFIX was also classified as the second nuclear DNA sensor and was shown to restrict HSV-1 replication. IFI16 interactions from this study and from additional IP-MS studies using an HSV-1 strain that lacks the ability to degrade IFI16 highlighted associations important for immune signaling, such as with ND10 body components (Diner et al., 2015b). An MS-based approach was also used to confirm the interaction of IFI16 with cGAS in multiple cell types (Orzalli et al., 2015). Domain-specific IP-MS analyses were later performed to assign different IFI16 interactions to its pyrin or HIN domains (Orzalli et al., 2015; Diner et al., 2015b; Diner et al., 2016). More recently, a broader cGAS protein interaction network during HSV-1 infection in primary fibroblasts was also characterized using IP-MS analyses

(Lum et al., 2018). In support of previous findings that HSV-1 pUL37 and pUL49 directly antagonize cGAS-mediated production of 2',3'-cGAMP, this study also identified these viral protein associations with cGAS, along with additional HSV-1 proteins, providing a future resource to study hypothesis-driven cGAS regulation and antagonism by HSV-1. Additionally, a cGAS interaction was also identified the 2'-5'-oligoadenylate synthase-like protein OASL, a protein associated with RNA sensing. Further investigation in this study showed that OASL suppresses cGAS-mediated IFN response, suggesting its role as a regulatory negative feedback loop for controlling cytokine induction.

Proteome alterations associated with immune response

Knowledge of PRRs, downstream effector pathways, and their protein interactions were also placed in a broader proteome context during infection to provide a clearer understanding of immune regulatory changes throughout the virus replication cycle. Kulej, *et al.* applied an epigenomics-proteomics approach to generate a resource of global protein abundance changes in both cellular and viral proteins, as well as phosphorylation sites and chromatin-bound histone acetylations during multiple time points of HSV-1 infection in primary human fibroblasts (Kulej et al., 2017). At different time points across HSV-1 infection, fibroblast lysates were sequentially digested with LysC and trypsin. Phosphorylated peptides were enriched using a titanium dioxide-based chromatographic resin, and then subjected to MS analysis. To identify histone acetylations, histones were purified from the lysates of infected cells, and assessed by MS.

Temporal proteome alterations during HSV-1 infection were further explored in the context of cGAS-mediated immune signaling. Lum, *et al.* placed their identified cGAS interactions in the context of the respective endogenous protein abundances during HSV-1 infection (Lum et al., 2018). This was accomplished via whole proteome analysis of primary human fibroblasts using isobaric tag-based quantitative mass spectrometry. Targeted MS using parallel reaction monitoring further validated several protein abundance changes throughout infection, and enabled the assessment of proteins with low endogenous abundances. Providing an additional layer of insight into immune regulation, proteome changes were compared between cells infected with wild type or mutant HSV-1 strains that induced varying degrees of cytokines and apoptosis.

Discovery of PTMs

The contribution of proteomics has extended beyond the identification of DNA sensors and protein-protein interactions. As highlighted in the above sections, PTMs (i.e., phosphorylation, acetylation, deamidation, and ubiquitination) are indispensable for the temporal and spatial modulation of both host immune signal transduction and effective circumvention of this defense by alphaherpesviruses. In addition to investigating global PTM mapping of specific modifications in cells, as mentioned above in the study by Kulej, *et al*, IP-MS has also been used to identify multiple PTMs on immune signaling proteins. For example, Li, *et al*. used a combinatorial IP-MS strategy to identify PTMs on endogenous IFI16 in primary fibroblasts and CEM T cells by digesting immunoisolated samples separately with the endoproteinases, trypsin and GluC (Li et al., 2012). This strategy facilitated the detection of multiple acetylation and phosphorylation sites on IFI16, including the aforementioned acetylation within its NLS, which the authors showed to serve as a toggle for nuclear-cytoplasmic localizations. Downstream of PRR function, two studies identified the capacity of HSV-1 pUS3 to hyperphosphorylate both IRF3 and NF-κB p65 via immunoisolations of epitope-tagged pUS3 and trypsin-based MS analyses (Wang et al., 2013b; Wang et al., 2014).

More recently, this approach was employed to identify both cellular and viral proteins that regulate the PTM state of cGAS, and thus regulate IFNβ induction, during HSV-1 infection. Seo, *et al.* first conducted IP-MS in cells expressing Flag-tagged cGAS and identified that the cellular protein kinase Akt promoted the addition of a phosphorylation on Ser305 of cGAS (Seo et al., 2015). This modification inhibited cGAS-mediated IFN expression and HSV-1 replication during infection upon exogenous Akt expression, suggesting that

Akt either functions as a regulator of cGAS activity, or is co-opted by HSV-1. Recently, IP-MS was used to identify cGAS deamidation by HSV-1 pUL37 (Zhang et al., 2018), emphasizing how this approach is amenable for exploring both anti-viral and pro-viral pathways.

Microscopy in determining the localization dynamics of DNA sensors

The implementation of advanced microscopy approaches has facilitated important insights into the dynamic localization of immune and viral proteins during infection. As an example, microscopy-based investigations into how and when IFI16 binds nuclear HSV-1 DNA have provided an elegant understanding of its mechanism of action with great temporal and spatial resolution. First via confocal microscopy, nuclear IFI16 was observed to re-localize from a diffuse state to distinct puncta asymmetrically at the nuclear periphery, coincident with viral genome-associated proteins (Cuchet-Lourenco et al., 2013). Imaging via *in vitro* single-molecule fluorescence, as well as via time-lapse total internal reflection fluorescence microscopy and Förster resonance energy transfer, fluorescently-labeled IFI16 was observed to track along biotinylated, euchromatinized DNA and assemble into oligomeric structures (Stratmann et al., 2015). This clustering was estimated to be comprised of eight to ten protomers. Subsequently, Diner, *et al.* conducted live cell imaging in primary fibroblasts infected with HSV-1 (Diner et al., 2016). Upon stably expressing fluorescently-tagged IFI16 in cells, IFI16 was observed to dynamically enrich first to incoming HSV-1 genomes within minutes of nucleocapsids docking at the nuclear periphery, and subsequently re-localize to diffuse nucleoplasmic puncta within hours after infection, and then result in loss of signal. These secondary localizations were consistent with previous indirect fluorescence microscopy reports of IFI16 localizing to nuclear puncta in an ICP0-dependent manner prior to its degradation during infection (Orzalli et al., 2012; Cuchet-Lourenco et al., 2013). Thus, microscopy techniques enhanced the understanding of IFI16-mediated signal transduction and antagonism by ICP0.

Concluding remarks and future trends

As we have seen, many alphaherpesvirus-sensing PRRs converge upon shared immune signaling pathways. With the ongoing characterization of PRRs in the cytosol and nucleus, our knowledge of how host defense components act in a synergistic fashion, in distinct immune signaling pathways, operate sequentially, or function redundantly is expected to grow in the coming years. We have also discussed several viral proteins with broad host substrate specificity, providing viruses with multiple opportunities to dismantle host immune signaling. Given the multifunctional capacity of many viral proteins, this ability to both evade immune induction and promote virus replication may explain why the host innate immune response contains such a complex network of PRRs and signaling cascades. As an example, while HSV-1 US3 is known to inactivate IRF3 and NF-κB via atypical phosphorylation, it also relieves the transcriptional repression of viral genes by hyperphosphorylating and preventing deacetylation of histone deacetylases, HDAC1 and HDAC2 (Poon et al., 2006; Walters et al., 2010; Shapira et al., 2016). Similarly, the homolog in VZV, ORF66p, was also shown to hyperphosphorylate HDAC1 and HDAC2 (Walters et al., 2009). In this way, it is perhaps not surprising that we continue to uncover additional functions for proteins involved in both cellular innate immunity and virus immune evasion.

Thus far, many intriguing examples of alphaherpesvirus-mediated IFN suppression have additionally been identified without defined mechanisms. For instance, in the central nervous system of mice, the HSV-1 kinase pUL13 was shown to subvert $CD8^+$ T cell infiltration by attenuating expression of the chemokine attractant CXCL9 (Koyanagi et al., 2017). The VZV immediate-early protein IE62 was found to prevent IFN-inducing IRF3 phosphorylation at serines 396, 398, 402, yet was not found to bind to either TBK-1 or IRF3 (Sen et al., 2010). In plasmacytoid dendritic cells infected with PRV, the glycoproteins gE and gI were observed to antagonize IFNα expression (Lamote et al., 2017). As more sensitive technology emerges, these techniques may be leveraged to provide more mechanistic insight into existing and as-of-yet discovered virus immune evasion strategies.

Altogether, the implementation of developing omic technology has significantly enriched our understanding of innate immunity and alphaherpesvirus immune evasion. Ongoing improvements in mass spectrometry instrumentation with unprecedented resolution and sensitivity provide a powerful tool to be leveraged for future investigations. The additional incorporation of biochemical and molecular tools (i.e., microscopy, chemical crosslinking, alpha screens) is expected to provide greater mechanistic insight and depth. We envisage that these emerging technologies will continue to facilitate exciting opportunities for future research in 1) identifying virus-host protein interactions; 2) uncovering dynamic PTMs on viral and cellular proteins across infection time points and subcellular locations; 3) quantifying global changes in protein abundance; 4) integrating information among multi-omics data (i.e., genomic, transcriptomic, lipidomic) for a systems-level view of alphaherpesvirus replication and host response under different biological contexts (i.e., alphaherpesvirus-induced disease states, variations in clinical isolates, mutant virus strains). These areas of research will be critical in the coming years to address more sophisticated questions regarding the modulation of innate immune signaling during alphaherpesvirus infection.

References

Aglipay, J.A., Lee, S.W., Okada, S., Fujiuchi, N., Ohtsuka, T., Kwak, J.C., Wang, Y., Johnstone, R.W., Deng, C.X., Qin, J., *et al.* (2003). A member of the Pyrin family, IFI16, is a novel BRCA1-associated protein involved in the p53-mediated apoptosis pathway. Oncogene 22, 8931-8938. https://dx.doi.org/10.1038/sj.onc.1207057

Arvin, A.M. (1996). Varicella-zoster virus. Clin Microbiol Rev 9, 361-+. https://dx.doi.org/Doi 10.1128/Cmr.9.3.361

Barbalat, R., Lau, L., Locksley, R.M., and Barton, G.M. (2009). Toll-like receptor 2 on inflammatory monocytes induces type I interferon in response to viral but not bacterial ligands. Nat Immunol 10, 1200-1207. https://dx.doi.org/10.1038/ni.1792

Boutell, C., Orr, A., and Everett, R.D. (2003). PML residue lysine 160 is required for the degradation of PML induced by herpes simplex virus type 1 regulatory protein ICP0. Journal of Virology 77, 8686-8694. https://dx.doi.org/10.1128/Jvi.77.16.8686-8694.2003

Burckstummer, T., Baumann, C., Bluml, S., Dixit, E., Durnberger, G., Jahn, H., Planyavsky, M., Bilban, M., Colinge, J., Bennett, K.L., *et al.* (2009). An orthogonal proteomic-genomic screen identifies AIM2 as a cytoplasmic DNA sensor for the inflammasome. Nature Immunology 10, 266-272. https://dx.doi.org/10.1038/ni.1702

Cai, M., Li, M., Wang, K., Wang, S., Lu, Q., Yan, J., Mossman, K.L., Lin, R., and Zheng, C. (2013). The herpes simplex virus 1-encoded envelope glycoprotein B activates NF-kappaB through the Toll-like receptor 2 and MyD88/TRAF6-dependent signaling pathway. PLoS One 8, e54586. https://dx.doi.org/10.1371/journal.pone.0054586

Carter-Timofte, M.E., Hansen, A.F., Christiansen, M., Paludan, S.R., and Mogensen, T.H. (2018). Mutations in RNA Polymerase III genes and defective DNA sensing in adults with varicella-zoster virus CNS infection. Genes and Immunity https://dx.doi.org/10.1038/s41435-018-0027-y

Chelbi-Alix, M.K., and de The, H. (1999). Herpes virus induced proteasome-dependent degradation of the nuclear bodies-associated PML and Sp100 proteins. Oncogene 18, 935-941. https://dx.doi.org/10.1038/sj.onc.1202366

Chiu, Y.H., MacMillan, J.B., and Chen, Z.J.J. (2009). RNA Polymerase III Detects Cytosolic DNA and Induces Type I Interferons through the RIG-I Pathway. Cell 138, 576-591. https://dx.doi.org/10.1016/j.cell.2009.06.015

Choi, M.K., Wang, Z., Ban, T., Yanai, H., Lu, Y., Koshiba, R., Nakaima, Y., Hangai, S., Savitsky, D., Nakasato, M., *et al.* (2009). A selective contribution of the RIG-I-like receptor pathway to type I interferon responses activated by cytosolic DNA. Proc Natl Acad Sci U S A 106, 17870-17875. https://dx.doi.org/10.1073/pnas.0909545106

Choubey, D., and Lengyel, P. (1992). Interferon Action - Nucleolar and Nucleoplasmic Localization of the Interferon-Inducible 72-Kd Protein That Is Encoded by the Ifi-204 Gene from the Gene-200 Cluster. J Cell Biol 116, 1333-1341. https://dx.doi.org/DOI 10.1083/jcb.116.6.1333

Christensen, M.H., Jensen, S.B., Miettinen, J.J., Luecke, S., Prabakaran, T., Reinert, L.S., Mettenleiter, T., Chen, Z.J., Knipe, D.M., Sandri-Goldin, R.M., *et al.* (2016). HSV-1 ICP27 targets the TBK1-activated STING signalsome to inhibit virus-induced type I IFN expression. Embo J 35, 1385-1399. https://dx.doi.org/10.15252/embj.201593458

Cristea, I.M., Moorman, N.J., Terhune, S.S., Cuevas, C.D., O'Keefe, E.S., Rout, M.P., Chait, B.T., and Shenk, T. (2010). Human cytomegalovirus pUL83 stimulates activity of the viral immediate-early promoter through its interaction with the cellular IFI16 protein. J Virol 84, 7803-7814. https://dx.doi.org/10.1128/JVI.00139-10

Crow, M.S., and Cristea, I.M. (2017). Human Antiviral Protein IFIX Suppresses Viral Gene Expression during Herpes Simplex Virus 1 (HSV-1) Infection and Is Counteracted by Virus-induced Proteasomal Degradation. Mol Cell Proteomics 16, S200-S214. https://dx.doi.org/10.1074/mcp.M116.064741

Cuchet-Lourenco, D., Anderson, G., Sloan, E., Orr, A., and Everett, R.D. (2013). The Viral Ubiquitin Ligase ICP0 Is neither Sufficient nor Necessary for Degradation of the Cellular DNA Sensor IFI16 during Herpes Simplex Virus 1 Infection. Journal of Virology 87, 13422-13432. https://dx.doi.org/10.1128/Jvi.02474-13

Cuchet, D., Sykes, A., Nicolas, A., Orr, A., Murray, J., Sirma, H., Heeren, J., Bartelt, A., and Everett, R.D. (2011). PML isoforms I and II participate in PML-dependent restriction of HSV-1 replication. J Cell Sci 124, 280-291. https://dx.doi.org/10.1242/jcs.075390

Danilchanka, O., and Mekalanos, J.J. (2013). Cyclic dinucleotides and the innate immune response. Cell 154, 962-970. https://dx.doi.org/10.1016/j.cell.2013.08.014

Davey, G.M., Wojtasiak, M., Proietto, A.I., Carbone, F.R., Heath, W.R., and Bedoui, S. (2010). Cutting Edge: Priming of CD8 T Cell Immunity to Herpes Simplex Virus Type 1 Requires Cognate TLR3 Expression In Vivo. Journal of Immunology 184, 2243-2246. https://dx.doi.org/10.4049/jimmunol.0903013

Deschamps, T., and Kalamvoki, M. (2017). Evasion of the STING DNA-Sensing Pathway by VP11/12 of Herpes Simplex Virus 1. J Virol 91 https://dx.doi.org/10.1128/JVI.00535-17

Diner, B.A., Li, T., Greco, T.M., Crow, M.S., Fuesler, J.A., Wang, J., and Cristea, I.M. (2015a). The functional interactome of PYHIN immune regulators reveals IFIX is a sensor of viral DNA. Mol Syst Biol 11, 787. https://dx.doi.org/10.15252/msb.20145808

Diner, B.A., Lum, K.K., Javitt, A., and Cristea, I.M. (2015b). Interactions of the Antiviral Factor Interferon Gamma-Inducible Protein 16 (IFI16) Mediate Immune Signaling and Herpes Simplex Virus-1 Immunosuppression. Mol Cell Proteomics 14, 2341-2356. https://dx.doi.org/10.1074/mcp.M114.047068

Diner, B.A., Lum, K.K., Toettcher, J.E., and Cristea, I.M. (2016). Viral DNA Sensors IFI16 and Cyclic GMP-AMP Synthase Possess Distinct Functions in Regulating Viral Gene Expression, Immune Defenses, and Apoptotic Responses during Herpesvirus Infection. MBio 7. https://dx.doi.org/10.1128/mBio.01553-16

Dobbs, N., Burnaevskiy, N., Chen, D.D., Gonugunta, V.K., Alto, N.M., and Yan, N. (2015). STING Activation by Translocation from the ER Is Associated with Infection and Autoinflammatory Disease. Cell Host Microbe 18, 157-168. https://dx.doi.org/10.1016/j.chom.2015.07.001

Dunphy, G., Flannery, S.M., Almine, J.F., Connolly, D.J., Paulus, C., Jonsson, K.L., Jakobsen, M.R., Nevels, M.M., Bowie, A.G., and Unterholzner, L. (2018). Non-canonical Activation of the DNA Sensing Adaptor STING by ATM and IFI16 Mediates NF-kappa B Signaling after Nuclear DNA Damage. Mol Cell 71, 745-+. https://dx.doi.org/10.1016/j.molcel.2018.07.034

Elgadi, M.M., Hayes, C.E., and Smiley, J.R. (1999). The herpes simplex virus vhs protein induces endoribonucleolytic cleavage of target RNAs in cell extracts. Journal of Virology 73, 7153-7164.

Everett, R.D. (2016). Dynamic Response of IFI16 and Promyelocytic Leukemia Nuclear Body Components to Herpes Simplex Virus 1 Infection. J Virol 90, 167-179. https://dx.doi.org/10.1128/JVI.02249-15

Everett, R.D., Rechter, S., Papior, P., Tavalai, N., Stamminger, T., and Orr, A. (2006). PML contributes to a cellular mechanism of repression of herpes simplex virus type 1 infection that is inactivated by ICP0. J Virol 80, 7995-8005. https://dx.doi.org/10.1128/JVI.00734-06

Gram, A.M., Frenkel, J., and Ressing, M.E. (2012). Inflammasomes and viruses: cellular defence versus viral offence. J Gen Virol 93, 2063-2075. https://dx.doi.org/10.1099/vir.0.042978-0

Hajjar, A.M., O'Mahony, D.S., Ozinsky, A., Underhill, D.M., Aderem, A., Klebanoff, S.J., and Wilson, C.B. (2001). Cutting edge: Functional interactions between toll-like receptor (TLR) 2 and TLR1 or TLR6 in response to phenol-soluble modulin. Journal of Immunology 166, 15-19. https://dx.doi.org/DOI 10.4049/jimmunol.166.1.15

Heine, J.W., Honess, R.W., Cassai, E., and Roizman, B. (1974). Proteins specified by herpes simplex virus. XII. The virion polypeptides of type 1 strains. J Virol 14, 640-651.

Horan, K.A., Hansen, K., Jakobsen, M.R., Holm, C.K., Soby, S., Unterholzner, L., Thompson, M., West, J.A., Iversen, M.B., Rasmussen, S.B., *et al.* (2013a). Proteasomal degradation of herpes simplex virus capsids in macrophages releases DNA to the cytosol for recognition by DNA sensors. J Immunol 190, 2311-2319. https://dx.doi.org/10.4049/jimmunol.1202749

Horan, K.A., Hansen, K., Jakobsen, M.R., Holm, C.K., Soby, S., Unterholzner, L., Thompson, M., West, J.A., Iversen, M.B., Rasmussen, S.B., *et al.* (2013b). Proteasomal Degradation of Herpes Simplex Virus Capsids in Macrophages Releases DNA to the Cytosol for Recognition by DNA Sensors. J Immunol 190, 2311-2319. https://dx.doi.org/10.4049/jimmunol.1202749

Huang, J., You, H.J., Su, C.H., Li, Y.X., Chen, S.H., and Zheng, C.F. (2018). Herpes Simplex Virus 1 Tegument Protein VP22 Abrogates cGAS/STING-Mediated Antiviral Innate Immunity. Journal of Virology 92. https://dx.doi.org/10.1128/JVI.00841-18

Ishikawa, H., and Barber, G.N. (2008). STING is an endoplasmic reticulum adaptor that facilitates innate immune signalling. Nature 455, 674-U674. https://dx.doi.org/10.1038/nature07317

Ishikawa, H., Ma, Z., and Barber, G.N. (2009). STING regulates intracellular DNA-mediated, type I interferon-dependent innate immunity. Nature 461, 788-U740. https://dx.doi.org/10.1038/nature08476

Jacquemont, B., and Roizman, B. (1975). Ribonucleic-Acid Synthesis in Cells Infected with Herpes-Simplex Virus - Characterization of Viral High

Molecular-Weight Nuclear-Rna. Journal of General Virology 29, 155-165. https://dx.doi.org/Doi 10.1099/0022-1317-29-2-155

Johnson, K.E., Chikoti, L., and Chandran, B. (2013). Herpes simplex virus 1 infection induces activation and subsequent inhibition of the IFI16 and NLRP3 inflammasomes. J Virol 87, 5005-5018. https://dx.doi.org/10.1128/JVI.00082-13

Karasneh, G.A., and Shukla, D. (2011). Herpes simplex virus infects most cell types in vitro: clues to its success. J Virol 8. https://dx.doi.org/10.1186/1743-422x-8-481

Karin, M., and Ben-Neriah, Y. (2000). Phosphorylation meets ubiquitination: the control of NF-[kappa]B activity. Annu Rev Immunol 18, 621-663. https://dx.doi.org/10.1146/annurev.immunol.18.1.621

Kawai, T., and Akira, S. (2006). Innate immune recognition of viral infection. Nat Immunol 7, 131-137. https://dx.doi.org/10.1038/ni1303

Kawai, T., and Akira, S. (2011). Toll-like Receptors and Their Crosstalk with Other Innate Receptors in Infection and Immunity. Immunity 34, 637-650. https://dx.doi.org/10.1016/j.immuni.2011.05.006

Kerur, N., Veettil, M.V., Sharma-Walia, N., Bottero, V., Sadagopan, S., Otageri, P., and Chandran, B. (2011). IFI16 Acts as a Nuclear Pathogen Sensor to Induce the Inflammasome in Response to Kaposi Sarcoma-Associated Herpesvirus Infection. Cell Host Microbe 9, 363-375. https://dx.doi.org/10.1016/j.chom.2011.04.008

Khalil, M.I., Sommer, M.H., Hay, J., Ruyechan, W.T., and Arvin, A.M. (2015). Varicella-zoster virus (VZV) origin of DNA replication oriS influences origin-dependent DNA replication and flanking gene transcription. Virology 481, 179-186. https://dx.doi.org/10.1016/j.virol.2015.02.049

Kim, J.A., Park, S.K., Seo, S.W., Lee, C.H., and Shin, O.S. (2017). STING Is Involved in Antiviral Immune Response against VZV Infection via the Induction of Type I and III IFN Pathways. J Invest Dermatol 137, 2101-2109. https://dx.doi.org/10.1016/j.jid.2017.03.041

Koyanagi, N., Imai, T., Shindo, K., Sato, A., Fujii, W., Ichinohe, T., Takemura, N., Kakuta, S., Uematsu, S., Kiyono, H., *et al.* (2017). Herpes simplex virus-1 evasion of CD8(+) T cell accumulation contributes to viral encephalitis. J Clin Invest 127, 3784-3795. https://dx.doi.org/10.1172/Jci92931

Kulej, K., Avgousti, D.C., Sidoli, S., Herrmann, C., Della Fera, A.N., Kim, E.T., Garcia, B.A., and Weitzman, M.D. (2017). Time-resolved Global and Chromatin Proteomics during Herpes Simplex Virus Type 1 (HSV-1) Infection. Mol Cell Proteomics 16, S92-S107. https://dx.doi.org/10.1074/mcp.M116.065987

Kurt-Jones, E.A., Chan, M., Zhou, S., Wang, J., Reed, G., Bronson, R., Arnold, M.M., Knipe, D.M., and Finberg, R.W. (2004). Herpes simplex virus 1 interaction with Toll-like receptor 2 contributes to lethal encephalitis. Proc Natl Acad Sci U S A 101, 1315-1320. https://dx.doi.org/10.1073/pnas.0308057100

Lahaye, X., Gentili, M., Silvin, A., Conrad, C., Picard, L., Jouve, M., Zueva, E., Maurin, M., Nadalin, F., Knott, G.J., *et al.* (2018). NONO Detects the Nuclear HIV Capsid to Promote cGAS-Mediated Innate Immune Activation. Cell 175, 488-501 e422. https://dx.doi.org/10.1016/j.cell.2018.08.062

Lamote, J.A.S., Kestens, M., Van Waesberghe, C., Delva, J., De Pelsmaeker, S., Devriendt, B., and Favoreel, H.W. (2017). The Pseudorabies Virus Glycoprotein gE/gI Complex Suppresses Type I Interferon Production by Plasmacytoid Dendritic Cells. Journal of Virology 91. https://dx.doi.org/10.1128/JVI.02276-16

Lecker, S.H., Goldberg, A.L., and Mitch, W.E. (2006). Protein degradation by the ubiquitin-proteasome pathway in normal and disease states. J Am Soc Nephrol 17, 1807-1819. https://dx.doi.org/10.1681/ASN.2006010083

Leoni, V., Gianni, T., Salvioli, S., and Campadelli-Fiume, G. (2012). Herpes simplex virus glycoproteins gH/gL and gB bind Toll-like receptor 2, and soluble gH/gL is sufficient to activate NF-kappaB. J Virol 86, 6555-6562. https://dx.doi.org/10.1128/JVI.00295-12

Leslie, J., Rixon, F.J., and McLauchlan, J. (1996). Overexpression of the herpes simplex virus type 1 tegument protein VP22 increases its incorporation into virus particles. Virology 220, 60-68. https://dx.doi.org/10.1006/viro.1996.0286

Lester, S.N., and Li, K. (2014). Toll-like receptors in antiviral innate immunity. J Mol Biol 426, 1246-1264. https://dx.doi.org/10.1016/j.jmb.2013.11.024

Li, T., Chen, J., and Cristea, I.M. (2013a). Human cytomegalovirus tegument protein pUL83 inhibits IFI16-mediated DNA sensing for immune evasion. Cell Host Microbe 14, 591-599. https://dx.doi.org/10.1016/j.chom.2013.10.007

Li, T., Diner, B.A., Chen, J., and Cristea, I.M. (2012). Acetylation modulates cellular distribution and DNA sensing ability of interferon-inducible protein IFI16. Proc Natl Acad Sci U S A 109, 10558-10563. https://dx.doi.org/10.1073/pnas.1203447109

Li, X.D., Wu, J., Gao, D., Wang, H., Sun, L., and Chen, Z.J. (2013b). Pivotal roles of cGAS-cGAMP signaling in antiviral defense and immune adjuvant effects. Science 341, 1390-1394. https://dx.doi.org/10.1126/science.1244040

Lien, E., Sellati, T.J., Yoshimura, A., Flo, T.H., Rawadi, G., Finberg, R.W., Carroll, J.D., Espevik, T., Ingalls, R.R., Radolf, J.D., *et al.* (1999). Toll-like receptor 2 functions as a pattern recognition receptor for diverse bacterial products. J Biol Chem 274, 33419-33425. https://dx.doi.org/DOI 10.1074/jbc.274.47.33419

Liu, H., Zhang, H., Wu, X., Ma, D., Wu, J., Wang, L., Jiang, Y., Fei, Y., Zhu, C., Tan, R., *et al.* (2018). Nuclear cGAS suppresses DNA repair and promotes tumorigenesis. Nature 563, 131-136. https://dx.doi.org/10.1038/s41586-018-0629-6

Lum, K.K., Howard, T.R., Pan, C., and Cristea, I.M. (2019). Charge-Mediated Pyrin Oligomerization Nucleates Antiviral IFI16 Sensing of Herpesvirus DNA. mBio 10. https://dx.doi.org/10.1128/mBio.01428-19

Lum, K.K., Song, B., Federspiel, J.D., Diner, B.A., Howard, T., and Cristea, I.M. (2018). Interactome and Proteome Dynamics Uncover Immune Modulatory Associations of the Pathogen Sensing Factor cGAS. Cell Syst 7, 627-642 e626. https://dx.doi.org/10.1016/j.cels.2018.10.010

Lund, J., Sato, A., Akira, S., Medzhitov, R., and Iwasaki, A. (2003). Toll-like receptor 9-mediated recognition of Herpes simplex virus-2 by plasmacytoid dendritic cells. J Exp Med 198, 513-520. https://dx.doi.org/10.1084/jem.20030162

Lundberg, P., Welander, P., Hang, X., and Cantin, E. (2003). Herpes simplex virus type 1 DNA is immunostimulatory in vitro and in vivo. Journal of Virology 77, 11158-11169. https://dx.doi.org/10.1128/Jvi.77.20.11158-11169.2003

Maruzuru, Y., Ichinohe, T., Sato, R., Miyake, K., Okano, T., Suzuki, T., Koshiba, T., Koyanagi, N., Tsuda, S., Watanabe, M., *et al.* (2018). Herpes Simplex Virus 1 VP22 Inhibits AIM2-Dependent Inflammasome Activation to Enable Efficient Viral Replication. Cell Host Microbe 23, 254-265 e257. https://dx.doi.org/10.1016/j.chom.2017.12.014

McGeoch, D.J., and Cook, S. (1994). Molecular phylogeny of the alphaherpesvirinae subfamily and a proposed evolutionary timescale. J Mol Biol 238, 9-22. https://dx.doi.org/10.1006/jmbi.1994.1264

McGeoch, D.J., Cook, S., Dolan, A., Jamieson, F.E., and Telford, E.A. (1995). Molecular phylogeny and evolutionary timescale for the family of mammalian herpesviruses. J Mol Biol 247, 443-458. https://dx.doi.org/10.1006/jmbi.1995.0152

Medzhitov, R. (2007). Recognition of microorganisms and activation of the immune response. Nature 449, 819-826. https://dx.doi.org/10.1038/nature06246

Melchjorsen, J., Rintahaka, J., Soby, S., Horan, K.A., Poltajainen, A., Ostergaard, L., Paludan, S.R., and Matikainen, S. (2010). Early Innate Recognition of Herpes Simplex Virus in Human Primary Macrophages Is Mediated via the MDA5/MAVS-Dependent and MDA5/MAVS/RNA Polymerase III-Independent Pathways. Journal of Virology 84, 11350-11358. https://dx.doi.org/10.1128/Jvi.01106-10

Melroe, G.T., Silva, L., Schaffer, P.A., and Knipe, D.M. (2007). Recruitment of activated IRF-3 and CBP/p300 to herpes simplex virus ICP0 nuclear foci: Potential role in blocking IFN-beta induction. Virology 360, 305-321. https://dx.doi.org/10.1016/j.virol.2006.10.028

Mohr, I., and Gluzman, Y. (1996). A herpesvirus genetic element which affects translation in the absence of the viral GADD34 function. Embo J 15, 4759-4766. https://dx.doi.org/10.1002/j.1460-2075.1996.tb00853.x

Morrone, S.R., Wang, T., Constantoulakis, L.M., Hooy, R.M., Delannoy, M.J., and Sohn, J. (2014). Cooperative assembly of IFI16 filaments on dsDNA provides insights into host defense strategy. Proc Natl Acad Sci U S A 111, E62-71. https://dx.doi.org/10.1073/pnas.1313577111

Mulvey, M., Poppers, J., Ladd, A., and Mohr, I. (1999). A herpesvirus ribosome-associated, RNA-binding protein confers a growth advantage upon mutants deficient in a GADD34-related function. Journal of Virology 73, 3375-3385.

Nour, A.M., Reichelt, M., Ku, C.C., Ho, M.Y., Heineman, T.C., and Arvin, A.M. (2011). Varicella-zoster virus infection triggers formation of an interleukin-1beta (IL-1beta)-processing inflammasome complex. J Biol Chem 286, 17921-17933. https://dx.doi.org/10.1074/jbc.M110.210575

Ogunjimi, B., Zhang, S.Y., Sorensen, K.B., Skipper, K.A., Carter-Timofte, M., Kerner, G., Luecke, S., Prabakaran, T., Cai, Y., Meester, J., *et al.* (2017). Inborn errors in RNA polymerase III underlie severe varicella zoster virus infections. J Clin Invest 127, 3543-3556. https://dx.doi.org/10.1172/Jci92280

Oroskar, A.A., and Read, G.S. (1987). A Mutant of Herpes-Simplex Virus Type-1 Exhibits Increased Stability of Immediate-Early (Alpha) Messenger-Rnas. Journal of Virology 61, 604-606.

Oroskar, A.A., and Read, G.S. (1989). Control of Messenger-Rna Stability by the Virion Host Shutoff Function of Herpes-Simplex Virus. Journal of Virology 63, 1897-1906.

Orzalli, M.H., Broekema, N.M., Diner, B.A., Hancks, D.C., Elde, N.C., Cristea, I.M., and Knipe, D.M. (2015). cGAS-mediated stabilization of IFI16 promotes innate signaling during herpes simplex virus infection. P Natl Acad Sci USA 112, E1773-E1781. https://dx.doi.org/10.1073/pnas.1424637112

Orzalli, M.H., Broekema, N.M., and Knipe, D.M. (2016). Relative Contributions of Herpes Simplex Virus 1 ICP0 and vhs to Loss of Cellular IFI16 Vary in Different Human Cell Types. J Virol 90, 8351-8359. https://dx.doi.org/10.1128/JVI.00939-16

Orzalli, M.H., Conwell, S.E., Berrios, C., DeCaprio, J.A., and Knipe, D.M. (2013). Nuclear interferon-inducible protein 16 promotes silencing of herpesviral and transfected DNA. Proc Natl Acad Sci U S A 110, E4492-4501. https://dx.doi.org/10.1073/pnas.1316194110

Orzalli, M.H., DeLuca, N.A., and Knipe, D.M. (2012). Nuclear IFI16 induction of IRF-3 signaling during herpesviral infection and degradation of IFI16 by the viral ICP0 protein. Proc Natl Acad Sci U S A 109, E3008-3017. https://dx.doi.org/10.1073/pnas.1211302109

Paludan, S.R., Bowie, A.G., Horan, K.A., and Fitzgerald, K.A. (2011). Recognition of herpesviruses by the innate immune system. Nature Reviews Immunology 11, 143-154. https://dx.doi.org/10.1038/nri2937

Parker, Z.M., Murphy, A.A., and Leib, D.A. (2015). Role of the DNA Sensor STING in Protection from Lethal Infection following Corneal and Intracerebral Challenge with Herpes Simplex Virus 1. Journal of Virology 89, 11080-11091. https://dx.doi.org/10.1128/Jvi.00954-15

Parkinson, J., Lees-Miller, S.P., and Everett, R.D. (1999). Herpes simplex virus type 1 immediate-early protein Vmw110 induces the proteasome-dependent degradation of the catalytic subunit of DNA-dependent protein kinase. Journal of Virology 73, 650-657.

Peri, P., Mattila, R.K., Kantola, H., Broberg, E., Karttunen, H.S., Waris, M., Vuorinen, T., and Hukkanen, V. (2008). Herpes Simplex Virus Type 1 Us3 Gene Deletion Influences Toll-like Receptor Responses in Cultured Monocytic Cells. Virol J 5. https://dx.doi.org/10.1186/1743-422x-5-140

Peters, G.A., Khoo, D., Mohr, I., and Sen, G.C. (2002). Inhibition of PACT-mediated activation of PKR by the herpes simplex virus type 1 Us11 protein. J Virol 76, 11054-11064.

Platanias, L.C. (2005). Mechanisms of type-I- and type-II-interferon-mediated signalling. Nat Rev Immunol 5, 375-386. https://dx.doi.org/10.1038/nri1604

Pomerantz, J.L., and Baltimore, D. (1999). NF-kappa B activation by a signaling complex containing TRAF2, TANK and TBK1, a novel IKK-related kinase. Embo J 18, 6694-6704. https://dx.doi.org/DOI 10.1093/emboj/18.23.6694

Poon, A.P., Gu, H., and Roizman, B. (2006). ICP0 and the US3 protein kinase of herpes simplex virus 1 independently block histone deacetylation to enable gene expression. Proc Natl Acad Sci U S A 103, 9993-9998. https://dx.doi.org/10.1073/pnas.0604142103

Read, G.S., and Frenkel, N. (1983). Herpes-Simplex Virus Mutants Defective in the Virion-Associated Shutoff of Host Polypeptide-Synthesis and Exhibiting Abnormal Synthesis of Alpha-(Immediate Early) Viral Polypeptides. Journal of Virology 46, 498-512.

Reske, A., Pollara, G., Krummenacher, C., Chain, B.M., and Katz, D.R. (2007). Understanding HSV-1 entry glycoproteins. Rev Med Virol 17, 205-215. https://dx.doi.org/10.1002/rmv.531

Rivas, H.G., Schmaling, S.K., and Gaglia, M.M. (2016). Shutoff of Host Gene Expression in Influenza A Virus and Herpesviruses: Similar Mechanisms and Common Themes. Viruses 8, 102. https://dx.doi.org/10.3390/v8040102

Roers, A., Hiller, B., and Hornung, V. (2016). Recognition of Endogenous Nucleic Acids by the Innate Immune System. Immunity 44, 739-754. https://dx.doi.org/10.1016/j.immuni.2016.04.002

Romano, P.R., Garcia-Barrio, M.T., Zhang, X.L., Wang, Q.Z., Taylor, D.R., Zhang, F., Herring, C., Mathews, M.B., Qin, J., and Hinnebusch, A.G. (1998). Autophosphorylation in the activation loop is required for full kinase activity in vivo of human and yeast eukaryotic initiation factor 2 alpha kinases PKR and GCN2. Mol Cell Biol 18, 2282-2297. https://dx.doi.org/10.1128/Mcb.18.4.2282

Saira, K., Zhou, Y., and Jones, C. (2007). The infected cell protein 0 encoded by bovine herpesvirus 1 (bICP0) induces degradation of interferon response factor 3 and, consequently, inhibits beta interferon promoter activity. J Virol 81, 3077-3086. https://dx.doi.org/10.1128/JVI.02064-06

Saitoh, T., Fujita, N., Hayashi, T., Takahara, K., Satoh, T., Lee, H., Matsunaga, K., Kageyama, S., Omori, H., Noda, T., *et al.* (2009). Atg9a controls dsDNA-driven dynamic translocation of STING and the innate immune response. P Natl Acad Sci USA 106, 20842-20846. https://dx.doi.org/10.1073/pnas.0911267106

Sato, A., Linehan, M.M., and Iwasaki, A. (2006). Dual recognition of herpes simplex viruses by TLR2 and TLR9 in dendritic cells. Proc Natl Acad Sci U S A 103, 17343-17348. https://dx.doi.org/10.1073/pnas.0605102103

Sato, R., Kato, A., Chimura, T., Saitoh, S.I., Shibata, T., Murakami, Y., Fukui, R., Liu, K.W., Zhang, Y., Arii, J., *et al.* (2018). Combating herpesvirus encephalitis by potentiating a TLR3-mTORC2 axis. Nature Immunology 19, 1071-+. https://dx.doi.org/10.1038/s41590-018-0203-2

Schlieker, C., Korbel, G.A., Kattenhorn, L.M., and Ploegh, H.L. (2005). A deubiquitinating activity is conserved in the large tegument protein of the herpesviridae. J Virol 79, 15582-15585. https://dx.doi.org/10.1128/JVI.79.24.15582-15585.2005

Schneider, W.M., Chevillotte, M.D., and Rice, C.M. (2014). Interferon-stimulated genes: a complex web of host defenses. Annu Rev Immunol 32, 513-545. https://dx.doi.org/10.1146/annurev-immunol-032713-120231

Schoggins, J.W., Wilson, S.J., Panis, M., Murphy, M.Y., Jones, C.T., Bieniasz, P., and Rice, C.M. (2011). A diverse range of gene products are effectors of the type I interferon antiviral response. Nature 472, 481-485. https://dx.doi.org/10.1038/nature09907

Schott, J., and Stoecklin, G. (2010). Networks controlling mRNA decay in the immune system. Wires Rna 1, 432-456. https://dx.doi.org/10.1002/wrna.13

Schwerk, J., and Sayan, R. (2015). Translating the Untranslated Region. Journal of Immunology 195, 2963-2971. https://dx.doi.org/10.4049/jimmunol.1500756

Sen, N., Sommer, M., Che, X.B., White, K., Ruyechan, W.T., and Arvin, A.M. (2010). Varicella-Zoster Virus Immediate-Early Protein 62 Blocks Interferon Regulatory Factor 3 (IRF3) Phosphorylation at Key Serine Residues: a Novel Mechanism of IRF3 Inhibition among Herpesviruses. Journal of Virology 84, 9240-9253. https://dx.doi.org/10.1128/Jvi.01147-10

Seo, G.J., Yang, A., Tan, B., Kim, S., Liang, Q., Choi, Y., Yuan, W., Feng, P., Park, H.S., and Jung, J.U. (2015). Akt Kinase-Mediated Checkpoint of cGAS DNA Sensing Pathway. Cell Rep 13, 440-449. https://dx.doi.org/10.1016/j.celrep.2015.09.007

Shapira, L., Ralph, M., Tomer, E., Cohen, S., and Kobiler, O. (2016). Histone Deacetylase Inhibitors Reduce the Number of Herpes Simplex Virus-1 Genomes Initiating Expression in Individual Cells. Front Microbiol 7, 1970. https://dx.doi.org/10.3389/fmicb.2016.01970

Shen, G., Wang, K., Wang, S., Cai, M., Li, M.L., and Zheng, C. (2014). Herpes simplex virus 1 counteracts viperin via its virion host shutoff protein UL41. J Virol 88, 12163-12166. https://dx.doi.org/10.1128/JVI.01380-14

Shu, M., Taddeo, B., and Roizman, B. (2015). Tristetraprolin Recruits the Herpes Simplex Virion Host Shutoff RNase to AU-Rich Elements in Stress

Response mRNAs To Enable Their Cleavage. J Virol 89, 5643-5650. https://dx.doi.org/10.1128/JVI.00091-15

Sorensen, L.N., Reinert, L.S., Malmgaard, L., Bartholdy, C., Thomsen, A.R., and Paludan, S.R. (2008). TLR2 and TLR9 synergistically control herpes simplex virus infection in the brain. J Immunol 181, 8604-8612.

Stratmann, S.A., Morrone, S.R., van Oijen, A.M., and Sohn, J. (2015). The innate immune sensor IFI16 recognizes foreign DNA in the nucleus by scanning along the duplex. Elife 4. https://dx.doi.org/10.7554/eLife.11721

Su, C., Zhang, J., and Zheng, C. (2015). Herpes simplex virus 1 UL41 protein abrogates the antiviral activity of hZAP by degrading its mRNA. Virol J 12, 203. https://dx.doi.org/10.1186/s12985-015-0433-y

Su, C., and Zheng, C. (2017). Herpes Simplex Virus 1 Abrogates the cGAS/STING-Mediated Cytosolic DNA-Sensing Pathway via Its Virion Host Shutoff Protein, UL41. J Virol 91. https://dx.doi.org/10.1128/JVI.02414-16

Sun, H., Zhang, Q., Jing, Y.Y., Zhang, M., Wang, H.Y., Cai, Z., Liuyu, T., Zhang, Z.D., Xiong, T.C., Wu, Y., *et al.* (2017). USP13 negatively regulates antiviral responses by deubiquitinating STING. Nat Commun 8, 15534. https://dx.doi.org/10.1038/ncomms15534

Sun, L.J., Wu, J.X., Du, F.H., Chen, X., and Chen, Z.J.J. (2013). Cyclic GMP-AMP Synthase Is a Cytosolic DNA Sensor That Activates the Type I Interferon Pathway. Science 339, 786-791. https://dx.doi.org/10.1126/science.1232458

Sun, W.X., Li, Y., Chen, L., Chen, H.H., You, F.P., Zhou, X., Zhou, Y., Zhai, Z.H., Chen, D.Y., and Jiang, Z.F. (2009). ERIS, an endoplasmic reticulum IFN stimulator, activates innate immune signaling through dimerization. P Natl Acad Sci USA 106, 8653-8658. https://dx.doi.org/10.1073/pnas.0900850106

Takeda, K., Kaisho, T., and Akira, S. (2003). Toll-like receptors. Annu Rev Immunol 21, 335-376. https://dx.doi.org/10.1146/annurev.immunol.21.120601.141126

Tanaka, Y., and Chen, Z.J.J. (2012). STING Specifies IRF3 Phosphorylation by TBK1 in the Cytosolic DNA Signaling Pathway. Sci Signal 5. https://dx.doi.org/10.1126/scisignal.2002521

Tian, J., Liu, Y.X., Liu, X.X., Sun, X., Zhang, J.K., and Qu, L.D. (2018). Feline Herpesvirus 1 US3 Blocks the Type I Interferon Signal Pathway by Targeting Interferon Regulatory Factor 3 Dimerization in a Kinase-Independent Manner. Journal of Virology 92. https://dx.doi.org/10.1128/JVI.00047-18

Unterholzner, L., Keating, S.E., Baran, M., Horan, K.A., Jensen, S.B., Sharma, S., Sirois, C.M., Jin, T.C., Latz, E., Xiao, T.S., *et al.* (2010). IFI16 is an innate immune sensor for intracellular DNA. Nature Immunology 11, 997-U942. https://dx.doi.org/10.1038/ni.1932

van Lint, A.L., Murawski, M.R., Goodbody, R.E., Severa, M., Fitzgerald, K.A., Finberg, R.W., Knipe, D.M., and Kurt-Jones, E.A. (2010). Herpes Simplex Virus Immediate-Early ICP0 Protein Inhibits Toll-Like Receptor 2-Dependent Inflammatory Responses and NF-kappa B Signaling. Journal of Virology 84, 10802-10811. https://dx.doi.org/10.1128/Jvi.00063-10

Vandevenne, P., Lebrun, M., El Mjiyad, N., Ote, I., Di Valentin, E., Habraken, Y., Dortu, E., Piette, J., and Sadzot-Delvaux, C. (2011). The Varicella-Zoster Virus ORF47 Kinase Interferes with Host Innate Immune Response by Inhibiting the Activation of IRF3. Plos One 6. https://dx.doi.org/10.1371/journal.pone.0016870

Verpooten, D., Ma, Y.J., Hou, S.W., Yan, Z.P., and He, B. (2009). Control of TANK-binding Kinase 1-mediated Signaling by the gamma(1)34.5 Protein of Herpes Simplex Virus 1. J Biol Chem 284, 1097-1105. https://dx.doi.org/10.1074/jbc.M805905200

von Roretz, C., Di Marco, S., Mazroui, R., and Gallouzi, I.E. (2011). Turnover of AU-rich-containing mRNAs during stress: a matter of survival. Wiley Interdiscip Rev RNA 2, 336-347. https://dx.doi.org/10.1002/wrna.55

Walters, M.S., Erazo, A., Kinchington, P.R., and Silverstein, S. (2009). Histone Deacetylases 1 and 2 Are Phosphorylated at Novel Sites during Varicella-Zoster Virus Infection. Journal of Virology 83, 11502-11513. https://dx.doi.org/10.1128/Jvi.01318-09

Walters, M.S., Kinchington, P.R., Banfield, B.W., and Silverstein, S. (2010). Hyperphosphorylation of histone deacetylase 2 by alphaherpesvirus US3 kinases. J Virol 84, 9666-9676. https://dx.doi.org/10.1128/JVI.00981-10

Wang, J.P., Kurt-Jones, E.A., Shin, O.S., Manchak, M.D., Levin, M.J., and Finberg, R.W. (2005). Varicella-zoster virus activates inflammatory cytokines in human monocytes and macrophages via Toll-like receptor 2. J Virol 79, 12658-12666. https://dx.doi.org/10.1128/JVI.79.20.12658-12666.2005

Wang, K.Z., Ni, L.W., Wang, S., and Zheng, C.F. (2014). Herpes Simplex Virus 1 Protein Kinase US3 Hyperphosphorylates p65/RelA and Dampens NF-kappa B Activation. Journal of Virology 88, 7941-7951. https://dx.doi.org/10.1128/Jvi.03394-13

Wang, L., Wen, M., and Cao, X. (2019). Nuclear hnRNPA2B1 initiates and amplifies the innate immune response to DNA viruses. Science 365. https://dx.doi.org/10.1126/science.aav0758

Wang, S., Wang, K.Z., Li, J., and Zheng, C.F. (2013a). Herpes Simplex Virus 1 Ubiquitin-Specific Protease UL36 Inhibits Beta Interferon Production by Deubiquitinating TRAF3. Journal of Virology 87, 11851-11860. https://dx.doi.org/10.1128/Jvi.01211-13

Wang, S., Wang, K.Z., Lin, R.T., and Zheng, C.F. (2013b). Herpes Simplex Virus 1 Serine/Threonine Kinase US3 Hyperphosphorylates IRF3 and Inhibits Beta Interferon Production. Journal of Virology 87, 12814-12827. https://dx.doi.org/10.1128/Jvi.02355-13

Weber, F., Wagner, V., Rasmussen, S.B., Hartmann, R., and Paludan, S.R. (2006). Double-stranded RNA is produced by positive-strand RNA viruses and DNA viruses but not in detectable amounts by negative-strand RNA viruses. J Virol 80, 5059-5064. https://dx.doi.org/10.1128/JVI.80.10.5059-5064.2006

West, J.A., Gregory, S.M., and Damania, B. (2012). Toll-like receptor sensing of human herpesvirus infection. Front Cell Infect Microbiol 2, 122. https://dx.doi.org/10.3389/fcimb.2012.00122

Willmann, O., Ahmad-Nejad, P., Neumaier, M., Hennerici, M.G., and Fatar, M. (2010). Toll-Like Receptor 3 Immune Deficiency May Be Causative for HSV-2-Associated Mollaret Meningitis. Eur Neurol 63, 249-251. https://dx.doi.org/10.1159/000287585

Wu, J.X., Sun, L.J., Chen, X., Du, F.H., Shi, H.P., Chen, C., and Chen, Z.J.J. (2013). Cyclic GMP-AMP Is an Endogenous Second Messenger in Innate

Immune Signaling by Cytosolic DNA. Science 339, 826-830. https://dx.doi.org/10.1126/science.1229963

Xing, J.J., Ni, L.W., Wang, S., Wang, K.Z., Lin, R.T., and Zheng, C.F. (2013). Herpes Simplex Virus 1-Encoded Tegument Protein VP16 Abrogates the Production of Beta Interferon (IFN) by Inhibiting NF-kappa B Activation and Blocking IFN Regulatory Factor 3 To Recruit Its Coactivator CBP. Journal of Virology 87, 9788-9801. https://dx.doi.org/10.1128/Jvi.01440-13

Xing, J.J., Wang, S., Lin, R.T., Mossman, K.L., and Zheng, C.F. (2012). Herpes Simplex Virus 1 Tegument Protein US11 Downmodulates the RLR Signaling Pathway via Direct Interaction with RIG-I and MDA-5. Journal of Virology 86, 3528-3540. https://dx.doi.org/10.1128/Jvi.06713-11

Ye, R.J., Su, C.H., Xu, H.Y., and Zheng, C.F. (2017). Herpes Simplex Virus 1 Ubiquitin-Specific Protease UL36 Abrogates NF-kappa B Activation in DNA Sensing Signal Pathway. Journal of Virology 91. https://dx.doi.org/10.1128/JVI.02417-16

Yu, H.R., Huang, H.C., Kuo, H.C., Sheen, J.M., Ou, C.Y., Hsu, T.Y., and Yang, K.D. (2011). IFN-alpha production by human mononuclear cells infected with varicella-zoster virus through TLR9-dependent and -independent pathways. Cell Mol Immunol 8, 181-188. https://dx.doi.org/10.1038/cmi.2010.84

Yuan, H., You, J., You, H., and Zheng, C.F. (2018). Herpes Simplex Virus 1 UL36USP Antagonizes Type I Interferon-Mediated Antiviral Innate Immunity. Journal of Virology 92. https://dx.doi.org/10.1128/JVI.01161-18

Zhang, D., Su, C., and Zheng, C. (2016). Herpes Simplex Virus 1 Serine Protease VP24 Blocks the DNA-Sensing Signal Pathway by Abrogating Activation of Interferon Regulatory Factor 3. J Virol 90, 5824-5829. https://dx.doi.org/10.1128/JVI.00186-16

Zhang, J., Wang, K., Wang, S., and Zheng, C. (2013). Herpes simplex virus 1 E3 ubiquitin ligase ICP0 protein inhibits tumor necrosis factor alpha-induced NF-kappaB activation by interacting with p65/RelA and p50/NF-kappaB1. J Virol 87, 12935-12948. https://dx.doi.org/10.1128/JVI.01952-13

Zhang, J., Zhao, J., Xu, S., Li, J., He, S., Zeng, Y., Xie, L., Xie, N., Liu, T., Lee, K., *et al.* (2018). Species-Specific Deamidation of cGAS by Herpes Simplex

Virus UL37 Protein Facilitates Viral Replication. Cell Host Microbe 24, 234-248 e235. https://dx.doi.org/10.1016/j.chom.2018.07.004

Zhang, R., Xu, A., Qin, C., Zhang, Q., Chen, S., Lang, Y., Wang, M., Li, C., Feng, W., Zhang, R., *et al.* (2017). Pseudorabies Virus dUTPase UL50 Induces Lysosomal Degradation of Type I Interferon Receptor 1 and Antagonizes the Alpha Interferon Response. J Virol 91. https://dx.doi.org/10.1128/JVI.01148-17

Zhang, S.Y., Jouanguy, E., Ugolini, S., Smahi, A., Elain, G., Romero, P., Segal, D., Sancho-Shimizu, V., Lorenzo, L., Puel, A., *et al.* (2007). TLR3 deficiency in patients with herpes simplex encephalitis. Science 317, 1522-1527. https://dx.doi.org/10.1126/science.1139522

Zhang, Z., Yuan, B., Bao, M., Lu, N., Kim, T., and Liu, Y.J. (2011). The helicase DDX41 senses intracellular DNA mediated by the adaptor STING in dendritic cells. Nat Immunol 12, 959-965. https://dx.doi.org/10.1038/ni.2091

Zhao, J., Zeng, Y., Xu, S., Chen, J., Shen, G., Yu, C., Knipe, D., Yuan, W., Peng, J., Xu, W., *et al.* (2016). A Viral Deamidase Targets the Helicase Domain of RIG-I to Block RNA-Induced Activation. Cell Host Microbe 20, 770-784. https://dx.doi.org/10.1016/j.chom.2016.10.011

Zhong, B., Yang, Y., Li, S., Wang, Y.Y., Li, Y., Diao, F.C., Lei, C.Q., He, X., Zhang, L., Tien, P., *et al.* (2008). The Adaptor Protein MITA Links Virus-Sensing Receptors to IRF3 Transcription Factor Activation. Immunity 29, 538-550. https://dx.doi.org/10.1016/j.immuni.2008.09.003

Zhu, H., Zheng, C., Xing, J., Wang, S., Li, S., Lin, R., and Mossman, K.L. (2011). Varicella-zoster virus immediate-early protein ORF61 abrogates the IRF3-mediated innate immune response through degradation of activated IRF3. J Virol 85, 11079-11089. https://dx.doi.org/10.1128/JVI.05098-11

Chapter 8

Nuclear Egress

Elizabeth B. Draganova, Michael K. Thorsen and Ekaterina E. Heldwein*

Department of Molecular Biology and Microbiology, Tufts University School of Medicine, Boston, MA, USA

*katya.heldwein@tufts.edu

DOI: https://doi.org/10.21775/9781913652555.08

Abstract

During viral replication, herpesviruses utilize a unique strategy, termed nuclear egress, to translocate capsids from the nucleus into the cytoplasm. This initial budding step transfers a newly formed capsid from within the nucleus, too large to fit through nuclear pores, through the inner nuclear membrane to the perinuclear space. The perinuclear enveloped virion must then fuse with the outer nuclear membrane to be released into the cytoplasm for further maturation, undergoing budding once again at the *trans*-Golgi network or early endosomes, and ultimately exit the cell non-lytically to spread infection. This first budding process is mediated by two conserved viral proteins, UL31 and UL34, that form a heterodimer called the nuclear egress complex (NEC). This chapter focuses on what we know about how the NEC mediates capsid transport to the perinuclear space, including steps prior to and after this budding event. Additionally, we discuss the involvement of other viral proteins in this process and how NEC-mediated budding may be regulated during infection.

Introduction: overview of nuclear egress

Herpesviruses replicate their double-stranded-DNA genomes and package them into capsids within the host cell nucleus. These progeny capsids must then exit the nucleus to form mature virions within the cytoplasm (Figure 1). The nucleus is surrounded by the nuclear envelope - a double membrane bilayer consisting of the inner nuclear membrane (INM) and the outer nuclear membrane (ONM) - and most transport in and out of the nucleus occurs through nuclear pores, which have an inner channel diameter of ~39-nm (Pante and Kann, 2002). But at ~125 nm in diameter, herpesviral capsids are too large to fit through the nuclear pores and must use a more complex route, termed nuclear egress, to exit the nucleus. Multiple lines of evidence support the model whereby capsids first become enveloped at the INM, pinching off into the perinuclear space (PNS) (primary envelopment) [reviewed in (Bigalke and Heldwein, 2016; Crump, 2018; Roller and Baines, 2017a)]. These perinuclear enveloped virions (PEVs) then fuse with the ONM, releasing the capsids into the cytoplasm (de-envelopment). The capsids then undergo another, final envelopment, gaining tegument proteins (layer between capsid and the envelope) and a final lipid envelope. The mature virion then hijacks the secretory pathway and leaves the cell by exocytosis.

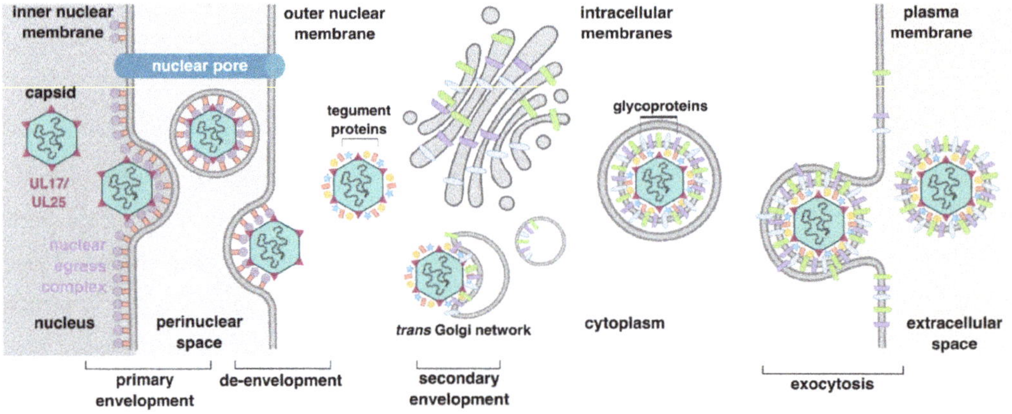

Figure 1. Overview of herpesvirus egress. Capsids undergo primary envelopment (nuclear budding) at the inner nuclear membrane where they pinch off into the perinuclear space in a process mediated by the nuclear egress complex. The resulting perinuclear enveloped vesicles fuse with the outer nuclear membrane and release the capsid into the cytoplasm (de-envelopment). The resulting virion gains necessary tegument proteins before undergoing secondary envelopment during which it acquires a glycoprotein envelope derived from the trans Golgi network or early endosomes. The mature virion then hijacks the host secretory pathway and is released into the extracellular space via exocytosis to spread infection.

This envelopment/de-envelopment model of nuclear egress is the most widely accepted model for capsid escape from the nucleus and is backed up by substantial data. First, enveloped virions in the PNS have been observed in cells infected with herpesvirus for several decades (Fuchs et al., 2002; Granzow et al., 1997; Reynolds et al., 2002; Stackpole, 1969). The phenotypes of viral gene deletions likewise provide compelling evidence in favor of the envelopment/de-envelopment model. Deletion of the US3 kinase gene from either HSV-1 or from the closely related pseudorabies virus (PRV) results in the accumulation of PEVs (Klupp et al., 2001; Reynolds et al., 2002; Ryckman and Roller, 2004; Sehl et al., 2020), which points to budding into the PNS being an intermediate stage in nuclear egress. Second, PEVs differ from mature virions in their morphology and protein composition. For example, UL31 and UL34, two viral proteins critical for nuclear egress, are absent from mature HSV-1 and PRV virions (Fuchs et al., 2002; Reynolds et al., 2002). Conversely, most tegument proteins found in mature virions are not found in the nucleus or in PEVs (Gershon et al., 1994; Granzow et al., 2001; Klupp et al., 2000; Skepper et al., 2001). Furthermore, mature virions are studded with glycoprotein spikes and have a thick tegument while PEVs have a smooth envelope and thin tegument (Gershon et al., 1994; Granzow et al., 2001), consistent with the two types of particles being formed during different budding events. Third, deletions of particular tegument proteins, such as HSV-1 UL36 and UL37, or the glycoproteins gD and gE from either HSV-1 or PRV result in unenveloped capsids accumulating in the cytoplasm (Brack et al., 2000; Desai et al., 2001; Desai, 2000; Farnsworth et al., 2003; Leege et al., 2009; Roberts et al., 2009) suggesting that a secondary envelopment step in the cytoplasm is necessary for egress. Lastly, the envelopes of mature viral particles from both HSV-1 and varicella-zoster virus (VZV), another alphaherpesvirus, are derived from cytoplasmic membranes, rather than nuclear membranes, as shown by immunogold electron microscopy and transmission electron microscopy radioautography experiments, respectively (Gershon et al., 1994; Skepper et al., 2001). Collectively, these observations make a strong case for a model of nuclear egress in which nuclear capsids gain their initial, temporary envelope at the INM, escape into the cytoplasm as unenveloped capsids, and then acquire the final envelope at the trans Golgi network (Henaff et al., 2012) or early endosomes (Hollinshead et al., 2012) (see Chapter 9).

The nuclear egress complex

The nuclear egress complex is essential for nuclear egress

Nuclear egress is mediated by two conserved viral proteins, UL31 and UL34, that form the nuclear egress complex (NEC) (Fuchs et al., 2002; Reynolds et al., 2001; Roller et al., 2000) [reviewed in (Bigalke and Heldwein, 2016; Hellberg et al., 2016; Roller and Baines, 2017a)]. This complex is conserved across all members of the Herpesviridae family (see Chapter 3), and both the UL31 and UL34 proteins are essential for this process (Fuchs et al., 2002) (Bubeck et al., 2004; Farina et al., 2005; Reynolds et al., 2001). While UL31 and UL34 homologs are known under different names in other herpesviruses, throughout this chapter, alphaherpesvirus nomenclature will be used in keeping with the subject of this book. UL31 is a soluble nuclear phosphoprotein that is 306-aa long in HSV-1 (Chang and Roizman, 1993; Reynolds et al., 2001). UL34, which is 275-aa long in HSV-1, has a single-spanning C-terminal transmembrane (TM) region that anchors the complex to the INM such that the NEC faces the nucleoplasm (Reynolds et al., 2001; Shiba et al., 2000).

UL31 homologs have nuclear localization signals that target the protein from its site of synthesis in the cytosol to the nucleus (Funk et al., 2015; Li et al., 2015; Passvogel et al., 2015). In the absence of HSV-1 UL34, UL31 is diffusely distributed throughout the nucleoplasm (Reynolds et al., 2001; Zhu et al., 1999). In the absence of UL31, UL34 localizes to the ER membranes, probably, because most UL34 homologs do not have a nuclear localization signal. UL31 and UL34 localize to the INM only when both proteins are present, either during infection or in uninfected cells expressing UL31 and UL34, which suggests that proper localization of the two proteins at the INM requires formation of the NEC (Fuchs et al., 2002; Funk et al., 2015; Reynolds et al., 2001) [reviewed in (Hellberg et al., 2016; Roller and Baines, 2017a)].

In the absence of either UL31 or UL34, viral replication is reduced by three to four orders of magnitude and capsids accumulate within the nucleus (Fuchs et al., 2002) (Bubeck et al., 2004; Chang and Roizman, 1993; Klupp et al., 2000; Reynolds et al., 2001; Roller et al., 2000), which indicates that both proteins are required for efficient nuclear egress. Moreover, PRV NEC expressed in uninfected cells results in the formation of capsidless perinuclear vesicles, demonstrating that UL31 and UL34 are the only viral proteins necessary for nuclear budding (Klupp et al., 2007). Formation of capsidless perinuclear vesicles was also observed for NEC homologs from gammaherpesviruses Epstein-Barr virus (EBV) (Farina et al., 2005; Gonnella et al., 2005) and Kaposi's sarcoma-associated herpesvirus (KSHV) (Desai et al., 2012; Luitweiler et al., 2013) expressed in insect cells using recombinant baculoviruses. Together, these findings implicated the NEC in the trafficking of capsids out of the nucleus through a nuclear budding mechanism.

High-resolution structures of the NEC

Within the last few years, five crystal structures of the NEC from subfamilies of alphaherpesviruses and betaherpesviruses have been determined: one from HSV-1 (PDB ID: 4ZXS) (Bigalke and Heldwein, 2015), two from PRV (PDB IDs: 4Z3U and 5E8C) (Bigalke and Heldwein, 2015; Zeev-Ben-Mordehai et al., 2015), and two from the human cytomegalovirus (HCMV) (PDB IDs: 5DOB and 5D5N) (Lye et al., 2015; Walzer et al., 2015). The NEC has an elongated cylindrical shape (Figure 2). UL34, located at the base of the complex, forms a globular pedestal for UL31. UL31, located at the top of the complex, is composed of a globular core that rests on UL34 and an N-terminal hook-like extension that projects downwards and wraps around UL34. Both UL31 and UL34 have unique folds according to Dali structural similarity algorithms (Holm and Rosenstrom, 2010). In all of the structures determined thus far, UL31 and UL34 form an extensive interface that can be subdivided into two regions: one that involves the globular core of UL31 (interface 1) and the other formed by the hook-like extension of UL31 (interface 2) (Figure 2). Prior to structural studies, genetic studies performed using in-frame deletion mutants pinpointed regions of UL34 from HSV-1 (Liang and Baines, 2005) and UL31 homolog from HCMV (Sam et al., 2009; Schnee et al., 2012) necessary for complex formation. While these regions, indeed, participate in UL31/UL34 interactions, they map to different interfaces and do not contact one another within the structure (Bigalke and Heldwein, 2015). Additionally, point mutations within HSV-1, PRV, and HCMV UL34 suggested specific residues necessary for complex formation (Bjerke et al., 2003; Bubeck et al., 2004; Milbradt et al., 2012; Passvogel et al., 2014; Passvogel et al., 2015; Passvogel et al., 2013; Roller et al., 2010).

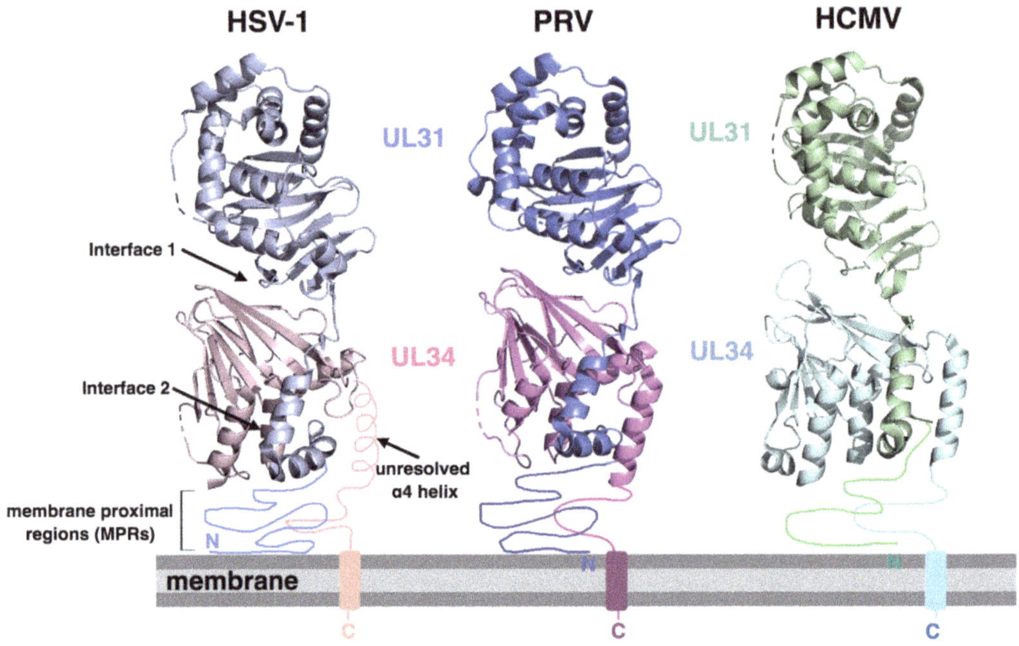

Figure 2. Structures of the HSV-1 (PDB: 4ZXS), PRV (PDB: 4Z3U), and HCMV (PDB: 5D5N) nuclear egress complexes. UL31 (purple and green) and UL34 (pink and light blue) form an elongated cylindrical molecule with the hook-like extension of UL31 wrapping around UL34 at the base of the molecules. The unresolved membrane proximal regions (MPRs) and transmembrane regions are drawn in schematically, with the MPRs located closest to the membrane. The approximate location of the unresolved α4 helix is schematically drawn for HSV-1. Interfaces are indicated by arrows. The space above the membrane is the nucleoplasm.

All five NEC structures lack the N terminus of UL31 and the C terminus of UL34, which are predicted to be unstructured and were left out of the crystallized constructs to promote crystallization (Bigalke and Heldwein, 2015; Lye et al., 2015; Walzer et al., 2015; Zeev-Ben-Mordehai et al., 2015). These segments contain sequences necessary for HSV-1 NEC interaction with membranes (Bigalke et al., 2014), along with the TM anchor of UL34, and are expected to be located near the membrane. These so-called membrane-proximal regions (MPRs) map to the base of the NEC complex (Figure 2), which places UL34 at the membrane-proximal and UL31 at the membrane-distal end of the NEC cylinder.

The NEC homologs have remarkably similar structures despite low sequence identity. The biggest difference among the five structures is the conserved C-terminal alpha helix α4 that was unresolved in the HSV-1 UL34 structure (Figure 2) (Bigalke and Heldwein, 2015), possibly due to differences in the construct boundaries, because the HSV-1 UL34 used for crystallization is five amino acid residues shorter than the PRV UL34. Truncated helix α4 in HSV-1 UL34 may be less stable and, as a result, higher flexibility of this region would cause it to be unresolved in the structure. No structural information is yet available on the NEC homologs from gammaherpesviruses such as EBV and KSHV. However, from secondary structure predictions, they are expected to share similar folds with their counterparts from alpha and betaherpesviruses.

A recent molecular dynamics simulation using the available NEC crystal structures investigated the relative flexibilities of NEC heterodimers from HSV-1 and HCMV (Diewald et al., 2018). For these studies, the PRV α4 helix was docked into the HSV-1 structure corresponding to the missing HSV-1 α4 helix region, prior to simulations. Molecular dynamics suggests that isolated NEC heterodimers have different flexibilities, with HCMV NEC being more flexible than HSV-1 NEC. Regions of increased flexibility map to the UL31/UL34 interface 1 (Figure 2), with UL31 capable of twisting around UL34. The differences in flexibility likely stem from low conservation of the interface regions, and it is tempting to speculate that the increased rotational freedom at this UL31/UL34 interface of the HCMV heterodimer could play a role in the nuclear budding process of betaherpesviruses.

Stages in primary envelopment

During primary envelopment, the first step in nuclear egress, capsids bud at the INM to form PEVs in the PNS. This process, at a minimum, requires the following events to take place: the nuclear lamina is dissolved, capsid docks at the INM, membrane is deformed around the capsid to form a nascent bud, and the bud is pinched off the INM releasing the PEV into the perinuclear space. Nuclear budding is a complex process that involves several viral and cellular participants, some of which have not yet been identified. It is also subject to regulation to prevent budding in the absence of the capsid and to ensure that the mature, DNA-containing C-capsids, rather than the immature procapsids, the empty A-capsids, or the scaffold-containing B-capsids, described in more detail below, preferentially undergo budding. These undercharacterized regulatory mechanisms ensure quality control and conservation of resources.

Nuclear lamina alteration

Nuclear lamina is a dense filamentous protein network underneath the INM that provides shape and mechanical stability to the nuclear envelope. Nuclear lamina is primarily composed of two types of intermediate filament proteins called lamins A and B that oligomerize into filaments [reviewed in (Dechat et al., 2008; Gruenbaum and Foisner, 2015)]. The A-type lamins, A and C, which are the products of alternative splicing of the LMNA gene, are expressed mainly in differentiated cells whereas the B-type lamins, B1 and B2, are expressed in all cells (Gruenbaum and Foisner, 2015). The B-type lamins tether the lamina to the INM by binding integral membrane proteins such as the lamin B receptor, lamina-associated proteins (LAPs) and emerin [reviewed in (Goldberg et al., 2008; Schirmer and Foisner, 2007; Wilson and Foisner, 2010)]. The A-type lamins, A and C, add additional stiffness to the nuclear envelope [reviewed in (Dechat et al., 2008; Gruenbaum and Foisner, 2015)].

The dense, overlapping filamentous networks of B-type and A-type lamins create a barrier to capsid docking at the INM and would probably impede membrane deformation. Efficient capsid budding thus requires that the nuclear lamina be disassembled. Complete lamina disassembly in uninfected cells occurs during mitosis [reviewed in (Hetzer, 2010)] and is achieved by phosphorylation of lamins and LAPs by cellular mitotic kinases CDK1 (cyclin dependent kinase 1), PKA (protein kinase A), PKC and MAPK (mitogen-activated protein kinase) (Fields and Thompson, 1995; Goss et al., 1994; Guttinger et al., 2009; Heald and McKeon, 1990; Likhacheva and Bogachev, 2001; Margalit et al., 2005; Peter et al., 1992). However, during herpesvirus infection, no massive disassembly of the lamina is observed, likely, because it is important for the virus to preserve the integrity of the nucleus since this is where viral gene transcription and genome replication take place. Instead, the lamina is disassembled locally (Granzow et al., 1997) through phosphorylation of lamins and LAPs by the viral US3 kinase and likely by cellular kinases that are different from the mitotic ones (Hertel, 2011; Lv et al., 2019a, b; Roller and Baines, 2017b) [reviewed in (Roller and Baines, 2017a)].

Lamins A/C are phosphorylated by PKC which is recruited to the INM, along with the host protein facilitator of nuclear egress p32, by the HSV-1 gamma(1)34.5 protein (Wang et al., 2014; Wu et al., 2016). PKC is also involved in hyperphosphorylation of emerin, along with US3, which also contributes to lamina disassembly (Leach et al., 2007; Morris et al., 2007). UL13 is another viral kinase that phosphorylates lamina in HSV-2 (Cano-Monreal et al., 2009) and HCMV (Marschall et al., 2005; Sharma et al., 2014) but in HSV-1, instead, appears to regulate the distribution of UL31 and UL34 by phosphorylating US3 (Kato et al., 2006). Further mechanistic studies are necessary to delineate the exact contributions of both viral and host proteins facilitating lamina disruption during nuclear egress.

Capsid docking to the INM

Herpesviral capsids are icosahedral, with T = 16, and consist of the major capsid protein, VP5 (UL19) in alphaherpesviruses, arranged into 11 pentons and 150 hexons, plus additional proteins UL18, UL38, and UL35 (Dai and Zhou, 2018) (see Chapter 3). In each capsid, one of the 12 vertices contains a single portal composed of the portal protein UL6 (McElwee et al., 2018) through which viral DNA is encapsidated during assembly (Newcomb et al., 2001) and, presumably, ejected early in infection (Newcomb et al., 2007). The capsid vertex specific complex (CVSC)/capsid-associated tegument complex (CATC) consists of the capsid proteins UL17 and UL25 and ensures proper DNA retention in the mature capsid (Dai and Zhou, 2018; Salmon et al., 1998; Stow, 2001) likely by stabilizing the capsid to help it withstand the internal pressure of several atmospheres from the encapsidated DNA genome (Bauer et al., 2013).

During assembly, a spherical scaffold-containing procapsid forms initially and matures into genome-containing icosahedral C-capsids [reviewed in (Baines, 2011)]. Two additional icosahedral capsid types observed in infected cells are the empty A-capsids and the scaffold-containing B-capsids, which are thought to be the defective by-products of capsid assembly [reviewed in (Baines, 2011)].

How the capsids reach the INM is yet unclear, but they likely dock to the lamina-free, exposed areas of the INM by binding to the UL31 component of the NEC (Funk et al., 2015; Leelawong et al., 2011). UL31 likely interacts with the CVSC/CATC at the capsid vertices because CVSC/CATC components UL25 and UL17 bind UL31 (Yang and Baines, 2011). Furthermore, UL25 mediates interactions between capsids purified from HSV-1 infected cells and NEC expressed *in vitro* (Takeshima et al., 2019). A putative capsid-binding site of UL31 has been mapped to a conserved surface patch at the membrane-distal end composed of residues (HSV-1 D275-D282, PRV D238-D245) (Bigalke and Heldwein, 2015). In HSV-1, residues R281 and D282 of UL31 have been implicated in NEC/capsid binding (Takeshima et al., 2019) whereas in PRV, residue K242 (HSV-1 K279) is important for the primary capsid envelopment (Ronfeldt et al., 2020; Ronfeldt et al., 2017) but may not interact with the capsid directly (Ronfeldt et al., 2020). Although capsids isolated from UL25-null HSV-1 or PRV can interact with UL31 (Leelawong et al., 2011; Yang et al., 2014), this interaction is insufficient for nuclear egress because in the absence of UL25, HSV-1 and PRV capsids can only dock at the INM but cannot exit the nucleus (Klupp et al., 2006; Kuhn et al., 2008). Taken together, these findings suggest that while UL25/UL31 interaction is not required for capsid docking at the INM, it is necessary for egress.

Although several capsid types are present in the nucleus, primarily C-capsids undergo primary envelopment at the INM whereas A- and B-capsids typically do not (Klupp et al., 2011; Roizman and Furlong, 1974; Stackpole, 1969). These observations imply the existence of a capsid selection mechanism that ensures only mature, DNA-filled capsids exit the nucleus, which could, in principle, increase the yield of infectious virions by reducing unproductive budding of the A- and B-capsids. The nature of this selection mechanism has not yet been identified but likely involve interactions between the CVSC/CATC and the NEC. Although CVSC/CATC is present on all three capsid types, A- and B- capsids have fewer CVSC/CATC copies on the capsid surface (Newcomb et al., 2006). It is also possible that the CVSC/CATC on C-capsids adopts a conformation distinct from those on A- or B-capsids (Newcomb et al., 2017). Multiple NEC binding sites on a mature C-capsid could produce avidity effects, providing the driving force for creating a PEV.

Membrane deformation and budding by the NEC

The NEC is a self-contained budding machine

While the studies discussed above suggested the NEC is necessary for capsid budding at the INM and sufficient for vesiculation of the INM, they left unclear the exact function of the NEC during nuclear egress. Was the NEC mediating the budding process at the INM directly or recruiting yet unknown host proteins? To answer this question, recombinant soluble HSV-1 NEC (lacking the C-terminal TM) was purified and incubated *in vitro* with synthetic lipid membranes in the form of giant unilamellar vesicles (Bigalke et al., 2014). The NEC vesiculated the membranes, forming intraluminal vesicles within the GUVs, in the absence of any other factors or ATP. This observation was subsequently recapitulated with recombinant PRV NEC anchored to the membranes through the TM of UL34 (Lorenz et al., 2015). These findings demonstrated that the NEC is a self-contained membrane budding machine. The fact that soluble HSV-1 NEC mediates budding shows that, at least, *in vitro,* the TM is dispensable for budding. Moreover, this region can be replaced with a heterologous TM without any ill effects on egress *in vivo* (Schuster et al., 2012). This suggested that while the TM anchors the NEC to the INM *in vivo*, other regions of the NEC are responsible for membrane interactions required for budding.

HSV-1 NEC-mediated budding *in vitro* requires the presence of acidic lipids (Bigalke et al., 2014), suggesting the potential involvement of electrostatic interactions. In the case of PRV NEC, *in-vitro* budding required the presence of sphingomyelin or cholesterol (Lorenz et al., 2015). Thus, lipid composition appears to play an important role in budding.

NEC-mediated membrane budding proceeds rapidly *in vitro* (Bigalke et al., 2014). In addition to generating a scaffold on the inner surface of the nascent bud, the NEC also mediates scission of its neck. The ability of the NEC to mediate the entire budding process is a unique property of this protein complex. Other enveloped viruses encode proteins that mediate budding, yet most must also recruit cellular ESCRT [endosomal sorting complexes required for transport (ESCRT)] proteins [reviewed in (Alonso et al., 2016; Hurley, 2015; Votteler and Sundquist, 2013)]. For example, the HIV matrix protein Gag forms a scaffold on the membranes of buds and recruits ESCRT proteins (Carlson and Hurley, 2012; Flower et al., 2020) [reviewed in (Hurley and Cada, 2018)], notably ESCRT-III proteins that mediate scission by assembling a spiral-like polymer on the inward face of the membrane neck and constricting it (Effantin et al., 2013; Nguyen et al., 2020). Likewise, in Hepatitis C and yellow fever viruses from the *Flaviviridae* family, membrane deformation is driven by the formation of an external viral glycoprotein coat, but scission requires ESCRT machinery (Barouch-Bentov et al., 2016; Carpp et al., 2011; Corless et al., 2010). Not all enveloped viruses, however, recruit ESCRT proteins for membrane budding. Influenza A virus, for example, solely uses virally encoded proteins M1 and M2 that mediate scaffolding and neck scission, respectively (Rossman et al., 2010) [reviewed in (Rossman and Lamb, 2011, 2013)]. How NEC accomplishes neck scission is unknown, but according to one proposed scenario (Bigalke et al., 2014), formation of the hexagonal NEC lattice could narrow the neck to the point where it could undergo spontaneous scission. Alternatively, the NEC could alter local lipid composition to membrane curvature conducive to scission, as has been proposed for the Influenza M2 protein (Rossman et al., 2010). Additional experiments are needed to elucidate the mechanism behind NEC scission.

HSV-1 nuclear egress is thought to be independent of the ESCRT-III machinery in HEK293 cells because it is insensitive to a dominant negative mutation of Vps4, an ATPase required for ESCRT-III-mediated scission (Crump et al., 2007), considered to be "a gold standard" for ESCRT-III involvement (Alonso et al., 2016). However, a more recent report challenged that assertion by showing that nuclear egress was reduced in the presence of the Vps4 dominant-negative mutant in HeLa cells, rather than in HEK293 cells (Arii et al., 2018). Moreover, depletion of CHMP4B, an ESCRT-III protein, resulted in accumulation of PEVs, apparently, due to impaired scission (Arii et al., 2018). Thus, while HSV-1 NEC is capable of mediating scission *in vitro*, it may recruit ESCRT-III machinery *in vivo* to increase the efficiency of scission. A similar concept is at play in Ebola virus, where the viral matrix protein VP40 mediates membrane budding *in vitro* (Soni and Stahelin,

2014) yet recruits ESCRT machinery *in vivo* (Licata et al., 2003; Silvestri et al., 2007) [reviewed in (Gordon et al., 2019)]. Due to the discordant findings regarding the role of ESCRTs in HSV-1 nuclear egress, whether the ESCRT-III machinery is universally important for nuclear egress remains unclear. Future studies with different viral strains and host cell types are needed to resolve this question.

NEC assembly into a hexagonal lattice is important for budding

Cryogenic electron microscopy and tomography (cryo-EM/T) demonstrated that during budding *in vitro*, the soluble HSV-1 NEC heterodimer forms hexagonal "honeycomb" coats on the inner surface of lipid vesicles (Figure 3B) (Bigalke et al., 2014). Given that the NEC cannot cross the membrane bilayer, these budded vesicles appear to be the products of budding and scission, because this would be the only way for the protein to get inside the vesicles. NEC also forms hexagonal coats on the inner surface of capsidless perinuclear vesicles formed in uninfected cells expressing PRV NEC (Figure 3C) (Hagen et al., 2015). Hexagonal symmetry evident in the coats formed as the result of budding both *in vitro* and *in vivo* suggested that the hexagonal arrangement is biologically relevant.

A high-resolution view of interactions involved in NEC oligomerization was afforded by the crystal structure of HSV-1 NEC (Bigalke and Heldwein, 2015). In the crystals, NEC assembled into a hexagonal lattice built from ~110-Å hexameric rings (Figure 3A). (Bigalke and Heldwein, 2015). The symmetry and the dimensions of the hexagonal crystal lattice are very similar to those of the hexagonal NEC coats formed *in vitro* (Bigalke et al., 2014) and *in vivo* (Figures 3B and C) (Hagen et al., 2015). Therefore, NEC interactions within the crystal lattice model interactions within the hexagonal coats enabling a systematic exploration of the role of oligomerization in the NEC budding mechanism. Additionally, HCMV NEC also packed into a hexagonal lattice in one of the crystal forms (Walzer et al., 2015), which suggests that the NEC from alpha- and betaherpesviruses use similar strategies for coat formation and budding. Although no structural information is yet available for NEC from any gammaherpesviruses, these principles likely apply to them as well.

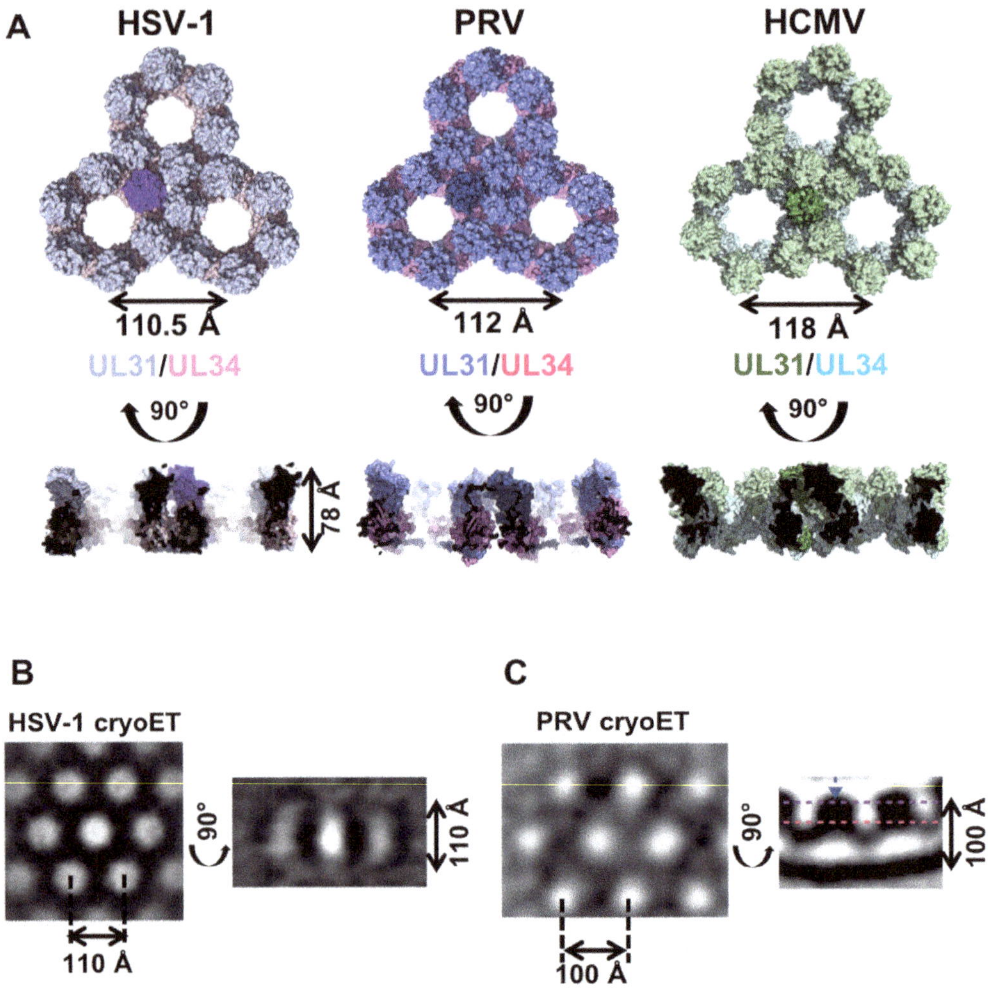

Figure 3. Comparison of NEC hexagonal lattice formed by HSV-1 (crystal lattice), PRV (model of cryo-ET coat) and HCMV (crystal lattice). A) Three hexameric rings shown in top and side views with UL31 (purple/green) sitting on top of UL34 (pink/blue). One NEC heterodimer is highlighted in each lattice. The HSV-1 NEC crystal lattice is 78 Å thick. The PRV lattice in the side view is curved since it was determined from a membrane coat and not a planar crystal lattice. Both the HSV-1 NEC (B) and the PRV NEC (C) cryo-ET coats are shown for comparison. The dimensions of the PRV coat are slightly smaller due to curvature. The missing ~30 Å from the crystal lattice is present in the cryo-ET coat. The HSV-1 NEC cryo-ET coat is reprinted from (Bigalke and Heldwein, 2015). The PRV NEC cryo-ET coat is reprinted from (Hagen et al., 2015).

The hexagonal lattice is formed by two types of oligomeric interactions: those within the hexameric rings (intra-hexameric) and those between adjacent hexamers (inter-hexameric) (Bigalke and Heldwein, 2015). Mutations disrupting oligomeric interfaces reduce budding both *in vitro* (Bigalke and Heldwein, 2015) and *in vivo* (Arii et al., 2019; Bjerke et al., 2003; Roller et al., 2010). For example, the dominant negative (DN) mutation in the UL34 (D35A/E37A) that blocks capsid budding *in vivo* (Roller et al., 2010), maps to the hexameric interface. Accordingly, this mutant is defective in both budding and coat formation *in vitro* (Bigalke et al., 2014). Therefore, the NEC oligomerization into a hexagonal coat is not only necessary for efficient budding, and ultimately nuclear egress, but is also the driving force behind the process.

Formation of spherical NEC coats

The NEC crystal lattice is flat while the coat formed on vesicles is curved. While the dimensions of the flat lattice are similar to those observed in the coat, the coat must allow a degree of flexibility that would permit the formation of a curved and ultimately, spherical array. Cryo-ET in combination with cryosectioning (CEMOVIS) and focused ion beam cryomilling (cryo-FIB) were used to analyze the structure of the NEC coats in capsidless perinuclear vesicles formed in uninfected cells expressing PRV NEC (Hagen et al., 2015). The PRV NEC-coated perinuclear vesicles were relatively uniform in size (~100-117 nm in diameter). Interestingly, budded intraluminal vesicles formed by soluble HSV-1 NEC *in vitro* range widely in size (Bigalke et al., 2014), suggesting that the C terminus of UL34 containing the TM (absent from the HSV- 1 NEC construct used in *in vitro* studies) may somehow determine the curvature of the NEC scaffold and, ultimately, the size of the budded vesicle. The PRV NEC-coated perinuclear vesicles formed in uninfected cells (Hagen et al., 2015) were smaller than the capsid [~125 nm in diameter (Liu et al., 2017a)] or the PEVs isolated from cells infected with HSV-1 US3-null mutant (Newcomb et al., 2017). This suggests that while the NEC oligomerization favors formation of a scaffold of relatively uniform curvature, the capsid defines the size of the NEC scaffold formed in infected cells.

Aside from flexibility, a purely hexagonal array is flat. Curvature can be achieved by the regular inclusion of another shape - typically a pentagon - within the hexagonal array, which leads to the formation of a polyhedral structure. Consider, for example, a soccer ball, in which a spherical shape is achieved by the inclusion of 12 pentagons within an otherwise hexagonal arrangement, a structure referred to as an icosahedron. Many viruses, including herpesviruses, employ this strategy to form their capsids, and it is reasonable to expect that the NEC coat may form an icosahedral structure. Although icosahedral symmetry was not observed in NEC coats formed either *in vitro* or in NEC-expressing cells (Bigalke et al., 2014; Hagen et al., 2015), these coats were formed in the absence of capsids, leaving open the possibility that formation of an icosahedral NEC coat requires the presence of an icosahedral capsid. Ideally, this would require *in situ* imaging of PEVs in infected cells at a sufficiently high resolution, which has not yet been achievable. Instead, NEC coats formed in the presence of capsids were visualized in PEVs isolated from cells infected with an HSV-1 US3-null virus (Newcomb et al., 2017). Deletion of the US3, a viral kinase, results in an accumulation of PEVs within the perinuclear space (Klupp et al., 2001; Reynolds et al., 2002; Ryckman and Roller, 2004). No icosahedral symmetry was evident in reconstructions of NEC coats in PEVs. However, these PEVs were formed in the absence of US3, which could have affected the coat structure. Therefore, further studies with wild-type HSV-1 or PRV are required to conclusively rule out the icosahedral symmetry in the NEC coats.

In the absence of icosahedral arrangement, a curved coat can alternatively be formed through the inclusion of irregular defects into the hexagonal NEC lattice. Such non-icosahedral arrangement has been observed for the immature HIV-1 capsid formed by the Gag matrix protein (Briggs et al., 2009; Schur et al., 2015) and for the early poxvirus envelope (Heuser, 2005) formed by D13 protein (Hyun et al., 2011). Although such irregular defects have not yet been reported for the HSV-1 NEC coats, molecular dynamics simulations of the hexagonal NEC lattice show it to be very flexible (Diewald et al., 2018). Such flexibility may allow for effortless incorporation of irregular defects within the coat. Ultimately, determining the presence of irregular defects within the NEC coat awaits high-resolution cryo-ET reconstructions.

Regulation of NEC budding

HSV-1 and PRV NEC-mediated budding is robust *in vitro* (Bigalke et al., 2014; Lorenz et al., 2015) and in uninfected cells expressing the PRV NEC (Klupp et al., 2007). However, during infection, capsidless perinuclear vesicles are rarely observed, which suggests that NEC-mediated budding at the INM is negatively regulated to prevent unproductive budding in the absence of C-capsids [reviewed in (Hellberg et al., 2016; Roller and Baines, 2017a)]. One way to regulate budding could be by maintaining the NEC at the INM in an inactive form prior to contact with the capsid. Phosphorylation of UL31 could be one potential mechanism of downregulating the NEC budding activity. The N terminus of HSV-1 UL31 contains six serines phosphorylated by the US3 kinase (Mou et al., 2009). Replacement of these UL31 serine residues with glutamates, which mimic phosphorylated serines, prevented capsid budding (Mou et al., 2009). By contrast, replacing these serines with alanines, which mimics a constitutively dephosphorylated state, resulted in the accumulation of the PEVs in the perinuclear space (Mou et al., 2009). Deletion or inactivation of the US3 kinase results in a similar phenotype (Bjerke et al., 2003; Klupp et al., 2007; Klupp et al., 2001; Reynolds et al., 2002; Ryckman and Roller, 2004), which suggests that PEVs accumulate whenever US3 cannot phosphorylate UL31. While this could mean that phosphorylation of UL31 is important for the de-envelopment process (Mou et al., 2009), the alternative possibility is that in the absence of negative regulation of budding leads to overproduction of the PEVs.

If phosphorylation of UL31, indeed, inhibits the NEC activity prior to C-capsid docking, dephosphorylation would then be required to relieve inhibition. Herpesviral genomes do not encode any phosphatases (Bernard Roizman, 2013) but could, potentially, recruit host phosphatases for this purpose. Indeed, PRV US3 activates host phosphatases to rearrange cytoplasmic actin (Jacob et al., 2013) (see Chapter 3). Altogether, these findings implicate US3 in nuclear egress regulation.

A number of tegument proteins, described later in this Chapter, have been implicated in nuclear egress and could, in principle, regulate the NEC activity either positively or negatively. Additionally, the nonstructural ICP22 (Maruzuru et al., 2014) protein interacts with the NEC but is not critical for egress. Another nonstructural HSV-1 protein, UL24, has also been implicated in contributing to nuclear egress (Lymberopoulos et al., 2011), yet not much is known about this phenotype.

De-envelopment

During de-envelopment, the PEVs lose their envelopes and NEC coats at the ONM, and the naked capsids are released into the cytoplasm. This poorly understood process must include fusion of the envelope with the ONM as well as the disassembly of the NEC scaffold around the capsid, which could occur concomitantly or sequentially. To ensure efficient nuclear egress, PEVs must preferentially fuse with the ONM rather than the INM, but how this selection is achieved or whether backfusion events occur is yet unknown. In uninfected cells expressing UL31 and UL34 from HSV-1 or PRV, de-envelopment of the capsidless perinuclear vesicles either does not take place or is very inefficient because PEVs accumulate in herniations (Klupp et al., 2007). This suggests that de-envelopment requires viral proteins besides UL31 and UL34.

Defects in de-envelopment typically manifest as the accumulation of PEVs in the PNS, with a concomitant reduction in viral titer. Several proteins, as described below, have been implicated in de-envelopment. Yet, proteins that mediate membrane fusion during de-envelopment have not yet been conclusively identified. The conserved viral glycoproteins gB and gH, both of which are essential for membrane fusion during herpesviral entry, have been reported as important for efficient de-envelopment in HSV-1 because deletion of both gB and gH impedes nuclear egress in HSV-1 (Farnsworth et al., 2007). Intact fusion loops of gB are important for the de-envelopment (Wright et al., 2009). However, single deletions of gB or gH have minimal effects on de-envelopment (Farnsworth et al., 2007) implying functional redundancy, which is inconsistent with both glycoproteins being essential for fusion during HSV-1 entry (see Chapter 2). This raises the possibility that instead of mediating fusion, gB and gH instead contribute to de-envelopment through a different mechanism. In any case, gB and gH are simultaneously dispensable for de-envelopment in the closely related PRV (Granzow et al., 2001; Klupp et al., 2008). Overexpression of another viral glycoprotein gK from HSV-1, which localizes to the ER during infection, results in the accumulation of PEVs (Hutchinson and Johnson, 1995), but its precise role in the de-envelopment remains unclear.

An outer tegument protein, UL51, has also been implicated in de-envelopment in HSV-1 (Nozawa et al., 2005). However, UL51 localizes to the Golgi and probably contributes to de-envelopment indirectly. Treatment of PRV-infected cells with the inhibitor of Golgi-to-ER transport, brefeldin A, causes the accumulation of PEVs (Whealy et al., 1991), indicating that UL51 could potentially facilitate de-envelopment indirectly by helping traffic yet unidentified membrane fusogen(s) to ER and the ONM. Nonetheless, UL51 is dispensable for de-envelopment in HCMV, PRV and Bovine Herpesvirus 1 (Klupp et al., 2005c; Raza et al., 2016; Schauflinger et al., 2011; Womack and Shenk, 2010). The identification of viral or cellular fusogens mediating de-envelopment and the elucidation of their mechanisms is an important area for future investigation.

During de-envelopment, the hexagonal NEC coats within the PEVs need to be disassembled to expose the surface of the capsid for subsequent interactions with the tegument proteins in the cytoplasm. The fate of the NEC following nuclear egress remains unclear. What triggers disassembly is also unknown, but phosphorylation of UL31 by US3 may play a role in the disassembly of the NEC coat. HSV-1 US3 is present in the PEVs (Reynolds et al., 2002), and it is tempting to speculate that phosphorylation of UL31 within the PEVs could lead to structural rearrangements that would disrupt the NEC interactions within the hexagonal lattice and promote the disassembly of the NEC coat. However, the N terminal region of UL31, targeted by US3, is thought to be located near the membrane due to its apparent requirement for NEC/membrane interactions, and it is unclear how US3 would access this region sandwiched between the NEC coat and the membrane. Therefore, disassembly of the NEC coat could, instead, be facilitated by perturbations to the envelope during fusion with the ONM.

Phosphorylation of UL31 by US3 could alternatively serve as a selection mechanism for mature C-capsids. According to this scenario, in the absence of phosphorylation, A- and B-capsids could undergo primary envelopment, along with

C-capsids, yet be incapable of undergoing de-envelopment and, therefore, accumulate in the PNS as PEVs.

gB, another target of US3, contains only one US3 phosphorylation site located within the gB cytoplasmic domain (Wisner et al., 2009). Mutation of this phosphorylation site results in the accumulation of PEVs when gH is absent (Wisner et al., 2009). The phosphorylation of the cytoplasmic domain could potentially regulate the function of gB by controlling its oligomerization state or by triggering a fusogenic conformational change, with the caveat that gB may not be mediating fusion during de-envelopment (described above).

Collectively, these findings point to the involvement of US3 kinase activity in the de-envelopment process. Nevertheless, the modest defect in viral growth, only 10-fold, observed in viruses lacking US3 kinase (Mou et al., 2009) argues that while US3 kinase activity is important for efficient de-envelopment, it is not essential, at least under the conditions of cell culture.

Tegument and nuclear egress

The tegument is a multiprotein layer sandwiched between the capsid and the envelope, which is a unique feature of herpesvirus particles. Although tegument proteins are thought to be recruited to cytoplasmic capsids, some tegument proteins localize to the nucleus and a few of these, UL36, UL47, UL16, UL11 and UL21, are implicated in nuclear egress.

At ~300-kDa, UL36 is the largest protein conserved across the Herpesviridae family. The best studied role of UL36 is its interaction with cellular motor proteins dynein and kinesin that is essential for directed trafficking of HSV-1 and PRV capsids post-entry and during egress (Shanda and Wilson, 2008; Zaichick et al., 2013) (see Chapters 3 and 9). In both HSV-1 and PRV, UL36 is dispensable for nuclear egress (Desai, 2000; Luxton et al., 2006; Roberts et al., 2009), but in PRV, the absence, capsids egress from the nucleus is less efficient in the absence of UL36 (Luxton et al., 2006). Infections with PRV UL36 dually labeled with fluorescent proteins at both N and C termini showed that nuclear UL36 is proteolytically cleaved (Leelawong et al., 2012), and only the C-terminal fragment of PRV UL36 localizes to the nucleus and co-purifies with nuclear C-capsids (Leelawong et al., 2012). The C-terminal fragment of PRV UL36 is also sufficient for increased levels of nuclear egress compared to a UL36-null virus (Leelawong et al., 2012). Association of the C-terminal fragment of UL36 with nuclear capsids has also been reported for HSV-1 (Radtke et al., 2010). Indeed, UL36 is a component of the aforementioned CVSC/CATC located at the capsid vertices. In addition to two copies of UL25 and one copy of UL17, each CVSC/CATC also contains two copies of the C-terminal helix of UL36 (Dai and Zhou, 2018; Liu et al., 2017b). Large numbers of B-capsids were observed in the cytoplasm of cells infected with UL36-null HSV-1 (Kharkwal et al., 2016), which suggests that UL36 contributes to the selection of mature C-capsids for nuclear egress.

Several outer tegument proteins have also been implicated in nuclear egress. UL11, UL16, and UL21 are important for efficient cytoplasmic budding of capsids that leads to the formation of infectious virions but during infection also localize to the nucleus. UL16 is a conserved, ~40-kDa tegument protein that colocalizes with HSV-1 capsids within the nucleus of infected cells but is not found on HSV-1 nor HSV-2 capsids after purification (Meckes and Wills, 2007; Nalwanga et al., 1996; Oshima et al., 1998), potentially due to a weak association between UL16 and the capsid. Infections with HSV-1 or HSV-2 lacking UL16 increases the proportion of A- and B-capsids relative to C-capsids within the nucleus, suggesting a defect in viral DNA packaging (Gao et al., 2018; Oshima et al., 1998). While both HSV-1 and HSV-2 lacking UL16 have reduced ability to generate C-capsids, they differ in their replicative capacity. In HSV-1 strains F and KOS, viral titers were reduced 10-fold when most of the UL16 gene was deleted (Gao et al., 2018) whereas in several HSV-2 strains, viral titers were reduced 100-fold (Gao et al., 2017; Gao et al., 2018), and in PRV, UL16 is dispensable for replication (Klupp et al., 2005b). These findings suggest a species-specific role for UL16 during infection such that the differences in replication defects could reflect varying roles of UL16 in the viral replication life cycle. In the absence of UL16, HSV-1 C-capsids relocate into the cytoplasm whereas HSV-2 C-capsids are retained in the nucleus (Gao et al., 2017; Gao et al., 2018). Perhaps, UL16 deletion has a lesser effect on the HSV-1 titers because more C-capsids can enter the maturation pathway.

UL11 is the smallest among conserved tegument proteins. During infection, HSV-1 UL11 localizes to both nuclear and cytoplasmic membranes but does not interact with intranuclear capsids, as assessed by immunogold electron microscopy (Baines et al., 1995). Nonetheless, a C-terminal deletion of ~2/3 of HSV-1 UL11 results in lower viral titers and a ~10% increase in capsid retention at the INM (Baines and Roizman, 1992). The deletion does not affect UL11/16 interactions indicating that this phenotype is due to a loss of some other function of UL11 (Loomis et al., 2003b). To date, this phenotype appears specific to HSV-1 because UL11-null PRV has no obvious defects in nuclear egress (Klupp et al., 2005b) (Klupp et al., 2005a; Kopp et al., 2003).

UL21 is a ~61-kDa capsid-associated tegument protein that interacts with UL11 and UL16 (de Wind et al., 1992; Harper et al., 2010; Klupp et al., 2005b; Loomis et al., 2003a). In HSV-2 strain 186, deletion of UL21 results in capsid retention within the nucleus and a defect in cell-to-cell spread (Finnen and Banfield, 2018; Le Sage et al., 2013). Some studies examining HSV-1 UL21 found it to be dispensable for replication (Baines et al., 1994; Mbong et al., 2012) while another reported a 100-fold decrease in replication, albeit due to the presence of A- and B-capsids in the cytoplasm rather than a defect in nuclear egress (Sarfo et al., 2017). While all PRV strains exhibit a 10-fold decrease in replication, PRV strain NIA-3 lacking UL21 results in A- and B-capsids representing 50% of cytoplasmic capsids whereas the Kaplan strain exhibits no defects in viral maturation (de Wind et al., 1992; Klopfleisch et al., 2006; Klupp et al., 2005b; Klupp et al., 1995; Wagenaar et al., 2001). Strain-specific differences could be the result of differences among the strains, the presence of additional unknown mutations in the virus, or due to differences in the cell lines used.

While mainly studied in HSV-1, UL47 may play an important role during nuclear egress. UL47-null HSV-1 results in a 50-fold reduction in PEVs and accumulation of capsids in the nucleus (Liu et al., 2014). The block in nuclear egress may be the result of disrupting a capsid-NEC interaction, as UL47 associates with both UL31/UL34 as well as UL17, which is found on capsids (Liu et al., 2014; Scholtes et al., 2010). UL47 homologs are not required for nuclear egress in PRV (Kopp et al., 2002) or bovine herpes virus 1 (Lobanov et al., 2010).

The gamma(1)34.5 protein interacts with the NEC (Jing et al., 2004; Wu et al., 2016), and its deletion results in drastically reduced viral replication due to a defect in nuclear lamina dissolution (described above) that ultimately prevents capsid egress (Wu et al., 2016).

Nuclear membrane alterations

When properly assembled, the NEC functions as a complete membrane budding nanomachine, forming perfectly sized vesicles encapsulating herpesvirus capsids in infected cells. However, when the proper assembly of the NEC is disrupted by mutations or when the capsid is absent, the perturbation of the membrane budding activity of the NEC results in oddly shaped membrane structures.

For example, tightly packed multilayered membranous structures at the nuclear envelope were observed in insect cells heterologously expressing EBV BFRF1, a homolog of HSV-1 UL34 (Gonnella et al., 2005). These structures were also observed in insect cells co-expressing BFRF1 and BFLF2, a homolog of HSV-1 UL31. However, some of the structures are less tightly packed and "were in concentric whorls" (Gonnella et al., 2005). The increased spacing between the INM and ONM was also seen in Vero cells infected with a baculovirus encoding only HSV-1 UL34 (Ye et al., 2000). The mechanism of UL34-mediated membrane alterations could be due to alternations of the nuclear lamina, which provides structural support to the nuclear envelope (Reynolds et al., 2004). While expression of UL34 alone resulted in nuclear membrane alterations, aberrant membrane vesiculation has been observed during infections where interaction between UL31 and UL34 were perturbed by mutagenesis (Roller et al., 2011). Disruption of the UL31/UL34 interaction also resulted in poor co-localization to the nuclear rim (Roller et al., 2011). These findings suggest that while UL34 alone is capable of altering the structure of the nuclear envelope, only the NEC can mediate productive budding. Tightly packed multilayered membranes have also been observed upon infection with PRV lacking UL11, UL16, and UL21 (Klupp et al., 2005b), although the mechanism behind this phenotype is yet unknown.

Nuclear budding in uninfected cells

Although the process of nuclear egress was once thought to be unique to herpesviruses, we now know that a topologically similar process is also utilized during the nuclear export of large cellular synaptic ribonucleoproteins (RNPs) in *Drosophila* (Speese et al., 2012) [reviewed in (Fradkin and Budnik, 2016)]. These RNPs cannot exit the nucleus through the nuclear pores and, instead, bud at the INM to form RNP cargo containing vesicles in the perinuclear space that subsequently fuse at the ONM like herpesviruses. Torsin ATPases are essential for nuclear budding in uninfected cells, and in their absence, vesicles formed in the perinuclear space remain attached to the INM (Goodchild et al., 2005; Jokhi et al., 2013; Kim et al., 2010). Abundant PEVs containing granules that resemble RNPs, some in the process of fusion with the ONM have also been observed in sea urchin gastrula, suggesting that nuclear egress may be involved in nucleocytoplasmic transfer during sea urchin development (LaMassa et al., 2018). Thus, the nuclear egress may be more common in host cells than previously thought.

The existence of the nuclear egress pathway in uninfected cells may imply that herpesviruses have hijacked it. However, the absence of both isoforms of torsin has only a modest effect on the efficiency of nuclear egress in HSV-1 (Turner et al., 2015). While this does not rule out the involvement of other host proteins in herpesvirus nuclear egress, it suggests that herpesviruses utilize a distinct nuclear egress pathway. Further studies are clearly necessary to delineate the contributions of host factors to nuclear egress of herpesviruses vs. large RNPs and mechanistic similarities and differences between these two processes.

Alternative routes of nuclear egress: nuclear envelope breakdown and nuclear pore enlargement

The NEC is essential for nuclear egress, and deletions of either UL31 or UL34 reduce titers at least 1000-fold (Fuchs et al., 2002; Reynolds et al., 2001; Roller et al., 2000). Nevertheless, some infectious virus can be recovered even in the absence of UL31 or UL34. This suggests that while the envelopment/de-envelopment route is preferred, other, far less efficient nuclear escape routes also exist. Two such routes have been proposed: the nuclear envelope breakdown

(NEBD) whereby the nuclear envelope is ruptured (Grimm et al., 2012; Klupp et al., 2011; Maric et al., 2014; Stackpole, 1969) and the dilation of nuclear pores (Leuzinger et al., 2005; Wild et al., 2009), either of which would allow the capsid to access the cytoplasm directly.

The NEBD phenomenon was discovered when upon passaging in culture, PRV lacking UL31 or UL34 acquired the ability to induce NEBD (Grimm et al., 2012; Klupp et al., 2011) by manipulating mitosis-related processes (Grimm et al., 2012). NEBD, along with reduced viral titers, was also observed during HSV-1 infections in cells lacking the cellular ATPase torsinA (Maric et al., 2014).

Studies using a combination of scanning and transmission electron microscopy of HSV-1 infected Vero and Hela cells showed instances of enlarged nuclear pores (~100 nm) (Wild et al., 2009), yet biochemical studies by other groups using fluorescent microscopy (Nagel et al., 2008) and live-cell imaging (Hofemeister and O'Hare, 2008) were unable to corroborate these findings. This suggests that while there may be instances of nuclear pore enlargement under certain experimental conditions, this route of capsid egress from the nucleus appears atypical.

It is unlikely that either of these alternative nuclear exit routes contribute significantly to nuclear egress under normal circumstances because they would eliminate an important checkpoint in the capsid quality control. NEC-mediated budding as an important quality control mechanism ensuring that primarily DNA-filled C-capsids undergo maturation and are released from the cells.

Many viruses that replicate in the nucleus, e.g., adenoviruses, parvoviruses, and polyomaviruses, escape the nucleus by rupturing the nuclear envelope (Cohen et al., 2011; Raghava et al., 2013; Tollefson et al., 1996) [reviewed in (Kobiler et al., 2012; Mettenleiter, 2015)]. By contrast, herpesviruses clearly favor the envelopment/de-envelopment pathway and exit by an alternative route only when this preferred route of nuclear egress is disrupted.

Discussion and future trends

The unusual nuclear escape route used by all herpesviruses, which involves capsids budding at the INM (primary envelopment) followed by fusion of the resulting perinuclear enveloped virions with the ONM (de-envelopment), is supported by substantial evidence. Yet, our mechanistic understanding of this process remains in infancy. The significant advances made in the field over the last decade have clarified some aspects of the nuclear egress mechanism. For example, we now know that the key player in nuclear egress, the NEC, is a membrane budding machine that scaffolds the membrane by forming the striking hexagonal coats on its surface and interacts with the capsid. At the same time, the number of proteins involved in nuclear egress has increased, and with it, the complexity of this process. In contrast, we know very little about the mechanism by which perinuclear enveloped virions fuse with the ONM or the proteins involved in this process. Therefore, we envision clarification of the mechanisms that control capsid budding and NEC activity as one major direction of inquiry for the next decade while another is the elucidation of the mechanism of the subsequent membrane fusion step. Technological advances in *in situ* imaging of infected cells by cryo-EM and cryo-ET will be essential for visualizing the process of nuclear egress and the interactions that shape it at high resolution.

Web Resources

1. HSV-1 NEC PDB ID: 4ZXS (Bigalke and Heldwein, 2015)
2. PRV NEC PDB ID: 4Z3U (Bigalke and Heldwein, 2015)
3. PRV NEC PDB ID: 5E8C (Zeev-Ben-Mordehai et al., 2015)
4. HCMV NEC PDB ID: 5DOB (Lye et al., 2015)
5. HCMV NEC PDB ID: 5D5N (Walzer et al., 2015)

References

Alonso, Y.A.M., Migliano, S.M., and Teis, D. (2016). ESCRT-III and Vps4: a dynamic multipurpose tool for membrane budding and scission. FEBS J *283*, 3288-3302.

Arii, J., Takeshima, K., Maruzuru, Y., Koyanagi, N., Kato, A., and Kawaguchi, Y. (2019). Roles of the Interhexamer Contact Site for Hexagonal Lattice Formation of the Herpes Simplex Virus 1 Nuclear Egress Complex in Viral Primary Envelopment and Replication. J Virol *93*.

Arii, J., Watanabe, M., Maeda, F., Tokai-Nishizumi, N., Chihara, T., Miura, M., Maruzuru, Y., Koyanagi, N., Kato, A., and Kawaguchi, Y. (2018). ESCRT-III mediates budding across the inner nuclear membrane and regulates its integrity. Nat Commun *9*, 3379.

Baines, J.D. (2011). Herpes simplex virus capsid assembly and DNA packaging: a present and future antiviral drug target. Trends Microbiol *19*, 606-613.

Baines, J.D., Jacob, R.J., Simmerman, L., and Roizman, B. (1995). The herpes simplex virus 1 UL11 proteins are associated with cytoplasmic and nuclear membranes and with nuclear bodies of infected cells. J Virol *69*, 825-833.

Baines, J.D., Koyama, A.H., Huang, T., and Roizman, B. (1994). The UL21 gene products of herpes simplex virus 1 are dispensable for growth in cultured cells. Journal of virology *68*, 2929-2936.

Baines, J.D., and Roizman, B. (1992). The UL11 gene of herpes simplex virus 1 encodes a function that facilitates nucleocapsid envelopment and egress from cells. J Virol *66*, 5168-5174.

Barouch-Bentov, R., Neveu, G., Xiao, F., Beer, M., Bekerman, E., Schor, S., Campbell, J., Boonyaratanakornkit, J., Lindenbach, B., Lu, A., *et al.* (2016). Hepatitis C Virus Proteins Interact with the Endosomal Sorting Complex Required for Transport (ESCRT) Machinery via Ubiquitination To Facilitate Viral Envelopment. MBio *7*.

Bauer, D.W., Huffman, J.B., Homa, F.L., and Evilevitch, A. (2013). Herpes virus genome, the pressure is on. J Am Chem Soc *135*, 11216-11221.

Bernard Roizman, D.M.K., and Richard J. Whitley (2013). Herpes Simplex Viruses. In Fields Virology, P.H. David M. Knipe ed. (Wolters Kluwer Lippincott Williams and Wilkins).

Bigalke, J.M., and Heldwein, E.E. (2015). Structural basis of membrane budding by the nuclear egress complex of herpesviruses. EMBO J *34*, 2921-2936.

Bigalke, J.M., and Heldwein, E.E. (2016). Nuclear Exodus: Herpesviruses Lead the Way. Annu Rev Virol *3*, 387-409.

Bigalke, J.M., Heuser, T., Nicastro, D., and Heldwein, E.E. (2014). Membrane deformation and scission by the HSV-1 nuclear egress complex. Nat Commun *5*, 4131.

Bjerke, S.L., Cowan, J.M., Kerr, J.K., Reynolds, A.E., Baines, J.D., and Roller, R.J. (2003). Effects of charged cluster mutations on the function of herpes simplex virus type 1 UL34 protein. J Virol *77*, 7601-7610.

Brack, A.R., Klupp, B.G., Granzow, H., Tirabassi, R., Enquist, L.W., and Mettenleiter, T.C. (2000). Role of the cytoplasmic tail of pseudorabies virus glycoprotein E in virion formation. J Virol *74*, 4004-4016.

Briggs, J.A., Riches, J.D., Glass, B., Bartonova, V., Zanetti, G., and Krausslich, H.G. (2009). Structure and assembly of immature HIV. Proc Natl Acad Sci U S A *106*, 11090-11095.

Bubeck, A., Wagner, M., Ruzsics, Z., Lotzerich, M., Iglesias, M., Singh, I.R., and Koszinowski, U.H. (2004). Comprehensive mutational analysis of a herpesvirus gene in the viral genome context reveals a region essential for virus replication. J Virol *78*, 8026-8035.

Cano-Monreal, G.L., Wylie, K.M., Cao, F., Tavis, J.E., and Morrison, L.A. (2009). Herpes simplex virus 2 UL13 protein kinase disrupts nuclear lamins. Virology *392*, 137-147.

Carlson, L.A., and Hurley, J.H. (2012). In vitro reconstitution of the ordered assembly of the endosomal sorting complex required for transport at membrane-bound HIV-1 Gag clusters. Proceedings of the National Academy of Sciences of the United States of America *109*, 16928-16933.

Carpp, L.N., Galler, R., and Bonaldo, M.C. (2011). Interaction between the yellow fever virus nonstructural protein NS3 and the host protein Alix contributes to the release of infectious particles. Microbes Infect *13*, 85-95.

Chang, Y.E., and Roizman, B. (1993). The product of the UL31 gene of herpes simplex virus 1 is a nuclear phosphoprotein which partitions with the nuclear matrix. J Virol *67*, 6348-6356.

Cohen, S., Marr, A.K., Garcin, P., and Pante, N. (2011). Nuclear envelope disruption involving host caspases plays a role in the parvovirus replication cycle. J Virol *85*, 4863-4874.

Corless, L., Crump, C.M., Griffin, S.D., and Harris, M. (2010). Vps4 and the ESCRT-III complex are required for the release of infectious hepatitis C virus particles. J Gen Virol *91*, 362-372.

Crump, C. (2018). Virus Assembly and Egress of HSV. Adv Exp Med Biol *1045*, 23-44.

Crump, C.M., Yates, C., and Minson, T. (2007). Herpes simplex virus type 1 cytoplasmic envelopment requires functional Vps4. Journal of virology *81*, 7380-7387.

Dai, X., and Zhou, Z.H. (2018). Structure of the herpes simplex virus 1 capsid with associated tegument protein complexes. Science *360*.

de Wind, N., Wagenaar, F., Pol, J., Kimman, T., and Berns, A. (1992). The pseudorabies virus homology of the herpes simplex virus UL21 gene product is a capsid protein which is involved in capsid maturation. J Virol *66*, 7096-7103.

Dechat, T., Pfleghaar, K., Sengupta, K., Shimi, T., Shumaker, D.K., Solimando, L., and Goldman, R.D. (2008). Nuclear lamins: major factors in the structural organization and function of the nucleus and chromatin. Genes Dev *22*, 832-853.

Desai, P., Sexton, G.L., McCaffery, J.M., and Person, S. (2001). A null mutation in the gene encoding the herpes simplex virus type 1 UL37 polypeptide abrogates virus maturation. J Virol *75*, 10259-10271.

Desai, P.J. (2000). A null mutation in the UL36 gene of herpes simplex virus type 1 results in accumulation of unenveloped DNA-filled capsids in the cytoplasm of infected cells. J Virol *74*, 11608-11618.

Desai, P.J., Pryce, E.N., Henson, B.W., Luitweiler, E.M., and Cothran, J. (2012). Reconstitution of the Kaposi's sarcoma-associated herpesvirus nuclear egress complex and formation of nuclear membrane vesicles by coexpression of ORF67 and ORF69 gene products. J virol *86*, 594-598.

Diewald, B., Socher, E., Soldner, C.A., and Sticht, H. (2018). Conformational Dynamics of Herpesviral NEC Proteins in Different Oligomerization States. International journal of molecular sciences *19*.

Effantin, G., Dordor, A., Sandrin, V., Martinelli, N., Sundquist, W.I., Schoehn, G., and Weissenhorn, W. (2013). ESCRT-III CHMP2A and CHMP3 form variable helical polymers in vitro and act synergistically during HIV-1 budding. Cellular microbiology *15*, 213-226.

Farina, A., Feederle, R., Raffa, S., Gonnella, R., Santarelli, R., Frati, L., Angeloni, A., Torrisi, M.R., Faggioni, A., and Delecluse, H.J. (2005). BFRF1 of Epstein-Barr virus is essential for efficient primary viral envelopment and egress. J Virol *79*, 3703-3712.

Farnsworth, A., Goldsmith, K., and Johnson, D.C. (2003). Herpes simplex virus glycoproteins gD and gE/gI serve essential but redundant functions during acquisition of the virion envelope in the cytoplasm. J Virol *77*, 8481-8494.

Farnsworth, A., Wisner, T.W., Webb, M., Roller, R., Cohen, G., Eisenberg, R., and Johnson, D.C. (2007). Herpes simplex virus glycoproteins gB and gH function in fusion between the virion envelope and the outer nuclear membrane. Proc Natl Acad Sci U S A *104*, 10187-10192.

Fields, A.P., and Thompson, L.J. (1995). The regulation of mitotic nuclear envelope breakdown: a role for multiple lamin kinases. Prog Cell Cycle Res *1*, 271-286.

Finnen, R.L., and Banfield, B.W. (2018). CRISPR/Cas9 Mutagenesis of UL21 in Multiple Strains of Herpes Simplex Virus Reveals Differential Requirements for pUL21 in Viral Replication. Viruses *10*.

Flower, T.G., Takahashi, Y., Hudait, A., Rose, K., Tjahjono, N., Pak, A.J., Yokom, A.L., Liang, X., Wang, H.G., Bouamr, F., *et al.* (2020). A helical assembly of human ESCRT-I scaffolds reverse-topology membrane scission. Nat Struct Mol Biol *27*, 570-580.

Fradkin, L.G., and Budnik, V. (2016). This bud's for you: mechanisms of cellular nucleocytoplasmic trafficking via nuclear envelope budding. Curr Opin Cell Biol *41*, 125-131.

Fuchs, W., Klupp, B.G., Granzow, H., Osterrieder, N., and Mettenleiter, T.C. (2002). The interacting UL31 and UL34 gene products of pseudorabies virus are involved in egress from the host-cell nucleus and represent components of primary enveloped but not mature virions. J Virol *76*, 364-378.

Funk, C., Ott, M., Raschbichler, V., Nagel, C.H., Binz, A., Sodeik, B., Bauerfeind, R., and Bailer, S.M. (2015). The Herpes Simplex Virus Protein pUL31 Escorts Nucleocapsids to Sites of Nuclear Egress, a Process Coordinated by Its N-Terminal Domain. PLoS Pathog *11*, e1004957.

Gao, J., Hay, T.J.M., and Banfield, B.W. (2017). The product of the herpes simplex virus 2 UL16 gene Is critical for the egress of capsids from the nuclei of infected cells. J Virol *91*.

Gao, J., Yan, X., and Banfield, B.W. (2018). Comparative Analysis of UL16 Mutants Derived from Multiple Strains of Herpes Simplex Virus 2 (HSV-2) and HSV-1 Reveals Species-Specific Requirements for the UL16 Protein. J Virol *92*, e00629-00618.

Gershon, A.A., Sherman, D.L., Zhu, Z., Gabel, C.A., Ambron, R.T., and Gershon, M.D. (1994). Intracellular transport of newly synthesized varicella-zoster virus: final envelopment in the trans-Golgi network. J Virol *68*, 6372-6390.

Goldberg, M.W., Fiserova, J., Huttenlauch, I., and Stick, R. (2008). A new model for nuclear lamina organization. Biochem Soc Trans *36*, 1339-1343.

Gonnella, R., Farina, A., Santarelli, R., Raffa, S., Feederle, R., Bei, R., Granato, M., Modesti, A., Frati, L., Delecluse, H.J., *et al.* (2005). Characterization and intracellular localization of the Epstein-Barr virus protein BFLF2: interactions with BFRF1 and with the nuclear lamina. J Virol *79*, 3713-3727.

Goodchild, R.E., Kim, C.E., and Dauer, W.T. (2005). Loss of the dystonia-associated protein torsinA selectively disrupts the neuronal nuclear envelope. Neuron *48*, 923-932.

Gordon, T.B., Hayward, J.A., Marsh, G.A., Baker, M.L., and Tachedjian, G. (2019). Host and Viral Proteins Modulating Ebola and Marburg Virus Egress. Viruses *11*.

Goss, V.L., Hocevar, B.A., Thompson, L.J., Stratton, C.A., Burns, D.J., and Fields, A.P. (1994). Identification of nuclear beta II protein kinase C as a mitotic lamin kinase. J Biol Chem *269*, 19074-19080.

Granzow, H., Klupp, B.G., Fuchs, W., Veits, J., Osterrieder, N., and Mettenleiter, T.C. (2001). Egress of alphaherpesviruses: comparative ultrastructural study. J Virol *75*, 3675-3684.

Granzow, H., Weiland, F., Jons, A., Klupp, B.G., Karger, A., and Mettenleiter, T.C. (1997). Ultrastructural analysis of the replication cycle of pseudorabies virus in cell culture: a reassessment. J Virol *71*, 2072-2082.

Grimm, K.S., Klupp, B.G., Granzow, H., Muller, F.M., Fuchs, W., and Mettenleiter, T.C. (2012). Analysis of viral and cellular factors influencing herpesvirus-induced nuclear envelope breakdown. J Virol *86*, 6512-6521.

Gruenbaum, Y., and Foisner, R. (2015). Lamins: nuclear intermediate filament proteins with fundamental functions in nuclear mechanics and genome regulation. Annu Rev Biochem *84*, 131-164.

Guttinger, S., Laurell, E., and Kutay, U. (2009). Orchestrating nuclear envelope disassembly and reassembly during mitosis. Nat Rev Mol Cell Biol *10*, 178-191.

Hagen, C., Dent, K.C., Zeev-Ben-Mordehai, T., Grange, M., Bosse, J.B., Whittle, C., Klupp, B.G., Siebert, C.A., Vasishtan, D., Bauerlein, F.J., *et al.* (2015).

Structural Basis of Vesicle Formation at the Inner Nuclear Membrane. Cell *163*, 1692-1701.

Harper, A.L., Meckes, D.G., Jr., Marsh, J.A., Ward, M.D., Yeh, P.C., Baird, N.L., Wilson, C.B., Semmes, O.J., and Wills, J.W. (2010). Interaction domains of the UL16 and UL21 tegument proteins of herpes simplex virus. J Virol *84*, 2963-2971.

Heald, R., and McKeon, F. (1990). Mutations of phosphorylation sites in lamin A that prevent nuclear lamina disassembly in mitosis. Cell *61*, 579-589.

Hellberg, T., Passvogel, L., Schulz, K.S., Klupp, B.G., and Mettenleiter, T.C. (2016). Nuclear Egress of Herpesviruses: The Prototypic Vesicular Nucleocytoplasmic Transport. Adv Virus Res *94*, 81-140.

Henaff, D., Radtke, K., and Lippe, R. (2012). Herpesviruses exploit several host compartments for envelopment. Traffic *13*, 1443-1449.

Hertel, L. (2011). Herpesviruses and intermediate filaments: close encounters with the third type. Viruses *3*, 1015-1040.

Hetzer, M.W. (2010). The nuclear envelope. Cold Spring Harb Perspect Biol *2*, a000539.

Heuser, J. (2005). Deep-etch EM reveals that the early poxvirus envelope is a single membrane bilayer stabilized by a geodetic "honeycomb" surface coat. J Cell Biol *169*, 269-283.

Hofemeister, H., and O'Hare, P. (2008). Nuclear pore composition and gating in herpes simplex virus-infected cells. J Virol *82*, 8392-8399.

Hollinshead, M., Johns, H.L., Sayers, C.L., Gonzalez-Lopez, C., Smith, G.L., and Elliott, G. (2012). Endocytic tubules regulated by Rab GTPases 5 and 11 are used for envelopment of herpes simplex virus. EMBO J *31*, 4204-4220.

Holm, L., and Rosenstrom, P. (2010). Dali server: conservation mapping in 3D. Nucleic acids research *38*, W545-549.

Hurley, J.H. (2015). ESCRTs are everywhere. EMBO J *34*, 2398-2407.

Hurley, J.H., and Cada, A.K. (2018). Inside job: how the ESCRTs release HIV-1 from infected cells. Biochem Soc Trans *46*, 1029-1036.

Hutchinson, L., and Johnson, D.C. (1995). Herpes simplex virus glycoprotein K promotes egress of virus particles. J Virol *69*, 5401-5413.

Hyun, J.K., Accurso, C., Hijnen, M., Schult, P., Pettikiriarachchi, A., Mitra, A.K., and Coulibaly, F. (2011). Membrane remodeling by the double-barrel scaffolding protein of poxvirus. PLoS Pathog *7*, e1002239.

Jacob, T., Van den Broeke, C., van Troys, M., Waterschoot, D., Ampe, C., and Favoreel, H.W. (2013). Alphaherpesviral US3 kinase induces cofilin dephosphorylation to reorganize the actin cytoskeleton. J Virol *87*, 4121-4126.

Jing, X., Cerveny, M., Yang, K., and He, B. (2004). Replication of herpes simplex virus 1 depends on the gamma 134.5 functions that facilitate virus response to interferon and egress in the different stages of productive infection. J Virol *78*, 7653-7666.

Jokhi, V., Ashley, J., Nunnari, J., Noma, A., Ito, N., Wakabayashi-Ito, N., Moore, M.J., and Budnik, V. (2013). Torsin mediates primary envelopment of large ribonucleoprotein granules at the nuclear envelope. Cell Rep *3*, 988-995.

Kato, A., Yamamoto, M., Ohno, T., Tanaka, M., Sata, T., Nishiyama, Y., and Kawaguchi, Y. (2006). Herpes simplex virus 1-encoded protein kinase UL13 phosphorylates viral Us3 protein kinase and regulates nuclear localization of viral envelopment factors UL34 and UL31. J Virol *80*, 1476-1486.

Kharkwal, H., Smith, C.G., and Wilson, D.W. (2016). Herpes Simplex Virus Capsid Localization to ESCRT-VPS4 Complexes in the Presence and Absence of the Large Tegument Protein UL36p. J Virol *90*, 7257-7267.

Kim, C.E., Perez, A., Perkins, G., Ellisman, M.H., and Dauer, W.T. (2010). A molecular mechanism underlying the neural-specific defect in torsinA mutant mice. Proc Natl Acad Sci U S A *107*, 9861-9866.

Klopfleisch, R., Klupp, B.G., Fuchs, W., Kopp, M., Teifke, J.P., and Mettenleiter, T.C. (2006). Influence of pseudorabies virus proteins on neuroinvasion and neurovirulence in mice. J Virol *80*, 5571-5576.

Klupp, B., Altenschmidt, J., Granzow, H., Fuchs, W., and Mettenleiter, T.C. (2008). Glycoproteins required for entry are not necessary for egress of pseudorabies virus. J Virol *82*, 6299-6309.

Klupp, B.G., Bottcher, S., Granzow, H., Kopp, M., and Mettenleiter, T.C. (2005a). Complex formation between the UL16 and UL21 tegument proteins of pseudorabies virus. J Virol *79*, 1510-1522.

Klupp, B.G., Bottcher, S., Granzow, H., Kopp, M., and Mettenleiter, T.C. (2005b). Complex formation between the UL16 and UL21 tegument proteins of pseudorabies virus. J Virol *79*, 1510-1522.

Klupp, B.G., Granzow, H., Fuchs, W., Keil, G.M., Finke, S., and Mettenleiter, T.C. (2007). Vesicle formation from the nuclear membrane is induced by coexpression of two conserved herpesvirus proteins. Proc Natl Acad Sci U S A *104*, 7241-7246.

Klupp, B.G., Granzow, H., Keil, G.M., and Mettenleiter, T.C. (2006). The capsid-associated UL25 protein of the alphaherpesvirus pseudorabies virus is nonessential for cleavage and encapsidation of genomic DNA but is required for nuclear egress of capsids. J Virol *80*, 6235-6246.

Klupp, B.G., Granzow, H., Klopfleisch, R., Fuchs, W., Kopp, M., Lenk, M., and Mettenleiter, T.C. (2005c). Functional analysis of the pseudorabies virus UL51 protein. J Virol *79*, 3831-3840.

Klupp, B.G., Granzow, H., and Mettenleiter, T.C. (2000). Primary envelopment of pseudorabies virus at the nuclear membrane requires the UL34 gene product. J Virol *74*, 10063-10073.

Klupp, B.G., Granzow, H., and Mettenleiter, T.C. (2001). Effect of the pseudorabies virus US3 protein on nuclear membrane localization of the UL34 protein and virus egress from the nucleus. J Gen Virol *82*, 2363-2371.

Klupp, B.G., Granzow, H., and Mettenleiter, T.C. (2011). Nuclear envelope breakdown can substitute for primary envelopment-mediated nuclear egress of herpesviruses. Journal of virology *85*, 8285-8292.

Klupp, B.G., Lomniczi, B., Visser, N., Fuchs, W., and Mettenleiter, T.C. (1995). Mutations affecting the UL21 gene contribute to avirulence of pseudorabies virus vaccine strain Bartha. Virology *212*, 466-473.

Kobiler, O., Drayman, N., Butin-Israeli, V., and Oppenheim, A. (2012). Virus strategies for passing the nuclear envelope barrier. Nucleus *3*, 526-539.

Kopp, M., Granzow, H., Fuchs, W., Klupp, B.G., Mundt, E., Karger, A., and Mettenleiter, T.C. (2003). The pseudorabies virus UL11 protein is a virion component involved in secondary envelopment in the cytoplasm. J Virol *77*, 5339-5351.

Kopp, M., Klupp, B.G., Granzow, H., Fuchs, W., and Mettenleiter, T.C. (2002). Identification and characterization of the pseudorabies virus tegument proteins

UL46 and UL47: role for UL47 in virion morphogenesis in the cytoplasm. J Virol *76*, 8820-8833.

Kuhn, J., Leege, T., Klupp, B.G., Granzow, H., Fuchs, W., and Mettenleiter, T.C. (2008). Partial functional complementation of a pseudorabies virus UL25 deletion mutant by herpes simplex virus type 1 pUL25 indicates overlapping functions of alphaherpesvirus pUL25 proteins. J Virol *82*, 5725-5734.

LaMassa, N., Arenas-Mena, C., and Phillips, G.R. (2018). Electron microscopic characterization of nuclear egress in the sea urchin gastrula. J Morphol *279*, 609-615.

Le Sage, V., Jung, M., Alter, J.D., Wills, E.G., Johnston, S.M., Kawaguchi, Y., Baines, J.D., and Banfield, B.W. (2013). The herpes simplex virus 2 UL21 protein is essential for virus propagation. J Virol *87*, 5904-5915.

Leach, N., Bjerke, S.L., Christensen, D.K., Bouchard, J.M., Mou, F., Park, R., Baines, J., Haraguchi, T., and Roller, R.J. (2007). Emerin is hyperphosphorylated and redistributed in herpes simplex virus type 1-infected cells in a manner dependent on both UL34 and US3. J Virol *81*, 10792-10803.

Leege, T., Granzow, H., Fuchs, W., Klupp, B.G., and Mettenleiter, T.C. (2009). Phenotypic similarities and differences between UL37-deleted pseudorabies virus and herpes simplex virus type 1. J Gen Virol *90*, 1560-1568.

Leelawong, M., Guo, D., and Smith, G.A. (2011). A physical link between the pseudorabies virus capsid and the nuclear egress complex. Journal of virology *85*, 11675-11684.

Leelawong, M., Lee, J.I., and Smith, G.A. (2012). Nuclear Egress of Pseudorabies Virus Capsids Is Enhanced by a Subspecies of the Large Tegument Protein That Is Lost upon Cytoplasmic Maturation. Journal of virology *86*, 6303-6314.

Leuzinger, H., Ziegler, U., Schraner, E.M., Fraefel, C., Glauser, D.L., Heid, I., Ackermann, M., Mueller, M., and Wild, P. (2005). Herpes simplex virus 1 envelopment follows two diverse pathways. J Virol *79*, 13047-13059.

Li, G., Nguyen, C.C., Ryckman, B.J., Britt, W.J., and Kamil, J.P. (2015). A viral regulator of glycoprotein complexes contributes to human cytomegalovirus cell tropism. Proc Natl Acad Sci U S A *112*, 4471-4476.

Liang, L., and Baines, J.D. (2005). Identification of an essential domain in the herpes simplex virus 1 UL34 protein that is necessary and sufficient to interact with UL31 protein. J Virol *79*, 3797-3806.

Licata, J.M., Simpson-Holley, M., Wright, N.T., Han, Z., Paragas, J., and Harty, R.N. (2003). Overlapping motifs (PTAP and PPEY) within the Ebola virus VP40 protein function independently as late budding domains: involvement of host proteins TSG101 and VPS-4. J Virol *77*, 1812-1819.

Likhacheva, E.V., and Bogachev, S.S. (2001). Lamins and their functions in cell cycle. Membr Cell Biol *14*, 565-577.

Liu, Y.T., Jiang, J., Bohannon, K.P., Dai, X., Gant Luxton, G.W., Hui, W.H., Bi, G.Q., Smith, G.A., and Zhou, Z.H. (2017a). A pUL25 dimer interfaces the pseudorabies virus capsid and tegument. J Gen Virol *98*, 2837-2849.

Liu, Y.T., Jiang, J., Bohannon, K.P., Dai, X., Gant Luxton, G.W., Hui, W.H., Bi, G.Q., Smith, G.A., and Zhou, Z.H. (2017b). A pUL25 dimer interfaces the pseudorabies virus capsid and tegument. J Gen Virol *98*, 2837-2849.

Liu, Z., Kato, A., Shindo, K., Noda, T., Sagara, H., Kawaoka, Y., Arii, J., and Kawaguchi, Y. (2014). Herpes simplex virus 1 UL47 interacts with viral nuclear egress factors UL31, UL34, and Us3 and regulates viral nuclear egress. J Virol *88*, 4657-4667.

Lobanov, V.A., Maher-Sturgess, S.L., Snider, M.G., Lawman, Z., Babiuk, L.A., and van Drunen Littel-van den Hurk, S. (2010). A UL47 gene deletion mutant of bovine herpesvirus type 1 exhibits impaired growth in cell culture and lack of virulence in cattle. J Virol *84*, 445-458.

Loomis, J.S., Courtney, R.J., and Wills, J.W. (2003a). Binding partners for the UL11 tegument protein of herpes simplex virus type 1. Journal of virology *77*, 11417-11424.

Loomis, J.S., Courtney, R.J., and Wills, J.W. (2003b). Binding partners for the UL11 tegument protein of herpes simplex virus type 1. J Virol *77*, 11417-11424.

Lorenz, M., Vollmer, B., Unsay, J.D., Klupp, B.G., Garcia-Saez, A.J., Mettenleiter, T.C., and Antonin, W. (2015). A single herpesvirus protein can mediate vesicle formation in the nuclear envelope. J Biol Chem *290*, 6962-6974.

Luitweiler, E.M., Henson, B.W., Pryce, E.N., Patel, V., Coombs, G., McCaffery, J.M., and Desai, P.J. (2013). Interactions of the Kaposi's Sarcoma-Associated Herpesvirus Nuclear Egress Complex: ORF69 Is a Potent Factor for Remodeling Cellular Membranes. J virol *87*, 3915-3929.

Luxton, G.W., Lee, J.I., Haverlock-Moyns, S., Schober, J.M., and Smith, G.A. (2006). The pseudorabies virus VP1/2 tegument protein is required for intracellular capsid transport. J Virol *80*, 201-209.

Lv, Y., Zhou, S., Gao, S., and Deng, H. (2019a). Remodeling of host membranes during herpesvirus assembly and egress. Protein Cell *10*, 315-326.

Lv, Y., Zhou, S., Gao, S., and Deng, H. (2019b). Remodeling of host membranes during herpesvirus assembly and egress. Protein Cell *10*, 315-326.

Lye, M.F., Sharma, M., El Omari, K., Filman, D.J., Schuermann, J.P., Hogle, J.M., and Coen, D.M. (2015). Unexpected features and mechanism of heterodimer formation of a herpesvirus nuclear egress complex. EMBO J *34*, 2937-2952.

Lymberopoulos, M.H., Bourget, A., Ben Abdeljelil, N., and Pearson, A. (2011). Involvement of the UL24 protein in herpes simplex virus 1-induced dispersal of B23 and in nuclear egress. Virology *412*, 341-348.

Margalit, A., Liu, J., Fridkin, A., Wilson, K.L., and Gruenbaum, Y. (2005). A lamin-dependent pathway that regulates nuclear organization, cell cycle progression and germ cell development. Novartis Found Symp *264*, 231-240; discussion 240-235.

Maric, M., Haugo, A.C., Dauer, W., Johnson, D., and Roller, R.J. (2014). Nuclear envelope breakdown induced by herpes simplex virus type 1 involves the activity of viral fusion proteins. Virology *460-461*, 128-137.

Marschall, M., Marzi, A., aus dem Siepen, P., Jochmann, R., Kalmer, M., Auerochs, S., Lischka, P., Leis, M., and Stamminger, T. (2005). Cellular p32 recruits cytomegalovirus kinase pUL97 to redistribute the nuclear lamina. J Biol Chem *280*, 33357-33367.

Maruzuru, Y., Shindo, K., Liu, Z., Oyama, M., Kozuka-Hata, H., Arii, J., Kato, A., and Kawaguchi, Y. (2014). Role of herpes simplex virus 1 immediate early protein ICP22 in viral nuclear egress. J Virol *88*, 7445-7454.

Mbong, E.F., Woodley, L., Frost, E., Baines, J.D., and Duffy, C. (2012). Deletion of UL21 causes a delay in the early stages of the herpes simplex virus 1 replication cycle. J Virol *86*, 7003-7007.

McElwee, M., Vijayakrishnan, S., Rixon, F., and Bhella, D. (2018). Structure of the herpes simplex virus portal-vertex. PLoS biology *16*, e2006191.

Meckes, D.G., Jr., and Wills, J.W. (2007). Dynamic interactions of the UL16 tegument protein with the capsid of herpes simplex virus. J Virol *81*, 13028-13036.

Mettenleiter, T.C. (2015). Breaching the Barrier-The Nuclear Envelope in Virus Infection. J Mol Biol.

Milbradt, J., Auerochs, S., Sevvana, M., Muller, Y.A., Sticht, H., and Marschall, M. (2012). Specific residues of a conserved domain in the N terminus of the human cytomegalovirus pUL50 protein determine its intranuclear interaction with pUL53. J Biol Chem *287*, 24004-24016.

Morris, J.B., Hofemeister, H., and O'Hare, P. (2007). Herpes simplex virus infection induces phosphorylation and delocalization of emerin, a key inner nuclear membrane protein. J Virol *81*, 4429-4437.

Mou, F., Wills, E., and Baines, J.D. (2009). Phosphorylation of the U(L)31 protein of herpes simplex virus 1 by the U(S)3-encoded kinase regulates localization of the nuclear envelopment complex and egress of nucleocapsids. J Virol *83*, 5181-5191.

Nagel, C.H., Dohner, K., Fathollahy, M., Strive, T., Borst, E.M., Messerle, M., and Sodeik, B. (2008). Nuclear egress and envelopment of herpes simplex virus capsids analyzed with dual-color fluorescence HSV1(17+). J Virol *82*, 3109-3124.

Nalwanga, D., Rempel, S., Roizman, B., and Baines, J.D. (1996). The UL 16 gene product of herpes simplex virus 1 is a virion protein that colocalizes with intranuclear capsid proteins. Virology *226*, 236-242.

Newcomb, W.W., Booy, F.P., and Brown, J.C. (2007). Uncoating the herpes simplex virus genome. Journal of molecular biology *370*, 633-642.

Newcomb, W.W., Fontana, J., Winkler, D.C., Cheng, N., Heymann, J.B., and Steven, A.C. (2017). The Primary Enveloped Virion of Herpes Simplex Virus 1: Its Role in Nuclear Egress. MBio *8*.

Newcomb, W.W., Homa, F.L., and Brown, J.C. (2006). Herpes simplex virus capsid structure: DNA packaging protein UL25 is located on the external surface of the capsid near the vertices. Journal of virology *80*, 6286-6294.

Newcomb, W.W., Juhas, R.M., Thomsen, D.R., Homa, F.L., Burch, A.D., Weller, S.K., and Brown, J.C. (2001). The UL6 gene product forms the portal for entry of DNA into the herpes simplex virus capsid. Journal of virology *75*, 10923-10932.

Nguyen, H.C., Talledge, N., McCullough, J., Sharma, A., Moss, F.R., 3rd, Iwasa, J.H., Vershinin, M.D., Sundquist, W.I., and Frost, A. (2020). Membrane constriction and thinning by sequential ESCRT-III polymerization. Nat Struct Mol Biol *27*, 392-399.

Nozawa, N., Kawaguchi, Y., Tanaka, M., Kato, A., Kato, A., Kimura, H., and Nishiyama, Y. (2005). Herpes simplex virus type 1 UL51 protein is involved in maturation and egress of virus particles. J Virol *79*, 6947-6956.

Oshima, S., Daikoku, T., Shibata, S., Yamada, H., Goshima, F., and Nishiyama, Y. (1998). Characterization of the UL16 gene product of herpes simplex virus type 2. Arch Virol *143*, 863-880.

Pante, N., and Kann, M. (2002). Nuclear pore complex is able to transport macromolecules with diameters of about 39 nm. Mol Biol Cell *13*, 425-434.

Passvogel, L., Janke, U., Klupp, B.G., Granzow, H., and Mettenleiter, T.C. (2014). Identification of conserved amino acids in pUL34 which are critical for function of the pseudorabies virus nuclear egress complex. J Virol *88*, 6224-6231.

Passvogel, L., Klupp, B.G., Granzow, H., Fuchs, W., and Mettenleiter, T.C. (2015). Functional characterization of nuclear trafficking signals in pseudorabies virus pUL31. J Virol *89*, 2002-2012.

Passvogel, L., Trube, P., Schuster, F., Klupp, B.G., and Mettenleiter, T.C. (2013). Mapping of Sequences in Pseudorabies Virus pUL34 That Are Required for Formation and Function of the Nuclear Egress Complex. Journal of virology *87*, 4475-4485.

Peter, M., Sanghera, J.S., Pelech, S.L., and Nigg, E.A. (1992). Mitogen-activated protein kinases phosphorylate nuclear lamins and display sequence specificity overlapping that of mitotic protein kinase p34cdc2. Eur J Biochem *205*, 287-294.

Radtke, K., Kieneke, D., Wolfstein, A., Michael, K., Steffen, W., Scholz, T., Karger, A., and Sodeik, B. (2010). Plus- and minus-end directed microtubule motors bind simultaneously to herpes simplex virus capsids using different inner tegument structures. PLoS Pathog *6*, e1000991.

Raghava, S., Giorda, K.M., Romano, F.B., Heuck, A.P., and Hebert, D.N. (2013). SV40 late protein VP4 forms toroidal pores to disrupt membranes for viral release. Biochemistry *52*, 3939-3948.

Raza, S., Deng, M., Shahin, F., Yang, K., Hu, C., Chen, Y., Chen, H., and Guo, A. (2016). A bovine herpesvirus 1 pUL51 deletion mutant shows impaired viral growth in vitro and reduced virulence in rabbits. Oncotarget *7*, 12235-12253.

Reynolds, A.E., Liang, L., and Baines, J.D. (2004). Conformational changes in the nuclear lamina induced by herpes simplex virus type 1 require genes U(L)31 and U(L)34. J Virol *78*, 5564-5575.

Reynolds, A.E., Ryckman, B.J., Baines, J.D., Zhou, Y., Liang, L., and Roller, R.J. (2001). U(L)31 and U(L)34 proteins of herpes simplex virus type 1 form a complex that accumulates at the nuclear rim and is required for envelopment of nucleocapsids. J Virol *75*, 8803-8817.

Reynolds, A.E., Wills, E.G., Roller, R.J., Ryckman, B.J., and Baines, J.D. (2002). Ultrastructural localization of the herpes simplex virus type 1 UL31, UL34, and US3 proteins suggests specific roles in primary envelopment and egress of nucleocapsids. J Virol *76*, 8939-8952.

Roberts, A.P., Abaitua, F., O'Hare, P., McNab, D., Rixon, F.J., and Pasdeloup, D. (2009). Differing roles of inner tegument proteins pUL36 and pUL37 during entry of herpes simplex virus type 1. J Virol *83*, 105-116.

Roizman, B., and Furlong, D. (1974). The Replication of Herpesviruses. In Comprehensive Virology, H. Fraenkel-Conrat, and R.R. Wagner, eds. (Boston, MA: Springer), pp. 229-403.

Roller, R.J., and Baines, J.D. (2017a). Herpesvirus Nuclear Egress. Adv Anat Embryol Cell Biol *223*, 143-169.

Roller, R.J., and Baines, J.D. (2017b). Herpesvirus Nuclear Egress. In Advances in Anatomy, Embryology, and Cell Biology, K. Osteirrieder, ed. (Springer), pp. 143-169.

Roller, R.J., Bjerke, S.L., Haugo, A.C., and Hanson, S. (2010). Analysis of a charge cluster mutation of herpes simplex virus type 1 UL34 and its extragenic suppressor suggests a novel interaction between pUL34 and pUL31 that is necessary for membrane curvature around capsids. J Virol *84*, 3921-3934.

Roller, R.J., Haugo, A.C., and Kopping, N.J. (2011). Intragenic and extragenic suppression of a mutation in herpes simplex virus 1 UL34 that affects both nuclear envelope targeting and membrane budding. J Virol *85*, 11615-11625.

Roller, R.J., Zhou, Y., Schnetzer, R., Ferguson, J., and DeSalvo, D. (2000). Herpes simplex virus type 1 U(L)34 gene product is required for viral envelopment. J Virol *74*, 117-129.

Ronfeldt, S., Franzke, K., Holper, J.E., Klupp, B.G., and Mettenleiter, T.C. (2020). Mutational Functional Analysis of the Pseudorabies Virus Nuclear Egress Complex-Nucleocapsid Interaction. J Virol *94*.

Ronfeldt, S., Klupp, B.G., Franzke, K., and Mettenleiter, T.C. (2017). Lysine 242 within Helix 10 of the Pseudorabies Virus Nuclear Egress Complex pUL31 Component Is Critical for Primary Envelopment of Nucleocapsids. J Virol *91*.

Rossman, J.S., Jing, X., Leser, G.P., and Lamb, R.A. (2010). Influenza virus M2 protein mediates ESCRT-independent membrane scission. Cell *142*, 902-913.

Rossman, J.S., and Lamb, R.A. (2011). Influenza virus assembly and budding. Virology *411*, 229-236.

Rossman, J.S., and Lamb, R.A. (2013). Viral membrane scission. Annu Rev Cell Dev Biol *29*, 551-569.

Ryckman, B.J., and Roller, R.J. (2004). Herpes simplex virus type 1 primary envelopment: UL34 protein modification and the US3-UL34 catalytic relationship. J Virol *78*, 399-412.

Salmon, B., Cunningham, C., Davison, A.J., Harris, W.J., and Baines, J.D. (1998). The herpes simplex virus type 1 U(L)17 gene encodes virion tegument proteins that are required for cleavage and packaging of viral DNA. J Virol *72*, 3779-3788.

Sam, M.D., Evans, B.T., Coen, D.M., and Hogle, J.M. (2009). Biochemical, biophysical, and mutational analyses of subunit interactions of the human cytomegalovirus nuclear egress complex. J Virol *83*, 2996-3006.

Sarfo, A., Starkey, J., Mellinger, E., Zhang, D., Chadha, P., Carmichael, J., and Wills, J.W. (2017). The UL21 Tegument Protein of Herpes Simplex Virus 1 Is Differentially Required for the Syncytial Phenotype. J Virol *91*.

Schauflinger, M., Fischer, D., Schreiber, A., Chevillotte, M., Walther, P., Mertens, T., and von Einem, J. (2011). The tegument protein UL71 of human cytomegalovirus is involved in late envelopment and affects multivesicular bodies. J Virol *85*, 3821-3832.

Schirmer, E.C., and Foisner, R. (2007). Proteins that associate with lamins: many faces, many functions. Exp Cell Res *313*, 2167-2179.

Schnee, M., Wagner, F.M., Koszinowski, U.H., and Ruzsics, Z. (2012). A cell free protein fragment complementation assay for monitoring the core interaction of the human cytomegalovirus nuclear egress complex. Antiviral Res *95*, 12-18.

Scholtes, L.D., Yang, K., Li, L.X., and Baines, J.D. (2010). The capsid protein encoded by U(L)17 of herpes simplex virus 1 interacts with tegument protein VP13/14. J Virol *84*, 7642-7650.

Schur, F.K., Hagen, W.J., Rumlova, M., Ruml, T., Muller, B., Krausslich, H.G., and Briggs, J.A. (2015). Structure of the immature HIV-1 capsid in intact virus particles at 8.8 A resolution. Nature *517*, 505-508.

Schuster, F., Klupp, B.G., Granzow, H., and Mettenleiter, T.C. (2012). Structural determinants for nuclear envelope localization and function of pseudorabies virus pUL34. J Virol *86*, 2079-2088.

Sehl, J., Portner, S., Klupp, B.G., Granzow, H., Franzke, K., Teifke, J.P., and Mettenleiter, T.C. (2020). Roles of the Different Isoforms of the Pseudorabies Virus Protein Kinase pUS3 in Nuclear Egress. J Virol *94*.

Shanda, S.K., and Wilson, D.W. (2008). UL36p is required for efficient transport of membrane-associated herpes simplex virus type 1 along microtubules. J Virol *82*, 7388-7394.

Sharma, M., Kamil, J.P., Coughlin, M., Reim, N.I., and Coen, D.M. (2014). Human cytomegalovirus UL50 and UL53 recruit viral protein kinase UL97, not protein kinase C, for disruption of nuclear lamina and nuclear egress in infected cells. J Virol *88*, 249-262.

Shiba, C., Daikoku, T., Goshima, F., Takakuwa, H., Yamauchi, Y., Koiwai, O., and Nishiyama, Y. (2000). The UL34 gene product of herpes simplex virus type 2 is a tail-anchored type II membrane protein that is significant for virus envelopment. J Gen Virol *81*, 2397-2405.

Silvestri, L.S., Ruthel, G., Kallstrom, G., Warfield, K.L., Swenson, D.L., Nelle, T., Iversen, P.L., Bavari, S., and Aman, M.J. (2007). Involvement of vacuolar protein sorting pathway in Ebola virus release independent of TSG101 interaction. The Journal of infectious diseases *196 Suppl 2*, S264-270.

Skepper, J.N., Whiteley, A., Browne, H., and Minson, A. (2001). Herpes simplex virus nucleocapsids mature to progeny virions by an envelopment --> deenvelopment --> reenvelopment pathway. J Virol *75*, 5697-5702.

Soni, S.P., and Stahelin, R.V. (2014). The Ebola virus matrix protein VP40 selectively induces vesiculation from phosphatidylserine-enriched membranes. J Biol Chem *289*, 33590-33597.

Speese, S.D., Ashley, J., Jokhi, V., Nunnari, J., Barria, R., Li, Y., Ataman, B., Koon, A., Chang, Y.T., Li, Q., *et al.* (2012). Nuclear Envelope Budding Enables Large Ribonucleoprotein Particle Export during Synaptic Wnt Signaling. Cell *149*, 832-846.

Stackpole, C.W. (1969). Herpes-type virus of the frog renal adenocarcinoma. I. Virus development in tumor transplants maintained at low temperatures. J Virol *4*, 75-93.

Stow, N.D. (2001). Packaging of genomic and amplicon DNA by the herpes simplex virus type 1 UL25-null mutant KUL25NS. J Virol *75*, 10755-10765.

Takeshima, K., Arii, J., Maruzuru, Y., Koyanagi, N., Kato, A., and Kawaguchi, Y. (2019). Identification of the Capsid Binding Site in the Herpes Simplex Virus 1 Nuclear Egress Complex and Its Role in Viral Primary Envelopment and Replication. J Virol *93*.

Tollefson, A.E., Scaria, A., Hermiston, T.W., Ryerse, J.S., Wold, L.J., and Wold, W.S. (1996). The adenovirus death protein (E3-11.6K) is required at very late stages of infection for efficient cell lysis and release of adenovirus from infected cells. J Virol *70*, 2296-2306.

Turner, E.M., Brown, R.S., Laudermilch, E., Tsai, P.L., and Schlieker, C. (2015). The Torsin Activator LULL1 Is Required for Efficient Growth of Herpes Simplex Virus 1. J Virol *89*, 8444-8452.

Votteler, J., and Sundquist, W.I. (2013). Virus Budding and the ESCRT Pathway. Cell Host Microbe *14*, 232-241.

Wagenaar, F., Pol, J.M., de Wind, N., and Kimman, T.G. (2001). Deletion of the UL21 gene in Pseudorabies virus results in the formation of DNA-deprived capsids: an electron microscopy study. Vet Res *32*, 47-54.

Walzer, S.A., Egerer-Sieber, C., Sticht, H., Sevvana, M., Hohl, K., Milbradt, J., Muller, Y.A., and Marschall, M. (2015). Crystal Structure of the Human Cytomegalovirus pUL50-pUL53 Core Nuclear Egress Complex Provides Insight into a Unique Assembly Scaffold for Virus-Host Protein Interactions. J Biol Chem *290*, 27452-27458.

Wang, Y., Yang, Y., Wu, S., Pan, S., Zhou, C., Ma, Y., Ru, Y., Dong, S., He, B., Zhang, C., *et al.* (2014). p32 is a novel target for viral protein ICP34.5 of herpes simplex virus type 1 and facilitates viral nuclear egress. J Biol Chem *289*, 35795-35805.

Whealy, M.E., Card, J.P., Meade, R.P., Robbins, A.K., and Enquist, L.W. (1991). Effect of brefeldin A on alphaherpesvirus membrane protein glycosylation and virus egress. J Virol *65*, 1066-1081.

Wild, P., Senn, C., Manera, C.L., Sutter, E., Schraner, E.M., Tobler, K., Ackermann, M., Ziegler, U., Lucas, M.S., and Kaech, A. (2009). Exploring the nuclear envelope of herpes simplex virus 1-infected cells by high-resolution microscopy. J Virol *83*, 408-419.

Wilson, K.L., and Foisner, R. (2010). Lamin-binding Proteins. Cold Spring Harb Perspect Biol *2*, a000554.

Wisner, T.W., Wright, C.C., Kato, A., Kawaguchi, Y., Mou, F., Baines, J.D., Roller, R.J., and Johnson, D.C. (2009). Herpesvirus gB-induced fusion between the virion envelope and outer nuclear membrane during virus egress is regulated by the viral US3 kinase. J Virol *83*, 3115-3126.

Womack, A., and Shenk, T. (2010). Human cytomegalovirus tegument protein pUL71 is required for efficient virion egress. MBio *1*.

Wright, C.C., Wisner, T.W., Hannah, B.P., Eisenberg, R.J., Cohen, G.H., and Johnson, D.C. (2009). Fusion between perinuclear virions and the outer nuclear membrane requires the fusogenic activity of herpes simplex virus gB. J Virol *83*, 11847-11856.

Wu, S., Pan, S., Zhang, L., Baines, J., Roller, R., Ames, J., Yang, M., Wang, J., Chen, D., Liu, Y., *et al.* (2016). Herpes Simplex Virus 1 Induces Phosphorylation and Reorganization of Lamin A/C through the gamma134.5 Protein That Facilitates Nuclear Egress. J Virol *90*, 10414-10422.

Yang, K., and Baines, J.D. (2011). Selection of HSV capsids for envelopment involves interaction between capsid surface components pUL31, pUL17, and pUL25. Proc Natl Acad Sci U S A *108*, 14276-14281.

Yang, K., Wills, E., Lim, H.Y., Zhou, Z.H., and Baines, J.D. (2014). Association of herpes simplex virus pUL31 with capsid vertices and components of the capsid vertex-specific complex. J Virol *88*, 3815-3825.

Ye, G.J., Vaughan, K.T., Vallee, R.B., and Roizman, B. (2000). The herpes simplex virus 1 U(L)34 protein interacts with a cytoplasmic dynein intermediate chain and targets nuclear membrane. J Virol *74*, 1355-1363.

Zaichick, S.V., Bohannon, K.P., Hughes, A., Sollars, P.J., Pickard, G.E., and Smith, G.A. (2013). The Herpesvirus VP1/2 Protein Is an Effector of Dynein-Mediated Capsid Transport and Neuroinvasion. Cell host and microbe *13*, 193-203.

Zeev-Ben-Mordehai, T., Weberruss, M., Lorenz, M., Cheleski, J., Hellberg, T., Whittle, C., El Omari, K., Vasishtan, D., Dent, K.C., Harlos, K., *et al.* (2015). Crystal Structure of the Herpesvirus Nuclear Egress Complex Provides Insights into Inner Nuclear Membrane Remodeling. Cell Rep *13*, 2645-2652.

Zhu, H.Y., Yamada, H., Jiang, Y.M., Yamada, M., and Nishiyama, Y. (1999). Intracellular localization of the UL31 protein of herpes simplex virus type 2. Arch Virol *144*, 1923-1935.

Chapter 9

Tegument Assembly, Secondary Envelopment and Exocytosis

Ian B. Hogue*

Center for Immunotherapy, Vaccines, and Virotherapy, Biodesign Institute and School of Life Sciences, Arizona State University, Tempe, AZ 85287, USA

*ihogue@asu.edu

DOI: https://doi.org/10.21775/9781913652555.09

Abstract

Alphaherpesvirus tegument assembly, secondary envelopment, and exocytosis processes are understood in broad strokes, but many of the individual steps in this pathway, and their molecular and cell biological details, remain unclear. Viral tegument and membrane proteins form an extensive and robust protein interaction network, such that essentially any structural protein can be deleted, yet particles are still assembled, enveloped, and released from infected cells. We conceptually divide the tegument proteins into three groups: conserved inner and outer teguments that participate in nucleocapsid and membrane contacts, respectively; and "middle" tegument proteins, consisting of some of the most abundant tegument proteins that serve as central hubs in the protein interaction network, yet which are unique to the alphaherpesviruses. We then discuss secondary envelopment, reviewing the tegument-membrane contacts and cellular factors that drive this process. We place this viral process in the context of cell biological processes, including the endocytic pathway, ESCRT machinery, autophagy, secretory pathway, intracellular transport, and exocytosis mechanisms. Finally, we speculate about potential relationships between cellular defenses against oligomerizing or aggregating membrane proteins and the envelopment and egress of viruses.

Introduction

The alphaherpesviruses include important human pathogens herpes simplex viruses 1 and 2 (HSV-1 and -2), varicella-zoster virus (VZV), and various veterinary viruses. The envelopment and egress processes are understood in broad strokes, but many of the individual steps, and their molecular and cell biological details, remain unclear. A virus replication cycle begins as virus particles enter by fusing their virion envelope with cellular membranes, undergo post-entry transport to the nucleus, and release their viral DNA genomes into the nucleus (see Chapters 2, 3, and 4). At this point, the virus may begin a lytic replication cycle, producing progeny virus particles and ultimately killing its host cell, or in peripheral nervous system neurons, alphaherpesviruses may establish latency, and persist for the lifetime of the host (see Chapter 6).

During lytic replication in non-neuronal cell types, or following reactivation from latency in neurons, DNA replication, capsid assembly, and genome packaging occurs in the host cell nucleus. Nucleocapsids then exit the nucleus via the process of primary envelopment, whereby nucleocapsids transiently acquire an envelope to shuttle across the nuclear membranes (see Chapter 8).

Meanwhile, newly-translated viral tegument proteins begin to assemble in the cytoplasm, and newly-translated viral transmembrane proteins traffic within the host endomembrane system. Nucleocapsids, tegument complexes, and viral membrane proteins then converge on intracellular membranes, thought to be trans-Golgi or endosomal organelles, and interactions between these sub-assemblies drive the process of virion envelopment - called "secondary envelopment" within the herpesvirus field to distinguish it from the process of primary envelopment/nuclear egress. Secondary envelopment, and possibly subsequent membrane trafficking steps, produces an enveloped infectious virion in the lumen of a secretory vesicle. This virion-in-a-vesicle then traffics to the cell periphery, and the surrounding secretory vesicle fuses with the plasma membrane, a process called exocytosis. This deposits the virion outside of the infected cell where it can infect another host cell, or the virion can be shed to the environment to infect another host organism.

A multitude of viral mechanisms, host cell biological mechanisms, and virus-host interactions drive tegument assembly, secondary envelopment, intracellular trafficking, and exocytosis of progeny virus particles from the infected cell. This chapter will focus on the viral tegument and membrane proteins, and what is known of the molecular and cell biology of these final steps in the virus replication cycle.

Tegument

The herpesvirus tegument is analogous to the matrix proteins of many other enveloped viruses in that they link the viral nucleocapsid to the viral envelope. Accordingly, the herpesvirus tegument, as a whole, plays a central role in the process of secondary envelopment. But, in contrast to typical viral matrix proteins, the herpesvirus tegument layer is much larger and more complex, consisting of around 24 different viral proteins (Kramer et al., 2011; Loret et al., 2008), forming a highly robust and redundant protein interaction network.

Identification of proteins incorporated into the tegument is not always straightforward. Whether detected by radiolabeling, immuno-assays, or mass spectrometry, contaminating infected cell material can present false positives, and proteins with very low abundance may not be detected (Engel et al., 2015; Leroy et al., 2016; Loret et al., 2008). Many host proteins are also incorporated into the tegument and envelope, but whether they are incorporated specifically and what functional roles they may play, if any, is not known (Kramer et al., 2011; Leroy et al., 2016; Loret et al., 2008).

Most of the roughly 24 viral tegument proteins do not appear to affect tegument assembly and secondary envelopment processes (Crump, 2018; Owen et al., 2015). Instead, many of these proteins have other well-studied enzymatic or regulatory roles, such as protein kinases (UL13 and US3), thymidine kinase (UL23), RNA endonuclease responsible for viral host shutoff (vhs/UL41), dUTPase (UL50), countermeasures against intrinsic cell defenses (ICP34.5, and US11), E3 ubiquitin ligase (ICP0), and transcription regulator (ICP4). While these proteins are not required for assembly and secondary envelopment, they may be specifically incorporated into the tegument because they are delivered to the cytoplasm during entry and serve to modulate viral and host functions during the earliest stages of infection (see Chapters 2, 3, and 4).

The functions of several other tegument proteins, namely UL55 (Nash and Spivack, 1994), US2 (Kang et al., 2013), and US10 (Nishiyama et al., 1993; Yamada et al., 1997) warrant further study, but these proteins are generally dispensable for virus replication *in vitro*, and so they are not thought to play major roles in tegument assembly or envelopment (Crump, 2018; Owen et al., 2015).

A large body of literature, discussed and cited in detail below, has identified at least 12 viral tegument proteins that form a protein interaction network (Figure 1) that drives tegument assembly and secondary envelopment. This network is highly robust with functional redundancy, so most individual tegument proteins and protein-protein contacts are not necessary for tegument assembly, secondary envelopment, or even production of infectious virions. In addition to this complexity and functional redundancy, the study of tegument proteins that drive assembly and secondary envelopment is complicated by the fact that many of these proteins also have enzymatic or regulatory roles during other stages of infection. As one example, in addition to its role in tegument assembly, VP16 (UL48) serves as a transactivator of viral gene transcription and interacts with vhs (UL41) to modulate its mRNA degradation activity. Deletion of VP16 (UL48) therefore impacts virus production in multiple ways: a direct effect on tegument assembly and indirect effects due to changes in viral transcription and unconstrained vhs (UL41) activity blocking viral and host gene expression (Mossman et al., 2000). Loss-of-function approaches, such as deletions in viral genes, are frequently difficult to interpret due to the

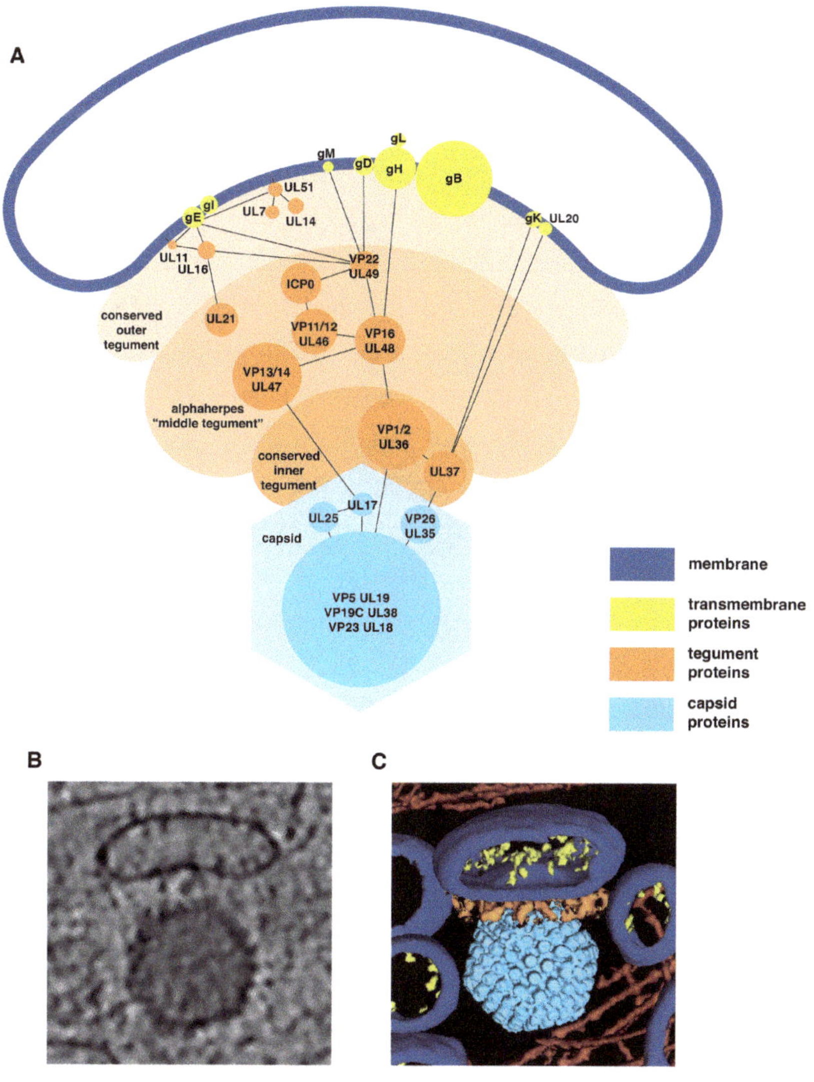

Figure 1. Alphaherpesvirus tegument and membrane protein interaction network drives secondary envelopment. (A) Tegument proteins (orange) interact with capsid proteins (light blue), viral membrane proteins (yellow), and directly with membranes (dark blue) during secondary envelopment. We conceptually divide the tegument proteins into 3 groups: Conserved inner tegument proteins, which are the major capsid-interacting proteins; "Middle" tegument proteins, which are among the most abundant tegument proteins, but are unique to the *alphaherpesvirinae*; Conserved outer tegument proteins, which participate in many membrane contacts. Each circle, representing a structural protein, is scaled by the molecular weight and estimated abundance (Loret et al., 2008) of the indicated protein, to represent the approximate relative mass in a typical HSV-1 virion. (B) CryoEM tomography slice showing an HSV-1 nucleocapsid interacting with an intracellular vesicle, possibly undergoing secondary envelopment. (C) Surface rendering depicting the nucleocapsid (light blue), tegument (orange), membrane protein densities (yellow) in the lumen of the vesicle (dark blue). Data in panels B-C kindly provided by Kay Grünewald (Ibiricu et al., 2011) and reproduced under the terms of the Creative Commons Attribution License.

multifunctional nature of most viral proteins, the ability of cloned virus strains to rapidly regenerate genetic diversity (Parsons et al., 2015), and therefore the tendency of mutant viruses to rapidly acquire compensatory mutations in other viral genes (Haugo et al., 2011).

Inner Versus Outer Tegument

In much of the literature, tegument proteins are conceptually divided into "inner tegument proteins", which are more tightly associated with the capsid, and "outer tegument proteins", which form the bulk of the tegument. The largest tegument protein, VP1/2 (UL36), its binding partner UL37, and protein kinase US3 are typically considered inner tegument proteins. This classification is mainly supported by biochemical fractionation of virions and studies showing that outer tegument proteins dissociate from the capsid during entry (Radtke et al., 2010; Wolfstein et al., 2006). However, this classification between inner and outer tegument proteins has been challenged by several notable studies using fluorescence-based methods to estimate the stoichiometry and radial distribution of select tegument proteins within individual virus particles (Bilali et al., 2017; Bohannon et al., 2013; Laine et al., 2015). While canonical inner tegument proteins VP1/2 (UL36) (Bohannon et al., 2013) and UL37 (Bilali et al., 2017; Bohannon et al., 2013) appeared to be incorporated with little variance, similarly to the stoichiometric incorporation of capsid proteins, US3 was highly variable (Bohannon et al., 2013). Canonical outer tegument proteins exhibited a wide range of variation, from stoichiometric incorporation to seemingly random incorporation (Bilali et al., 2017; Bohannon et al., 2013). Furthermore, super-resolution fluorescence microscopy showed that both the inner tegument protein UL37 and outer tegument protein VP16 (UL48) were located at nearly the same average radius from the capsid (Laine et al., 2015). Thus, the historical distinction between "inner" and "outer" tegument proteins does not necessarily reflect how individual tegument proteins are incorporated or distributed within the virion.

Given our collective knowledge of the alphaherpesvirus tegument, we find it more useful to think about tegument proteins in terms of their connectivity within the protein interaction network, the requirement for assembly and secondary envelopment, and degree of conservation among herpesvirus subfamilies. As detailed below, we conceptually distinguish between a conserved "inner tegument" that makes contacts with the nucleocapsid, a conserved "outer tegument" that mediates many membrane contacts, and a "middle tegument" that links the two. Paradoxically, while these "middle tegument" proteins are among the most abundant in the virion and form central hubs within the tegument protein interaction network, they are unique to the *alphaherpesvirinae* and do not appear to be absolutely necessary for production of infectious virus particles. Therefore, this chapter will also consider, from an evolutionary perspective, what conserved viral protein interactions and host cell biological mechanisms may be minimally required for tegument assembly, secondary envelopment, and production of infectious virions.

Conserved Inner Tegument Protein VP1/2 (UL36) is the Lynchpin of Capsid-Tegument Interactions

VP1/2 (UL36), the largest tegument protein, directly links the capsid to the tegument. Previous studies identified capsid-binding sites within the C-terminal third of VP1/2 (UL36), and more recently, cryoEM studies have shown that VP1/2 (UL36), together with capsid-associated proteins UL17 and UL25, form the capsid vertex specific component (CVSC), also called the capsid-associated tegument complex (CATC) (Dai and Zhou, 2018; Fan et al., 2015; McElwee et al., 2018). The CVSC is a pentameric protein complex composed of two VP1/2 (UL36) molecules, two UL25 molecules, and one UL17 molecule. Five CVSC complexes bind with 5-fold symmetry around each of the 11 penton vertices. However, while UL17 and UL25 are required for primary envelopment and nuclear egress, VP1/2 (UL36) is not. This contradiction has been clarified by a study (Fan et al., 2015) showing that while VP1/2 (UL36) is required for formation of the CVSC density in cryoEM reconstructions, in the absence of VP1/2 (UL36), UL17 and UL25 can still bind to capsids, likely in a more disordered fashion, and promote nuclear egress (see Chapter 8). In addition to its role at the penton vertices, the CVSC, including VP1/2 (UL36), contributes to formation of the portal vertex-associated tegument (PVAT) (McElwee et al., 2018). However, additional densities within the PVAT remain unaccounted for, and whether there are additional contributions from VP1/2 (UL36) or other tegument proteins remains to be determined (McElwee et al., 2018; Schmid et al., 2012).

The N-terminal third of VP1/2 (UL36) projects away from the capsid (Laine et al., 2015; Newcomb and Brown, 2010; Scrima et al., 2015) and binds other tegument proteins, namely the inner tegument protein UL37 (Bucks et al., 2007; Klupp et al., 2002; Vittone et al., 2005) and the outer tegument protein VP16 (UL48) (Ko et al., 2010; Svobodova et al., 2012), linking capsids to the rest of the tegument. UL37, in turn, also appears to bind the small capsid protein VP26 (UL35) to provide another capsid-tegument interaction (Lee et al., 2008), and may also interact with viral envelope proteins (see "*Tegument/ Membrane Interactions in Secondary Envelopment*" below). In the absence of VP1/2 (UL36), capsids still associate with cellular membranes to some degree (Kharkwal et al., 2015), arguing for the existence of other capsid-tegument-membrane interactions; however, these capsids accumulate in the cytoplasm, fail to recruit a substantial, electron-dense tegument layer, and do not undergo secondary envelopment (Desai, 2000; Schipke et al., 2012).

Importantly, VP1/2 (UL36) and its binding partners, UL37 and VP16 (UL48), are the only three tegument proteins that are necessary for production of infectious virions by HSV-1 (Desai et al., 2001; Weinheimer et al., 1992). In contrast, pseudorabies virus (PRV) containing mutations disrupting VP16 (UL48) or UL37 can be propagated, although these viruses are severely attenuated (Fuchs et al., 2002a; Leege et al., 2009). These observations underscore the critical role of VP1/2 (UL36) as the lynchpin of capsid-tegument interactions. Since VP1/2 (UL36) and UL37 are core genes conserved among the herpesviruses, they likely represent an ancestral inner tegument that links nucleocapsids to the rest of the tegument.

"Middle" Tegument Proteins

VP11/12 (UL46), VP13/14 (UL47), VP16 (UL48), and VP22 (UL49), the most abundant tegument proteins in viral particles, are unique to the *alphaherpesvirinae*, suggesting that there is significant divergence in tegument structure and function between herpesvirus subfamilies (Mocarski, 2007). VP16 (UL48), described above as a VP1/2 (UL36) binding partner, also interacts with each of the other middle tegument proteins, VP11/12 (UL46), VP13/14 (UL47), and VP22 (UL49), making it a central hub of tegument interactions (Elliott et al., 1995; Kato et al., 2000; Svobodova et al., 2012; Vittone et al., 2005). VP22 (UL49) is also a hub of protein-protein interactions, as it is reported to interact with VP16 (UL48) and UL16 (Maringer et al., 2012; Starkey et al., 2014). VP22 (UL49) also interacts with several viral membrane proteins and may directly bind to the membrane bilayer via charged amino acids (see "*Tegument/Membrane Interactions in Secondary Envelopment*" below). VP11/12 (UL46) and VP13/14 (UL47) both bind VP16 (UL48), and VP13/14 (UL47) is reported to bind the capsid protein UL17, which may provide another link between capsid and tegument (Scholtes et al., 2010). VP22 (UL49) and VP11/12 (UL46) may also interact indirectly via their common binding partner ICP0 (Lin et al., 2013; Maringer and Elliott, 2010; Maringer et al., 2012). Numerous other potential interactions have been detected using yeast two-hybrid, protein cross-linking, and co-immunoprecipitation/mass spectrometry approaches, but most of these have not yet been validated using other methods.

In HSV-1 and PRV, three of these middle tegument proteins, VP11/12 (UL46), VP13/14 (UL47), and VP22 (UL49) are not necessary for virus replication (del Rio et al., 2002; Elliott et al., 2005; Kopp et al., 2002; Mossman et al., 2000; Pomeranz and Blaho, 2000; Weinheimer et al., 1992; Zhang and McKnight, 1993; Zhang et al., 1991). VP16 (UL48) is necessary for HSV-1 replication, but it is not necessary for replication of VZV and PRV (Cohen and Seidel, 1994; Fuchs et al., 2002a). Strikingly, simultaneous deletion of all four middle tegument proteins does not fully block production of infectious virions in PRV, although these viruses are severely attenuated (Fuchs et al., 2003). Thus, while these tegument proteins are present at high abundance and form central hubs in the tegument protein interaction network, they are not absolutely required for tegument assembly and secondary envelopment. Since these proteins are also not conserved between the herpesvirus subfamilies, it seems unlikely that they represent the ancestral herpesvirus tegument. Probably other tegument protein interactions exist between the conserved inner and outer tegument proteins that are capable driving tegument assembly and secondary envelopment in the absence of these "middle" tegument proteins, but few of these putative interactions have been discovered.

Conserved Outer Tegument Proteins

In addition to conserved inner tegument proteins VP1/2 (UL36) and UL37, several outer tegument proteins are conserved among the herpesvirus subfamilies and participate in tegument assembly and secondary envelopment. Conserved proteins UL11 and UL16, together with the alphaherpesvirus-specific protein UL21, are reported to form a tripartite complex (Han et al., 2012; Harper et al., 2010; Lee et al., 2008; Loomis et al., 2003; Sadaoka et al., 2014; Vittone et al., 2005; Yeh et al., 2008), and deletion of UL11 and UL16 reduces, but does not fully eliminate secondary envelopment (Baines and Roizman, 1992; Fulmer et al., 2007; Kopp et al., 2003; Starkey et al., 2014). UL16 interacts with VP22 (UL49), one of the central hubs of the tegument protein interaction network (Starkey et al., 2014). UL11 and UL16 provide multiple links between tegument and membrane (see "*Tegument/Membrane Interactions in Secondary Envelopment",* below), and UL16 and UL21 are also reported to interact with nucleocapsids (de Wind et al., 1992; Meckes and Wills, 2007; Meckes et al., 2010; Oshima et al., 1998), possibly providing direct links between nucleocapsids and membranes during secondary envelopment.

UL51, UL7, and UL14 are also conserved among the herpes subfamilies, and deletion in any of these genes causes impaired secondary envelopment (Albecka et al., 2017; Fuchs et al., 2005; Klupp et al., 2005; Nozawa et al., 2002; 2005; Oda et al., 2016). Both UL7 and UL14 bind UL51 (Albecka et al., 2017; Oda et al., 2016; Roller and Fetters, 2015), so these proteins may also form a tripartite complex. Like UL11, UL51 provides multiple membrane contacts (see "*Tegument/Membrane Interactions in Secondary Envelopment*", below).

Notably, although these tegument proteins are conserved among the herpesvirus subfamilies, and many of them are considered necessary for replication of the betaherpesvirus HCMV, these proteins are not strictly necessary for tegument assembly and secondary envelopment in HSV-1 and PRV (Mocarski, 2007). These proteins may represent an ancestral herpesvirus outer tegument that may drive tegument assembly and secondary envelopment in the absence of the middle tegument proteins, so further research into the molecular and structural details of these tegument interactions may be particularly valuable.

Tegument/Membrane Interactions in Secondary Envelopment

Just as the alphaherpesvirus tegument is far more complex than the matrix of other viruses, so too are the alphaherpesvirus membrane proteins. HSV-1 incorporates 9-12 glycoproteins into virions: gB, gC, gD, gE, gG, gH, gI, gL, and gM, and it is ambiguous whether gJ, gK and gN are incorporated into virions (Loret et al., 2008). In addition, HSV-1 incorporates at least 4 other non-glycosylated transmembrane proteins into mature virions: UL20, UL45, UL56, and US9, and it is ambiguous whether UL43 is incorporated (Loret et al., 2008). In addition, several tegument proteins, such as UL11, UL51, and possibly VP22 (UL49) (detailed below), contain membrane-binding motifs and are peripherally associated with the membrane. Like the tegument protein interaction network, which is robust with redundancy, the interactions between tegument and intracellular membranes during secondary envelopment is also robust and redundant.

The conserved tegument proteins UL11 and UL51 (see "*Conserved Outer Tegument Proteins*", above) are acylated - UL51 is palmitoylated, and UL11 is both myristoylated and palmitoylated - allowing them to directly interact with cellular membranes (Loomis et al., 2001; MacLean et al., 1989; Nozawa et al.,

2003). UL11 and its binding partner UL16 are both reported to also bind the cytosolic tail of gE (Farnsworth et al., 2007; Han et al., 2012; 2011; Yeh et al., 2011), and UL51 is also reported to interact with gE (Roller et al., 2014). More recent proteomics approaches have identified many other viral proteins that may also be palmitoylated, including tegument proteins UL24, US2, and US3, and membrane proteins gE, gI, gK, gG, gH, and UL56 (Serwa et al., 2015).

Cellular enzymes methionine aminopeptidase 2 and N-myristoyltransferase 1/2 add myristoyl moieties during protein translation, and this modification is irreversible (Udenwobele et al., 2017). Palmitoylation is driven by a large number of cellular protein acyl-transferases with different substrate affinities. Unlike myristoylation, palmitoylation is reversible, and de-palmitoylation is mediated by cellular acylprotein thioesterase 1/2 (Ko and Dixon, 2018). Some cellular proteins cycle between palmitoylated and de-palmitoylated states on a timescale of seconds, allowing dynamic modulation of their membrane association, biophysical properties in the membrane, and intracellular trafficking and localization (Ko and Dixon, 2018). Therefore, dynamic palmitoylation/de-palmitoylation may serve as a molecular switch that regulates the function of these proteins during tegument assembly and secondary envelopment.

VP22 (UL49) is a central hub of both the tegument protein interaction network (see "*'Middle' Tegument Proteins*", above) and also of tegument-membrane interactions. VP22 (UL49) is reported to bind the cytosolic tails of gE, gM, and gD (Farnsworth et al., 2007; Fuchs et al., 2002b; Maringer et al., 2012). VP22 (UL49) also binds to UL16, which in turn binds UL11 and gE, providing another contact to membranes (Starkey et al., 2014). Furthermore, VP22 (UL49) may also directly associate with membranes via basic residues binding to acidic cellular phospholipids (Brignati et al., 2003). Triple deletions of gE, gI, and either gM or gD blocks secondary envelopment (Brack et al., 1999; Farnsworth et al., 2003), whereas individual deletions of these glycoproteins or VP22 (UL49) have minimal effects on secondary envelopment, indicating that these protein interactions function in a redundant manner. VP16 (UL48), another central hub of tegument protein interactions, is reported to bind gH (Gross et al., 2003; Kamen et al., 2005), although this interaction may not be conserved with PRV (Omar et al., 2013).

Finally, the capsid protein UL37 may also participate directly in secondary envelopment as it may share some structural homology with cellular proteins that regulate intracellular membrane trafficking (Pitts et al., 2014) and is reported to interact with viral envelope proteins gK and UL20 (Jambunathan et al., 2014). gK and UL20 form a complex, and deletion of either protein impairs secondary envelopment (Foster et al., 2004; Jayachandra et al., 1997; Melancon et al., 2005). However, the secondary envelopment defects of these mutants may not be solely due to loss of particular tegument-envelope contacts because gK and UL20 (similarly to gM) influence the intracellular trafficking of viral envelope glycoproteins gD, gH, and gL, as well as host membrane proteins (Crump et al., 2004; Lau and Crump, 2015; Ren et al., 2012).

Interestingly, the only conserved tegument-membrane interactions reported so far are the conserved outer tegument proteins UL11 and UL51 binding directly to the membrane via acyl modifications. All of the other reported interactions between tegument proteins and viral transmembrane proteins are unique to the *alphaherpesvirinae* because many membrane contacts are either mediated by the "middle" tegument proteins, which are not conserved, or involve the cytosolic domains of gD, gE, gK, and UL20, which are all unique to the alphaherpesvirus subfamily (Mocarski, 2007). Therefore, just as there appear to be significant differences in the "middle" tegument, there also appears to be significant divergence between the herpesvirus subfamilies in the tegument-membrane interactions that drive secondary envelopment.

L-Particle Envelopment

In addition to assembly of infectious virions, the alphaherpesviruses produce a great number of noninfectious "L-particles", or "light particles" (Rixon et al., 1992; Szilágyi and Cunningham, 1991). The functions of L-particles in viral infection and pathogenesis are not entirely clear, but have been discussed in recent reviews (Bello-Morales and López-Guerrero, 2018; Heilingloh and Krawczyk, 2017; Kalamvoki and Deschamps, 2016). L-particles are similar in size to virions and contain viral tegument and envelope proteins, but do not contain viral capsids or genomes. Dense tegument structures, which closely resemble the contents of extracellular L-particles, can be observed budding into cytoplasmic membranes in the absence of capsids (Alemañ et al., 2003; Granzow et al., 2001). Therefore, L-particle assembly and envelopment appear to share mechanistic similarities with assembly and secondary envelopment of infectious virions. Moreover, following envelopment, it has also been reported that L-particles traffic to the plasma membrane using the same secretory pathway Rab GTPases as virions (Hogue et al., 2016).

Viruses lacking tegument proteins VP1/2 (UL36), UL37, and VP16 (UL48) still produce copious L-particles (Desai et al., 2001; Fuchs et al., 2002a; 2004). Deletion of VP16 (UL48) blocks incorporation of the conserved inner tegument proteins VP1/2 (UL36) and UL37 into L-particles (Fuchs et al., 2002a), supporting the conclusion that VP16 (UL48) is an important "middle" tegument link between inner and outer tegument. L-particles are assembled even when viral DNA replication is blocked, indicating that viral proteins expressed with true-late (γ2) kinetics (e.g. VP1/2 (UL36), VP13/14 (UL47), and gC) are not required (Dargan et al., 1995).

Only simultaneously disrupting multiple membrane-tegument interactions blocks L-particle envelopment. In particular, combinations of deletions that include the conserved viral membrane protein gM or the conserved outer tegument protein UL11 are reported to block L-particle envelopment (Brack et al., 1999; Kopp et al., 2004; Mettenleiter, 2006), suggesting that some conserved membrane-tegument interactions may be minimally required for envelopment of tegument complexes into L-particles. However, the subsets of tegument-membrane interactions necessary for L-particle envelopment have not been investigated systematically.

It is tempting to speculate that production of L-particles is related to the evolutionary path these viruses followed as they diverged from an ancestral herpesvirus. Over the course of alphaherpesvirus evolution, the ancestral tegument and tegument-membrane interactions were gradually replaced by the alphaherpesvirus-specific "middle" tegument and membrane proteins. Perhaps in this course of this process, conserved outer tegument proteins, such as UL11 and its interacting partners, and conserved viral membrane proteins, such as gM, acquired the ability to drive envelopment more promiscuously, providing the flexibility to gradually evolve new alphaherpesvirus-specific "middle" tegument and membrane protein interactions.

Regardless of what role they play in viral infection and pathogenesis, L-particles may be useful, even if only as a tool to better understand secondary envelopment. Further work to identify the minimal requirements for envelopment of tegument complexes into L-particles may provide a reductionist system to clarify the mechanisms of membrane curvature during budding/wrapping, and recruitment of cellular factors that contribute to secondary envelopment and subsequent membrane trafficking steps.

Which cellular membranes/organelles serve as sites of envelopment?

There is strong evidence that secondary envelopment occurs at the trans-Golgi network, based for the most part on colocalization with cellular trans-Golgi markers (Henaff et al., 2012). Moreover, a variety of cellular factors involved in trafficking secretory vesicles from the trans-Golgi to the plasma membrane have been implicated in HSV-1 and PRV egress. For example, in uninfected cells, protein kinase D (PKD) is recruited to the trans-Golgi network and initiates a cascade of protein and lipid-based signaling that culminates in the fission of secretory vesicles from the trans-Golgi membranes. Inhibiting PKD and downstream factors reduces HSV-1 egress, and results in the accumulation of virus particles at the trans-Golgi (Rémillard-Labrosse et al., 2009; Roussel and Lippé, 2018). Several studies indicate that HSV-1 and PRV use Rab6 and Rab8-positive cellular organelles for viral glycoprotein trafficking prior to secondary envelopment and trafficking of enveloped particles to the plasma membrane after secondary envelopment (Hogue et al., 2016; 2014; Johns et al., 2014). In uninfected cells, these Rab GTPases mediate trafficking of secretory vesicles from the trans-Golgi to plasma membrane (Fourriere et al., 2019; Grigoriev et al., 2011; Miserey-Lenkei et al., 2010; Noordstra and Akhmanova, 2017).

However, several lines of evidence suggest that endocytic membranes also contribute to secondary envelopment. Many viral glycoproteins contain endocytic sorting motifs (Favoreel, 2006), indicating that glycoproteins may traffic to the site of secondary envelopment via the plasma membrane. Studies using surface biotinylation (Maresova et al., 2005), antibodies against viral glycoproteins (Albecka et al., 2016), or horseradish peroxidase added to the extracellular medium (Hollinshead et al., 2012) efficiently labeled cellular membranes involved in secondary envelopment or the resulting progeny virions. Loss-of-function experiments disrupting the endocytic pathway, by disrupting dynamin and endocytic Rab GTPases, Rab5, Rab7, and Rab11, showed that the endocytic trafficking of viral glycoproteins is required for efficient secondary envelopment of HSV-1 (Albecka et al., 2016; Hollinshead et al., 2012; Johns et al., 2014). Following secondary envelopment, PRV particle exocytosis colocalizes with the canonical recycling endosomal Rab protein, Rab11 in non-neuronal cells (Hogue et al., 2014; 2016), although Rab11 also participates in many other intracellular trafficking events. Altogether, these data indicate that at least some fraction of the viral membrane proteins and cellular membranes traffic via the plasma membrane and endocytic pathway prior to their involvement in secondary envelopment. It is well established that uninfected cells possess retrograde endomembrane trafficking mechanisms that sort endocytic cargoes into the trans-Golgi. This process is mediated by cellular retromer and sorting nexin proteins (Gallon and Cullen, 2015; Wang et al., 2018), and is exploited by several viruses and toxins for entry into cells; e.g. (Campos, 2017; Williams and Tsai, 2016). If the trans-Golgi is the site of secondary envelopment, this trafficking pathway may explain how endocytosed membrane proteins and endocytic tracers are routed to the trans-Golgi.

Alternatively, by analogy to the betaherpesviruses, it is possible that alphaherpesviruses form a unique virus-induced membrane compartment derived from both trans-Golgi and endocytic cellular membranes. HCMV causes a dramatic reorganization of cellular membranes, leading to the formation of a "virus assembly compartment" (Alwine, 2012; Henaff et al., 2012; Johnson and Baines, 2011; Tandon and Mocarski, 2012). This juxtanuclear assemblage of membranes is organized in concentric layers, with the microtubule organizing center (MTOC) and endosomal markers in the center, surrounded by trans-Golgi, Golgi, and ER markers (Das and Pellett, 2011; Das et al., 2007). Both secretory and endocytic pathway markers colocalize in the HCMV assembly compartment, and are incorporated into HCMV and HHV-6 particles (Cepeda et al., 2010; Mori et al., 2008). Although the alphaherpesviruses do not cause such an overt rearrangement of intracellular organelles, it is possible that viral infection alters intracellular membrane traffic and protein sorting to blur the distinction between trans-Golgi and endocytic organelles.

What is the role of autophagy in secondary envelopment and egress?

Another ambiguity in alphaherpesvirus secondary envelopment concerns the role of autophagy (Grose et al., 2016; 2015; Lussignol and Esclatine, 2017). Autophagy is typically a degradative pathway in which cytosolic contents are wrapped and enclosed within cellular membranes, called autophagosomes. Subsequently, autophagosomes can fuse with endosomal organelles to form amphisomes that are capable of subsequent intracellular trafficking, and/or fuse with lysosomes to form degradative autolysosomes (Morishita and Mizushima, 2019). Uninfected cells generally exhibit a constitutive homeostatic level of autophagy, but strongly upregulate autophagy in response to cell stress signals (which may include processes of viral infection). Autophagy was originally thought to be antiviral, since it could lead to lysosomal degradation of viral proteins and particles. Accordingly, several herpesviruses encode countermeasures that downregulate autophagy in infected cells. For example, HSV-1 ICP34.5 (Tallóczy et al., 2006) and US11 (Lussignol et al., 2013) are reported to block signaling pathways leading to autophagy. In contrast, varicelloviruses, like VZV, and even some primate simplexviruses that are more closely related to HSV-1, do not encode ICP34.5 or US11 homologs (Eberle and Jones-Engel, 2018; Grose et al., 2015; Severini et al., 2013). In VZV, upregulation of autophagy and numerous autophagosomes are readily detected in infected cells *in vivo* and in culture (Takahashi et al., 2009), and inhibiting autophagy reduces viral replication (Buckingham et al., 2014). Inhibiting autophagy appears to block VZV replication at the step of secondary envelopment (Girsch et al., 2019). Autophagosome markers together with endosome marker Rab11 were associated with VZV virions and secretory vesicles, suggesting that VZV trafficking and egress may occur in amphisomes (Buckingham et al., 2016). While PRV, like VZV, does not encode ICP34.5 or US11 homologs, there are mixed reports whether autophagy is induced and hijacked by PRV (Xu et al., 2018), or has antiviral effects and is inhibited by PRV protein kinase US3 (Sun et al., 2017). In summary, there appears to be fundamental differences in how different alphaherpesviruses evade or exploit cellular autophagy mechanisms; however, in some cases, autophagy mechanisms and autophagic membranes may contribute to secondary envelopment and egress of alphaherpesviruses.

Membrane Scission by ESCRT Complexes

The cellular ESCRT (endosomal sorting complex required for transport) machinery induces membrane curvature and scission "away" from the cytoplasm, into the extracellular or lumenal space. For example, ESCRT produces intralumenal vesicles inside multivesicular bodies, but also participates in many other membrane processes throughout the cell. ESCRT functions by cascade recruitment of several multi-subunit complexes, namely ESCRT-0, -I, -II, and -III. ESCRT-0 associates with cellular membranes by binding acidic phospholipids, and can initiate the cascade by recruiting ESCRT-I complexes. ESCRT-I complexes recruit ESCRT-II complexes, which, in turn, induce protein conformational changes to activate ESCRT-III proteins. The ESCRT accessory protein ALIX is also recruited by ESCRT-I and can also activate ESCRT-III proteins. Activation then nucleates the assembly of ESCRT-III proteins to form membrane-bound spiral filaments. The ATPase Vps4 then binds, remodels, and ultimately disassembles the ESCRT-III filaments, and the coordinated activity of ESCRT-III self-assembly and Vps4-mediated remodeling and disassembly are thought to drive membrane constriction and scission (Barnes and Wilson, 2019; Christ et al., 2017).

Since this budding and membrane scission is topologically equivalent to virus particle envelopment, it is not surprising that many enveloped viruses have evolved to exploit the cellular ESCRT machinery. Several studies have shown that ESCRT-III and Vps4 are required for secondary envelopment of HSV-1 and PRV (Arii et al., 2018; Crump et al., 2007; Kharkwal et al., 2014; Pawliczek and Crump, 2009), and their disruption using knockdown or dominant-negative mutant approaches leads to the appearance of partially-enveloped virus particles that appear to be arrested at the stage of membrane scission (Crump et al., 2007; Kharkwal et al., 2016).

It is not clear how alphaherpesviruses recruit ESCRT activity to sites of secondary envelopment. The structural proteins of many other viruses recruit ESCRT proteins using particular "late domain" peptide motifs, so named because they function in one of the latest steps in the production of infectious viral progeny. These motifs include P(T/S)AP motifs that recruit ESCRT-I protein TSG101, YPXL motifs that recruit the accessory protein ALIX, and PPXY motifs that recruit ESCRT-I and ALIX via NEDD4-like E3 ubiquitin ligases (Votteler and Sundquist, 2013). Many alphaherpesvirus tegument and envelope proteins contain such peptide motifs, suggesting that ESCRT-I and ALIX may be recruited in multiple redundant ways (Pawliczek and Crump, 2009).

In addition, ESCRT can be recruited by viral structural proteins via ubiquitination. For example, genetically fusing ubiquitin to a retrovirus Gag structural protein can functionally replace its late domain motifs (Joshi et al., 2008), and hepatitis C virus, which does not encode any known late domain motifs, recruits ESCRT-0 protein HRS via ubiquitination of its viral protein NS2 (Barouch-Bentov et al., 2016). ESCRT-I, -II, and ALIX also bind and sort ubiquitinated cargo (Christ et al., 2017). Mass spectrometry proteomics studies have found several ubiquitinated alphaherpesvirus structural proteins, including VP1/2 (UL36), VP13/14 (UL47), gB, and gH (Bell et al., 2013). Moreover, alphaherpesviruses may modulate ubiquitination directly via E3 ubiquitin ligase activity of ICP0 and deubiquitinase activity of VP1/2 (UL36), respectively (Boutell et al., 2002; Kattenhorn et al., 2005). Therefore, it is possible that dynamic ubiquitination of alphaherpesvirus structural proteins may also serve to recruit ESCRT during secondary envelopment. Interestingly, HSV-1 VP1/2 (UL36) has been shown to deubiquitinate ESCRT-I protein TSG101 (Caduco et al., 2013; Calistri et al., 2015), but the role of this in secondary envelopment is unclear.

All of the aforementioned methods of recruiting ESCRT - via late domain motifs and via ubiquitination - may contribute to secondary envelopment, but are not necessary. Knockdown or dominant-negative mutants of ESCRT-I protein TSG101, ESCRT-II protein EAP20/VPS25, or Bro1 domain proteins ALIX, HD-PTP, or BROX does not inhibit HSV-1 secondary envelopment (Barnes and Wilson, 2020; Pawliczek and Crump, 2009). Thus, while ESCRT appears to be critical cellular machinery driving membrane scission during alphaherpesvirus secondary envelopment, it is not known how ESCRT-III proteins are recruited, activated, or regulated in this process. As with the robustness and redundancy of tegument assembly and tegument-membrane contacts, it appears that ESCRT activity may be recruited in a robust and redundant manner.

Intracellular Transport and Exocytosis

The process of secondary envelopment creates enveloped virions in the lumen of intracellular organelles, but, as discussed above, the identity of these membranes is not clear. If the membrane of secondary envelopment is not itself a secretory vesicle (e.g. a post-Golgi secretory vesicle or plasma membrane-directed recycling endosome), the virion may require subsequent membrane trafficking steps to be sorted into a secretory organelle (e.g. sorting at the trans-Golgi network, or sorting through endosomal compartments). Whether there exist additional membrane sorting and trafficking steps between secondary envelopment and transport to the plasma membrane is not known. In either case, once inside a secretory organelle, this virion-in-a-vesicle transports to the plasma membrane, where it is released by exocytosis (Hogue et al., 2014).

Whether traveling relatively short distances in non-neuronal cells or very long distances in the axons and dendrites of neurons, cellular secretory vesicles, including those containing viral cargoes, are transported on microtubule tracks by cellular kinesin and dynein molecular motors (Miranda-Saksena et al., 2018). Recruitment of microtubule motors to cellular cargoes is mediated by many different cargo adaptor proteins, which can be transmembrane proteins, lipid-binding proteins, or associate with membranes via small GTPase proteins (Cross and Dodding, 2019). For example, the Rab GTPases implicated in alpha herpesvirus egress include: Rab3 (Miranda-Saksena et al., 2009), which recruits KIF1A and KIF1Bβ (kinesin-3) via cargo adaptor DENN/MADD (Niwa et al., 2008); Rab6 (Hogue et al., 2014; Johns et al., 2014; Stegen et al., 2013), which recruits dynein and KIF5A/B (kinesin-1) via cargo adaptors BICD1/2 (Grigoriev et al., 2007; Matanis et al., 2002; Short et al., 2002; Young et al., 2005), and dynein and KIF1C (kinesin-3) via cargo adaptor BICDR-1 (Schlager et al., 2010); Rab11 (Hogue et al., 2014; Hollinshead et al., 2012; Johns et al., 2014; Stegen et al., 2013), which recruits kinesin-1 via cargo adaptor FIP3 (Simon and Prekeris, 2008) and KIF3B (kinesin-2) via cargo adaptor Rip11/FIP5 (Schonteich et al., 2008); and Rab27 (Bello-Morales et al., 2012), which recruits KIF5 (kinesin-1) via cargo adaptors Slp3 or Slp1 and CRMP-2 (Arimura et al., 2009; Kurowska et al., 2012). Rab GTPases can also directly recruit microtubule motors, as in the case of Rab6, which may directly bind dynein motor complexes via their dynactin subunits (Short et al., 2002), KIF20A (kinesin-6) (Echard et al., 1998; Hill et al., 2000), and KIF1C (kinesin-3) (Lee et al., 2015). In addition to Rab GTPases, there are likely many other cellular factors present in the viral secretory vesicle that recruit cargo adaptors and microtubule motors. For example, HSV-1 particles co-traffic with amyloid precursor protein (APP) (Cheng et al., 2011), which recruits dynein and kinesin-1 via the cargo adaptor JIP1 (Fu and Holzbaur, 2013).

Viral proteins may also function as cargo adaptors to recruit microtubule motors to viral secretory vesicles. For example, conserved inner tegument proteins VP1/2 (UL36) and UL37 recruit dynein motors that mediate post-entry intracellular transport to the nucleus (Bearer et al., 2000; Döhner et al., 2002; McElwee et al., 2013; Richards et al., 2017; Sodeik et al., 1997; Zaichick et al., 2013). VP1/2 (UL36) also contains a conserved W-acidic motif that may recruit kinesin-1 (Ivanova et al., 2016). A consensus W-acidic motif, φ(D/E)W(D/E) (where φ represents a hydrophobic amino acid), present in many cargo adaptors, binds the tetratricopeptide repeat (TPR) domain of kinesin light chain 1 and 2 (KLC1 and KLC2) (Cross and Dodding, 2019). VP1/2 (UL36) and UL37 may also be present as part of tegument-glycoprotein complexes on the cytoplasmic face of viral secretory vesicles, where they may affect post-assembly transport of virus particles (Shanda and Wilson, 2008). Other simplex proteins, HSV-1 US11 (Diefenbach et al., 2002) and HSV-2 UL56 (Koshizuka et al., 2005) are reported to recruit kinesin-1 or kinesin-3, respectively. However, the significance of all of these putative motor recruitment activities specifically during secondary envelopment and egress is not clear.

Viral membrane proteins gE, gI, and US9 may also function as a kinesin motor recruitment complex. These proteins affect intracellular transport and egress in polarized cell types, including polarized epithelial cells and neurons (Johnson et al., 2001), and also in non-polarized cell types, like common transformed cell lines (Feutz et al., 2019). In particular, these membrane proteins are collectively required for virus particles to traffic into the axons of neurons (Draper et al., 2013; Howard et al., 2013; Kratchmarov et al., 2013a; LaVail et al., 2007; Lyman et al., 2007; Snyder et al., 2008), a necessary function for alpha herpesviruses to spread from peripheral nervous system neurons to epithelial tissues following reactivation from latency (Koyuncu et al., 2013; Miranda-Saksena et al., 2018).

In HSV-1, US9 is reported to interact with KIF5 (kinesin-1) (Diefenbach et al., 2015). The alpha herpesvirus US9 proteins contain a conserved Y-acidic motif, (D/E)(A/C)YY^{po}SE^{po}S (where ^{po}S represents phosphoserine residues), followed by 2-3 additional acidic residues, and a highly basic domain. The basic domain is necessary for KIF5 (kinesin-1) binding (Diefenbach et al., 2015), and may function by recruiting the viral protein kinase US3 to phosphorylate the serine residues (Kato et al., 2005). Alternatively, the serine residues may be phosphorylated by host casein kinase 2 (Kratchmarov et al., 2013b). Mutating the tyrosine, serine, or surrounding acidic residues reduces or eliminates axonal sorting and transport of PRV, HSV-1, and BHV-1 (Brideau et al., 2000; Chowdhury et al., 2011; Draper et al., 2013; Kratchmarov et al., 2013b). A consensus Y-acidic motif, DφYφ(D/E) (where φ represents a hydrophobic amino acid), is present in many cargo adaptors and selectively binds kinesin light chain 1 (KLC1) (Cross and Dodding, 2019). Since phosphoserine is similar in size and charge to aspartic or glutamic acid, US9 phosphorylation may create a Y-acidic motif to dynamically regulate kinesin-1 recruitment to viral secretory vesicles.

In PRV, a complex of gE, gI, and US9 are also reported to recruit KIF1A (kinesin-3) (Kramer et al., 2012; Kratchmarov et al., 2013a). KIF1A (kinesin-3) is highly expressed in the nervous system, where it is the primary motor transporting synaptic vesicle precursors down the axon (Hirokawa et al., 2009). KIF1A and KIF1Bβ (kinesin-3) contain a pleckstrin homology (PH) domain that binds acidic phosphatidylinositol-phosphate lipids (PIPs) on the cytosolic face of host membranes (Klopfenstein et al., 2002). Accordingly, the association between PRV gE/gI/US9 and KIF1A (kinesin-3) appears to depend on host membranes, since it is only detected in the nonionic detergent-resistant membrane fraction, and not when membranes are fully solubilized (Kramer et al., 2012; Lyman et al., 2008). Recruitment of KIF1A (kinesin-3) to secretory vesicles is also regulated by calmodulin and intracellular Ca^{2+} signaling (Stucchi et al., 2018), which may explain the correlation between intracellular Ca^{2+} concentration and efficient axonal transport and spread observed with PRV (Kramer and Enquist, 2012). Aside from these potential roles of PIPs and

Ca^{2+}, the mechanistic and structural details of kinesin-3 motor recruitment by gE/gI/US9 are yet to be determined.

Once viral secretory vesicles arrive at the plasma membrane, actin cytoskeleton motors, such as myosinVa, promotes virion egress, likely by facilitating viral secretory vesicle transport through cortical actin to reach the plasma membrane (Roberts and Baines, 2010). Cellular SNARE proteins mediate vesicle fusion with the plasma membrane to release secretory cargoes. SNARE proteins SNAP-25 and VAMP2, which mediate exocytosis at the plasma membrane, colocalize with virus particles in the axons of neurons (Antinone et al., 2010; Miranda-Saksena et al., 2009). These SNAREs, along with two additional SNAREs, Syntaxin6 and Vti1b, which are also reported to mediate secretion of cellular cargoes (Murray et al., 2005), were also enriched in membrane fractions containing viral membrane proteins (Kramer et al., 2012). However, the cellular factors involved in virus particle exocytosis have not been thoroughly investigated. Future work is needed to identify which viral membrane and tegument proteins may be present on the cytosolic face of viral secretory vesicles, and determine how these viral proteins interact with host cell biology to modulate virus particle egress, identify additional cell biological factors involved, and demonstrate the function of these viral and cellular factors in virion exocytosis.

Conclusions

As detailed in this chapter, a multitude of viral and host cell biological mechanisms drive tegument assembly, secondary envelopment, intracellular trafficking, and exocytosis of progeny virus particles. Because of their redundancy and robustness, disentangling the tegument and membrane protein interactions involved in these complex multistep processes is a challenge. Further work to identify individual molecular links and minimal subsets of tegument and membrane proteins required to drive envelopment and recruit cell biological factors may be particularly valuable.

Future Trends

Broader View: Do conserved cellular responses to aggregating and pore-forming membrane proteins contribute to the robustness of viral envelopment?

The plasma membrane is the defining feature of cellular life, which separates the cell from its environment. All cellular organisms express transmembrane proteins to be able to interact with their extracellular environment. Dimerization or oligomerization of membrane proteins is one of the common ways to transduce extracellular signals across the lipid bilayer. However, pathological oligomerization or aggregation of membrane proteins can also threaten the structural integrity of the membrane. Cells have evolved many defenses against membrane damage, including cell stress responses that are conserved between *Eukarya*, *Bacteria*, and *Archaea*, likely inherited from the last universal common ancestor of cellular organisms (Kültz, 2005). In eukaryotes, misfolding and aggregation of membrane proteins leads to a variety of interrelated defenses, including the unfolded protein response (UPR), ER-associated degradation (ERAD), autophagy, and ultimately, programmed cell death pathways (Qi et al., 2017; Ruggiano et al., 2014). Each of these stress response pathways are induced by alphaherpesvirus infection, but antagonized by viral countermeasures, and often exploited to promote viral replication (Grose et al., 2016).

This ability of multimeric membrane proteins to destabilize the membrane bilayer has been weaponized via evolutionary arms races to such an extent that pore-forming toxins represent the largest class of bacterial toxins, and are found in every kingdom of life (Dal Peraro and van der Goot, 2016). Studies of cellular responses to pore-forming toxins have described a common response to plasma membrane damage, sometimes termed "ectocytosis" or the "intrinsic repair pathway" (Romero et al., 2017). In this response, a plasma membrane lesion causes a rapid influx of Ca^{2+}, changes in cortical actin dynamics and multimerization of annexin proteins that stabilize the membrane and prevent expansion of the lesion, and recruitment of ESCRT machinery to drive blebbing and shedding of microvesicles from the plasma membrane (Jimenez et al., 2014). Alternatively, pore-forming toxins or protein aggregates can also be endocytosed, sequestered into intralumenal vesicles by ESCRT machinery and/or via autophagy, and subsequently exocytosed from the cell in the form of exosomes (Dal Peraro and van der Goot, 2016; Romero et al., 2017). These cellular responses rapidly remove the membrane destabilizing toxins or aggregates, allowing recovery of membrane integrity within seconds to minutes. Importantly, recent studies of mutant pore-forming toxins showed that oligomerization of the toxin protein is necessary and sufficient to induce shedding of microvesicles, even when pore formation and Ca^{2+} influx is blocked (Romero et al., 2017). This finding shows that cells possess some intrinsic mechanism to detect and respond to membrane protein oligomerization as a potential threat, even before pores or lesions are actually formed.

Studies from the retrovirus field suggest a possible link between membrane protein oligomerization, virus envelopment, and the cellular pathways producing exosomes/microvesicles (Figure 2). Many studies have noted striking similarities between retrovirus particles and cellular exosomes. Both exosomes and virus particles are capable of delivering biologically-active proteins and nucleic acids between cells, incorporate similar membrane lipids and cellular proteins, and cellular ESCRT machinery may be involved (although ESCRT-independent exosome/ microvesicle production has also been reported) (Booth et al., 2006; Nguyen et al., 2003; Nolte-'t Hoen et al., 2016; Pegtel and Gould, 2019). These similarities led several researchers to propose the "Trojan exosome" hypothesis, which posits that that viral structural proteins have evolved to direct themselves into pre-existing cellular exosome/ microvesicle pathways, and retrovirus particles are essentially exosomes/ microvesicles carrying infectious viral cargoes (Gould et al., 2003; Nolte-'t Hoen et al., 2016).

The structural polyprotein Gag drives assembly of retrovirus particles. Gag contains membrane-binding motifs, protein-protein interaction domains, RNA binding motifs, and late domain motifs that recruit cellular ESCRT complexes. The ability of Gag to drive assembly of virus-like particles is highly robust

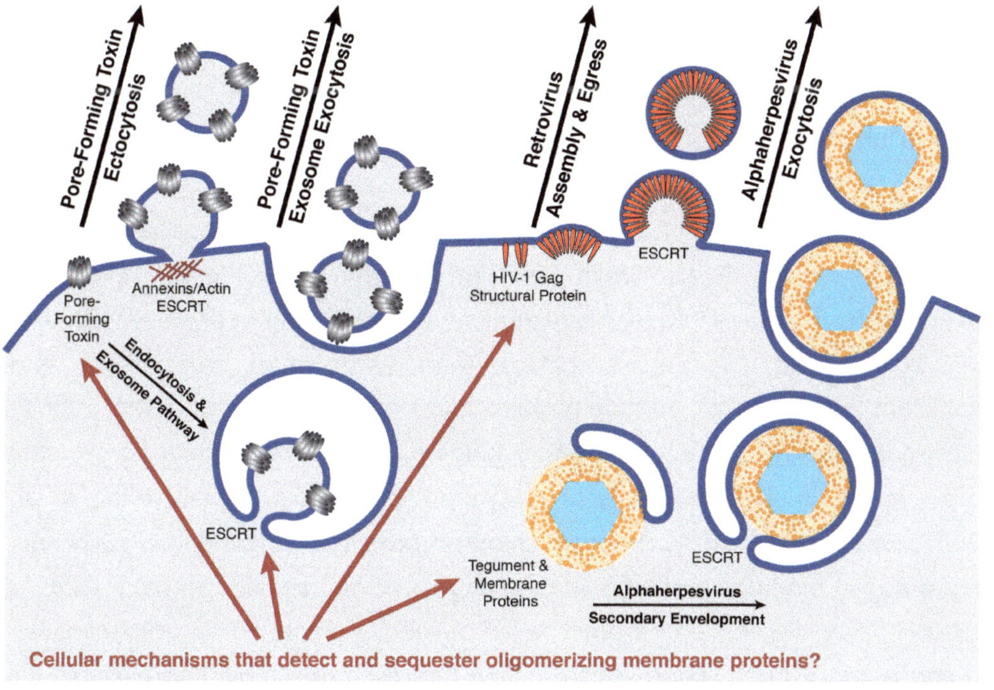

Figure 2. Common cellular responses to pore-forming toxins and other oligomerizing membrane proteins may contribute to virus envelopment. Pore-forming toxins and aggregated membrane proteins can be shed by ectocytosis, budding directly from the plasma membrane, or can be endocytosed and directed into the exosome pathway. The retrovirus Gag structural polyprotein may have evolved to mimic pathological oligomerizing membrane proteins to engage these cellular responses during assembly and egress (the "Trojan exosome" hypothesis). Does the alphaherpesvirus tegument and membrane protein interaction network engage cellular factors in this manner, and does this explain the robustness of alphaherpesvirus envelopment?

(O'Carroll et al., 2012). Essentially any part of Gag can be replaced by exogenous functional motifs, yet these chimeric Gag proteins still drive assembly and envelopment. The protein-protein interaction and RNA binding domains of Gag can be replaced by multimerizing leucine zipper motifs (Accola et al., 2000; Fang et al., 2007). The Gag membrane binding motifs can be replaced by exogenous acylation motifs or phospholipid binding domains (Chukkapalli et al., 2008; Jouvenet et al., 2006; Scholz et al., 2008; Urano et al., 2008). Indeed, HSV-1 UL11, the membrane-anchored conserved outer tegument protein that participates in herpes secondary envelopment, is capable of functionally replacing the membrane binding motifs of retroviral Gag, leading to the assembly and envelopment of virus-like particles on intracellular membranes (Bowzard et al., 2000).

Further studies exploring the relationship between retrovirus assembly and exosome/microvesicle pathways identified the minimal features sufficient to direct protein cargoes into exosomes/microvesicles (Fang et al., 2007; Shen et al., 2011; Yang and Gould, 2013). Antibody-mediated cross-linking was sufficient to redirect cell-surface proteins into microvesicles (Fang et al., 2007). Genetically fusing various membrane binding motifs onto oligomeric cytosolic proteins was also sufficient to redirect them into microvesicles (Fang et al., 2007; Shen et al., 2011). Even a chimeric protein consisting minimally of a membrane binding motif, a dimerizing leucine zipper domain, and a tetramerizing fluorescent protein was efficiently secreted in microvesicles (Fang et al., 2007). These studies show that membrane binding plus higher-order oligomerization are sufficient to direct proteins into exosomes/ microvesicles. This explains the robustness of retrovirus assembly and envelopment: essentially any part of Gag can be replaced by exogenous functional domains, as long as membrane binding and oligomerization functions are preserved. These observations may help to explain several curious features of alphaherpesvirus tegument assembly and secondary envelopment highlighted in this chapter. First, these processes are highly robust, such that essentially any viral structural protein can be deleted, yet particles (virions and/or L-particles) are still enveloped and released. Second, while herpesvirus proteins that mediate most major steps in the viral replication cycle, e.g., membrane fusion and entry, post-entry transport to the nucleus, DNA replication, capsid assembly and genome encapsidation, and nuclear egress, are conserved amongst all of the herpesviruses, the most abundant middle tegument proteins and many of the tegument-membrane protein interactions are not conserved.

Since cellular organisms have evolved a broad, common, and conserved suite of responses to defend against aggregation of membrane proteins and pore-forming toxins, it is likely that enveloped viruses have evolved to exploit these cellular pathways for assembly and egress (Figure 2). Is it possible that alphaherpesvirus tegument and envelope proteins exploit similar conserved cellular responses to drive secondary envelopment? Like retroviral Gag proteins, the alphaherpesvirus tegument and membrane protein interaction network may have evolved to use the very general properties of membrane binding plus higher-order oligomerization to engage such cellular mechanisms. Is it possible that engaging a common cellular pathway that senses and envelops membrane-bound oligomers has afforded the herpesviruses the robustness and flexibility to evolve novel subfamily-specific tegument and membrane proteins? Future work to identify cellular factors that sense the biochemical and biophysical changes in membrane dynamics caused by oligomerizing membrane proteins may shed light on these intriguing questions.

Web Resources

Alphaherpesvirinae orthologs of tegument proteins on UniProt:

VP1/2 (UL36): https://tinyurl.com/alphaherpesUL36

UL37: https://tinyurl.com/alphaherpesUL37

VP11/12 (UL46): https://tinyurl.com/alphaherpesUL46

VP13/14 (UL47): https://tinyurl.com/alphaherpesUL47

VP16 (UL48): https://tinyurl.com/alphaherpesUL48

VP22 (UL49): https://tinyurl.com/alphaherpesUL49

UL21: https://tinyurl.com/alphaherpesUL21

UL11: https://tinyurl.com/alphaherpesUL11

UL16: https://tinyurl.com/alphaherpesUL16

UL51: https://tinyurl.com/alphaherpesUL51

UL7: https://tinyurl.com/alphaherpesUL7

UL14: https://tinyurl.com/alphaherpesUL14

References

Accola, M.A., Strack, B., and Göttlinger, H.G. (2000). Efficient particle production by minimal Gag constructs which retain the carboxy-terminal domain of human immunodeficiency virus type 1 capsid-p2 and a late assembly domain. J. Virol. 74, 5395-5402. https://dx.doi.org/10.1128/jvi.74.12.5395-5402.2000

Albecka, A., Laine, R.F., Janssen, A.F.J., Kaminski, C.F., and Crump, C.M. (2016). HSV-1 Glycoproteins Are Delivered to Virus Assembly Sites Through Dynamin-Dependent Endocytosis. Traffic 17, 21-39. https://dx.doi.org/10.1111/ tra.12340

Albecka, A., Owen, D.J., Ivanova, L., Brun, J., Liman, R., Davies, L., Ahmed, M.F., Colaco, S., Hollinshead, M., Graham, S.C., et al. (2017). Dual Function of the pUL7-pUL51 Tegument Protein Complex in Herpes Simplex Virus 1 Infection. J. Virol. 91, 448. https://dx.doi.org/10.1128/JVI.02196-16

Alemañ, N., Quiroga, M.I., López-Peña, M., Vázquez, S., Guerrero, F.H., and Nieto, J.M. (2003). L-particle production during primary replication of pseudorabies virus in the nasal mucosa of swine. J. Virol. 77, 5657-5667. https://dx.doi.org/10.1128/JVI.77.10.5657-5667.2003

Alwine, J.C. (2012). The human cytomegalovirus assembly compartment: a masterpiece of viral manipulation of cellular processes that facilitates assembly and egress. PLoS Pathog. 8, e1002878. https://dx.doi.org/10.1371/journal.ppat. 1002878

Antinone, S.E., Zaichick, S.V., and Smith, G.A. (2010). Resolving the assembly state of herpes simplex virus during axon transport by live-cell imaging. J. Virol. 84, 13019-13030. https://dx.doi.org/10.1128/JVI.01296-10

Arimura, N., Kimura, T., Nakamuta, S., Taya, S., Funahashi, Y., Hattori, A., Shimada, A., Ménager, C., Kawabata, S., Fujii, K., et al. (2009). Anterograde transport of TrkB in axons is mediated by direct interaction with Slp1 and Rab27. Developmental Cell 16, 675-686. https://dx.doi.org/10.1016/j.devcel.2009.03.005

Baines, J.D., and Roizman, B. (1992). The UL11 gene of herpes simplex virus 1 encodes a function that facilitates nucleocapsid envelopment and egress from cells. J. Virol. 66, 5168-5174.

Barnes, J., and Wilson, D.W. (2020). The ESCRT-II Subunit EAP20/VPS25 and the Bro1 Domain Proteins HD-PTP and BROX Are Individually Dispensable for Herpes Simplex Virus 1 Replication. J. Virol. 94, e01641-19. https://dx.doi.org/ 10.1128/JVI.01641-19

Barouch-Bentov, R., Neveu, G., Xiao, F., Beer, M., Bekerman, E., Schor, S., Campbell, J., Boonyaratanakornkit, J., Lindenbach, B., Lu, A., et al. (2016). Hepatitis C Virus Proteins Interact with the Endosomal Sorting Complex Required for Transport (ESCRT) Machinery via Ubiquitination To Facilitate Viral Envelopment. MBio 7, 556. https://dx.doi.org/10.1128/mBio.01456-16

Bearer, E.L., Breakefield, X.O., Schuback, D., Reese, T.S., and LaVail, J.H. (2000). Retrograde axonal transport of herpes simplex virus: evidence for a single mechanism and a role for tegument. Proc. Natl. Acad. Sci. U.S.A. 97, 8146-8150. https://dx.doi.org/10.1073/pnas.97.14.8146

Bell, C., Desjardins, M., Thibault, P., and Radtke, K. (2013). Proteomics analysis of herpes simplex virus type 1-infected cells reveals dynamic changes of viral protein expression, ubiquitylation, and phosphorylation. J. Proteome Res. 12, 1820-1829. https://dx.doi.org/10.1021/pr301157j

Bello-Morales, R., and López-Guerrero, J.A. (2018). Extracellular Vesicles in Herpes Viral Spread and Immune Evasion. Front Microbiol 9, 2572. https://dx.doi.org/10.3389/fmicb.2018.02572

Bello-Morales, R., Crespillo, A.J., Fraile-Ramos, A., Tabarés, E., Alcina, A., and López-Guerrero, J.A. (2012). Role of the small GTPase Rab27a during herpes simplex virus infection of oligodendrocytic cells. BMC Microbiol. 12, 265. https://dx.doi.org/10.1186/1471-2180-12-265

Bilali, El, N., Duron, J., Gingras, D., and Lippé, R. (2017). Quantitative Evaluation of Protein Heterogeneity within Herpes Simplex Virus 1 Particles. J. Virol. 91, 8605. https://dx.doi.org/10.1128/JVI.00320-17

Bohannon, K.P., Jun, Y., Gross, S.P., and Smith, G.A. (2013). Differential protein partitioning within the herpesvirus tegument and envelope underlies a complex and variable virion architecture. Proc. Natl. Acad. Sci. U.S.A. 110, E1613-E1620. https://dx.doi.org/10.1073/pnas.1221896110

Booth, A.M., Fang, Y., Fallon, J.K., Yang, J.-M., Hildreth, J.E.K., and Gould, S.J. (2006). Exosomes and HIV Gag bud from endosome-like domains of the

T cell plasma membrane. The Journal of Cell Biology 172, 923-935. https://dx.doi.org/ 10.1083/jcb.200508014

Boutell, C., Sadis, S., and Everett, R.D. (2002). Herpes simplex virus type 1 immediate-early protein ICP0 and is isolated RING finger domain act as ubiquitin E3 ligases in vitro. J. Virol. 76, 841-850. https://dx.doi.org/10.1128/JVI.76.2.841-850.2002

Bowzard, J.B., Visalli, R.J., Wilson, C.B., Loomis, J.S., Callahan, E.M., Courtney, R.J., and Wills, J.W. (2000). Membrane targeting properties of a herpesvirus tegument protein-retrovirus Gag chimera. J. Virol. 74, 8692-8699. https://dx.doi.org/10.1128/jvi.74.18.8692-8699.2000

Brack, A.R., Dijkstra, J.M., Granzow, H., Klupp, B.G., and Mettenleiter, T.C. (1999). Inhibition of virion maturation by simultaneous deletion of glycoproteins E, I, and M of pseudorabies virus. J. Virol. 73, 5364-5372.

Brideau, A.D., Eldridge, M.G., and Enquist, L.W. (2000). Directional transneuronal infection by pseudorabies virus is dependent on an acidic internalization motif in the Us9 cytoplasmic tail. J. Virol. 74, 4549-4561. https://dx.doi.org/10.1128/ jvi.74.10.4549-4561.2000

Brignati, M.J., Loomis, J.S., Wills, J.W., and Courtney, R.J. (2003). Membrane association of VP22, a herpes simplex virus type 1 tegument protein. J. Virol. 77, 4888-4898. https://dx.doi.org/10.1128/JVI.77.8.4888-4898.2003

Buckingham, E.M., Carpenter, J.E., Jackson, W., and Grose, C. (2014). Autophagy and the effects of its inhibition on varicella-zoster virus glycoprotein biosynthesis and infectivity. J. Virol. 88, 890-902. https://dx.doi.org/10.1128/ JVI.02646-13

Buckingham, E.M., Jarosinski, K.W., Jackson, W., Carpenter, J.E., and Grose, C. (2016). Exocytosis of Varicella-Zoster Virus Virions Involves a Convergence of Endosomal and Autophagy Pathways. J. Virol. 90, 8673-8685. https://dx.doi.org/ 10.1128/JVI.00915-16

Bucks, M.A., O'Regan, K.J., Murphy, M.A., Wills, J.W., and Courtney, R.J. (2007). Herpes simplex virus type 1 tegument proteins VP1/2 and UL37 are associated with intranuclear capsids. Virology 361, 316-324. https://dx.doi.org/10.1016/ j.virol.2006.11.031

Caduco, M., Comin, A., Toffoletto, M., Munegato, D., Sartori, E., Celestino, M., Salata, C., Parolin, C., Palù, G., and Calistri, A. (2013). Tsg101 interacts with herpes simplex virus 1 VP1/2 and is a substrate of VP1/2 ubiquitin-specific protease domain activity. J. Virol. 87, 692-696. https://dx.doi.org/10.1128/ JVI. 01969-12

Calistri, A., Munegato, D., Toffoletto, M., Celestino, M., Franchin, E., Comin, A., Sartori, E., Salata, C., Parolin, C., and Palù, G. (2015). Functional Interaction Between the ESCRT-I Component TSG101 and the HSV-1 Tegument Ubiquitin Specific Protease. J. Cell. Physiol. 230, 1794-1806. https:// dx.doi.org/10.1002/ jcp.24890

Campos, S.K. (2017). Subcellular Trafficking of the Papillomavirus Genome during Initial Infection: The Remarkable Abilities of Minor Capsid Protein L2. Viruses 9, 370. https://dx.doi.org/10.3390/v9120370

Cepeda, V., Esteban, M., and Fraile-Ramos, A. (2010). Human cytomegalovirus final envelopment on membranes containing both trans-Golgi network and endosomal markers. Cell. Microbiol. 12, 386-404. https:// dx.doi.org/10.1111/ j.1462-5822.2009.01405.x

Cheng, S.-B., Ferland, P., Webster, P., and Bearer, E.L. (2011). Herpes simplex virus dances with amyloid precursor protein while exiting the cell. PLoS ONE 6, e17966. https://dx.doi.org/10.1371/journal.pone.0017966

Chowdhury, S.I., Brum, M.C.S., Coats, C., Doster, A., Wei, H., and Jones, C. (2011). The bovine herpesvirus type 1 envelope protein Us9 acidic domain is crucial for anterograde axonal transport. Vet. Microbiol. 152, 270-279. https:// dx.doi.org/10.1016/j.vetmic.2011.05.012

Christ, L., Raiborg, C., Wenzel, E.M., Campsteijn, C., and Stenmark, H. (2017). Cellular Functions and Molecular Mechanisms of the ESCRT Membrane-Scission Machinery. Trends Biochem. Sci. 42, 42-56. https://dx.doi.org/ 10.1016/ j.tibs.2016.08.016

Chukkapalli, V., Hogue, I.B., Boyko, V., Hu, W.-S., and Ono, A. (2008). Interaction between the human immunodeficiency virus type 1 Gag matrix domain and phosphatidylinositol-(4,5)-bisphosphate is essential for efficient gag membrane binding. J. Virol. 82, 2405-2417. https://dx.doi.org/10.1128/ JVI.01614-07

Cohen, J.I., and Seidel, K. (1994). Varicella-zoster virus (VZV) open reading frame 10 protein, the homolog of the essential herpes simplex virus protein VP16, is dispensable for VZV replication in vitro. J. Virol. 68, 7850-7858.

Cross, J.A., and Dodding, M.P. (2019). Motor-cargo adaptors at the organelle-cytoskeleton interface. Curr. Opin. Cell Biol. 59, 16-23. https://dx.doi.org/ 10.1016/j.ceb.2019.02.010

Crump, C. (2018). Virus Assembly and Egress of HSV. In Human Herpesviruses. Advances in Experimental Medicine and Biology, vol. 1045. Y. Kawaguchi, Y. Mori, H. Kimura, eds. (Singapore: Springer), pp. 23-44. https:// dx.doi.org/ 10.1007/978-981-10-7230-7_2

Crump, C.M., Bruun, B., Bell, S., Pomeranz, L.E., Minson, T., and Browne, H.M. (2004). Alphaherpesvirus glycoprotein M causes the relocalization of plasma membrane proteins. J. Gen. Virol. 85, 3517-3527. https://dx.doi.org/ 10.1099/ vir.0.80361-0

Crump, C.M., Yates, C., and Minson, T. (2007). Herpes simplex virus type 1 cytoplasmic envelopment requires functional Vps4. J. Virol. 81, 7380-7387. https://dx.doi.org/10.1128/JVI.00222-07

Dai, X., and Zhou, Z.H. (2018). Structure of the herpes simplex virus 1 capsid with associated tegument protein complexes. Science 360, eaao7298. https://dx.doi.org/10.1126/science.aao7298

Dal Peraro, M., and van der Goot, F.G. (2016). Pore-forming toxins: ancient, but never really out of fashion. Nat. Rev. Microbiol. 14, 77-92. https://:// dx.doi.org/ 10.1038/nrmicro.2015.3

Dargan, D.J., Patel, A.H., and Subak-Sharpe, J.H. (1995). PREPs: herpes simplex virus type 1-specific particles produced by infected cells when viral DNA replication is blocked. J. Virol. 69, 4924-4932.

Das, S., and Pellett, P.E. (2011). Spatial relationships between markers for secretory and endosomal machinery in human cytomegalovirus-infected cells versus those in uninfected cells. J. Virol. 85, 5864-5879. https://dx.doi.org/ 10.1128/JVI.00155-11

Das, S., Vasanji, A., and Pellett, P.E. (2007). Three-dimensional structure of the human cytomegalovirus cytoplasmic virion assembly complex includes a reoriented secretory apparatus. J. Virol. 81, 11861-11869. https://dx.doi.org/10.1128/JVI.01077-07

de Wind, N., Wagenaar, F., Pol, J., Kimman, T., and Berns, A. (1992). The pseudorabies virus homology of the herpes simplex virus UL21 gene product is a capsid protein which is involved in capsid maturation. J. Virol. 66, 7096-7103.

del Rio, T., Werner, H.C., and Enquist, L.W. (2002). The pseudorabies virus VP22 homologue (UL49) is dispensable for virus growth in vitro and has no effect on virulence and neuronal spread in rodents. J. Virol. 76, 774-782. https://dx.doi.org/ 10.1128/JVI.76.2.774-782.2002

Desai, P.J. (2000). A null mutation in the UL36 gene of herpes simplex virus type 1 results in accumulation of unenveloped DNA-filled capsids in the cytoplasm of infected cells. J. Virol. 74, 11608-11618.

Desai, P., Sexton, G.L., McCaffery, J.M., and Person, S. (2001). A null mutation in the gene encoding the herpes simplex virus type 1 UL37 polypeptide abrogates virus maturation. J. Virol. 75, 10259-10271. https://dx.doi.org/10.1128/ JVI.75.21.10259-10271.2001

Diefenbach, R.J., Davis, A., Miranda-Saksena, M., Fernandez, M.A., Kelly, B.J., Jones, C.A., LaVail, J.H., Xue, J., Lai, J., and Cunningham, A.L. (2015). The Basic Domain of Herpes Simplex Virus 1 pUS9 Recruits Kinesin-1 To Facilitate Egress from Neurons. J. Virol. 90, 2102-2111. https://dx.doi.org/10.1128/JVI.03041-15

Diefenbach, R.J., Miranda-Saksena, M., Diefenbach, E., Holland, D.J., Boadle, R.A., Armati, P.J., and Cunningham, A.L. (2002). Herpes simplex virus tegument protein US11 interacts with conventional kinesin heavy chain. J. Virol. 76, 3282-3291. https://dx.doi.org/10.1128/jvi.76.7.3282-3291.2002

Döhner, K., Wolfstein, A., Prank, U., Echeverri, C., Dujardin, D., Vallee, R., and Sodeik, B. (2002). Function of dynein and dynactin in herpes simplex virus capsid transport. Mol. Biol. Cell 13, 2795-2809. https://dx.doi.org/10.1091/mbc.01-07-0348

Draper, J.M., Huang, G., Stephenson, G.S., Bertke, A.S., Cortez, D.A., and LaVail, J.H. (2013). Delivery of herpes simplex virus to retinal ganglion cell axon is dependent on viral protein Us9. Invest. Ophthalmol. Vis. Sci. 54, 962-967. https://dx.doi.org/10.1167/iovs.12-11274

Eberle, R., and Jones-Engel, L. (2018). Questioning the Extreme Neurovirulence of Monkey B Virus (Macacine alphaherpesvirus 1). Adv Virol 2018, 5248420-17. https://dx.doi.org/10.1155/2018/5248420

Echard, A., Jollivet, F., Martinez, O., Lacapère, J.J., Rousselet, A., Janoueix-Lerosey, I., and Goud, B. (1998). Interaction of a Golgi-associated kinesin-like protein with Rab6. Science 279, 580-585. https://dx.doi.org/10.1126/science.279.5350.580

Elliott, G., Hafezi, W., Whiteley, A., and Bernard, E. (2005). Deletion of the herpes simplex virus VP22-encoding gene (UL49) alters the expression, localization, and virion incorporation of ICP0. J. Virol. 79, 9735-9745. https://dx.doi.org/10.1128/ JVI.79.15.9735-9745.2005

Elliott, G., Mouzakitis, G., and O'Hare, P. (1995). VP16 interacts via its activation domain with VP22, a tegument protein of herpes simplex virus, and is relocated to a novel macromolecular assembly in coexpressing cells. J. Virol. 69, 7932-7941.

Engel, E.A., Song, R., Koyuncu, O.O., and Enquist, L.W. (2015). Investigating the biology of alpha herpesviruses with MS-based proteomics. Proteomics 15, 1943-1956. https://dx.doi.org/10.1002/pmic.201400604

Fan, W.H., Roberts, A.P.E., McElwee, M., Bhella, D., Rixon, F.J., and Lauder, R. (2015). The large tegument protein pUL36 is essential for formation of the capsid vertex-specific component at the capsid-tegument interface of herpes simplex virus 1. J. Virol. 89, 1502-1511. https://dx.doi.org/10.1128/JVI.02887-14

Fang, Y., Wu, N., Gan, X., Yan, W., Morrell, J.C., and Gould, S.J. (2007). Higher-order oligomerization targets plasma membrane proteins and HIV gag to exosomes. PLoS Biol. 5, e158. https://dx.doi.org/10.1371/journal.pbio.0050158

Farnsworth, A., Goldsmith, K., and Johnson, D.C. (2003). Herpes simplex virus glycoproteins gD and gE/gI serve essential but redundant functions during acquisition of the virion envelope in the cytoplasm. J. Virol. 77, 8481-8494. https://dx.doi.org/10.1128/JVI.77.15.8481-8494.2003

Farnsworth, A., Wisner, T.W., and Johnson, D.C. (2007). Cytoplasmic residues of herpes simplex virus glycoprotein gE required for secondary envelopment and binding of tegument proteins VP22 and UL11 to gE and gD. J. Virol. 81, 319-331. https://dx.doi.org/10.1128/JVI.01842-06

Favoreel, H.W. (2006). The why's of Y-based motifs in alphaherpesvirus envelope proteins. Virus Res. 117, 202-208. https://dx.doi.org/10.1016/j.virusres. 2005.11.007

Feutz, E., McLeland-Wieser, H., Ma, J., and Roller, R.J. (2019). Functional interactions between herpes simplex virus pUL51, pUL7 and gE reveal cell-specific mechanisms for epithelial cell-to-cell spread. Virology 537, 84-96. https://dx.doi.org/10.1016/j.virol.2019.08.014

Foster, T.P., Melancon, J.M., Baines, J.D., and Kousoulas, K.G. (2004). The herpes simplex virus type 1 UL20 protein modulates membrane fusion events during cytoplasmic virion morphogenesis and virus-induced cell fusion. J. Virol. 78, 5347-5357. https://dx.doi.org/10.1128/JVI.78.10.5347-5357.2004

Fu, M.-M., and Holzbaur, E.L.F. (2013). JIP1 regulates the directionality of APP axonal transport by coordinating kinesin and dynein motors. The Journal of Cell Biology 202, 495-508. https://dx.doi.org/10.1083/jcb.201302078

Fuchs, W., Granzow, H., and Mettenleiter, T.C. (2003). A pseudorabies virus recombinant simultaneously lacking the major tegument proteins encoded by the UL46, UL47, UL48, and UL49 genes is viable in cultured cells. J. Virol. 77, 12891-12900. https://dx.doi.org/10.1128/JVI.77.23.12891-12900.2003

Fuchs, W., Granzow, H., Klopfleisch, R., Klupp, B.G., Rosenkranz, D., and Mettenleiter, T.C. (2005). The UL7 gene of pseudorabies virus encodes a nonessential structural protein which is involved in virion formation and egress. J. Virol. 79, 11291-11299. https://dx.doi.org/10.1128/JVI.79.17.11291-11299.2005

Fuchs, W., Granzow, H., Klupp, B.G., Kopp, M., and Mettenleiter, T.C. (2002a). The UL48 tegument protein of pseudorabies virus is critical for intracytoplasmic assembly of infectious virions. J. Virol. 76, 6729-6742. https://dx.doi.org/10.1128/ JVI.76.13.6729-6742.2002

Fuchs, W., Klupp, B.G., Granzow, H., and Mettenleiter, T.C. (2004). Essential function of the pseudorabies virus UL36 gene product is independent of its interaction with the UL37 protein. J. Virol. 78, 11879-11889. https://dx.doi.org/10.1128/JVI.78.21.11879-11889.2004

Fuchs, W., Klupp, B.G., Granzow, H., Hengartner, C., Brack, A., Mundt, A., Enquist, L.W., and Mettenleiter, T.C. (2002b). Physical interaction between envelope glycoproteins E and M of pseudorabies virus and the major tegument protein UL49. J. Virol. 76, 8208-8217. https://dx.doi.org/10.1128/JVI.76.16.8208-8217.2002

Fulmer, P.A., Melancon, J.M., Baines, J.D., and Kousoulas, K.G. (2007). UL20 protein functions precede and are required for the UL11 functions of herpes simplex virus type 1 cytoplasmic virion envelopment. J. Virol. 81, 3097-3108. https://dx.doi.org/10.1128/JVI.02201-06

Gallon, M., and Cullen, P.J. (2015). Retromer and sorting nexins in endosomal sorting. Biochem. Soc. Trans. 43, 33-47. https://dx.doi.org/10.1042/BST20140290

Girsch, J.H., Walters, K., Jackson, W., and Grose, C. (2019). Progeny Varicella-Zoster Virus Capsids Exit the Nucleus but Never Undergo Secondary Envelopment during Autophagic Flux Inhibition by Bafilomycin A1. J. Virol. 93, 23. https://dx.doi.org/10.1128/JVI.00505-19

Gould, S.J., Booth, A.M., and Hildreth, J.E.K. (2003). The Trojan exosome hypothesis. Proc. Natl. Acad. Sci. U.S.A. 100, 10592-10597. https://dx.doi.org/ 10.1073/pnas.1831413100

Granzow, H., Klupp, B.G., Fuchs, W., Veits, J., Osterrieder, N., and Mettenleiter, T.C. (2001). Egress of alphaherpesviruses: comparative ultrastructural study. J. Virol. 75, 3675-3684. https://dx.doi.org/10.1128/JVI.75.8.3675-3684.2001

Grigoriev, I., Splinter, D., Keijzer, N., Wulf, P.S., Demmers, J., Ohtsuka, T., Modesti, M., Maly, I.V., Grosveld, F., and Hoogenraad, C.C. (2007). Rab6 Regulates Transport and Targeting of Exocytotic Carriers. Developmental Cell 13, 305-314. https://dx.doi.org/10.1016/j.devcel.2007.06.010

Grose, C., Buckingham, E.M., Carpenter, J.E., and Kunkel, J.P. (2016). Varicella-Zoster Virus Infectious Cycle: ER Stress, Autophagic Flux, and Amphisome-Mediated Trafficking. Pathogens 5, 67. https://dx.doi.org/10.3390/ pathogens5040067

Grose, C., Buckingham, E.M., Jackson, W., and Carpenter, J.E. (2015). The pros and cons of autophagic flux among herpesviruses. Autophagy 11, 716-717. https://dx.doi.org/10.1080/15548627.2015.1017223

Gross, S.T., Harley, C.A., and Wilson, D.W. (2003). The cytoplasmic tail of Herpes simplex virus glycoprotein H binds to the tegument protein VP16 in vitro and in vivo. Virology 317, 1-12.

Han, J., Chadha, P., Meckes, D.G., Baird, N.L., and Wills, J.W. (2011). Interaction and interdependent packaging of tegument protein UL11 and glycoprotein e of herpes simplex virus. J. Virol. 85, 9437-9446. https://dx.doi.org/10.1128/ JVI.05207-11

Han, J., Chadha, P., Starkey, J.L., and Wills, J.W. (2012). Function of glycoprotein E of herpes simplex virus requires coordinated assembly of three tegument proteins on its cytoplasmic tail. Proc. Natl. Acad. Sci. U.S.A. 109, 19798-19803. https://dx.doi.org/10.1073/pnas.1212900109

Harper, A.L., Meckes, D.G., Marsh, J.A., Ward, M.D., Yeh, P.-C., Baird, N.L., Wilson, C.B., Semmes, O.J., and Wills, J.W. (2010). Interaction domains of the UL16 and UL21 tegument proteins of herpes simplex virus. J. Virol. 84, 2963-2971. https://dx.doi.org/10.1128/JVI.02015-09

Haugo, A.C., Szpara, M.L., Parsons, L., Enquist, L.W., and Roller, R.J. (2011). Herpes simplex virus 1 pUL34 plays a critical role in cell-to-cell spread of virus in addition to its role in virus replication. J. Virol. 85, 7203-7215. https://dx.doi.org/ 10.1128/JVI.00262-11

Heilingloh, C.S., and Krawczyk, A. (2017). Role of L-Particles during Herpes Simplex Virus Infection. Front Microbiol 8, 2565. https://dx.doi.org/10.3389/ fmicb.2017.02565

Henaff, D., Radtke, K., and Lippé, R. (2012). Herpesviruses exploit several host compartments for envelopment. Traffic 13, 1443-1449. https://dx.doi.org/ 10.1111/j.1600-0854.2012.01399.x

Hill, E., Clarke, M., and Barr, F.A. (2000). The Rab6-binding kinesin, Rab6-KIFL, is required for cytokinesis. Embo J. 19, 5711-5719. https://dx.doi.org/ 10.1093/ emboj/19.21.5711

Hirokawa, N., Noda, Y., Tanaka, Y., and Niwa, S. (2009). Kinesin superfamily motor proteins and intracellular transport. Nat. Rev. Mol. Cell Biol. 10, 682-696. https://dx.doi.org/10.1038/nrm2774

Hogue, I.B., Bosse, J.B., Hu, J.-R., Thiberge, S.Y., and Enquist, L.W. (2014). Cellular mechanisms of alpha herpesvirus egress: live cell fluorescence microscopy of pseudorabies virus exocytosis. PLoS Pathog. 10, e1004535. https://dx.doi.org/10.1371/journal.ppat.1004535

Hogue, I.B., Scherer, J., and Enquist, L.W. (2016). Exocytosis of Alphaherpesvirus Virions, Light Particles, and Glycoproteins Uses Constitutive Secretory Mechanisms. MBio 7, e00820-16. https://dx.doi.org/ 10.1128/ mBio.00820-16

Hollinshead, M., Johns, H.L., Sayers, C.L., Gonzalez-Lopez, C., Smith, G.L., and Elliott, G. (2012). Endocytic tubules regulated by Rab GTPases 5 and 11 are used for envelopment of herpes simplex virus. Embo J. 31, 4204-4220. https://dx.doi.org/10.1038/emboj.2012.262

Howard, P.W., Howard, T.L., and Johnson, D.C. (2013). Herpes simplex virus membrane proteins gE/gI and US9 act cooperatively to promote transport of capsids and glycoproteins from neuron cell bodies into initial axon segments. J. Virol. 87, 403-414. https://dx.doi.org/10.1128/JVI.02465-12

Ibiricu, I., Huiskonen, J.T., Döhner, K., Bradke, F., Sodeik, B., Grünewald, K. (2011). Cryo Electron Tomography of Herpes Simplex Virus during Axonal Transport and Secondary Envelopment in Primary Neurons. PLoS Pathog. 7, e1002406. https://dx.doi.org/10.1371/journal.ppat.1002406

Ivanova, L., Buch, A., Döhner, K., Pohlmann, A., Binz, A., Prank, U., Sandbaumhüter, M., Bauerfeind, R., and Sodeik, B. (2016). Conserved Tryptophan Motifs in the Large Tegument Protein pUL36 Are Required for Efficient Secondary Envelopment of Herpes Simplex Virus Capsids. J. Virol. 90, 5368-5383. https://dx.doi.org/10.1128/JVI.03167-15

Jambunathan, N., Chouljenko, D., Desai, P., Charles, A.-S., Subramanian, R., Chouljenko, V.N., and Kousoulas, K.G. (2014). Herpes simplex virus 1 protein UL37 interacts with viral glycoprotein gK and membrane protein UL20 and functions in cytoplasmic virion envelopment. J. Virol. 88, 5927-5935. https://dx.doi.org/10.1128/JVI.00278-14

Jayachandra, S., Baghian, A., and Kousoulas, K.G. (1997). Herpes simplex virus type 1 glycoprotein K is not essential for infectious virus production in actively replicating cells but is required for efficient envelopment and translocation of infectious virions from the cytoplasm to the extracellular space. J. Virol. 71, 5012-5024.

Jimenez, A.J., Maiuri, P., Lafaurie-Janvore, J., Divoux, S., Piel, M., and Perez, F. (2014). ESCRT machinery is required for plasma membrane repair. Science 343, 1247136-1247136. https://dx.doi.org/10.1126/science.1247136

Johns, H.L., Gonzalez-Lopez, C., Sayers, C.L., Hollinshead, M., and Elliott, G. (2014). Rab6 dependent post-Golgi trafficking of HSV1 envelope proteins to sites of virus envelopment. Traffic 15, 157-178. https://dx.doi.org/10.1111/tra.12134

Johnson, D.C., Webb, M., Wisner, T.W., and Brunetti, C. (2001). Herpes simplex virus gE/gI sorts nascent virions to epithelial cell junctions, promoting virus spread. J. Virol. 75, 821-833. https://dx.doi.org/10.1128/JVI.75.2.821-833.2001

Johnson, D.C., and Baines, J.D. (2011). Herpesviruses remodel host membranes for virus egress. Nat. Rev. Microbiol. 9, 382-394. https://dx.doi.org/10.1038/ nrmicro2559

Joshi, A., Munshi, U., Ablan, S.D., Nagashima, K., and Freed, E.O. (2008). Functional replacement of a retroviral late domain by ubiquitin fusion. Traffic 9, 1972-1983. https://dx.doi.org/10.1111/j.1600-0854.2008.00817.x

Jouvenet, N., Neil, S.J.D., Bess, C., Johnson, M.C., Virgen, C.A., Simon, S.M., and Bieniasz, P.D. (2006). Plasma membrane is the site of productive HIV-1 particle assembly. PLoS Biol. 4, e435. https://dx.doi.org/10.1371/journal.pbio.0040435

Kalamvoki, M., and Deschamps, T. (2016). Extracellular vesicles during Herpes Simplex Virus type 1 infection: an inquire. Virol. J. 13, 63. https://dx.doi.org/10.1186/s12985-016-0518-2

Kamen, D.E., Gross, S.T., Girvin, M.E., and Wilson, D.W. (2005). Structural basis for the physiological temperature dependence of the association of VP16 with the cytoplasmic tail of herpes simplex virus glycoprotein H. J. Virol. 79, 6134-6141. https://dx.doi.org/10.1128/JVI.79.10.6134-6141.2005

Kang, M.-H., Roy, B.B., Finnen, R.L., Le Sage, V., Johnston, S.M., Zhang, H., and Banfield, B.W. (2013). The Us2 gene product of herpes simplex virus 2 is a membrane-associated ubiquitin-interacting protein. J. Virol. 87, 9590-9603. https://dx.doi.org/10.1128/JVI.00994-13

Kato, K., Daikoku, T., Goshima, F., Kume, H., Yamaki, K., and Nishiyama, Y. (2000). Synthesis, subcellular localization and VP16 interaction of the herpes simplex virus type 2 UL46 gene product. Arch. Virol. 145, 2149-2162.

Kato, A., Yamamoto, M., Ohno, T., Kodaira, H., Nishiyama, Y., Kawaguchi, Y. (2005). Identification of proteins phosphorylated directly by the Us3 protein kinase encoded by herpes simplex virus 1. J. Virol. 79, 9325-9331. https://dx.doi.org/ 10.1128/JVI.79.14.9325-9331.2005

Kattenhorn, L.M., Korbel, G.A., Kessler, B.M., Spooner, E., and Ploegh, H.L. (2005). A deubiquitinating enzyme encoded by HSV-1 belongs to a family of cysteine proteases that is conserved across the family Herpesviridae. Mol. Cell 19, 547-557. https://dx.doi.org/10.1016/j.molcel.2005.07.003

Kharkwal, H., Furgiuele, S.S., Smith, C.G., and Wilson, D.W. (2015). Herpes Simplex Virus Capsid-Organelle Association in the Absence of the Large Tegument Protein UL36p. J. Virol. 89, 11372-11382. https://dx.doi.org/ 10.1128/ JVI.01893-15

Kharkwal, H., Smith, C.G., and Wilson, D.W. (2016). Herpes Simplex Virus Capsid Localization to ESCRT-VPS4 Complexes in the Presence and Absence of the Large Tegument Protein UL36p. J. Virol. 90, 7257-7267. https://dx.doi.org/ 10.1128/JVI.00857-16

Klupp, B.G., Fuchs, W., Granzow, H., Nixdorf, R., and Mettenleiter, T.C. (2002). Pseudorabies virus UL36 tegument protein physically interacts with the UL37 protein. J. Virol. 76, 3065-3071. https://dx.doi.org/10.1128/JVI.76.6.3065-3071.2002

Klupp, B.G., Granzow, H., Klopfleisch, R., Fuchs, W., Kopp, M., Lenk, M., and Mettenleiter, T.C. (2005). Functional analysis of the pseudorabies virus UL51 protein. J. Virol. 79, 3831-3840. https://dx.doi.org/10.1128/JVI.79.6.3831-3840.2005

Ko, D.H., Cunningham, A.L., and Diefenbach, R.J. (2010). The major determinant for addition of tegument protein pUL48 (VP16) to capsids in herpes simplex virus type 1 is the presence of the major tegument protein pUL36 (VP1/2). J. Virol. 84, 1397-1405. https://dx.doi.org/10.1128/JVI.01721-09

Ko, P.-J., and Dixon, S.J. (2018). Protein palmitoylation and cancer. EMBO Rep. 19, e46666. https://dx.doi.org/10.15252/embr.201846666

Kopp, M., Granzow, H., Fuchs, W., Klupp, B.G., Mundt, E., Karger, A., and Mettenleiter, T.C. (2003). The pseudorabies virus UL11 protein is a virion component involved in secondary envelopment in the cytoplasm. J. Virol. 77, 5339-5351. https://dx.doi.org/10.1128/JVI.77.9.5339-5351.2003

Kopp, M., Klupp, B.G., Granzow, H., Fuchs, W., and Mettenleiter, T.C. (2002). Identification and characterization of the pseudorabies virus tegument proteins UL46 and UL47: role for UL47 in virion morphogenesis in the cytoplasm. J. Virol. 76, 8820-8833. https://dx.doi.org/10.1128/JVI.76.17.8820-8833.2002

Koshizuka, T., Kawaguchi, Y., and Nishiyama, Y. (2005). Herpes simplex virus type 2 membrane protein UL56 associates with the kinesin motor protein KIF1A. J. Gen. Virol. 86, 527-533. https://dx.doi.org/10.1099/vir.0.80633-0

Koyuncu, O.O., Hogue, I.B., and Enquist, L.W. (2013). Virus infections in the nervous system. Cell Host Microbe 13, 379-393. https://dx.doi.org/10.1016/j.chom.2013.03.010

Kramer, T., and Enquist, L.W. (2012). Alphaherpesvirus infection disrupts mitochondrial transport in neurons. Cell Host Microbe 11, 504-514. https://dx.doi.org/10.1016/j.chom.2012.03.005

Kramer, T., Greco, T.M., Enquist, L.W., and Cristea, I.M. (2011). Proteomic characterization of pseudorabies virus extracellular virions. J. Virol. 85, 6427-6441. https://dx.doi.org/10.1128/JVI.02253-10

Kramer, T., Greco, T.M., Taylor, M.P., Ambrosini, A.E., Cristea, I.M., and Enquist, L.W. (2012). Kinesin-3 mediates axonal sorting and directional transport of alphaherpesvirus particles in neurons. Cell Host Microbe 12, 806-814. https://dx.doi.org/10.1016/j.chom.2012.10.013

Kratchmarov, R., Kramer, T., Greco, T.M., Taylor, M.P., Ch'ng, T.H., Cristea, I.M., and Enquist, L.W. (2013a). Glycoproteins gE and gI are required for efficient KIF1A-dependent anterograde axonal transport of alphaherpesvirus particles in neurons. J. Virol. 87, 9431-9440. https://dx.doi.org/10.1128/JVI.01317-13

Kratchmarov, R., Taylor, M.P., and Enquist, L.W. (2013b). Role of Us9 phosphorylation in axonal sorting and anterograde transport of pseudorabies virus. PLoS ONE 8, e58776. https://dx.doi.org/10.1371/journal.pone.0058776

Kurowska, M., Goudin, N., Nehme, N.T., Court, M., Garin, J., Fischer, A., de Saint Basile, G., and Ménasché, G. (2012). Terminal transport of lytic granules to the immune synapse is mediated by the kinesin-1/Slp3/Rab27a complex. Blood 119, 3879-3889. https://dx.doi.org/10.1182/blood-2011-09-382556

Kültz, D. (2005). Molecular and evolutionary basis of the cellular stress response. Annu. Rev. Physiol. 67, 225-257. https://dx.doi.org/10.1146/annurev.physiol. 67.040403.103635

Laine, R.F., Albecka, A., van de Linde, S., Rees, E.J., Crump, C.M., and Kaminski, C.F. (2015). Structural analysis of herpes simplex virus by optical super-resolution imaging. Nat Comms 6, 5980. https://dx.doi.org/10.1038/ncomms6980

Lau, S.-Y.K., and Crump, C.M. (2015). HSV-1 gM and the gK/pUL20 complex are important for the localization of gD and gH/L to viral assembly sites. Viruses 7, 915-938. https://dx.doi.org/10.3390/v7030915

LaVail, J.H., Tauscher, A.N., Sucher, A., Harrabi, O., and Brandimarti, R. (2007). Viral regulation of the long distance axonal transport of herpes simplex virus nucleocapsid. Neuroscience 146, 974-985. https://dx.doi.org/10.1016/ j.neuroscience.2007.02.010

Lee, J.H., Vittone, V., Diefenbach, E., Cunningham, A.L., and Diefenbach, R.J. (2008). Identification of structural protein-protein interactions of herpes simplex virus type 1. Virology 378, 347-354. https://dx.doi.org/10.1016/j.virol.2008.05.035

Lee, P.L., Ohlson, M.B., and Pfeffer, S.R. (2015). Rab6 regulation of the kinesin family KIF1C motor domain contributes to Golgi tethering. Elife 4, 917. https://dx.doi.org/10.7554/eLife.06029

Leege, T., Granzow, H., Fuchs, W., Klupp, B.G., and Mettenleiter, T.C. (2009). Phenotypic similarities and differences between UL37-deleted pseudorabies virus and herpes simplex virus type 1. J. Gen. Virol. 90, 1560-1568. https://dx.doi.org/ 10.1099/vir.0.010322-0

Leroy, B., Gillet, L., Vanderplasschen, A., and Wattiez, R. (2016). Structural Proteomics of Herpesviruses. Viruses 8, 50. https://dx.doi.org/10.3390/v8020050

Lin, A.E., Greco, T.M., Döhner, K., Sodeik, B., and Cristea, I.M. (2013). A proteomic perspective of inbuilt viral protein regulation: pUL46 tegument protein is targeted for degradation by ICP0 during herpes simplex virus type 1 infection. Mol. Cell Proteomics 12, 3237-3252. https://dx.doi.org/10.1074/mcp.M113.030866

Loomis, J.S., Bowzard, J.B., Courtney, R.J., and Wills, J.W. (2001). Intracellular trafficking of the UL11 tegument protein of herpes simplex virus type 1. J. Virol. 75, 12209-12219. https://dx.doi.org/10.1128/JVI.75.24.12209-12219.2001

Loomis, J.S., Courtney, R.J., and Wills, J.W. (2003). Binding partners for the UL11 tegument protein of herpes simplex virus type 1. J. Virol. 77, 11417-11424. https://dx.doi.org/10.1128/JVI.77.21.11417-11424.2003

Loret, S., Guay, G., and Lippé, R. (2008). Comprehensive characterization of extracellular herpes simplex virus type 1 virions. J. Virol. 82, 8605-8618. https://dx.doi.org/10.1128/JVI.00904-08

Lussignol, M., and Esclatine, A. (2017). Herpesvirus and Autophagy: "All Right, Everybody Be Cool, This Is a Robbery!." Viruses 9, 372. https://dx.doi.org/10.3390/v9120372

Lussignol, M., Queval, C., Bernet-Camard, M.-F., Cotte-Laffitte, J., Beau, I., Codogno, P., and Esclatine, A. (2013). The herpes simplex virus 1 Us11 protein inhibits autophagy through its interaction with the protein kinase PKR. J. Virol. 87, 859-871. https://dx.doi.org/10.1128/JVI.01158-12

Lyman, M.G., Feierbach, B., Curanovic, D., Bisher, M., and Enquist, L.W. (2007). Pseudorabies virus Us9 directs axonal sorting of viral capsids. J. Virol. 81, 11363-11371. https://dx.doi.org/10.1128/JVI.01281-07

Lyman, M.G., Curanović, D., and Enquist, L.W. (2008). Targeting of pseudorabies virus structural proteins to axons requires association of the viral Us9 protein with lipid rafts. PLoS Pathog. 4, e1000065. https://dx.doi.org/10.1371/ journal.ppat.1000065

MacLean, C.A., Clark, B., and McGeoch, D.J. (1989). Gene UL11 of herpes simplex virus type 1 encodes a virion protein which is myristylated. J. Gen. Virol. 70, 3147-3157. https://dx.doi.org/10.1099/0022-1317-70-12-3147

Maresova, L., Pasieka, T.J., Homan, E., Gerday, E., and Grose, C. (2005). Incorporation of three endocytosed varicella-zoster virus glycoproteins, gE, gH, and gB, into the virion envelope. J. Virol. 79, 997-1007. https://dx.doi.org/10.1128/JVI.79.2.997-1007.2005

Maringer, K., and Elliott, G. (2010). Recruitment of herpes simplex virus type 1 immediate-early protein ICP0 to the virus particle. J. Virol. 84, 4682-4696. https://dx.doi.org/10.1128/JVI.00126-10

Maringer, K., Stylianou, J., and Elliott, G. (2012). A network of protein interactions around the herpes simplex virus tegument protein VP22. J. Virol. 86, 12971-12982. https://dx.doi.org/10.1128/JVI.01913-12

Matanis, T., Akhmanova, A., Wulf, P., Del Nery, E., Weide, T., Stepanova, T., Galjart, N., Grosveld, F., Goud, B., De Zeeuw, C.I., et al. (2002). Bicaudal-D regulates COPI-independent Golgi-ER transport by recruiting the dynein-dynactin motor complex. Nat. Cell Biol. 4, 986-992. https://dx.doi.org/10.1038/ncb891

McElwee, M., Beilstein, F., Labetoulle, M., Rixon, F.J., and Pasdeloup, D. (2013). Dystonin/BPAG1 promotes plus-end-directed transport of herpes simplex virus 1 capsids on microtubules during entry. J. Virol. 87, 11008-11018. https://dx.doi.org/10.1128/JVI.01633-13

McElwee, M., Vijayakrishnan, S., Rixon, F., and Bhella, D. (2018). Structure of the herpes simplex virus portal-vertex. PLoS Biol. 16, e2006191. https://dx.doi.org/10.1371/journal.pbio.2006191

Meckes, D.G., and Wills, J.W. (2007). Dynamic interactions of the UL16 tegument protein with the capsid of herpes simplex virus. J. Virol. 81, 13028-13036. https://dx.doi.org/10.1128/JVI.01306-07

Meckes, D.G., Marsh, J.A., and Wills, J.W. (2010). Complex mechanisms for the packaging of the UL16 tegument protein into herpes simplex virus. Virology 398, 208-213. https://dx.doi.org/10.1016/j.virol.2009.12.004

Melancon, J.M., Luna, R.E., Foster, T.P., and Kousoulas, K.G. (2005). Herpes simplex virus type 1 gK is required for gB-mediated virus-induced cell fusion, while neither gB and gK nor gB and UL20p function redundantly in virion de-envelopment. J. Virol. 79, 299-313. https://dx.doi.org/10.1128/JVI.79.1.299-313.2005

Miranda-Saksena, M., Boadle, R.A., Aggarwal, A., Tijono, B., Rixon, F.J., Diefenbach, R.J., and Cunningham, A.L. (2009). Herpes simplex virus utilizes the large secretory vesicle pathway for anterograde transport of tegument and envelope proteins and for viral exocytosis from growth cones of human fetal axons. J. Virol. 83, 3187-3199. https://dx.doi.org/10.1128/JVI.01579-08

Miranda-Saksena, M., Denes, C.E., Diefenbach, R.J., and Cunningham, A.L. (2018). Infection and Transport of Herpes Simplex Virus Type 1 in Neurons: Role of the Cytoskeleton. Viruses 10, 92. https://dx.doi.org/10.3390/v10020092

Mocarski, E.S. (2007). Comparative analysis of herpesvirus-common proteins. In Human Herpesviruses: Biology, Therapy, and Immunoprophylaxis. A.M. Arvin, G. Campadelli-Fiume, E.S. Mocarski, P.S. Moore, B. Roizman, et al., eds. (Cambridge, UK: Cambridge University Press).

Mori, Y., Koike, M., Moriishi, E., Kawabata, A., Tang, H., Oyaizu, H., Uchiyama, Y., and Yamanishi, K. (2008). Human herpesvirus-6 induces MVB formation, and virus egress occurs by an exosomal release pathway. Traffic 9, 1728-1742. https://dx.doi.org/10.1111/j.1600-0854.2008.00796.x

Morishita, H., and Mizushima, N. (2019). Diverse Cellular Roles of Autophagy. Annu. Rev. Cell Dev. Biol. 35. https://dx.doi.org/10.1146/annurev-cellbio-100818-125300

Mossman, K.L., Sherburne, R., Lavery, C., Duncan, J., and Smiley, J.R. (2000). Evidence that herpes simplex virus VP16 is required for viral egress downstream of the initial envelopment event. J. Virol. 74, 6287-6299.

Murray, R.Z., Wylie, F.G., Khromykh, T., Hume, D.A., and Stow, J.L. (2005). Syntaxin 6 and Vti1b form a novel SNARE complex, which is up-regulated in activated macrophages to facilitate exocytosis of tumor necrosis Factor-alpha. J. Biol. Chem. 280, 10478-10483. https://dx.doi.org/10.1074/jbc.M414420200

Nash, T.C., and Spivack, J.G. (1994). The UL55 and UL56 genes of herpes simplex virus type 1 are not required for viral replication, intraperitoneal virulence, or establishment of latency in mice. Virology 204, 794-798. https://dx.doi.org/10.1006/viro.1994.1595

Newcomb, W.W., and Brown, J.C. (2010). Structure and capsid association of the herpesvirus large tegument protein UL36. J. Virol. 84, 9408-9414. https://dx.doi.org/10.1128/JVI.00361-10

Nguyen, D.G., Booth, A., Gould, S.J., and Hildreth, J.E.K. (2003). Evidence that HIV budding in primary macrophages occurs through the exosome release pathway. J. Biol. Chem. 278, 52347-52354. https://dx.doi.org/10.1074/ jbc.M309009200

Nishiyama, Y., Kurachi, R., Daikoku, T., and Umene, K. (1993). The US 9, 10, 11, and 12 genes of herpes simplex virus type 1 are of no importance for its neurovirulence and latency in mice. Virology 194, 419-423. https://dx.doi.org/10.1006/viro.1993.1279

Niwa, S., Tanaka, Y., and Hirokawa, N. (2008). KIF1Bbeta- and KIF1A-mediated axonal transport of presynaptic regulator Rab3 occurs in a GTP-dependent manner through DENN/MADD. Nat. Cell Biol. 10, 1269-1279. https://dx.doi.org/ 10.1038/ncb1785

Nolte-'t Hoen, E., Cremer, T., Gallo, R.C., and Margolis, L.B. (2016). Extracellular vesicles and viruses: Are they close relatives? Proc. Natl. Acad. Sci. U.S.A. 113, 9155-9161. https://dx.doi.org/10.1073/pnas.1605146113

Nozawa, N., Daikoku, T., Koshizuka, T., Yamauchi, Y., Yoshikawa, T., and Nishiyama, Y. (2003). Subcellular localization of herpes simplex virus type 1 UL51 protein and role of palmitoylation in Golgi apparatus targeting. J. Virol. 77, 3204-3216. https://dx.doi.org/10.1128/JVI.77.5.3204-3216.2003

Nozawa, N., Daikoku, T., Yamauchi, Y., Takakuwa, H., Goshima, F., Yoshikawa, T., and Nishiyama, Y. (2002). Identification and characterization of the UL7 gene product of herpes simplex virus type 2. Virus Genes 24, 257-266.

Nozawa, N., Kawaguchi, Y., Tanaka, M., Kato, A., Kato, A., Kimura, H., and Nishiyama, Y. (2005). Herpes simplex virus type 1 UL51 protein is involved in maturation and egress of virus particles. J. Virol. 79, 6947-6956. https://dx.doi.org/10.1128/JVI.79.11.6947-6956.2005

O'Carroll, I.P., Crist, R.M., Mirro, J., Harvin, D., Soheilian, F., Kamata, A., Nagashima, K., and Rein, A. (2012). Functional redundancy in HIV-1 viral particle assembly. J. Virol. 86, 12991-12996. https://dx.doi.org/10.1128/JVI.06287-11

Oda, S., Arii, J., Koyanagi, N., Kato, A., and Kawaguchi, Y. (2016). The Interaction between Herpes Simplex Virus 1 Tegument Proteins UL51 and UL14 and Its Role in Virion Morphogenesis. J. Virol. 90, 8754-8767. https://dx.doi.org/10.1128/ JVI.01258-16

Omar, O.S., Simmons, A.J., Andre, N.M., Wilson, D.W., and Gross, S.T. (2013). Pseudorabies virus and herpes simplex virus type 1 utilize different tegument-glycoprotein interactions to mediate the process of envelopment. Intervirology 56, 50-54. https://dx.doi.org/10.1159/000339467

Oshima, S., Daikoku, T., Shibata, S., Yamada, H., Goshima, F., and Nishiyama, Y. (1998). Characterization of the UL16 gene product of herpes simplex virus type 2. Arch. Virol. 143, 863-880.

Owen, D.J., Crump, C.M., and Graham, S.C. (2015). Tegument Assembly and Secondary Envelopment of Alphaherpesviruses. Viruses 7, 5084-5114. https://dx.doi.org/10.3390/v7092861

Parsons, L.R., Tafuri, Y.R., Shreve, J.T., Bowen, C.D., Shipley, M.M., Enquist, L.W., and Szpara, M.L. (2015). Rapid genome assembly and comparison decode intrastrain variation in human alphaherpesviruses. MBio 6, 1043. https://dx.doi.org/10.1128/mBio.02213-14

Pawliczek, T., and Crump, C.M. (2009). Herpes simplex virus type 1 production requires a functional ESCRT-III complex but is independent of TSG101 and ALIX expression. J. Virol. 83, 11254-11264. https://dx.doi.org/10.1128/JVI.00574-09

Pegtel, D.M., and Gould, S.J. (2019). Exosomes. Annu. Rev. Biochem. 88, 487-514. https://dx.doi.org/10.1146/annurev-biochem-013118-111902

Pitts, J.D., Klabis, J., Richards, A.L., Smith, G.A., and Heldwein, E.E. (2014). Crystal structure of the herpesvirus inner tegument protein UL37 supports its essential role in control of viral trafficking. J. Virol. 88, 5462-5473. https://dx.doi.org/10.1128/JVI.00163-14

Pomeranz, L.E., and Blaho, J.A. (2000). Assembly of infectious Herpes simplex virus type 1 virions in the absence of full-length VP22. J. Virol. 74, 10041-10054.

Qi, L., Tsai, B., and Arvan, P. (2017). New Insights into the Physiological Role of Endoplasmic Reticulum-Associated Degradation. Trends Cell Biol. 27, 430-440. https://dx.doi.org/10.1016/j.tcb.2016.12.002

Radtke, K., Kieneke, D., Wolfstein, A., Michael, K., Steffen, W., Scholz, T., Karger, A., and Sodeik, B. (2010). Plus- and minus-end directed microtubule motors bind simultaneously to herpes simplex virus capsids using different inner tegument structures. PLoS Pathog. 6, e1000991. https://dx.doi.org/10.1371/journal.ppat.1000991

Ren, Y., Bell, S., Zenner, H.L., Lau, S.-Y.K., and Crump, C.M. (2012). Glycoprotein M is important for the efficient incorporation of glycoprotein H-L into herpes simplex virus type 1 particles. Journal of General Virology 93, 319-329. https://dx.doi.org/10.1099/vir.0.035444-0

Rémillard-Labrosse, G., Mihai, C., Duron, J., Guay, G., and Lippé, R. (2009). Protein kinase D-dependent trafficking of the large Herpes simplex virus type 1 capsids from the TGN to plasma membrane. Traffic 10, 1074-1083. https://dx.doi.org/10.1111/j.1600-0854.2009.00939.x

Richards, A.L., Sollars, P.J., Pitts, J.D., Stults, A.M., Heldwein, E.E., Pickard, G.E., and Smith, G.A. (2017). The pUL37 tegument protein guides alpha-herpesvirus retrograde axonal transport to promote neuroinvasion. PLoS Pathog. 13, e1006741. https://dx.doi.org/10.1371/journal.ppat.1006741

Rixon, F.J., Addison, C., and McLauchlan, J. (1992). Assembly of enveloped tegument structures (L particles) can occur independently of virion maturation in herpes simplex virus type 1-infected cells. J. Gen. Virol. 73, 277-284.

Roberts, K.L., and Baines, J.D. (2010). Myosin Va enhances secretion of herpes simplex virus 1 virions and cell surface expression of viral glycoproteins. J. Virol. 84, 9889-9896. https://dx.doi.org/10.1128/JVI.00732-10

Roller, R.J., and Fetters, R. (2015). The herpes simplex virus 1 UL51 protein interacts with the UL7 protein and plays a role in its recruitment into the virion. J. Virol. 89, 3112-3122. https://dx.doi.org/10.1128/JVI.02799-14

Roller, R.J., Haugo, A.C., Yang, K., and Baines, J.D. (2014). The herpes simplex virus 1 UL51 gene product has cell type-specific functions in cell-to-cell spread. J. Virol. 88, 4058-4068. https://dx.doi.org/10.1128/JVI.03707-13

Romero, M., Keyel, M., Shi, G., Bhattacharjee, P., Roth, R., Heuser, J.E., and Keyel, P.A. (2017). Intrinsic repair protects cells from pore-forming toxins by microvesicle shedding. Cell Death Differ. 24, 798-808. https://dx.doi.org/ 10.1038/ cdd.2017.11

Roussel, É., and Lippé, R. (2018). Cellular Protein Kinase D Modulators Play a Role during Multiple Steps of Herpes Simplex Virus 1 Egress. J. Virol. 92, 785. https://dx.doi.org/10.1128/JVI.01486-18

Ruggiano, A., Foresti, O., and Carvalho, P. (2014). Quality control: ER-associated degradation: protein quality control and beyond. The Journal of Cell Biology 204, 869-879. https://dx.doi.org/10.1083/jcb.201312042

Sadaoka, T., Serada, S., Kato, J., Hayashi, M., Gomi, Y., Naka, T., Yamanishi, K., and Mori, Y. (2014). Varicella-zoster virus ORF49 functions in the efficient production of progeny virus through its interaction with essential tegument protein ORF44. J. Virol. 88, 188-201. https://dx.doi.org/10.1128/JVI.02245-13

Schipke, J., Pohlmann, A., Diestel, R., Binz, A., Rudolph, K., Nagel, C.-H., Bauerfeind, R., and Sodeik, B. (2012). The C terminus of the large tegument protein pUL36 contains multiple capsid binding sites that function differently during assembly and cell entry of herpes simplex virus. J. Virol. 86, 3682-3700. https://dx.doi.org/10.1128/JVI.06432-11

Schlager, M.A., Kapitein, L.C., Grigoriev, I., Burzynski, G.M., Wulf, P.S., Keijzer, N., de Graaff, E., Fukuda, M., Shepherd, I.T., Akhmanova, A., et al. (2010). Pericentrosomal targeting of Rab6 secretory vesicles by Bicaudal-D-related protein 1 (BICDR-1) regulates neuritogenesis. Embo J. 29, 1637-1651. https://dx.doi.org/10.1038/emboj.2010.51

Schmid, M.F., Hecksel, C.W., Rochat, R.H., Bhella, D., Chiu, W., and Rixon, F.J. (2012). A tail-like assembly at the portal vertex in intact herpes simplex type-1 virions. PLoS Pathog. 8, e1002961. https://dx.doi.org/10.1371/ journal.ppat. 1002961

Scholtes, L.D., Yang, K., Li, L.X., and Baines, J.D. (2010). The capsid protein encoded by U(L)17 of herpes simplex virus 1 interacts with tegument protein VP13/14. J. Virol. 84, 7642-7650. https://dx.doi.org/10.1128/JVI.00277-10

Scholz, I., Still, A., Dhenub, T.C., Coday, K., Webb, M., and Barklis, E. (2008). Analysis of human immunodeficiency virus matrix domain replacements. Virology 371, 322-335. https://dx.doi.org/10.1016/j.virol.2007.10.010

Schonteich, E., Wilson, G.M., Burden, J., Hopkins, C.R., Anderson, K., Goldenring, J.R., and Prekeris, R. (2008). The Rip11/Rab11-FIP5 and kinesin II complex regulates endocytic protein recycling. Journal of Cell Science 121, 3824-3833. https://dx.doi.org/10.1242/jcs.032441

Scrima, N., Lepault, J., Boulard, Y., Pasdeloup, D., Bressanelli, S., and Roche, S. (2015). Insights into herpesvirus tegument organization from structural analyses of the 970 central residues of HSV-1 UL36 protein. J. Biol. Chem. 290, 8820-8833. https://dx.doi.org/10.1074/jbc.M114.612838

Serwa, R.A., Abaitua, F., Krause, E., Tate, E.W., and O'Hare, P. (2015). Systems Analysis of Protein Fatty Acylation in Herpes Simplex Virus-Infected Cells Using Chemical Proteomics. Chemistry and Biology 22, 1008-1017. https://dx.doi.org/ 10.1016/j.chembiol.2015.06.024

Severini, A., Tyler, S.D., Peters, G.A., Black, D., and Eberle, R. (2013). Genome sequence of a chimpanzee herpesvirus and its relation to other primate alphaherpesviruses. Arch. Virol. 158, 1825-1828. https://dx.doi.org/ 10.1007/ s00705-013-1666-y

Shanda, S.K., and Wilson, D.W. (2008). UL36p is required for efficient transport of membrane-associated herpes simplex virus type 1 along microtubules. J. Virol. 82, 7388-7394. https://dx.doi.org/10.1128/JVI.00225-08

Shen, B., Wu, N., Yang, J.-M., and Gould, S.J. (2011). Protein targeting to exosomes/microvesicles by plasma membrane anchors. J. Biol. Chem. 286, 14383-14395. https://dx.doi.org/10.1074/jbc.M110.208660

Short, B., Preisinger, C., Schaletzky, J., Kopajtich, R., and Barr, F.A. (2002). The Rab6 GTPase regulates recruitment of the dynactin complex to Golgi membranes. Curr. Biol. 12, 1792-1795. https://dx.doi.org/10.1016/ s0960-9822(02)01221-6

Simon, G.C., and Prekeris, R. (2008). Mechanisms regulating targeting of recycling endosomes to the cleavage furrow during cytokinesis. Biochem. Soc. Trans. 36, 391-394. https://dx.doi.org/10.1042/BST0360391

Snyder, A., Polcicova, K., and Johnson, D.C. (2008). Herpes simplex virus gE/gI and US9 proteins promote transport of both capsids and virion glycoproteins in neuronal axons. J. Virol. 82, 10613-10624. https://dx.doi.org/10.1128/JVI.01241-08

Sodeik, B., Ebersold, M.W., and Helenius, A. (1997). Microtubule-mediated transport of incoming herpes simplex virus 1 capsids to the nucleus. The Journal of Cell Biology 136, 1007-1021. https://dx.doi.org/10.1083/jcb.136.5.1007

Starkey, J.L., Han, J., Chadha, P., Marsh, J.A., and Wills, J.W. (2014). Elucidation of the block to herpes simplex virus egress in the absence of tegument protein UL16 reveals a novel interaction with VP22. J. Virol. 88, 110-119. https://dx.doi.org/10.1128/JVI.02555-13

Stegen, C., Yakova, Y., Henaff, D., Nadjar, J., Duron, J., and Lippé, R. (2013). Analysis of virion-incorporated host proteins required for herpes simplex virus type 1 infection through a RNA interference screen. PLoS ONE 8, e53276. https://dx.doi.org/10.1371/journal.pone.0053276

Stucchi, R., Plucińska, G., Hummel, J.J.A., Zahavi, E.E., Guerra San Juan, I., Klykov, O., Scheltema, R.A., Altelaar, A.F.M., and Hoogenraad, C.C. (2018). Regulation of KIF1A-Driven Dense Core Vesicle Transport: Ca2+/CaM Controls DCV Binding and Liprin-α/TANC2 Recruits DCVs to Postsynaptic Sites. Cell Rep 24, 685-700. https://dx.doi.org/10.1016/j.celrep.2018.06.071

Sun, M., Hou, L., Tang, Y.-D., Liu, Y., Wang, S., Wang, J., Shen, N., An, T., Tian, Z., and Cai, X. (2017). Pseudorabies virus infection inhibits autophagy in permissive cells in vitro. Sci Rep 7, 39964. https://dx.doi.org/10.1038/srep39964

Svobodova, S., Bell, S., and Crump, C.M. (2012). Analysis of the interaction between the essential herpes simplex virus 1 tegument proteins VP16 and VP1/2. J. Virol. 86, 473-483. https://dx.doi.org/10.1128/JVI.05981-11

Szilágyi, J.F., and Cunningham, C. (1991). Identification and characterization of a novel non-infectious herpes simplex virus-related particle. J. Gen. Virol. 72, 661-668.

Takahashi, M.-N., Jackson, W., Laird, D.T., Culp, T.D., Grose, C., Haynes, J.I., and Benetti, L. (2009). Varicella-zoster virus infection induces autophagy in both cultured cells and human skin vesicles. J. Virol. 83, 5466-5476. https://dx.doi.org/ 10.1128/JVI.02670-08

Tallóczy, Z., Virgin, H.W., and Levine, B. (2006). PKR-dependent autophagic degradation of herpes simplex virus type 1. Autophagy 2, 24-29. https://dx.doi.org/10.4161/auto.2176

Tandon, R., and Mocarski, E.S. (2012). Viral and host control of cytomegalovirus maturation. Trends Microbiol. 20, 392-401. https://dx.doi.org/ 10.1016/ j.tim.2012.04.008

Udenwobele, D.I., Su, R.-C., Good, S.V., Ball, T.B., Varma Shrivastav, S., and Shrivastav, A. (2017). Myristoylation: An Important Protein Modification in the Immune Response. Front Immunol 8, 751. https://dx.doi.org/10.3389/ fimmu. 2017.00751

Urano, E., Aoki, T., Futahashi, Y., Murakami, T., Morikawa, Y., Yamamoto, N., and Komano, J. (2008). Substitution of the myristoylation signal of human immunodeficiency virus type 1 Pr55Gag with the phospholipase C-delta1 pleckstrin homology domain results in infectious pseudovirion production. J. Gen. Virol. 89, 3144-3149. https://dx.doi.org/10.1099/vir.0.2008/004820-0

Vittone, V., Diefenbach, E., Triffett, D., Douglas, M.W., Cunningham, A.L., and Diefenbach, R.J. (2005). Determination of interactions between tegument proteins of herpes simplex virus type 1. J. Virol. 79, 9566-9571. https://dx.doi.org/10.1128/JVI.79.15.9566-9571.2005

Votteler, J., and Sundquist, W.I. (2013). Virus budding and the ESCRT pathway. Cell Host Microbe 14, 232-241. https://dx.doi.org/10.1016/j.chom. 2013.08.012

Wang, J., Fedoseienko, A., Chen, B., Burstein, E., Jia, D., and Billadeau, D.D. (2018). Endosomal receptor trafficking: Retromer and beyond. Traffic 19, 578-590. https://dx.doi.org/10.1111/tra.12574

Weinheimer, S.P., Boyd, B.A., Durham, S.K., Resnick, J.L., and O'Boyle, D.R. (1992). Deletion of the VP16 open reading frame of herpes simplex virus type 1. J. Virol. 66, 258-269.

Williams, J.M., and Tsai, B. (2016). Intracellular trafficking of bacterial toxins. Curr. Opin. Cell Biol. 41, 51-56. https://dx.doi.org/10.1016/j.ceb.2016.03.019

Wolfstein, A., Nagel, C.-H., Radtke, K., Döhner, K., Allan, V.J., and Sodeik, B. (2006). The inner tegument promotes herpes simplex virus capsid motility along microtubules in vitro. Traffic 7, 227-237. https://dx.doi.org/10.1111/j.1600-0854.2005.00379.x

Xu, C., Wang, M., Song, Z., Wang, Z., Liu, Q., Jiang, P., Bai, J., Li, Y., and Wang, X. (2018). Pseudorabies virus induces autophagy to enhance viral replication in mouse neuro-2a cells in vitro. Virus Res. 248, 44-52. https://dx.doi.org/10.1016/j.virusres.2018.02.004

Yamada, H., Daikoku, T., Yamashita, Y., Jiang, Y.M., Tsurumi, T., and Nishiyama, Y. (1997). The product of the US10 gene of herpes simplex virus type 1 is a capsid/tegument-associated phosphoprotein which copurifies with the nuclear matrix. J. Gen. Virol. 78, 2923-2931. https://dx.doi.org/10.1099/0022-1317-78-11-2923

Yang, J.-M., and Gould, S.J. (2013). The cis-acting signals that target proteins to exosomes and microvesicles. Biochem. Soc. Trans. 41, 277-282. https://dx.doi.org/10.1042/BST20120275

Yeh, P.-C., Han, J., Chadha, P., Meckes, D.G., Ward, M.D., Semmes, O.J., and Wills, J.W. (2011). Direct and specific binding of the UL16 tegument protein of herpes simplex virus to the cytoplasmic tail of glycoprotein E. J. Virol. 85, 9425-9436. https://dx.doi.org/10.1128/JVI.05178-11

Yeh, P.-C., Meckes, D.G., and Wills, J.W. (2008). Analysis of the interaction between the UL11 and UL16 tegument proteins of herpes simplex virus. J. Virol. 82, 10693-10700. https://dx.doi.org/10.1128/JVI.01230-08

Young, J., Stauber, T., Del Nery, E., Vernos, I., Pepperkok, R., and Nilsson, T. (2005). Regulation of microtubule-dependent recycling at the trans-Golgi network by Rab6A and Rab6A'. Mol. Biol. Cell 16, 162-177. https://dx.doi.org/10.1091/ mbc.e04-03-0260

Zaichick, S.V., Bohannon, K.P., Hughes, A., Sollars, P.J., Pickard, G.E., and Smith, G.A. (2013). The Herpesvirus VP1/2 Protein Is an Effector of Dynein-Mediated Capsid Transport and Neuroinvasion. Cell Host Microbe 13, 193-203. https://dx.doi.org/10.1016/j.chom.2013.01.009

Zhang, Y., and McKnight, J.L. (1993). Herpes simplex virus type 1 UL46 and UL47 deletion mutants lack VP11 and VP12 or VP13 and VP14, respectively, and exhibit altered viral thymidine kinase expression. J. Virol. 67, 1482-1492.

Zhang, Y., Sirko, D.A., and McKnight, J.L. (1991). Role of herpes simplex virus type 1 UL46 and UL47 in alpha TIF-mediated transcriptional induction: characterization of three viral deletion mutants. J. Virol. 65, 829-841.

Chapter 10

Alphaherpesvirus Vaccines

Clare Burn Aschner[1] and Betsy C. Herold[1,2*]

[1]Department of Microbiology-Immunology and [2]Pediatrics, Albert Einstein College of Medicine, Bronx, NY, USA

*betsy.herold@einsteinmed.org

DOI: https://doi.org/10.21775/9781913652555.10

Abstract

Prophylactic and therapeutic vaccines for the alphaherpesviruses including varicella zoster virus (VZV) and herpes simplex virus types 1 and 2 have been the focus of enormous preclinical and clinical research. A live viral vaccine for prevention of chickenpox and a subunit therapeutic vaccine to prevent zoster are highly successful. In contrast, progress towards the development of effective prophylactic or therapeutic vaccines against HSV-1 and HSV-2 has met with limited success. This chapter provides an overview of the successes and failures, the different types of immune responses elicited by various vaccine modalities, and the need to reconsider the preclinical models and immune correlates of protection against HSV.

Background

The *Alphaherpesvirinae* subfamily of the *Herpesviridae* are large, enveloped double-stranded DNA viruses that include several important human and animal pathogens. The majority of the world's population is infected with at least one or more of the three major human *Alphaherpesvirinae*: herpes simplex virus type 1 (HSV-1), HSV type 2 (HSV-2) and human herpesvirus 3, more commonly known as varicella zoster virus (VZV). This subfamily is differentiated from the beta and gammaherpesviruses by genetic and biological properties including a relatively short reproductive cycle, lytic infection in cell culture, and ability to universally establish latency primarily in sensory nerve ganglia (see Chapter 6). Viral latency is characterized by episodes of subclinical or, less commonly, clinical reactivation, which further fuel transmission and contribute to disease burden.

HSV-1 infects over 3.7 billion people worldwide and is the primary cause of oral HSV disease, as well as a leading cause of infectious corneal blindness and fatal sporadic encephalitis (Roberts et al. 2003). HSV-1 has also emerged as a leading cause of primary genital herpes in the United States and other developed nations (Roberts et al. 2003). HSV-2, which infects approximately 400 million people worldwide, primarily causes genital disease and is associated with an increased risk of HIV acquisition and transmission (Looker et al. 2015; Looker et al. 2017). There is significant geographic overlap when comparing the prevalence of HIV and HSV-2, and mathematical modeling indicates that a prophylactic HSV-2 vaccine would have a major impact on the HIV epidemic (Looker et al. 2017). HSV-2 is also associated with recurrent aseptic meningitis, known as Mollaret's syndrome (Whitley 2006). Perinatal transmission of both HSV-1 and HSV-2, the incidence of which is estimated to be approximately 1 in 3200 live births in the United States, is associated with substantial morbidity even with appropriate antiviral therapy (James and Kimberlin 2015).

VZV causes the common, but usually self-limited primary childhood exanthemous illness, chickenpox. Clinical reactivation of VZV results in shingles or zoster, is typically restricted to one or a few dermatomes, and is more common in older or immunocompromised individuals (Ljungman et al. 1986). Less commonly, VZV can cause ocular disease including keratitis, uveitis or zoster opthalmicus. The latter is linked to VZV reactivation within the trigeminal ganglion and may lead to blindness. VZV may also cause a vasculopathy, which is associated with stroke and other vascular complications. Viral genetic studies have identified five VZV clades; clades 1, 3 and 5 are of European origin, clade 2 dominates in Asia, and clade 4 in Africa (Schmidt-Chanasit and Sauerbrei 2011). The enormous global health and economic burden of alphaherpesvirus infections has generated extensive vaccine development efforts over the past decades. However, while there has been great success in developing vaccines to prevent varicella and to reduce the risk of zoster, there is to date no effective vaccine for HSV-1 or HSV-2.

Varicella zoster vaccines

Live attenuated varicella vaccine

A live attenuated varicella vaccine (LAVV) was initially developed in Japan in 1974 with the goal of reducing disease burden in children with leukemia. The vaccine was attenuated by serially passaging a varicella strain isolated from a child whose last name was Oka (Oka strain) in guinea pig embryo cultures and then subsequently propagating the virus in human diploid cells (Takahashi et al. 1974). The vaccine strain (vOka) is attenuated primarily for replication in the skin but less so in T cells or trigeminal ganglia (Moffat et al. 1998). It also establishes latency, albeit less efficiently than primary infection (Moffat et al. 1998). Zoster attributable to reactivation of vOka occurs much less frequently (~7 fold less in children with leukemia) than reactivation of wild-type virus (Krause and Klinman 2000).

Although there were initial concerns about the safety of a live, attenuated viral vaccine, the extensive data in children with leukemia in remission (Hattori et al. 1976; Brunell et al. 1982) and subsequent studies in healthy children (Asano and Takahashi 1977; Weibel et al. 1985; White et al. 1991; White 1996) led to licensure of LAVV as a single-dose, universal vaccination of all healthy children 1-13 years of age and healthy adults in the United States in 1995 (Gershon 1980; White et al. 1991). However, numerous varicella outbreaks were reported among immunized school-aged children in the 1990s, which, combined with documentation of lower than initially anticipated seroconversion rates, led to the adoption in 2006 of a routine two-dose schedule with initial vaccination between 12-15 months and a booster at 4-6 years (Watson et al. 1995; Arvin and Gershon 2006). This revised schedule has been associated with a greater than 95% decline in primary varicella disease in healthy children and has provided herd immunity for immunocompromised children for whom a live viral vaccine is not indicated (Shapiro et al. 2011). There are two licensed vaccine formulations: VARIVAX®, the single-antigen varicella vaccine, and ProQuad®, a combination measles, mumps, rubella, and varicella (MMRV) vaccine (Merck and Co).

VOka was sequenced years after initial licensure and is comprised of a mixture of VZV Oka genotypes and differs from wild-type strains by at least 224 single nucleotide polymorphisms (SNPs) (Argaw et al. 2000; Faga et al. 2001; Quinlivan and Breuer 2006). Precisely which ones are responsible for the attenuating phenotype is unknown, although studies indicate that mutations in ORF62, a critical viral transcription factor and homolog of HSV-1 ICP4, are important for attenuation (Kwak et al. 2017). The sequencing data allows for the differentiation of the vaccine strain from the wild-type virus circulating in the United States, since the majority of clade 2 VZV viruses have Bgl1 and Pst1 restriction sites as well as several SNPs that differentiate them from vOka (Quinlivan et al. 2012; Peters et al. 2012). This genetic information is used when differentiating vaccine-associated disease from breakthrough varicella caused by infections with the wild-type virus.

Correlates of immune protection

The precise correlates of immune protection against natural disease elicited by the VZV vaccine are not fully defined. Both cell-mediated and humoral immune responses are important, as evidenced by the observations that patients with impaired T cell function are at risk for severe disease and that administration of high dose anti-VZV immune serum (zoster immune globulin) may prevent or modify disease manifestations (Arvin 2008). The gold standard for measuring antibody responses is the fluorescent antibody to membrane antigen assay (FAMA) against live VZV-infected cells. However, difficulties with assay reproducibility and automation have led to the development of alternatives including a glycoprotein (gp) based ELISA, which uses a lectin affinity purified preparation of viral glycoproteins from VZV-infected cells. FAMA VZV antibody titers ≥ 1:4 or gpELISA titers ≥ 5 gpELISA units/ml correlate with protection (Gershon et al. 1994). Limited studies demonstrate that natural infection and vaccination induce both neutralizing and antibody-dependent cell-mediated cytotoxicity (ADCC) immune responses, and FAMA titers correlate with both (Kamiya et al. 1982). Notably, in one study, ADCC responses were detected earlier than neutralizing responses, suggesting that ADCC may be involved in the early stages of recovery from VZV disease (Ihara et al. 1994). Neutralizing antibodies target several viral glycoproteins, including gE, gB, and the gH/gL heterodimer; the targets of ADCC responses have not been delineated.

Zoster vaccines

The success of vaccination with the vOka strain against primary varicella suggested that the live, attenuated vaccine could also be used to boost immunity in older adults to prevent zoster. The rationale for this stemmed from clinical observations by Hope-Simpson, which suggested that exposure to exogenous (chickenpox in the community) and/or endogenous (reactivation) VZV stimulated memory T cell responses (Hope-Simpson 1965; Guzzetta et al. 2013). This notion was supported by the recognition that second episodes of zoster were uncommon. Building on these observations, a double-blind, placebo-controlled trial in heathy adults ≥ 60 years of age was conducted to compare a single dose of high-potency (14x greater than primary LAVV dose) vOka compared to placebo (Oxman et al. 2008). This dose was selected based on cell-mediated immune responses. There were 38,546 healthy adults enrolled in the study who were randomized 1:1 to receive vaccine or placebo. There were 315 confirmed cases of zoster in vaccine recipients compared to 642 cases in placebo recipients; none of the zoster isolates were the vaccine strain. Based on these results, this high dose LAVV, marketed as Zostavax (Merck and Co) was licensed by the FDA in 2006 and recommended by the CDC for routine immunization of healthy adults aged > 50 for the prevention of zoster and post-herpetic neuralgia.

More recently, Shingrix, a liposome-based subunit vaccine containing recombinant glycoprotein E (gE), combined with adjuvants to stimulate strong

CD4+ T cell responses (AS01B, the Toll-like receptor agonist monophosphoryl lipid A combined with saponin QS21), was developed by GlaxoSmithKline and evaluated for prevention of zoster. gE is the most abundant glycoprotein on VZV-infected cells and is a target for neutralizing antibody responses. Two large Phase III placebo-controlled trials in healthy adults > 50 years or > 70 years using a 2-dose schedule of Shingrix demonstrated 97% protection against zoster compared to the 51% efficacy observed in response to Zostavax. Shingrix was approved by the FDA at the end of 2017 (Lal et al. 2015; Cunningham et al. 2016).

The specific correlates of immune protection for these vaccines are not yet fully defined. Both Zostavax and Shingrix boost cell-mediated immune responses but differ with respect to antigenic targets (Sei et al. 2015). T cell responses to Zostavax target IE63, IE62, gB and ORF9 more frequently than gE, whereas Shingrix, as expected, elicits gE-specific responses (Weinberg, Kroehl, et al. 2018). However, strong gE-specific CD4+ T cell responses are observed only after the second dose of Shingrix, suggesting that the majority of this response may be dependent on stimulation of otherwise naïve cells (Weinberg, Kroehl, et al. 2018; Weinberg, Popmihajlov, et al. 2018). In addition to the T cell responses, a 4-fold or greater rise in gE-specific antibody responses was observed in over 97% of Shingrix recipients.

In addition to greater immunogenicity and efficacy, an important advantage of Shingrix is that, as a subunit vaccine, it may be safely administered to immunocompromised hosts. Studies are ongoing in adult solid-organ transplant recipients (Stadtmauer et al. 2014; Winston et al. 2018) and in HIV-infected patients (Berkowitz et al. 2014). Additional questions that require future studies include the durability of the response to Shringrix, and whether patients who received Zostavax should also receive a booster dose of Shingrix. Whether there is any role for this subunit vaccine in preventing or boosting immunity to prevent primary chickenpox will also require future investigations.

Herpes simplex virus vaccines

Obstacles to HSV prevention

In contrast to the successes with VZV, an effective prophylactic or therapeutic vaccine for HSV has remained elusive. A live viral vaccine for varicella was developed in response to a specific population, children with leukemia, in whom the morbidity and mortality drove vaccine development. Only after the vaccine was shown to be safe and effective in this high-risk group, was it evaluated and approved for healthy children, providing herd immunity and thus further protection for high-risk populations. Although HSV also causes devastating disease in the subset of patients who develop encephalitis, overwhelming hepatic failure, dissemination to other organs, and recurrent corneal disease with blindness, the drive for a vaccine has been driven more by the morbidity and social stigma associated with HSV as a sexually transmitted infection and its link to HIV. Until very recently, the majority of vaccine efforts have focused on subunit protein strategies and neutralizing antibody responses, and this focus may have contributed to the lack of success.

One of the difficulties in designing a vaccine for any herpesvirus is the ability of all of these viruses (except HHV-8) to uniformly establish latency. However, subclinical and clinical reactivations are more common with HSV than VZV, reflecting a greater ability for the immune system to prevent or control latent VZV (Kronenberg et al. 2005; Papaevangelou et al. 2013). The term "leaky latency" has been proposed to better describe latent HSV (Wald et al. 1995; Wald et al. 2000; Mark et al. 2008). The natural immune response to HSV limits the number and severity of clinical recurrences and provides partial protection against heterologous (opposite serotype) challenge, but does not prevent viral reactivation. Thus, effective vaccines must elicit a more potent and/or different type of immune response compared to natural infection if the goals are to prevent the establishment of latency (prophylaxis) and/or provide greater protection against viral reactivation and transmission (therapeutic). Whether it is possible to safely eliminate the latent viral reservoir in seropositive individuals, a goal for HIV cure, is unknown. In addition, because of the high rate of HSV-1 seropositivity at a very young age, a vaccine ideally should be immunogenic in the setting of preexisting HSV-1 immunity (Roberts et al. 2003).

HSV-1 and HSV-2 have evolved a myriad of immune evasion strategies, many of which are unique to HSV and need to be considered in vaccine development. The ability of the virus to evade innate, humoral, and T cell immunity may contribute to the greater difficulty in developing an effective HSV vaccine. Examples of these immune evasion strategies are summarized below. Both HSV-1 and HSV-2 encode several proteins (e.g. ICP0 and ICP34.5) that inhibit interferon production and/or subsequent signal transduction to evade this major innate immune pathway (Chew et al. 2009). HSV also evades antibody responses through multiple mechanisms. First, the virus spreads directly from infected to neighboring uninfected cells across bridges, thus evading neutralizing antibodies (Sattentau 2008). Second, HSV glycoproteins E and I (gE/gI) act as an Fc receptor (FcR) analog, thereby shielding antibodies from engaging FcRs, which are required for non-neutralizing antibody dependent cell killing (Dubin et al. 1991; McNearney 1987). gE-gI may also promote clearance of viral antigen-Ab complexes and immune evasion through a process termed antibody bipolar bridging. The Fab fragment of IgG binds viral protein (e.g. gD) and the Fc fragment binds gE-gI, leading to endocytosis of the complex (Ndjamen et al. 2014). Third, HSV glycoprotein C (gC) interferes with the C3 complement cascade inhibiting complement-mediated lysis of virus and infected cells (Gerber et al. 1995; Friedman et al. 1984; Fries et al. 1986). Another potential immune evasion strategy is the binding of gD to herpesvirus entry mediator (HVEM) on immune cells, which competes with binding of its natural ligands (Spear 2004). HVEM binding to LIGHT (TNFSF14) (Mauri et al. 1998; Kwon et al. 1997) provides a stimulatory signal (Harrop, Reddy, et al. 1998; Harrop, McDonnell, et al. 1998; Tamada et al. 2000; Shi 2002), whereas binding to B and T lymphocyte attenuator (BTLA) (Šedý et al. 2004; Cheung et al. 2009) provides an inhibitory signal to T cells and gD may interfere with these responses to modulate immune responses (Sarrias et al. 2000; Carfi et al. 2001; Connolly et al. 2002; Carfi et al. 2002;

Connolly et al. 2003; Compaan et al. 2005; Stiles et al. 2010; T. L. Murphy and K. M. Murphy 2010). Moreover, HSV-2 gD induces the degradation of CD112/ nectin-2, a ligand for the NK-activating receptor DNAX accessory molecule 1 (DNAM-1). This impairs binding of DNAM-1 to the cell surface, thereby suppressing NK-mediated killing of HSV-2 infected cells (Grauwet et al. 2014). Finally, HSV also interferes with T cell responses. HSV ICP47 binds the human transporter associated with antigen presentation (TAP) and blocks the binding of antigenic peptides thus interfering with the major histocompatibility complex (MHC) class I antigen presentation pathway (Hill et al. 1995; Früh et al. 1995; Tomazin et al. 1996; Ahn et al. 1996).

VZV has also developed several immune evasion strategies, but these are more limited. For example, VZV interferes with interferon α and β production by epithelial cells by sequestering NF-κB proteins in the cytoplasm, but neighboring uninfected cells overcome this evasion strategy (Jones and Arvin 2006). The strongest immune evasion by VZV is linked to the T cell response. VZV down regulates MHC class I (Abendroth et al. 2001) and inhibits the up-regulation of interferon-γ-induced MHC class II expression (Abendroth et al. 2000). However, shielding of antibodies through FcR mimicry and interference with complement pathway have not been described for VZV.

HSV prophylactic subunit vaccine clinical trials

Two subunit vaccines progressed through Phase 3 clinical efficacy trials as prophylactic vaccines with limited success (Table 1). Both were predicated on the notion that high titer neutralizing Abs targeting gD and/or gB, which are the primary targets of neutralizing antibody responses during natural infection (in contrast to gE for VZV) would prove protective and were supported by studies in murine and guinea pig models. However, neither the recombinant gB/gD/ MF59 vaccine, which included a squalene-based adjuvant [Chiron], nor the recombinant gD-2/AS04 (*Simplirix™*) vaccine, which used an alum-monophosphoryl lipid A-based adjuvant [GSK], achieved their endpoints.

Table 1. Summary of clinical trial results with HSV vaccine candidates

Name	Platform	Antigens	Immune Response	Clinical Results
gB/gD/MF59 (Chiron)	Subunit MF59 adjuvant	gD and gB	Neutralizing Abs	Phase 3 Prophylaxis 9% (95% CI: -29%, 36%)
gD/AS04 (GSK)	Subunit AS04 adjuvant	gD	Neutralizing Abs CD4+ T cells	Phase 3 Prophylaxis 20% (95% CI: -29%, 50%)
gH-null (Cantab Pharmaceuticals)	Single-cycle live	Multiple	Neutralizing Abs	Phase 3 therapeutic No difference in recurrences or shedding
HSV529 (Sanofi-Pasteur)	Replication-defective	Multiple	Neutralizing Abs CD4, CD8 T cells	Phase 1 Safe but immunogenic in doubly-seronegatives only
Gen-003 (Genocea)	Subunit Saponin adjuvant	gD, ICP4	Neutralizing Abs Cytolytic T cells	Phase 2 therapeutic Dose variable decrease shedding
HerpV (Agenus)	Heat shock protein 70-HSV peptides QS-21 adjuvant	32 HSV peptides (gD, additional envelope, tegument, and other)	Cytotoxic T cells	Phase 2 therapeutic 17% reduction in shedding frequency
Vaxfectin (Vical)	DNA Vaxfectin adjuvant	gD, UL46/47	Neutralizing Abs	Phase 1/2 therapeutic Failed to reduce shedding
COR-1 (Admedus)	Codon optimized DNA	Codon optimized gD2 and ubiquitin-fused truncated gD2 to target the antigen to the proteasome	gD-specific Abs Cytotoxic T cells	Phase 2 therapeutic No difference in recurrences
ΔNLS (Rational Vaccines)	Deleted in ICP0 Replication competent Attenuated for latency	Multiple targets	Neutralizing Abs gD-specific Abs	Phase 1 non-FDA approved therapeutic study Self-reported reduction in recurrences

The gB/gD/MF59 vaccine was evaluated in two randomized, double-blind, placebo-controlled multicenter trials. One trial enrolled 531 HSV-2-seronegative partners of HSV-2-seropositive people, and the other enrolled 1862 persons attending sexually transmitted disease clinics. The vaccine induced high levels of neutralizing antibodies, but the overall vaccine efficacy was 9% (95% confidence interval, 29% to 36%) (Corey et al. 1999). The authors concluded that prevention will likely require more than high titer neutralizing antibodies. In subsequent studies, they found that the vaccine did not elicit ADCC antibodies and hypothesized that low ADCC responses may be associated with poor vaccine efficacy (Kohl et al. 2000).

Others speculated that the reason for the vaccine failure might be attributable to the need for higher neutralizing antibody titers and/or more potent T cell responses and focused their efforts on combining gD with different adjuvants. Thus, the gD-2/AS04 vaccine, composed of gD-2 with monophosphoryl lipid A (MPL) and aluminum hydroxide (Alum), was evaluated in three different Phase 3 clinical trials. The first two studies were conducted in serodiscordant couples. In one, 847 doubly seronegative participants (268 women) were enrolled and in the other, 1867 HSV-1 seropositive (710 women) were enrolled. Combining the data from both studies showed that that the vaccine was ~70% protective in doubly-seronegative women, but was not protective in HSV-1 seropositive women or in men regardless of their serostatus (Stanberry et al. 2002). The reasons for this sex discrepancy in protection in the HSV-1 seronegative participants were not understood.

The subpopulation results supported a third clinical field trial, which enrolled 8323 doubly-seronegative women 18 to 30 years of age with a primary endpoint of genital herpes disease caused by either serotype. The overall vaccine efficacy was only 20% (95% confidence interval [CI], 29% to 50%) against genital herpes disease; however, modest efficacy against HSV-1 genital disease (58%; 95% CI, 12% to 80%) was observed (Belshe et al. 2012). In a subsequent analysis of sera from 30 participants, the neutralizing antibody titers against HSV-1 were 3.5-fold higher than the titers against HSV-2, possibly reflecting serotype differences in the ability of the immune evading glycoproteins, gE and gC, to interfere with the neutralization activity of the gD-specific neutralizing antibodies (Awasthi, Belshe, et al. 2014). Overall, these disappointing clinical trial outcomes with adjuvanted gD subunit vaccines raised questions about the reliance on neutralizing antibodies as a sole correlate of immune protection and challenged the traditional preclinical models that were conducted with intravaginal challenges in female mice or guinea pigs with laboratory-adapted HSV-2 isolates at doses that typically do not exceed an LD_{90} (dose that causes lethal disease in 90% of control animals).

"Next generation" prophylactic subunit vaccine candidates

Building on the observation that HSV gE and gC are immune evasion molecules that might shield against neutralizing antibodies, Friedman and colleagues developed a trivalent subunit vaccine (THV) comprised of gD, gC and gE (Table 2) (Awasthi, Huang, et al. 2014). Preclinical data using the standard murine or guinea pig models suggest that the trivalent vaccine is more potent than recombinant gD alone. The THV protected 100% of mice following intravaginal challenge with a 3 x LD_{50} dose of the laboratory-adapted strain of HSV-2, MS, and 97% of mice were protected from the establishment of latent infection in the dorsal root ganglia (DRG). In a female guinea pig vaginal challenge model, the THV and recombinant gD-2 vaccines were both highly effective in preventing genital lesions (Awasthi, Hook, Shaw, Pahar, et

Table 2. HSV-2 vaccine candidates in preclinical development

Candidate	Platform	Antigenic Targets	Immune Response	Results in animal models
THV	Trivalent subunit	gD, gC, gE	Neutralizing Abs	Prophylactic in mice and guinea pigs Decreased shedding in guinea pigs
ΔgD-2	gD null Single cycle	Multiple targets excluding gD	FcγR-activating Abs CD4, CD8 responses Passive protection	100% protection against HSV-1 and HSV-2 clinical isolates in male and female mice with multiple challenge routes; Prevents establishment of latency Passive protection
gD27	gD-nectin binding impaired; Live attenuated	Multiple	Neutralizing Abs	Improved protection compared to gD-2 subunit
VC2	Deleted in gK and UL20 Defective in establishing latency; replication competent	Multiple	Neutralizing Abs T cell memory (gB, gD) T_{FH} and Th17 responses	100% protection in mice intravaginally challenged with HSV-1(McKrae) or HSV-2(G) Prevents latency

al. 2017). Shedding of replication-competent virus (i.e. virus capable of infecting cells in a co-culture model) occurred on fewer days in the THV compared to gD-2/AS04 immunized animals, although there was no difference between the two vaccines in HSV DNA shedding, and the gD-specific serum antibody and neutralization titers with and without complement were similar (Awasthi, Hook, Shaw and Friedman 2017). Challenges with higher doses, clinical isolates, or in male animals (skin challenge) have not yet been reported. More recently, the group compared the THV vaccine, a protein-based vaccine, to a nucleoside-modified mRNA vaccine containing sequences for the same 3 viral proteins and formulated in lipid nanoparticles. The trivalent mRNA vaccine was superior to the THV protein vaccine and elicited higher titer ELISA IgG antibodies, neutralizing gD-specific antibodies and $CD4^+$ T cell responses and was associated with decreased viral shedding following subclinical infections in mice and guinea pigs (Awasthi et al. 2019).

Live attenuated or single-cycle HSV prophylactic vaccine candidates

In contrast to subunit vaccines, live virus vaccines are likely to elicit polyfunctional immune responses targeting a greater breadth of antigens. In addition to the success of the live attenuated varicella vaccine, live virus vaccines have also had success against animal alphaherpesviruses. For example, a herpes turkey virus (HVT) vaccine has been used for years to immunize against the related Marek's Disease Virus (MDV) that causes an economically important neoplastic disease of chickens (Okazaki et al. 1970). Unfortunately, MDV has evolved to overcome the protection from the HVT vaccine, which has resulted in the need to modify the HVT strain to address the new outbreaks (Schat and Baranowski 2007). Live attenuated virus vaccines carrying deletions in specific viral proteins have also been used to protect against pseudorabies virus (PrV), the causative agent of Aujeszky's disease, another agriculturally and economically important infection that results in losses of swine populations (Bartha 1961). As with the vaccine for Marek's Disease, the PrV vaccines have to be modified to address changes in the circulating strains of PrV (Freuling et al. 2017).

Candidates that have been evaluated in clinical trials

The expansion of molecular technologies has facilitated the engineering of live virus vaccine candidates with specific deletions in gene(s) of known function. This defined approach has replaced the strategy of attenuation by serial passage in culture, which was used to develop the varicella vaccine. The first such HSV vaccine candidate to be evaluated in clinical trials lacked the gene for glycoprotein H (gH) (Table 1) (McLean et al. 1994). gH (along with gD, gB, and gL) is required for viral entry and cell-to-cell spread (see Chapter 2). Thus, the gH-null virus, propagated in gH-expressing complementing cells, is restricted to a single infectious cycle. The vaccine protected female guinea pigs from primary genital HSV-2, reduced the number of viral recurrences in the model, and elicited high serum neutralizing antibody responses (Boursnell et al. 1997). Based on these preclinical studies, this candidate was advanced into the clinic and evaluated as a therapeutic vaccine in volunteers with six or more annual genital herpes recurrences. The vaccine was safe, but there were no significant differences in the mean number of recurrences, time to healing, or days with viral shedding relative to placebo (de Bruyn et al. 2006). Specific immune responses to the vaccine, be they boosting of total, neutralizing, or other functional HSV-specific antibody or T cell responses, were not reported in the clinical trial. However, the lack of protection suggested that the preclinical models were not predictive of human clinical outcomes. The vaccine was not further pursued, and was never tested as a prophylactic.

Subsequently, several different replication-defective or single-cycle viruses have been engineered as candidate prophylactic vaccines. The first of these to enter clinical trial was an HSV-2 virus lacking the genes for UL5 (helicase-primase) and UL29 (DNA binding protein) and designated dl5-29 or HSV529 (Da Costa et al. 2000; Da Costa et al. 2001). The virus was grown on complementing V295 cells, which stably express the HSV-2 UL5 and UL29 genes. The salient findings in the preclinical studies were that the vaccine was safe, generated high titer neutralizing antibody and CD4 and CD8 T cell responses, and protected against acute or recurrent genital HSV-2 disease in naïve female mice as well as naïve or HSV-1 seropositive female guinea pigs using vaginal challenge models (Hoshino et al. 2008). It reduced, but did not completely prevent the establishment of latency (Hoshino et al. 2004; Hoshino et al. 2009; Bernard et al. 2015). Notably, efficacy was reduced if mice were challenged with a clinical African HSV-2 isolate, SD90, suggesting that there may be differences in protective epitopes among clinical strains and highlighting the need for preclinical testing against a wide array of isolates (Dudek et al. 2011). HSV529 was recently evaluated in a Phase 1 study (Table 1). The vaccine was safe and elicited neutralizing and gD-specific antibody responses in doubly seronegative participants. However, in contrast to the studies in guinea pigs, the vaccine did not boost neutralizing antibody responses in HSV-1 or HSV-2 seropositive participants (Dropulic et al. 2017; Dropulic et al. 2019). Efforts to increase the immunogenicity of this vaccine are currently being evaluated. One such approach is to replace HSV-2 UL41, which encodes for the virion host shutoff (vhs) function, with the heterologous gene from HSV-1. The rationale for this intertypic recombination is that vhs-1 is 40-fold less active than vhs-2, and thus, that HSV-2 encoding vhs-1 might elicit a stronger immune response (Reszka et al. 2010).

Candidates in preclinical development

A completely different candidate live single-cycle vaccine in preclinical development is an HSV-2 virus lacking the gD gene (ΔgD-2) (Table 2). This candidate vaccine, which is grown on HSV-1 gD-expressing complementing cells, generates high titer antibodies and T cell responses that recognize an array of HSV antigens exclusive of gD. The preclinical model testing was expanded to include challenges with multiple clinical isolates of both HSV-1 and HSV-2, male and female mice, and several different routes of infection including intravaginal, skin, intraocular and intranasal (Burn et al. 2017; Ramsey et al. 2020; Burn Aschner, Knipe, et al. 2020). Most other vaccine candidates were tested using laboratory-adapted isolates of HSV-2 in vaginally challenged female animals. Results of the studies with ΔgD-2 demonstrated complete protection against disease and latency following challenge with all clinical isolates tested to date, including the African clinical isolate SD90 (Petro et al. 2015; Petro et al. 2016; Burn et al. 2017). No latent virus was detected in dorsal root ganglia in vaccinated mice by quantitative PCR or *ex vivo* cultures, further distinguishing this candidate vaccine from the subunit vaccines and from HSV529. The vaccine was equally protective in female and male mice following challenge with doses as high as 100x LD90 (Burn et al. 2017; Burn Aschner, Knipe, et al. 2020). Passive transfer studies demonstrated that antibodies alone were sufficient to mediate complete protection. Notably, the antibodies were weakly neutralizing, but activated the Fcγ receptor IV (FcγRIV) to induce ADCC and antibody dependent cell mediated phagocytosis (ADCP) (Petro et al. 2015; Petro et al. 2016; Burn et al. 2017). Protection was lost in mice that did not express FcγRs and reduced in mice that did not express the Fc neonatal receptor.

In a direct comparison with an adjuvanted (alum and MPL) gD-2 subunit vaccine, ΔgD-2 provided significantly greater protection following high-dose skin challenges with clinical isolates of HSV-1 (strain B^3x1.1) and HSV-2 (strain SD90), a more robust total HSV-specific antibody response, and, in contrast to the gD-2 subunit vaccine, complete protection against the establishment of latency (Burn et al. 2017). While the gD-2 subunit vaccine elicited a predominantly IgG1 response, ΔgD-2 induced a predominantly IgG2 antibody response, the latter of which is associated with FcγRIV activation. Consistent with the isotype differences, adjuvanted gD-2 elicited higher neutralizing titers, whereas ΔgD-2 elicited significantly higher FcγR-activating responses (ADCC and ADCP). Although the neutralizing titer induced by gD-2 subunit vaccine was higher against SD90 compared to B^3x1.1, the subunit vaccine provided greater protection against the latter in the murine studies. Greater protection against HSV-1 versus HSV-2 was also observed in a study with cotton rats, and in the Phase III field trial of the gD-2/AS04 vaccine (Boukhvalova et al. 2015; Stanberry et al. 2002; Awasthi, Belshe, et al. 2014; Burn et al. 2017). It is unclear whether the serotype differences in vaccine efficacy for the recombinant gD-2 vaccine (but not ΔgD-2) reflect serotype differences in the immune response required to achieve protection.

The notion that ADCC antibodies are important in preventing HSV disease is supported by clinical studies showing that low levels of transplacentally acquired ADCC antibodies at the time of infection correlated with the severity of neonatal disease even after controlling for neutralizing antibodies (Kohl 1991; Kohl et al. 1989). Indeed, vaccinologists have begun to challenge the reliance on neutralizing antibodies as the benchmark for vaccine efficacy for multiple pathogens. Indeed, a study with ΔgD-2 illustrated that maternal vaccination protected pups from HSV challenge (Kao et al. 2019). Ongoing studies in influenza, HIV, CMV and dengue have also demonstrated the importance of Fc effector functions such as ADCC and ADCP in immune protection (Laoprasopwattana et al. 2007; Corey et al. 2015; Nelson et al. 2018).

Several other attenuated HSV vaccines have been developed (Table 2) but are unlikely to advance as prophylactic vaccines because, unlike HSV529 and ΔgD-2, they cause variable levels of disease in mice and/or establish latency, rendering them less suitable for clinical development, particularly for prophylaxis. Nonetheless, the preclinical studies may provide additional insights into correlates of immune protection. For example, multiple studies have been conducted with viruses deleted in the gene for thymidine kinase (tk). Tk-deletion viruses are attenuated in adult but not neonatal mice (Ben-Hur et al. 1983; Oram et al. 2000) and cause significant disease in humans (Wilson et al. 2009). For example, Iwasaki and colleagues found that the "prime-pull" strategy, in which protective CD8 T cells were recruited into the vaginal tract in response to the intravaginal introduction of chemokines, was required for protection with a tk-deletion virus vaccine.(Shin and Iwasaki 2012)

Another attenuated strain that has been evaluated is HSV-2-gD27, a genetically modified HSV-2 strain in which the nectin-binding domain of gD is altered (Wang et al. 2012). Compared to wild-type virus, gD27 is less neurovirulent, but retains the ability to establish latency, albeit at reduced levels. Intramuscular immunization with this attenuated strain provided greater protection than a recombinant gD-2 subunit vaccine despite eliciting lower neutralizing antibody titers. These findings suggest that non-neutralizing immune responses contribute to the increased protection relative to recombinant gD subunit vaccines, although quantification of different antibody functions (e.g. ADCC or complement activation), T cell responses, passive transfer, or depletion studies have not yet been published (Wang et al. 2015).

An attenuated HSV-2 strain lacking the glycoprotein K (gK) gene provided partial protection against HSV-1 (strain McKrae) and HSV-2 (strain G) intravaginal challenge and elicited T cell memory immune responses to specific gB and gD epitopes that were detected 7 months post vaccination (Iyer et al. 2013). However, the vaccine strain could establish latency, and, therefore, a second mutation in the membrane protein UL20 was introduced, resulting in the inability of the vaccine strain to establish latency in mice. This double gK and UL20 deletion strain (VC2) induced cross-protective neutralizing antibody responses in rhesus macaques as well as expansion of follicular T helper (T_{FH}) cells (Stanfield et al. 2017). In murine studies, VC2 provided 100% protection against HSV-1 (strain McKrae) and HSV-2 (strain G) intravaginal challenge and prevented the establishment of latency (Stanfield et al. 2014). In guinea pigs, the vaccine promoted a Th17 response to challenge, which was proposed to contribute to immune protection (Stanfield et al. 2018). Precisely why the double deletion prevented the establishment of latency and why it provided greater protection than the gK deletion virus is unclear.

In a related strategy designed to prevent the establishment of latency, an attenuated HSV-1 strain containing mutations in the UL37 tegument protein was found to be defective for retrograde axonal transport and unable to establish latent infections of the ganglia (Richards et al. 2017). In early preclinical studies, this vaccine (R2), administered intraocularly, protected mice from subsequent ocular challenge on the opposite side with HSV-1 strains F or McKrae. Pseudorabies virus (PRV) containing analogous mutations in the UL37 homolog provided protection from lethal challenge with wild-type PRV, suggesting that this strategy could be applied to both human and animal alphaherpesviruses (Richards et al. 2017).

An HSV-2 virus containing a deletion in ICP0 (ΔNLS) has also been evaluated as a candidate vaccine (Halford et al. 2010). ΔNLS is replication-competent, but interferon-sensitive, which contributes to its attenuated phenotype in animal models. Because ICP0 plays a major role in viral reactivation, the vaccine would likely be impaired in both the establishment and reactivation from latency, although this was not formally tested. ΔNLS provided greater protection against high-dose HSV-2 challenge compared to a gD-2 subunit vaccine in guinea pigs challenged intravaginally with HSV-2 (strain MS). It elicited polyantigenic immune responses including gD-2 specific and neutralizing antibodies, although the specific correlates of immune protection were not determined (Halford et al. 2011). The latter may be ascertained by performing passive transfer, antibody or T cell depletion and/or studies in defined knockout mice.

In a highly controversial step, Rational Vaccines circumvented the FDA and a safety Institutional Review Board (IRB) and tested ICP0 (ΔNLS) in a Phase 1 safety study in 20 HSV-1 and/or HSV-2 seropositive volunteers with a history of genital herpes. Participants self-reported a reduction in symptoms. A Phase 2 study was planned but was disbanded when it became known that the vaccine had been tested without FDA or IRB input and without the requisite preclinical safety studies. Moreover, several participants reported adverse side effects.

Therapeutic vaccines

The vaccines candidates described above were developed primarily as prophylactic vaccines with the primary goal of preventing HSV-2 (or both HSV-1 and HSV-2) disease in seronegative hosts and/or preventing HSV-2 in HSV-1 seropositive hosts. Whether the types of immune responses needed for therapeutic vaccines, which are designed to reduce the frequency and severity of clinical recurrences and/or asymptomatic shedding, differ from those required for prophylaxis is not known. The "gold standard" model for evaluating therapeutic vaccines is the female guinea pig because the virus spontaneously reactivates. However, despite several vaccines demonstrating efficacy in this model, the clinical trial results have been disappointing.

For example, a vaccine composed of HSV-2 gD and ICP4 combined with a saponin adjuvant was evaluated in a Phase 2 study at three different doses (Van Wagoner et al. 2018). The vaccine boosted neutralizing antibody titers and increased the cytolytic T cell response, but was associated with only a modest decrease in viral shedding (Genocea, Gen-003). Agenus Inc. developed a heat shock protein-based vaccine (HerpV) composed of a mixture of HSV-2 immunogenic peptides (including not only gD, but also other envelope, tegument, capsid, and regulatory viral proteins) and the saponin-based adjuvant QS-21. In a randomized, Phase 2, double-blind study, the majority of patients developed a robust anti-HSV cytotoxic T-cell immune response, but there was only a 17 % reduction in frequency of viral shedding despite a 75% reduction in viral load (Wald et al. 2011; Cohen 2017).

DNA may be more immunogenic than protein and, based on this notion, two other vaccines have been evaluated as candidate therapeutic vaccines. An HSV-2 DNA vaccine composed of gD and the tegument protein UL46 was advanced into the clinic after showing efficacy in the guinea pig recurrent disease model (Veselenak et al. 2012). However, the results were, once again, disappointing, and the vaccine failed to meet its clinical endpoint of annualized lesion recurrences during a minimum of nine months of surveillance. The Phase 2 study, in which volunteers were randomized 2:1 to vaccine or placebo, was conducted in 261 healthy HSV-2 seropositive adults with a self-reported history of 4 to 9 recurrences per year. Unfortunately, the annualized recurrence rate in the placebo group was lower than what was expected, and, thus, the trial did not have sufficient power to demonstrate efficacy (Vical 2018).

Another DNA-based vaccine based on a codon-optimized HSV-2 gD and the addition of a ubiquitin-encoding sequence to target the antigen to the proteasome for processing and to enhance cytotoxic T cell responses, was evaluated in a Phase 2 therapeutic vaccine trial after demonstrating that it induced high-titer gD-specific antibodies and T cell responses. The vaccine enhanced survival and a reduction in latent virus in a vaginal murine model (Dutton et al. 2013). However, while safe, the vaccine did not reduce the severity of clinical recurrences (Admedus 2018).

Vector vaccine strategies

An alternative to protein, DNA, or live attenuated or single-cycle vaccines is a vector vaccine strategy. Vector vaccines refer to the insertion of antigenic target genes from one pathogen into an unrelated vector "backbone". These have been evaluated, for example, as a strategy for HIV prevention (Draper and Heeney 2009). This approach has the potential to promote a T cell-dependent high-affinity class-switched response with memory, albeit to a more restricted set of antigens than a live virus virus. Depending on the vector backbone, this approach may be safer than a live vaccine, particularly in the case of HIV. However, the safety depends on the nature of the vector. The most commonly used vectors for HIV vaccines have been adeno-associated viruses, adenoviruses, and, more recently, a rhesus CMV vector (Barouch 2010; Hansen et al. 2011; Nieto and Salvetti 2014). These approaches may be limited by pre-existing immunity to the vector, which can decrease the efficacy. This was illustrated by the experience with a recombinant adenovirus serotype 5 (rAD5) HIV vaccine that may have enhanced HIV infection in the subpopulation of vaccinees with prior immunity to adenoviruses (STEP trial) (Buchbinder et al. 2008).

There have been several preclinical studies of vector vaccine strategies for the prevention or treatment of HSV that may provide insights for future development. For example, HSV-1 gB (gB-1) was introduced into a feline immunodeficiency virus (FIV) vector and tested against HSV-1 and HSV-2 vaginal challenges in mice. The vaccine elicited cross-neutralizing antibodies and cell-mediated responses that protected 100 and 75% animals from severe HSV-1- and HSV-2 disease, respectively, but not from infection (Chiuppesi et al. 2012). The authors proposed that the addition of other HSV viral genes into this vector might improve the immunogenicity and efficacy.

In a different study, HSV gB and/or gD, lacking their transmembrane domains, were inserted into human papilloma virus pseudoviruses (HPV PsV) as a potential mucosal vaccine strategy, with the goal of eliciting protective genital-resident T cell responses (Çuburu et al. 2014). The PsV secrete the HSV viral glycoproteins. Investigators found that secreted gB (gBsec)-expressing HPV PsVs induced gB-peptide-specific $CD8^+$ T cell responses. In subsequent studies, mucosal vaccination with PsVs expressing gBsec and gDsec were compared to intramuscular immunization with an adjuvanted (alum and MPL) recombinant gD-2 protein vaccine. Both provided a similar level of protection in mice following a lethal challenge (60% and 80%, respectively) (Çuburu et al. 2014). The PsV vaccine induced lower gD-specific and neutralizing antibody titers compared to recombinant gD-2 protein vaccine, but greater gB-specific CD8+ T cell responses, highlighting a potential role for gB-specific T cells.

A modified vaccinia Ankara (MVA) vaccine vector expressing HSV-2 gD (MVA-gD), used in combination with a gD-2-expressing plasmid vaccine, illustrated that the combination of the two delivery methods was effective in inducing robust anti-HSV IgG responses (Meseda et al. 2002; Meseda et al. 2006). The use of MVA-gD as either the prime and boost or as the prime followed by plasmid gD-2 as the boost induced a greater diversity of IgG isotype responses - shifting from an almost entirely IgG1 (predominantly neutralizing antibodies) to a combination of IgG1, IgG2a, and IgG2b responses (IgG2 antibodies have greater affinity for FcR and for complement). Both MVA-gD vector strategies resulted in greater survival following intraperitoneal challenge with HSV-2 as well as more IL-2 and IFN-γ production following stimulation of splenocytes with inactivated HSV (Meseda et al. 2002).

Another study compared immunogenicity and efficacy of an MVA vector expressing the HSV capsid VP5 (UL19) and tegument VP13/14 (UL47) proteins to adjuvanted gD and gC, and a combination of both the MVA vector and subunit proteins (Awasthi et al. 2015). The combination of MVA and subunit proteins elicited higher total and neutralizing antibody responses and greater protection compared to either component alone in mice or guinea pigs (Awasthi et al. 2015). There was also a greater reduction in the size of the latent reservoir in response the combination, compared to the gD and gC subunit vaccine, but no differences in viral shedding in the guinea pig model of recurrent disease. The results suggest that the MVA vector expressing viral antigens that are known to induce a T cell response provided a modest additional benefit compared to the subunit protein vaccine, although why this resulted in higher neutralizing titers is not known.

Lessons learned and future directions

The disappointing outcomes of clinical trials with HSV prophylactic and therapeutic vaccine candidates suggest several important lessons for future research. First, the established "gold-standard" preclinical small animal models (mice and guinea pigs) have not proven predictive of the human response. While the types of immune response (e.g. neutralizing antibody and/or T cell responses) observed in preclinical models were reproduced in humans, the magnitude of responses and the efficacy observed in mice or guinea pigs was not recapitulated. Studies conducted after the clinical trials were completed using higher dose challenges and a broader array of clinical isolates, provided a more predictive picture. For example, when mice were challenged with the clinical isolates HSV-2 (strain SD90) or HSV-1 (strain B^3x1.1), only modest protection was observed in mice immunized with adjuvanted recombinant gD protein vaccines, more closely recapitulating the clinical trial results (Burn et al. 2017). Studies with a broad range of clinical isolates have not been reported in the guinea pig model of recurrent disease, which was also not predictive of clinical trial outcomes. For example, a gD-2 subunit vaccine underwent several rounds of adjuvant optimization, and the optimized product (gD-2/AS04) (*Simplirix*™) provided nearly complete protection against primary and recurrent disease with either HSV-1 or HSV-2 in guinea pigs (Bourne et al. 2003; Bourne et al. 2005; Bernstein et al. 2009; Bernstein et al. 2010) and was also effective in HSV-1 seropositive guinea pigs (Hoshino et al. 2009), results that were not replicated in the clinical trials. Thus, modifications of the guinea pig model, perhaps including clinical isolates, should be explored.

Recent studies with a medroxyprogesterone (MPA)-treated cotton rat more closely recapitulated the outcomes of the *Simplirix*™ gD2 subunit vaccine. Cotton rats were primed and boosted intramuscularly with increasing doses of gD-2/AS04, a hepatitis B vaccine control (Fendrix), or PBS and subsequently challenged intravaginally with HSV-2(G) (5×10^4 pfu) or HSV-1(17) (2×10^6 pfu) (note that higher doses of HSV-1 were needed due to differences in virulence) (Boukhvalova et al. 2015). The gD/AS04 vaccine provided little or no protection at the lower doses (0.06 and 0.3 µg/ml) and partial protection from disease (2/5 cotton rats had no lesions; p=0.008 compared to PBS, log rank test) when administered at the highest dose of 2 µg per animal, which is ~75-fold higher than the human dose of 20 mg of gD. Moreover, there were no differences in amount of HSV-2 shed in vaginal washes in immunized cotton rats. However, the 2 µg gD-2/AS04 dose completely protected cotton rats from HSV-1 disease and reduced viral shedding by ~1.4 log10 in animals vaccinated with 2 µg gD-2/AS04. Consistent with clinical trial results, the gD-2/AS04 vaccine elicited gD-specific neutralizing antibodies in the serum and passive transfer of serum from the immunized cotton rats to naïve cotton rats, protecting the latter from HSV-1 but not HSV-2 disease (Boukhvalova et al. 2015). These results suggest that the cotton rat could be a more predictive model, but further studies are needed. Notably, and distinct from guinea pigs, reagents for studying immune responses in cotton rats are readily available and continue to expand (Niewiesk 2009).

The second major lesson suggested by the HSV vaccine experience is the recognition that non-neutralizing antibody responses may play an important role in immune protection. This was first suggested by the failure of the Chiron gD/gB subunit vaccine and is supported by limited clinical data showing that higher maternally-derived ADCC antibodies at the time of neonatal infection are associated with better clinical outcomes (Kohl 1991; Kohl et al. 1989). Preclinical studies comparing an adjuvanted gD-2 subunit vaccine, which elicits an almost exclusively neutralizing antibody response, to the ΔgD-2 single-cycle candidate vaccine, which elicits a predominantly ADCC response, further highlight the role of ADCC responses. The latter vaccine provided 100% protection against clinical isolates of either serotype in male and female mice and prevented the establishment of latency, whereas the former provided only partial protection against disease and no protection against latency (Burn et al. 2017). Moreover, the immune serum from ΔgD-2-immunized, but not gD subunit-immunized mice, provided complete protection when passively transferred to naïve mice. In addition, ΔgD-2 but not the adjuvanted gD subunit vaccine, protected HSV-1 seropositive mice from subsequent challenge with HSV-2 and this protection correlated with ADCC and not neutralizing antibody responses (Burn Aschner, Knipe, et al. 2020).

Precisely what antigenic and host factors determine whether a neutralizing and/or non-neutralizing FcγR-activating antibody response will be elicited to vaccination or infection is complex and requires further study. However, the history of failed HSV vaccine trials, which focused predominantly on generating neutralizing gD-specific antibody responses, and the ability of HSV to escape neutralizing antibodies by spreading directly from cell-to-cell, suggest that optimal protection will require polyfunctional non-neutralizing immune responses. The importance of these more complex immune responses is supported by ongoing vaccine efforts with other viruses, including CMV, HIV, and influenza, and by the success with VZV vaccines, which are not restricted to the generation of neutralizing antibody responses. The expanding ability to genetically engineer viruses and construct DNA, RNA, and protein vaccines with new adjuvants, in combination with new tools to evaluate immune responses, portends future success in developing vaccine(s) for HSV prevention and treatment.

Acknowledgements

BCH is supported by NIH NIAID and NICHD, X-Vax Technology and the Price Family Foundation. CBA is supported by a Howard Hughes Medical Institute International Student Research Fellowship.

Disclosures

Betsy C. Herold holds a patent for the ΔgD-2 vaccine.

References

Abendroth, A. et al., 2000. Modulation of major histocompatibility class II protein expression by varicella-zoster virus. *Journal of Virology*, 74(4), pp. 1900-1907.

Abendroth, A. et al., 2001. Varicella-zoster virus retains major histocompatibility complex class I proteins in the Golgi compartment of infected cells. *Journal of Virology*, 75(10), pp.4878-4888.

Admedus, 2018. ADMEDUS HSV 2 PHASE IIA RESULTS - MAY 2017.

Ahn, K. et al., 1996. Molecular mechanism and species specificity of TAP inhibition by herpes simplex virus ICP47. *The EMBO journal*, 15(13), pp. 3247-3255.

Argaw, T. et al., 2000. Nucleotide sequences that distinguish Oka vaccine from parental Oka and other varicella-zoster virus isolates. *Journal of Infectious Diseases*, 181(3), pp.1153-1157.

Arvin, A. and Gershon, A., 2006. Control of Varicella. *The Pediatric Infectious Disease Journal*, 25(6), pp.475-476.

Arvin, A.M., 2008. Humoral and Cellular Immunity to Varicella–Zoster Virus: An Overview. *Journal of Infectious Diseases*, 197(s2), pp.S58-S60.

Asano, Y. and Takahashi, M., 1977. Clinical and serologic testing of a live varicella vaccine and two-year follow-up for immunity of the vaccinated children. *PEDIATRICS*, 60(6), pp.810-814.

Awasthi, S. et al., 2015. A Dual-Modality Herpes Simplex Virus 2 Vaccine for Preventing Genital Herpes by Using Glycoprotein C and D Subunit Antigens To Induce Potent Antibody Responses and Adenovirus Vectors Containing Capsid and Tegument Proteins as T Cell Immunogens L. Hutt-Fletcher, ed. *Journal of Virology*, 89(16), pp.8497-8509.

Awasthi, S. et al., 2019. Nucleoside-modified mRNA encoding HSV-2 glycoproteins C, D, and E prevents clinical and subclinical genital herpes. *Science immunology*, 4(39), p.eaaw7083.

Awasthi, S., Belshe, R.B. and Friedman, H.M., 2014. Better Neutralization of Herpes Simplex Virus Type 1 (HSV-1) Than HSV-2 by Antibody From Recipients of GlaxoSmithKline HSV-2 Glycoprotein D2 Subunit Vaccine. *Journal of Infectious Diseases*, 210(4), pp.571-575.

Awasthi, S., Hook, L.M., Shaw, C.E. and Friedman, H.M., 2017. A trivalent subunit antigen glycoprotein vaccine as immunotherapy for genital herpes in the guinea pig genital infection model. *Human Vaccines and Immunotherapeutics*, 13(12), pp.2785-2793.

Awasthi, S., Hook, L.M., Shaw, C.E., Pahar, B., et al., 2017. An HSV-2 Trivalent Vaccine Is Immunogenic in Rhesus Macaques and Highly Efficacious in Guinea Pigs L. Hutt-Fletcher, ed. *PLoS Pathogens*, 13(1), pp.e1006141-24.

Awasthi, S., Huang, J., et al., 2014. Blocking Herpes Simplex Virus 2 Glycoprotein E Immune Evasion as an Approach To Enhance Efficacy of a Trivalent Subunit Antigen Vaccine for Genital Herpes. *Journal of Virology*, 88(15), pp.8421-8432.

Barouch, D.H., 2010. Novel adenovirus vector-based vaccines for HIV-1. *Current Opinion in HIV and AIDS*, 5(5), pp.386-390.

Bartha, A., 1961. Experimental reduction of virulence of Aujeszky's disease virus. *Mag. Allatorv. Lapja*, 16, pp.42-45.

Belshe, R.B. et al., 2012. Efficacy results of a trial of a herpes simplex vaccine. *New England Journal of Medicine*, 366(1), pp.34-43.

Ben-Hur, T. et al., 1983. Neurovirulence of herpes simplex virus type 1 depends on age in mice and thymidine kinase expression. *Archives of virology*, 78(3-4), pp.303-308.

Berkowitz, E.M. et al., 2014. Safety and Immunogenicity of an Adjuvanted Herpes Zoster Subunit Candidate Vaccine in HIV-Infected Adults: A Phase 1/2a Randomized, Placebo-Controlled Study. *Journal of Infectious Diseases*, 211(8), pp.1279-1287.

Bernard, M.-C. et al., 2015. Immunogenicity, Protective Efficacy, and Non-Replicative Status of the HSV-2 Vaccine Candidate HSV529 in Mice and Guinea Pigs L. BenMohamed, ed. *PLoS ONE*, 10(4), pp.e0121518-21.

Bernstein, D.I. et al., 2009. Potent adjuvant activity of cationic liposome-DNA complexes for genital herpes vaccines. *Clinical and vaccine immunology : CVI*, 16(5), pp.699-705.

Bernstein, D.I. et al., 2010. The adjuvant CLDC increases protection of a herpes simplex type 2 glycoprotein D vaccine in guinea pigs. *Vaccine*, 28(21), pp.3748-3753.

Boukhvalova, M. et al., 2015. Efficacy of HSV-2 Glycoprotein D/AS04 vaccine against genital HSV-2 and HSV-1 infection and disease in the cotton rat Sigmodon hispidus model. *Journal of Virology*.

Bourne, N. et al., 2003. Herpes simplex virus (HSV) type 2 glycoprotein D subunit vaccines and protection against genital HSV-1 or HSV-2 disease in guinea pigs. *Journal of Infectious Diseases*, 187(4), pp.542-549.

Bourne, N. et al., 2005. Impact of immunization with glycoprotein D2/AS04 on herpes simplex virus type 2 shedding into the genital tract in guinea pigs that become infected. *Journal of Infectious Diseases*, 192(12), pp.2117-2123.

Boursnell, M.E. et al., 1997. A genetically inactivated herpes simplex virus type 2 (HSV-2) vaccine provides effective protection against primary and recurrent HSV-2 disease. *Journal of Infectious Diseases*, 175(1), pp.16-25.

Brunell, P.A. et al., 1982. Administration of live varicella vaccine to children wtih leukaemia. *The Lancet*, 2(8307), pp.1069-1073.

Buchbinder, S.P. et al., 2008. Efficacy assessment of a cell-mediated immunity HIV-1 vaccine (the Step Study): a double-blind, randomised, placebo-controlled, test-of-concept trial. *Lancet (London, England)*, 372(9653), pp. 1881-1893.

Burn Aschner, C., Knipe, D.M. and Herold, B.C., 2020. Model of vaccine efficacy against HSV-2 superinfection of HSV-1 seropositive mice demonstrates protection by antibodies mediating cellular cytotoxicity. *npj Vaccines*, 5:35

Burn, C. et al., 2017. An HSV-2 single-cycle candidate vaccine deleted in glycoprotein D, ΔgD-2, protects male mice from lethal skin challenge with clinical isolates of HSV-1 and HSV-2. *Journal of Infectious Diseases*, pp.1-5.

Carfi, A. et al., 2002. Crystallization and preliminary diffraction studies of the ectodomain of the envelope glycoprotein D from herpes simplex virus 1 alone and in complex with the ectodomain of the human receptor HveA. *Acta crystallographica. Section D, Biological crystallography*, 58(Pt 5), pp.836-838.

Carfi, A. et al., 2001. Herpes simplex virus glycoprotein D bound to the human receptor HveA. *Molecular cell*, 8(1), pp.169-179.

Cheung, T.C. et al., 2009. Unconventional ligand activation of herpesvirus entry mediator signals cell survival. *Proceedings of the National Academy of Sciences*, 106(15), pp.6244-6249.

Chew, T., Taylor, K. and Mossman, K., 2009. Innate and Adaptive Immune Responses to Herpes Simplex Virus. *Viruses*, 1(3), pp.979-1002.

Chiuppesi, F. et al., 2012. A Lentiviral Vector-Based, Herpes Simplex Virus 1 (HSV-1) Glycoprotein B Vaccine Affords Cross-Protection against HSV-1 and HSV-2 Genital Infections. *Journal of Virology*, 86(12), pp.6563-6574.

Cohen, J.I., 2017. Vaccination to Reduce Reactivation of Herpes Simplex Virus 2. *Journal of Infectious Diseases*, pp.jix006-3.

Compaan, D.M. et al., 2005. Attenuating lymphocyte activity: the crystal structure of the BTLA-HVEM complex. *The Journal of biological chemistry*, 280(47), pp.39553-39561.

Connolly, S.A. et al., 2002. Structure-Based Analysis of the Herpes Simplex Virus Glycoprotein D Binding Site Present on Herpesvirus Entry Mediator HveA (HVEM). *Journal of Virology*, 76(21), pp.10894-10904.

Connolly, S.A. et al., 2003. Structure-based mutagenesis of herpes simplex virus glycoprotein D defines three critical regions at the gD-HveA/HVEM binding interface. *Journal of Virology*, 77(14), pp.8127-8140.

Corey, L. et al., 2015. Immune correlates of vaccine protection against HIV-1 acquisition. *Science Translational Medicine*, 7(310), pp.310rv7-310rv7.

Corey, L. et al., 1999. Recombinant glycoprotein vaccine for the prevention of genital HSV-2 infection: two randomized controlled trials. Chiron HSV Vaccine Study Group. *JAMA*, 282(4), pp.331-340.

Cunningham, A.L. et al., 2016. Efficacy of the Herpes Zoster Subunit Vaccine in Adults 70 Years of Age or Older. *New England Journal of Medicine*, 375(11), pp.1019-1032.

Çuburu, N. et al., 2014. Topical Herpes Simplex Virus 2 (HSV-2) Vaccination with Human Papillomavirus Vectors Expressing gB/gD Ectodomains Induces

Genital-Tissue-Resident Memory CD8 +T Cells and Reduces Genital Disease and Viral Shedding after HSV-2 Challenge R. M. Sandri-Goldin, ed. *Journal of Virology*, 89(1), pp.83-96.

Da Costa, X.J.E.A. et al., 2000. Construction, phenotypic analysis, and immunogenicity of a UL5/UL29 double deletion mutant of herpes simplex virus 2. *Journal of Virology*, 74(17), pp.7963-7971.

Da Costa, X.J.E.A., Morrison, L.A. and Knipe, D.M., 2001. Comparison of Different Forms of Herpes Simplex Replication-Defective Mutant Viruses as Vaccines in a Mouse Model of HSV-2 Genital Infection. *Virology*, 288(2), pp. 256-263.

de Bruyn, G. et al., 2006. A randomized controlled trial of a replication defective (gH deletion) herpes simplex virus vaccine for the treatment of recurrent genital herpes among immunocompetent subjects. *Vaccine*, 24(7), pp. 914-920.

Draper, S.J. and Heeney, J.L., 2009. Viruses as vaccine vectors for infectious diseases and cancer. *Nature Reviews Microbiology*, 8(1), pp.62-73.

Dropulic, L. et al., 2017. A Replication-Defective Herpes Simplex Virus (HSV)-2 Vaccine, HSV529, is Safe and Well-Tolerated in Adults with or without HSV Infection and Induces Significant HSV-2 Specific Antibody Responses in HSV Seronegative Individuals. *Open Forum Infectious Diseases*, (4), pp.S415-S416.

Dropulic, L.K. et al., 2019. A Randomized, Double-Blinded, Placebo-Controlled, Phase 1 Study of a Replication-Defective Herpes Simplex Virus (HSV) Type 2 Vaccine, HSV529, in Adults With or Without HSV Infection. *Journal of Infectious Diseases*, 220(6), pp.990-1000.

Dubin, G. et al., 1991. Herpes simplex virus type 1 Fc receptor protects infected cells from antibody-dependent cellular cytotoxicity. *Journal of Virology*, 65(12), pp.7046-7050.

Dudek, T.E. et al., 2011. Evidence for Differences in Immunologic and Pathogenesis Properties of Herpes Simplex Virus 2 Strains From the United States and South Africa. *Journal of Infectious Diseases*, 203(10), pp. 1434-1441.

Dutton, J.L. et al., 2013. A Novel DNA Vaccine Technology Conveying Protection against a Lethal Herpes Simplex Viral Challenge in Mice A. A. Ashkar, ed. *PLoS ONE*, 8(10), pp.e76407-10.

Faga, B. et al., 2001. Identification and Mapping of Single Nucleotide Polymorphisms in the Varicella-Zoster Virus Genome. *Virology*, 280(1), pp. 1-6.

Freuling, C.M., Müller, T.F. and Mettenleiter, T.C., 2017. Vaccines against pseudorabies virus (PrV). *Veterinary Microbiology*, 206, pp.3-9.

Friedman, H.M. et al., 1984. Glycoprotein C of herpes simplex virus 1 acts as a receptor for the C3b complement component on infected cells. *Nature*, 309(5969), pp.633-635.

Fries, L.F. et al., 1986. Glycoprotein C of herpes simplex virus 1 is an inhibitor of the complement cascade. *The Journal of Immunology*, 137(5), pp. 1636-1641.

Früh, K. et al., 1995. A viral inhibitor of peptide transporters for antigen presentation. *Nature*, 375(6530), pp.415-418.

Gerber, S.I., Belval, B.J. and Herold, B.C., 1995. Differences in the role of glycoprotein C of HSV-1 and HSV-2 in viral binding may contribute to serotype differences in cell tropism. *Virology*, 214(1), pp.29-39.

Gershon, A.A., 1980. Live attenuated Varicella-Zoster vaccine. *Reviews of infectious diseases*, 2(3), pp.393-407.

Gershon, A.A., LaRussa, P. and Steinberg, S., 1994. Detection of antibodies to varicella-zoster virus using a latex agglutination assay. *Clinical and diagnostic virology*, 2(4-5), pp.271-277.

Grauwet, K. et al., 2014. Modulation of CD112 by the alphaherpesvirus gD protein suppresses DNAM-1-dependent NK cell-mediated lysis of infected cells. *Proceedings of the National Academy of Sciences*, 111(45), pp. 16118-16123.

Guzzetta, G. et al., 2013. Hope-Simpson's Progressive Immunity Hypothesis as a Possible Explanation for Herpes Zoster Incidence Data. *American Journal of Epidemiology*, 177(10), pp.1134-1142.

Halford, W.P. et al., 2011. A Live-Attenuated HSV-2 ICP0− Virus Elicits 10 to 100 Times Greater Protection against Genital Herpes than a Glycoprotein D Subunit Vaccine R. Mosley, ed. *PLoS ONE*, 6(3), pp.e17748-18.

Halford, W.P., Püschel, R. and Rakowski, B., 2010. Herpes Simplex Virus 2 ICP0− Mutant Viruses Are Avirulent and Immunogenic: Implications for a Genital Herpes Vaccine D. P. Martin, ed. *PLoS ONE*, 5(8), pp.e12251-17.

Hansen, S.G. et al., 2011. Profound early control of highly pathogenic SIV by an effector memory T-cell vaccine. *Nature*, 473(7348), pp.523-527.

Harrop, J.A., McDonnell, P.C., et al., 1998. Herpesvirus entry mediator ligand (HVEM-L), a novel ligand for HVEM/TR2, stimulates proliferation of T cells and inhibits HT29 cell growth. *The Journal of biological chemistry*, 273(42), pp.27548-27556.

Harrop, J.A., Reddy, M., et al., 1998. Antibodies to TR2 (herpesvirus entry mediator), a new member of the TNF receptor superfamily, block T cell proliferation, expression of activation markers, and production of cytokines. *The Journal of Immunology*, 161(4), pp.1786-1794.

Hattori, A. et al., 1976. Use of live varicella vaccine in children with acute leukaemia or other malignancies. *The Lancet*, 2(7978), p.210.

Hill, A. et al., 1995. Herpes simplex virus turns off the TAP to evade host immunity. *Nature*, 375(6530), pp.411-415.

Hope-Simpson, R.E., 1965. The nature of herpes zoster: a long-term study and a new hypothesis. *Proceedings of the royal society of medicine*, 58, pp.9-20.

Hoshino, Y. et al., 2004. Comparative Efficacy and Immunogenicity of Replication-Defective, Recombinant Glycoprotein, and DNA Vaccines for Herpes Simplex Virus 2 Infections in Mice and Guinea Pigs. *Journal of Virology*, 79(1), pp.410-418.

Hoshino, Y. et al., 2008. Comparison of immunogenicity and protective efficacy of genital herpes vaccine candidates herpes simplex virus 2 dl5-29 and dl5-29-41L in mice and guinea pigs. *Vaccine*, 26(32), pp.4034-4040.

Hoshino, Y. et al., 2009. Protection from herpes simplex virus (HSV)-2 infection with replication-defective HSV-2 or glycoprotein D2 vaccines in HSV-1-seropositive and HSV-1-seronegative guinea pigs. *Journal of Infectious Diseases*, 200(7), pp.1088-1095.

Ihara, T. et al., 1994. Cytotoxicity against varicella zoster virus infected targets in children with acute leukemia. *Acta paediatrica Japonica : Overseas edition*, 36(1), pp.53-56.

Iyer, A.V. et al., 2013. Single dose of Glycoprotein K (gK)-deleted HSV-1 live-attenuated virus protects mice against lethal vaginal challenge with HSV-1 and HSV-2 and induces lasting T cell memory immune responses. *Virology journal*, 10(1), p.317.

James, S.H. and Kimberlin, D.W., 2015. Neonatal Herpes Simplex Virus Infection. *Infectious disease clinics of North America*, 29(3), pp.391-400.

Jones, J.O. and Arvin, A.M., 2006. Inhibition of the NF-kappaB pathway by varicella-zoster virus in vitro and in human epidermal cells in vivo. *Journal of Virology*, 80(11), pp.5113-5124.

Kamiya, H. et al., 1982. Antibody-dependent cell-mediated cytotoxicity against varicella-zoster virus-infected targets. *Infection and Immunity*, 38(2), pp. 554-557.

Kao, C.M. et al., 2019. Murine Model of Maternal Immunization Demonstrates Protective Role for Antibodies That Mediate Antibody-Dependent Cellular Cytotoxicity in Protecting Neonates From Herpes Simplex Virus Type 1 and Type 2. *Journal of Infectious Diseases*, 42, pp.47-10.

Kohl, S., 1991. Role of antibody-dependent cellular cytotoxicity in defense against herpes simplex virus infections. *Reviews of infectious diseases*, 13(1), pp.108-114.

Kohl, S. et al., 2000. Limited antibody-dependent cellular cytotoxicity antibody response induced by a herpes simplex virus type 2 subunit vaccine. *Journal of Infectious Diseases*, 181(1), pp.335-339.

Kohl, S. et al., 1989. Neonatal antibody-dependent cellular cytotoxic antibody levels are associated with the clinical presentation of neonatal herpes simplex virus infection. *Journal of Infectious Diseases*, 160(5), pp.770-776.

Krause, P.R. and Klinman, D.M., 2000. Varicella vaccination: evidence for frequent reactivation of the vaccine strain in healthy children. *Nature medicine*, 6(4), pp.451-454.

Kronenberg, A. et al., 2005. Retrospective analysis of varicella zoster virus (VZV) copy DNA numbers in plasma of immunocompetent patients with herpes zoster, of immunocompromised patients with disseminated VZV disease, and of asymptomatic solid organ transplant recipients. *Transplant infectious disease : an official journal of the Transplantation Society*, 7(3-4), pp.116-121.

Kwak, B.O. et al., 2017. Genotype analysis of ORF 62 identifies varicella-zoster virus infections caused by a vaccine strain in children. *Archives of virology*, 162(6), pp.1725-1730.

Kwon, B.S. et al., 1997. A newly identified member of the tumor necrosis factor receptor superfamily with a wide tissue distribution and involvement in lymphocyte activation. *The Journal of biological chemistry*, 272(22), pp. 14272-14276.

Lal, H. et al., 2015. Efficacy of an Adjuvanted Herpes Zoster Subunit Vaccine in Older Adults. *New England Journal of Medicine*, 372(22), pp.2087-2096.

Laoprasopwattana, K. et al., 2007. Antibody-Dependent Cellular Cytotoxicity Mediated by Plasma Obtained before Secondary Dengue Virus Infections: Potential Involvement in Early Control of Viral Replication. *Journal of Infectious Diseases*, 195(8), pp.1108-1116.

Ljungman, P. et al., 1986. Clinical and subclinical reactivations of varicella-zoster virus in immunocompromised patients. *Journal of Infectious Diseases*, 153(5), pp.840-847.

Looker, K.J. et al., 2017. Effect of HSV-2 infection on subsequent HIV acquisition: an updated systematic review and meta-analysis. *The Lancet Infectious ....*

Looker, K.J. et al., 2015. Global Estimates of Prevalent and Incident Herpes Simplex Virus Type 2 Infections in 2012. *PLoS ONE*, 10(1), pp.e114989-23.

Mark, K.E. et al., 2008. Rapidly Cleared Episodes of Herpes Simplex Virus Reactivation in Immunocompetent Adults. *Journal of Infectious Diseases*, 198(8), pp.1141-1149.

Mauri, D.N. et al., 1998. LIGHT, a new member of the TNF superfamily, and lymphotoxin alpha are ligands for herpesvirus entry mediator. *Immunity*, 8(1), pp.21-30.

McLean, C.S. et al., 1994. Protective vaccination against primary and recurrent disease caused by herpes simplex virus (HSV) type 2 using a genetically disabled HSV-1. *Journal of Infectious Diseases*, 170(5), pp.1100-1109.

McNearney, T.A., 1987. Herpes simplex virus glycoproteins gC-1 and gC-2 bind to the third component of complement and provide protection against complement- mediated neutralization of viral infectivity. *Journal of Experimental Medicine*, 166(5), pp.1525-1535.

Meseda, C.A. et al., 2002. Prime-boost immunization with DNA and modified vaccinia virus ankara vectors expressing herpes simplex virus-2 glycoprotein D elicits greater specific antibody and cytokine responses than DNA vaccine alone. *Journal of Infectious Diseases*, 186(8), pp.1065-1073.

Meseda, C.A., Stout, R.R. and Weir, J.P., 2006. Evaluation of a needle-free delivery platform for prime-boost immunization with DNA and modified vaccinia virus ankara vectors expressing herpes simplex virus 2 glycoprotein D. *Viral immunology*, 19(2), pp.250-259.

Moffat, J.F. et al., 1998. Attenuation of the vaccine Oka strain of varicella-zoster virus and role of glycoprotein C in alphaherpesvirus virulence demonstrated in the SCID-hu mouse. *Journal of Virology*, 72(2), pp.965-974.

Murphy, T.L. and Murphy, K.M., 2010. Slow Down and Survive: Enigmatic Immunoregulation by BTLA and HVEM. *Annual Review of Immunology*, 28(1), pp.389-411.

Ndjamen, B. et al., 2014. The herpes virus Fc receptor gE-gI mediates antibody bipolar bridging to clear viral antigens from the cell surface. *PLoS Pathogens*, 10(3), p.e1003961.

Nelson, C.S., Herold, B.C. and Permar, S.R., 2018. A new era in cytomegalovirus vaccinology: considerations for rational design of next-generation vaccines to prevent congenital cytomegalovirus infection. *npj Vaccines*, pp.1-9.

Nieto, K. and Salvetti, A., 2014. AAV Vectors Vaccines Against Infectious Diseases. *Frontiers in Immunology*, 5, p.5.

Niewiesk, S., 2009. Current animal models: cotton rat animal model. *Current topics in microbiology and immunology*, 330, pp.89-110.

Okazaki, W., Purchase, H.G. and Burmester, B.R., 1970. Protection against Marek's disease by vaccination with a herpesvirus of turkeys. *Avian diseases*, 14(2), pp.413-429.

Oram, R.J. et al., 2000. Characterization of an acyclovir-resistant herpes simplex virus type 2 strain isolated from a premature neonate. *Journal of Infectious Diseases*, 181(4), pp.1458-1461.

Oxman, M.N., Levin, M.J.Shingles Prevention Study Group, 2008. Vaccination against Herpes Zoster and Postherpetic Neuralgia. *Journal of Infectious Diseases*, 197(s2), pp.S228-S236.

Papaevangelou, V. et al., 2013. Subclinical VZV reactivation in immunocompetent children hospitalized in the ICU associated with prolonged fever duration. *Clinical microbiology and infection : the official publication of the European Society of Clinical Microbiology and Infectious Diseases*, 19(5), pp.E245-51.

Peters, G.A. et al., 2012. The Attenuated Genotype of Varicella-Zoster Virus Includes an ORF0 Transitional Stop Codon Mutation. *Journal of Virology*, 86(19), pp.10695-10703.

Petro, C. et al., 2015. Herpes simplex type 2 virus deleted in glycoprotein D protects against vaginal, skin and neural disease. *eLife*.

Petro, C.D. et al., 2016. HSV-2 ΔgD elicits FcγR-effector antibodies that protect against clinical isolates. *JCI Insight*, 1(12), pp.1-15.

Quinlivan, M. and Breuer, J., 2006. Molecular studies of Varicella zoster virus. *Reviews in Medical Virology*, 16(4), pp.225-250.

Quinlivan, M.L. et al., 2012. Novel Genetic Variation Identified at Fixed Loci in ORF62 of the Oka Varicella Vaccine and in a Case of Vaccine-Associated Herpes Zoster. *Journal of Clinical Microbiology*, 50(5), pp.1533-1538.

Ramsey, N.L.M. et al., 2020. A single-cycle glycoprotein D deletion viral vaccine candidate, ΔgD-2, elicits polyfunctional antibodies that protect against ocular herpes simplex virus. *Journal of Virology*.

Reszka, N.J., Dudek, T. and Knipe, D.M., 2010. Construction and properties of a herpes simplex virus 2 dl5-29 vaccine candidate strain encoding an HSV-1 virion host shutoff protein. *Vaccine*, 28(15), pp.2754-2762.

Richards, A.L. et al., 2017. The pUL37 tegument protein guides alpha-herpesvirus retrograde axonal transport to promote neuroinvasion D. A. Leib, ed. *PLoS Pathogens*, 13(12), pp.e1006741-32.

Roberts, C.M., Pfister, J.R. and Spear, S.J., 2003. Increasing Proportion of Herpes Simplex Virus Type 1 as a Cause of Genital Herpes Infection in College Students. *Sexually Transmitted Diseases*, 30(10), pp.797-800.

Sarrias, M.R. et al., 2000. The three HveA receptor ligands, gD, LT-alpha and LIGHT bind to distinct sites on HveA. *Molecular Immunology*, 37(11), pp. 665-673.

Sattentau, Q., 2008. Avoiding the void: cell-to-cell spread of human viruses. *Nature Reviews Microbiology*, 6(11), pp.815-826.

Schat, K.A. and Baranowski, E., 2007. Animal vaccination and the evolution of viral pathogens. *Revue scientifique et technique*, 26(2), pp.327-338.

Schmidt-Chanasit, J. and Sauerbrei, A., 2011. Evolution and world-wide distribution of varicella-zoster virus clades. *Infection, genetics and evolution : journal of molecular epidemiology and evolutionary genetics in infectious diseases*, 11(1), pp.1-10.

Sei, J.J. et al., 2015. Effector and Central Memory Poly-Functional CD4+ and CD8+ T Cells are Boosted upon ZOSTAVAX® Vaccination. *Frontiers in Immunology*, 6(Suppl 1), pp.675-15.

Shapiro, E.D. et al., 2011. Effectiveness of 2 Doses of Varicella Vaccine in Children. *Journal of Infectious Diseases*, 203(3), pp.312-315.

Shi, G., 2002. Mouse T cells receive costimulatory signals from LIGHT, a TNF family member. *Blood*, 100(9), pp.3279-3286.

Shin, H. and Iwasaki, A., 2012. A vaccine strategy that protects against genital herpes by establishing local memory T cells. *Nature*, 491(7424), pp.463-467.

Spear, P.G., 2004. Herpes simplex virus: receptors and ligands for cell entry. *Cellular Microbiology*, 6(5), pp.401-410.

Stadtmauer, E.A. et al., 2014. A phase 1/2 study of an adjuvanted varicella-zoster virus subunit vaccine in autologous hematopoietic cell transplant recipients. *Blood*, 124(19), pp.2921-2929.

Stanberry, L.R. et al., 2002. Glycoprotein-D-adjuvant vaccine to prevent genital herpes. *New England Journal of Medicine*, 347(21), pp.1652-1661.

Stanfield, B.A. et al., 2014. A single intramuscular vaccination of mice with the HSV-1 VC2 virus with mutations in the glycoprotein K and the membrane protein UL20 confers full protection against lethal intravaginal challenge with virulent HSV-1 and HSV-2 strains. *PLoS ONE*, 9(10), p.e109890.

Stanfield, B.A. et al., 2018. Intramuscular vaccination of guinea pigs with the live-attenuated human herpes simplex vaccine VC2 stimulates a transcriptional profile of vaginal Th17 and regulatory Tr1 responses. *Vaccine*, 36(20), pp.2842-2849.

Stanfield, B.A. et al., 2017. Vaccination of rhesus macaques with the live-attenuated HSV-1 vaccine VC2 stimulates the proliferation of mucosal T cells and germinal center responses resulting in sustained production of highly neutralizing antibodies. *Vaccine*, 35(4), pp.536-543.

Stiles, K.M. et al., 2010. Herpes Simplex Virus Glycoprotein D Interferes with Binding of Herpesvirus Entry Mediator to Its Ligands through Downregulation and Direct Competition. *Journal of Virology*, 84(22), pp.11646-11660.

Šedý, J.R. et al., 2004. B and T lymphocyte attenuator regulates T cell activation through interaction with herpesvirus entry mediator. *Nature immunology*, 6(1), pp.90-98.

Takahashi, M. et al., 1974. Live vaccine used to prevent the spread of varicella in children in hospital. *The Lancet*, 2(7892), pp.1288-1290.

Tamada, K. et al., 2000. LIGHT, a TNF-Like Molecule, Costimulates T Cell Proliferation and Is Required for Dendritic Cell-Mediated Allogeneic T Cell Response. *The Journal of Immunology*, 164(8), pp.4105-4110.

Tomazin, R. et al., 1996. Stable binding of the herpes simplex virus ICP47 protein to the peptide binding site of TAP. *The EMBO journal*, 15(13), pp. 3256-3266.

Van Wagoner, N. et al., 2018. Effects of Different Doses of GEN-003, a Therapeutic Vaccine for Genital Herpes Simplex Virus-2, on Viral Shedding and Lesions: Results of a Randomized Placebo-Controlled Trial. *Journal of Infectious Diseases*, 375, pp.1906-10.

Veselenak, R.L. et al., 2012. A Vaxfectin®-adjuvanted HSV-2 plasmid DNA vaccine is effective for prophylactic and therapeutic use in the guinea pig model of genital herpes. *Vaccine*, 30(49), pp.7046-7051.

Vical, 2018. Vical Reports Phase 2 Trial of HSV-2 Therapeutic Vaccine Did Not Meet Primary Endpoint.

Wald, A. et al., 2000. Reactivation of genital herpes simplex virus type 2 infection in asymptomatic seropositive persons. *New England Journal of Medicine*, 342(12), pp.844-850.

Wald, A. et al., 2011. Safety and immunogenicity of long HSV-2 peptides complexed with rhHsc70 in HSV-2 seropositive persons. *Vaccine*, 29(47), pp. 8520-8529.

Wald, A. et al., 1995. Virologic characteristics of subclinical and symptomatic genital herpes infections. *New England Journal of Medicine*, 333(12), pp. 770-775.

Wang, K. et al., 2015. A Herpes Simplex Virus 2 (HSV-2) gD Mutant Impaired for Neural Tropism Is Superior to an HSV-2 gD Subunit Vaccine To Protect Animals from Challenge with HSV-2 R. M. Longnecker, ed. *Journal of Virology*, 90(1), pp.562-574.

Wang, K. et al., 2012. A herpes simplex virus 2 glycoprotein D mutant generated by bacterial artificial chromosome mutagenesis is severely impaired for infecting neuronal cells and infects only Vero cells expressing exogenous HVEM. *Journal of Virology*, 86(23), pp.12891-12902.

Watson, B. et al., 1995. Safety and cellular and humoral immune responses of a booster dose of varicella vaccine 6 years after primary immunization. *Journal of Infectious Diseases*, 172(1), pp.217-219.

Weibel, R.E. et al., 1985. Live Oka/Merck varicella vaccine in healthy children. Further clinical and laboratory assessment. *JAMA*, 254(17), pp.2435-2439.

Weinberg, A., Kroehl, M.E., et al., 2018. Comparative Immune Responses to Licensed Herpes Zoster Vaccines. *Journal of Infectious Diseases*, 218(suppl_2), pp.S81-S87.

Weinberg, A., Popmihajlov, Z., et al., 2018. Persistence of Varicella-Zoster Virus Cell-Mediated Immunity After the Administration of a Second Dose of Live Herpes Zoster Vaccine. *Journal of Infectious Diseases*, 352, pp.2271-4.

White, C.J., 1996. Clinical trials of varicella vaccine in healthy children. *Infectious disease clinics of North America*, 10(3), pp.595-608.

White, C.J. et al., 1991. Varicella vaccine (VARIVAX) in healthy children and adolescents: results from clinical trials, 1987 to 1989. *PEDIATRICS*, 87(5), pp.604-610.

Whitley, R.J., 2006. Herpes simplex encephalitis: Adolescents and adults. *Antiviral Research*, 71(2-3), pp.141-148.

Wilson, S.S., Fakioglu, E. and Herold, B.C., 2009. Novel approaches in fighting herpes simplex virus infections. *Expert Review of Anti-infective Therapy*, 7(5), pp.559-568.

Winston, D.J. et al., 2018. Inactivated varicella zoster vaccine in autologous haemopoietic stem-cell transplant recipients: an international, multicentre, randomised, double-blind, placebo-controlled trial. *Lancet (London, England)*, 391(10135), pp.2116-2127.

Chapter 11

Oncolytic HSV Vectors and Anti-Tumor Immunity

Joseph C. Glorioso[1*], Justus B. Cohen[1], William F. Goins[1], Bonnie Hall[1], Joseph W. Jackson[1], Gary Kohanbash[2], Nduka Amankulor[2], Balveen Kaur[3], Michael A. Caligiuri[4], E. Antonio Chiocca[5], Eric C. Holland[6] and Christophe Quéva[7]

[1]Department of Microbiology and Molecular Genetics, University of Pittsburgh School of Medicine, Pittsburgh, PA , USA
[2]Department of Neurological Surgery, Hillman Cancer Center, University of Pittsburgh School of Medicine, Pittsburgh, PA, USA
[3]Department of Neurosurgery, McGovern Medical School at UTHealth, Houston, TX, USA
[4]City of Hope Medical Center, Duarte, CA, USA
[5]Department of Neurosurgery, Brigham and Women's Hospital, Harvard Medical School, Boston, MA, USA
[6]Fred Hutchinson Cancer Research Center, University of Washington, Seattle, WA, USA
[7]Oncorus Inc., Cambridge, MA, USA

*glorioso@pitt.edu

DOI: https://doi.org/10.21775/9781913652555.11

Abstract
The therapeutic promise of oncolytic viruses (OVs) rests on their ability to both selectively kill tumor cells and induce anti-tumor immunity. The potential of tumors to be recognized and eliminated by an effective anti-tumor immune

response has been spurred on by the discovery that immune checkpoint inhibition can overcome tumor-specific cytotoxic T cell (CTL) exhaustion and provide durable responses in multiple tumor indications. OV-mediated tumor destruction is now recognized as a powerful means to assist in the development of anti-tumor immunity for two important reasons: (i) OVs, through the elicitation of an anti-viral response and the production of type I interferon, are potent stimulators of inflammation and can be armed with transgenes to further enhance anti-tumor immune responses; and (ii) lytic activity can promote the release of tumor-associated antigens (TAAs) and tumor neoantigens that function as *in situ* tumor-specific vaccines to elicit adaptive immunity. Oncolytic herpes simplex viruses (oHSVs) are among the most widely studied OVs for the treatment of solid malignancies, and Amgen's oHSV Imlygic® for the treatment of melanoma is the only OV approved in major markets. Here we describe important biological features of HSV that make it an attractive OV, clinical experience with HSV-based vectors, and strategies to increase applicability to cancer treatment.

A. Introduction

Recent advances in cancer therapeutics include the use of tumor-specific lytic (oncolytic) viruses that both debulk tumors - reduce their size by eliminating cancer cells - and break down barriers to the induction of anti-tumor immunity. Oncolytic viruses have been derived from a wide variety of both DNA and RNA virus families, each having unique host ranges and replication mechanisms that make them attractive for attacking different tumor types. Oncolytic virus-mediated destruction of tumor cells leads to the production of danger signals by infected cells, a process referred to as immunogenic cell death, and to the release and presentation by antigen presenting cells (APCs) of virus and tumor-derived antigens. Tumor antigenic peptides are created by either an accumulation of mutations in cellular proteins, re-emergence of fetal/testis proteins without complete host tolerance, or by misfolded proteins to create novel processed peptides (Coulie et al., 2014; Kamran et al., 2016; Wang and Wang, 2017; Yarchoan et al., 2017). Robust immune responses to viral antigens can attract APCs to the tumor site wherein viral and tumor antigenic peptides can participate in cross-priming and the development of both anti-viral and anti-tumor immunity.

Oncolytic herpes simplex viruses (oHSVs) are among the most widely pursued for treatment of solid malignancies and have been validated with the commercial approval of Imlygic® (Talimogene laherparepvec, Amgen, Inc.) for the treatment of metastatic melanoma (Andtbacka et al., 2015; Pol et al., 2016). HSVs have several important features that contribute to their utility as tumor virolytics. These include a broad host cell range, which renders multiple types of tumors permissive for viral growth, and an ability to accommodate large or multiple transgene expression cassettes (referred to as large payload capacity) that enables oHSV-mediated delivery of multiple non-viral immunomodulatory products. Arming oHSVs with combinations of transgenes may increase oncolytic activity, help promote an immune-responsive tumor microenvironment and enhance the induction of innate and adaptive anti-tumor immune responses (Martuza et al., 1991; Andreansky et al., 1998; Chase et al., 1998). Advances in HSV engineering over the past decade and a half have produced oHSVs that have the full complement of viral genes, including virulence functions that resist cellular anti-viral responses and enable aggressive infection and destruction of tumors (Kambara et al., 2005; Gambini et al., 2012; Uchida et al., 2013). Strategies to enable uncompromised viral infection of tumor cells while preserving safety, to increase intratumoral vector spread, and to resist rapid elimination of replicating virus by counteracting anti-viral host responses should enhance oHSV potency and utility as anti-cancer agents. In this review, a brief history of the development of oHSVs, clinical outcomes in monotherapy and in combination with checkpoint inhibitors, and prospects for improved vector design are described.

B. oHSV design: curbing neurovirulence

1. Relevant biology of HSV

HSV-1 and HSV-2 are closely related neurotropic viruses that infect humans, their natural hosts. However, the virus can infect a wide variety of species, which has enabled the evaluation of oHSV vector designs in xenogeneic and syngeneic rodent tumor models. Some 70% of the adult human population is seropositive for HSV-1 (Kaufman et al., 2005), the most commonly used oncolytic HSV, and is likely to harbor latent virus in sensory neurons that innervate the site of initial infection. Latent virus reactivation can lead to virus spread to other susceptible individuals by direct contact with a virus-containing lesion. With the exception of herpes keratitis and encephalitis, HSV-1 infections of adults are mostly self-limiting, and latency is held in check largely by local anti-viral immunity (St Leger and Hendricks, 2011; St Leger et al., 2013).

The HSV genome is a double-stranded 152-kb DNA that encodes at least 84 gene products, the majority of which have multiple functions (Roizman, 2001; Roizman and Zhou, 2015). The genome consists of large and small unique segments flanked by repeat elements containing duplicate genes that contribute to virus replication, neurovirulence and latency. The repeat elements are responsible for genome isomerization such that 4 isomers are produced in infected cells. Approximately half of the viral genes encode essential products to complete the replication cycle in permissive cell cultures while the remaining genes contribute to the complex life cycle in the host. The viral genome is surrounded by an icosahedral capsid; a tegument composed of viral proteins involved in processes such as regulation of viral and host gene expression, host immune response, capsid trafficking and viral egress; and a host-derived lipid envelope studded with viral glycoproteins, many of which are involved in host-cell entry and virus spread to neighboring cells.

The basic biology of the virus has been exploited and modified in multiple ways in the design of oHSV. The 14-kb internal repeat region (IRR) can be removed without preventing virus replication since the full array of gene functions remains (Poffenberger et al., 1983; Meignier et al., 1988; Mazzacurati et al., 2015). IRR removal prevents genome isomerization, locking the viral genome into a single species and enhancing recombinant virus stability during production. With IRR removal, the HSV genome can accept ~25-kb of foreign DNA without straining the capsid packaging limits, allowing virus-mediated expression of one or more transgenes that can assist in the induction of anti-tumor immunity. In first-generation oHSVs, nonessential genes that overcome innate anti-viral responses, such as type I interferon (IFN) production, were deleted or modified to attenuate the virus in non-tumor cells and limit neurovirulence. More recent modifications have included the addition of microRNA response elements in the 3' UTR of essential genes to restrict replication to tumor cells (Lee et al., 2009; Mazzacurati et al., 2015) and modification of several viral envelope glycoproteins to allow only tumor-specific virus infection (Campadelli-Fiume et al., 2016). Inhibition of neuronal retrograde transport through specific mutation in the R2 domain of the tegument protein UL37 is a recently described alternative to prevent HSV-1 infection of the brain, enabling the design of HSV vaccines (Richards et al., 2017). Because HSV-1 is neurotropic, the initial work in developing oHSV vectors centered on the treatment of brain tumors by direct intratumoral injection (Markert et al., 2000; Rampling et al., 2000; Papanastassiou et al., 2002). Tumor cells of neuronal origin, such as glioma, are generally permissive for HSV mutants that are attenuated for replication in normal brain and do not elicit encephalitis. Although early clinical trials have thus far documented few complete responses using attenuated HSV, recent data from Todo and colleagues have shown promising activity and tolerability for G47Δ, an unarmed oHSV, in patients with recurrent or residual GBM (Todo, 2019), spurring a renewed interest into the treatment of GBM by oHSV.

2. oHSV vectors and treatment of neuronal tumors

a. Glioblastoma multiforme (GBM)

GBM (grade IV glioma) is the most common type of primary brain tumor in adults and is almost universally fatal despite aggressive therapies, including combinations of surgery, radiotherapy, and chemotherapy. The incidence of GBM is approximately 3.2 per 100,000 persons in the United States (Tamimi and Juweid, 2017), and GBM patients have a median survival of 12-18 months from initial diagnosis and 6-9 months after recurrence, with only 5% of patients surviving for 5 years (Wen and Kesari, 2008). The Standard of Care for management of these tumors includes tumor resection, followed by external beam radiotherapy with treatment doses of approximately 60 Gy (Desjardins et al., 2009; Thaker and Pollack, 2009). Partial responses to chemotherapy [e.g., temozolomide (Temodar)] are seen in approximately 30% of patients (Gilbert et al., 2013; Norden et al., 2013; Blumenthal et al., 2017), but overall survival is only improved by 2.5 months in newly diagnosed GBM in combination with radiotherapy (Stupp et al., 2005; Stupp et al., 2009; Grossman et al., 2010; Delgado-Lopez and Corrales-Garcia, 2016; Feng et al., 2017). Because tumor cells infiltrate the brain, tumor resection is not always complete. Even removal of the entire tumor-containing hemisphere (hemispherectomy) has been found to be non-curative (Chaichana et al., 2014). This is further compounded by the fact that the tumors are genetically and phenotypically heterogeneous, are poorly immunogenic, with scarce CD8+ T-cell infiltration (few T cells in mesenchymal and classical glioblastomas, essentially none in proneural GBM), and do not appear to respond to immune checkpoint inhibitors without induction of an inflammatory tumor microenvironment (Preusser et al., 2015; Reardon et al., 2016; Hodges et al., 2017). Due to the limited benefit of these therapeutic approaches, efforts to develop treatments that improve survival have included various biologic therapies, such as different types of oncolytic viruses, cell-based immunotherapy including chimeric antigen receptor (CAR)-T cells (Bagley et al., 2018), CAR-NK cells (Han et al., 2015), and gene therapy (Chiocca et al., 2019). Below is a brief history of the development of oHSVs for treatment of brain tumors.

b. Preclinical development

The first oncolytic HSV-1 described was dlsptk, a U_L23 [thymidine kinase (TK)] deletion mutant that displayed a promising therapeutic potential in the treatment of glioma in animal studies (Martuza et al., 1991). Replication of this mutant HSV-1 virus occurs only in mitotic tumor cells that upregulate their endogenous tk gene. Despite demonstrated tumor-specific replication in mouse models, there were serious concerns about vector safety because this strain of virus is insensitive to the most potent anti-herpetic agents, such as acyclovir, and because high titers of dlsptk caused neurotoxicity in mouse brain tumor models (Martuza et al., 1991). Hence, further development of HSV-1 vectors involved a search for other genes, the inactivation of which blocked virus replication in normal neurons but not in dividing cells.

One resulting vector was hrR3, an HSV-1 mutant with an in-frame insertion of the *lacZ* gene into the U_L39 (ICP6) gene encoding the large subunit of ribonucleotide reductase (RR), expressed as an ICP6-LacZ fusion protein that retained no RR activity. RR is a key enzyme for viral DNA synthesis in non-dividing cells but not in dividing cells where elevated levels of the host-cell RR can provide the deoxynucleotides needed as substrates for viral DNA synthesis. In tumor cells that are defective for p16 or the retinoblastoma (Rb) gene product, the transcription factor E2F is activated, resulting in increased endogenous RR activity (Hanson et al., 1994; DeGregori et al., 1995) that enables replication of HSV-1 mutants lacking virally encoded RR. Therefore, HSV-1 ICP6 mutants are relatively selective for cells with defects in the p16 tumor suppressor pathway (Aghi et al., 2008). Intratumoral inoculation of hrR3 into rats bearing malignant gliosarcomas improved animal survival (Fulci et al., 2007; Aghi et al., 2008), and this was further improved by ganciclovir (GCV) administration (Boviatsis et al., 1994). Viral TK converts GCV to a toxic product that acts as a chain terminator during DNA synthesis (Fulci and Chiocca, 2007), and the RR-defective phenotype conferred tumor-cell hypersensitivity to GCV (Luo et al., 2007).

Another approach to achieve tumor-restricted vector replication targets viral proteins involved in counteracting the innate immune response to HSV. Incoming HSV is sensed by a variety of pattern recognition receptors (PRRs), including cGAS, melanoma differentiation-associated protein 5 (MDA5), and Toll-like receptors (TLR-2, -3 and -7) (Mossman and Smiley, 2002; Leib et al., 2009; Paludan et al., 2011). These sensors, via multiple signaling pathways, induce the production of type 1 IFNs and the expression of over 2,000 IFN-stimulated gene products (ISGs) that mediate viral clearance and increase the expression of MHC class I molecules and the recruitment of innate lymphoid cells. Unsurprisingly, HSV expresses several gene products that counteract the IFN response, including ICP34.5 encoded by the γ34.5 gene. ICP34.5 is an inhibitor of NF-κB and STING (Pan et al., 2018), key mediators in the IFN pathway needed for induction of the ISGs. One well-characterized ISG, protein kinase R (PKR), shuts off host and viral protein synthesis by phosphorylating the translation initiation factor eIF2α. PKR is activated by binding to double-stranded RNA generated during viral infection and subsequent autophosphorylation and dimerization. While protein phosphatase-1α (PP1α) can inactivate PKR by dephosphorylation (Cheng et al., 2003), viral ICP34.5 selectively activates PP1α to mediate eIF2α dephosphorylation, allowing viral protein synthesis to proceed (Li et al., 2011). In addition to these functions, ICP34.5 can directly bind to and inhibit Beclin-1 to block cell death by autophagy (Orvedahl et al., 2007). Another viral protein, encoded by the Us11 gene, inhibits sensing of viral dsRNA by direct interaction with RIG-I and MDA-5 and works together with ICP34.5 to inhibit host shut-down of viral protein synthesis by the PKR pathway (Poppers et al., 2000; Xing et al., 2012).

While the type I IFN response induced upon virus infection represents an essential first line of defense against incoming virus in normal cells, tumors are often defective for components of the IFN response pathway and this can be exploited in the design of oHSV vectors. Eliminating HSV-1 gene products, such as ICP34.5, compromises virus growth in normal cells with an intact IFN response but allows growth in tumor cells. For example, HSV-1 strains expressing a mutant form of ICP34.5 fail to replicate in normal brain tissue and are non-pathogenic for highly HSV-susceptible primates such as *Aotus* monkeys following intracranial inoculation (Hunter et al., 1999; Todo et al., 2000; Varghese et al., 2001). Replication of HSV1716, a γ34.5 deleted oHSV, in ependymal cells in BALB/c mice is a notable exception (Kesari et al., 1998). In tumor cells with increased signaling through epidermal growth-factor receptor 1 (EGFR) or 2 (HER2/Neu) or platelet-derived growth-factor receptor (PDGFR), persistent Ras activation inhibits PKR activity. Many glioblastomas express a constitutively active mutant form of EGFR as part of the tumor phenotype that may contribute to effective replication of γ34.5 mutant HSV-1 vectors.

HSV-1 vectors deleted for γ34.5 function have thus been highly touted for utility as oncolytic vectors. R3616, a replication competent mutant of HSV-1 strain F, lacks most of the coding regions of the two γ34.5 genes (Kramm et al., 1997) (Figure 1A). R3616 is capable of replicating in glioma xenografts in nude mice, and tumor killing is enhanced by ionizing radiation (Advani et al., 1998; Bradley et al., 1999; Huszthy et al., 2008). R3616 has also been shown to be effective for treatment of chemotherapy-resistant tumors (Chahlavi et al., 1999). HSV1716, a mutant of HSV-1 strain 17, also lacks both γ34.5 genes (Figure 1B) (MacLean et al., 1991; Valyi-Nagy et al., 1994; Kesari et al., 1995) and shows a similar anti-tumor and safety profile (Randazzo et al., 1995; Rampling et al., 2000).

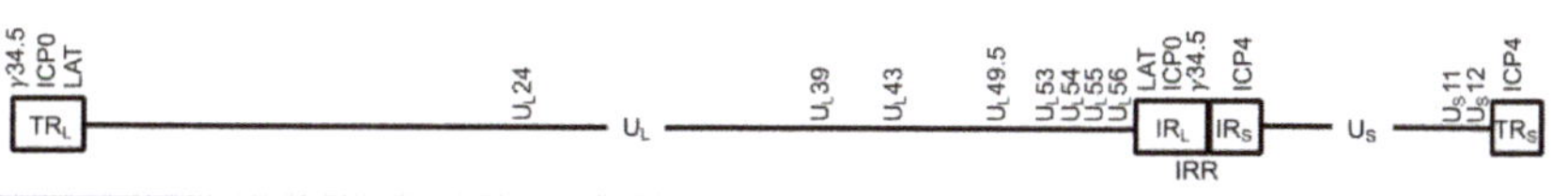

	Virus	Strain	γ34.5	U_L23 (TK)	U_L24	U_L39 (ICP6)	U_L43	$U_L49.5$ (gN)	U_L55/ U_L56	IRR	U_S11	U_S12 (ICP47)	Transgene(s)
A	R3616	F	Δγ34.5										
B	HSV1716	17	Δγ34.5										
C	G207	F	Δγ34.5			ΔICP6							ICP6p-LacZ [ΔICP6]
D	G47Δ	F	Δγ34.5			ΔICP6					U_S12p	ΔICP47	ICP6p-LacZ [ΔICP6]
E	T-VEC	JS1	Δγ34.5								U_S12p	ΔICP47	CMVp-GMCSF [Δγ34.5]
F	rQNestin34.5	F	Δγ34.5			ΔICP6							NestinEp-γ34.5 [ΔICP6]
G	C134	F	Δγ34.5										CMVp-CMVIRS1 [$U_L3/4$]
H	M032	F	Δγ34.5			ΔICP6							Egr1p-hIL12 [Δγ34.5] ICP6p-LacZ [ΔICP6]
J	NV1020[1]	F		ICP4p-TK	$ΔU_L24$				$ΔU_L56$	ΔIRR			
K	HF10[2]	HF					$ΔU_L43$	ΔgN	$ΔU_L55/56$	ΔLAT			
L	rRp450	KOS				ΔICP6							ICP6p-rCYP2B1 [ΔICP6]

[1]NV1020 is deleted for a 700-bp region encompassing the thymidine kinase (tk) locus and the promoter of the U_L24 gene, as well as a region encompassing U_L56 and the internal repeat region (IRR, containing a single copy of ICP0, LAT, γ34.5, and ICP4); the U_L/U_S junction carries a 3.7-kb duplication of $U_L5/6$, a 5.2-kb fragment of HSV-2 DNA, and a copy of the tk gene under the ICP4 promoter.
[2]HF10 naturally overexpresses U_L53 (gK) and U_L54 (ICP27).

Figure 1. Oncolytic HSV vectors employed in human clinical trials. Schematic of the wild-type HSV genome (top) with relevant genes indicated above. The unique long (U_L) and unique short (U_S) segments contain single copies of each gene, while the terminal (TR_S and TR_L) and internal repeats (IR_L and IR_S, together comprising the IRR) contain duplicate copies of ICP4, ICP0, γ34.5, and LAT. The genome modifications used to attenuate the oHSV vectors described in this review are summarized in the table underneath (descriptions in the text refer to the letters labeling each row). Transgenes expressed from the vectors, their promoters (p), and [insertion sites] in the viral genome are listed in the last column.

HSV1716 was among the first oHSVs to be taken into human clinical trials, along with G207, another γ34.5-deleted virus that additionally contains a *lacZ*-disrupted U_L39 RR (ICP6) gene (Figure 1C). In preclinical experiments, G207 extended the survival of mice with intracranial GBM tumors, and more recent studies have shown that cisplatin enhanced the effect of G207 without inhibiting viral replication (Chahlavi et al., 1999). Radiation has also been shown to be effective in combination with oncolytic vector therapy (Jorgensen et al., 2001; Blank et al., 2002; Stanziale et al., 2002). Additional viral genome modifications have been considered to improve oncolytic activity of HSV-1. γ34.5 mutants have been combined with deletion of the immediate-early U_S12 gene encoding ICP47, a protein that blocks the loading of HSV-1 antigenic peptides onto MHC class I surface molecules (York et al., 1994; Fruh et al., 1995; Goldsmith et al., 1998). In the absence of ICP47, both viral and tumor-related antigenic peptides can be presented to cytotoxic T cells (CTLs) to eliminate virus-infected and potentially uninfected tumor cells (York et al., 1994; Fruh et al., 1995; Goldsmith et al., 1998). Deletion of the U_S12 gene and the U_S11 gene promoter in G207-derivative G47Δ (Figure 1D) (Todo et al., 2001; Zeng et al., 2013) places the U_S11 gene, normally a true late gene, under the control of the U_S12 immediate-early gene promoter, resulting in compensation for γ34.5 deletion, increased virus yield and killing of GBM and other tumor lines in culture (Cassady et al., 1998; Mohr et al., 2001; Todo et al., 2001; Liu et al., 2003).

c. Clinical trials

The tumor-restricted replication and efficacy of γ34.5 mutant oHSVs in preclinical experiments led to multiple clinical trials with various HSV-1 strain backgrounds and mutation combinations (Figure 1). In a Phase-I clinical trial of 21 patients with non-histologically proven recurrent GBM, G207 (Figure 1C) was delivered by intra-tumoral injection at three-fold increasing doses from 1 x 10^6 to 3 x 10^9 pfu (Markert et al., 2000). Patients only experienced mild adverse events such as fever, malaise and other signs of viral infection that resolved after 3-4 days. This trial showed that inoculation of oHSV was relatively safe in human brain but did not provide an effective treatment. This was followed by a Phase-Ib trial in 6 patients with recurrent GBM using a dose of 1.15 x 10^9 pfu, showing radiographic evidence of tumor regression and the presence of an immune infiltrate (Markert et al., 2009). Another trial employing G207 at a dose of 3 x 10^9 pfu in combination with 5-Gy radiotherapy is currently recruiting children with recurrent GBM (ClinicalTrials.gov NCT0257845) (Waters et al., 2017; Friedman et al., 2018). In a Phase-I trial of the related G47Δ virus (Figure 1D) (UMIN000015995), doses of up to 1 x 10^9 pfu/injection for a total of 6 injections were found to be safe, and efficacy -- 92.3% of recurrent GBM patients remaining alive at 12 months -- was deemed sufficient to prompt the Japanese Ministry of Health, Labor and Welfare to designate G47Δ for fast-track approval (Fukuhara et al., 2016; Todo, 2019).

HSV1716, the strain 17 γ34.5 mutant (Figure 1B), was well tolerated at doses of up to 10^5 infectious units after stereotactic injection into recurrent malignant gliomas (Papanastassiou et al., 2002; Harrow et al., 2004), a dose much lower than the safe doses observed with G207 (Markert et al., 2000; Markert et al., 2009), which may be due to strain differences and the retention of ICP6 in HSV1716. Analysis of HSV1716-infected tumor explants revealed viral replication at 4-9 days after injection, and the amount of recovered virus exceeded the input dose in at least some patient samples. Radiographic evidence supported a reduction in tumor mass in some patients; yet, long-term survival was not altered, suggesting that more robust oHSV replication may be required.

Strategies employed to maintain vector safety while increasing efficacy include the expression of γ34.5 from a hybrid nestin enhancer-HSP68 minimal promoter in the ICP6-deleted oHSV mutant rQNestin34.5 (Figure 1F), an oHSV that was first shown to be highly active in a GBM model in nude mice (Kambara et al., 2005). A Phase-I trial safety/dose-escalation study using rQNestin34.5 alone or in combination with cyclophosphamide (CPA) for treatment of recurrent GBM patients is underway (NCT03152318). Another strategy is employed in the strain F-based vector C134 that expresses the human cytomegalovirus (CMV) IRS1 gene (Figure 1G) to enhance the oncolytic activity of γ34.5 mutant oHSV (Cassady et al., 2017). A Phase-I dose-escalation trial (NCT03657576) using C134 began enrolling patients in November 2018.

A new wave of pre-clinical studies, ultimately leading to further human clinical trials for GBM, now employs these and other oHSV backbones for expression of various transgenes to increase viral spread and penetration within the tumor mass (Dmitrieva et al., 2011; Kim et al., 2014; Jaime-Ramirez et al., 2017; Sette et al., 2019), block neo-vascularization of the tumor (Goodwin et al., 2012; Yoo et al., 2012; Fujii et al., 2013) or modify the immune response in the host tumor microenvironment via the expression of cytokines or chemokines (Liu et al., 2003; Roth et al., 2014; Meisen et al., 2015; Patel et al., 2016; Leoni et al., 2018b). For example, G207 was armed with human IL-12 expressed from the Egr1 promoter (Figure 1H). The resulting oHSV, M032, has been shown to be safe in non-human primates and only induce a transient neutrophil and white blood cell host immune response (Roth et al., 2014). It is currently being tested in a Phase-I dose-escalation trial (NCT02062827) in 36 patients with recurrent GBM (Patel et al., 2016). The results of this trial await publication and are likely to influence further human trials for treating GBM with armed oHSV (see section D below for a discussion of arming genes). Together, these Phase-I oHSV trials for GBM demonstrated that the vectors are safe both in HSV-naïve patients and in patients with pre-existing anti-HSV immunity, but they were not designed to examine long-term improvement in overall survival (OS) or progression free survival (PFS).

3. oHSV vectors and treatment of non-neuronal tumors

Despite the vast preclinical and clinical effort to develop oHSVs to treat GBM, the initial success of using oHSVs to treat patients has come about by treating a tumor of non-neuronal origin. Talimogene laherparepvec (T-VEC; Imlygic®), an oHSV developed by BioVex (under the name Oncovex$^{GM\text{-}CSF}$) and Amgen, has achieved regulatory approval for the treatment of metastatic melanoma (Pol et al., 2016). T-VEC is an intratumorally injected product; its clinical development for melanoma was facilitated by the accessibility of the malignant lesions and the possibility to perform repeated injections. Lesions distant to the sites of injection or visceral metastases, left uninjected, allow for the observation of an abscopal effect, i.e., the induction of anti-tumor activity affecting untreated tumor masses, likely through the generation of a systemic anti-tumor immune response. In addition, melanoma is well-known for its high degree of antigenicity and immunogenicity, a notably 'hot' tumor with expression of several prominent tumor antigens, such as MART-1, gp100, MAGE-1.A1, MAGE-3.A1, Melan-1, Mel-3 and tyrosinase, and a high degree of response to checkpoint inhibitors (Gujar et al., 2018a; Gujar et al., 2018b; Kamran et al., 2018). T-VEC (JS1/γ34.5$^-$:HCMVp-GM-CSF/ICP47$^-$; Figure 1E) was derived from an HSV-1 clinical isolate (JS1) and is deleted for both copies of the γ34.5 gene (Liu et al., 2003). The human GM-CSF gene is expressed from both of the deleted γ34.5 loci. In addition, the U_S12/ICP47 gene is deleted to release interference with transporter associated with antigen presentation (TAP)-dependent MHC class I antigen presentation. As in G47Δ discussed above, the deletion of ICP47 changes the temporal expression of the late gene U_S11 to that of an immediate early gene, at which time the U_S11 product can block PKR activation (Cassady et al., 1998; Mohr et al., 2001; Todo et al., 2001; Liu et al., 2003), thereby enhancing viral replication in tumor cells. The T-VEC Phase-I study in 30 patients established dose and regimen and showed that the intratumoral injection was well tolerated, with minor treatment-associated adverse events (AEs), including flu-like symptoms such as fatigue, pyrexia, chills, nausea, and focal cellulitis at the injection site. Notably, these events were more pronounced in previously HSV-seronegative patients, leading to the adoption of an initial dose of 1-4 x 10^6 pfu to seroconvert all patients before full therapeutic doses of 1-4 x 10^8 pfu every 2 weeks (Hu et al., 2006). T-VEC DNA was detected in injected tumors and in some neighboring uninjected lesions, but not in distal lesions, indicating that T-VEC did not spread to distal tumor sites.

The favorable results of this Phase-I trial supported a Phase-II trial with 50 melanoma patients, in which a similar safety profile was observed as in the Phase-I trial, an overall response rate (ORR; i.e., arrested tumor growth) of

26%, and a complete response rate (CR; tumor no longer detectable) of 16.9% (Senzer et al., 2009; Kaufman et al., 2010; Andtbacka et al., 2019). Regression of both injected local tumors as well as more distal tumors was observed, with overall survival (OS) rates of >50% at 1-2 years post-therapy. The immune response was evaluated and showed an increase in MART-1-specific T cells and a corresponding decrease in regulatory T cells (Tregs) and myeloid-derived suppressor cells (MDSCs) within the tumor microenvironment (TME) (Senzer et al., 2009; Kaufman et al., 2010). This prompted a large Phase-III OPTiM (Oncovex$^{GM\text{-}CSF}$ Pivotal Trial in Melanoma) study of 436 patients, in which T-VEC was compared to GM-CSF protein as a standard control therapy for melanoma. T-VEC decreased by >50% the size of 64% of the injected tumors, 34% of non-visceral tumors, and 15% of visceral tumors (Andtbacka et al., 2015). The durable response rate (DRR; no evidence of recurrent disease) was 16.3% for T-VEC, but only 2.1% for GM-CSF alone, while the ORR for T-VEC was 26.4% versus 5.7% for GM-CSF alone (Andtbacka et al., 2015; Harrington et al., 2016). These numbers were even better for the earlier-stage subgroup (57.1% of evaluated patients) with DRRs of 25.2% versus 1.2% and ORRs of 40.5% versus 2.3% (Andtbacka et al., 2015; Harrington et al., 2016). Together, these results warranted the first FDA approval of an oncolytic virus and gene therapy product in the U.S. for the treatment of a solid tumor, melanoma, in October of 2015.

Currently there are 37 active clinical trials employing T-VEC under its commercial name (Imlygic®), targeting melanoma, sarcomas, breast cancer, pancreatic cancer, colorectal cancer, head and neck squamous cell carcinoma (HNSCC), lymphoma, liver cancer, pleural and intraperitoneal cancer, and bladder cancer. Since its original success in melanoma treatment, T-VEC has been evaluated in combination with chemo- and radiotherapies. A Phase-I/II trial using T-VEC in HNSCC, in combination with 70 Gy radiotherapy and cisplatin, in 14 patients showed an 82.3% tumor response and OS of 82.4% at 2-years post therapy (Harrington et al., 2010). HSV-1 DNA was detected not only within the injected tumor but also in adjacent tumors, and the number of detected genomes was significantly higher than the injected dose, suggesting active virus replication in the TME. The fact that T-VEC treatment alone resulted in regression of local tumors but not distal and visceral tumors suggested that the immunosuppressive TME may be inhibiting the cytotoxic T lymphocyte (CTL) response required to effectively eliminate untreated distal and visceral lesions, pointing to a need for combination trials with immune checkpoint-inhibitory antibodies such as anti-CTLA-4 ipilimumab (Yervoy) and anti-PD-1 pembrolizumab (Keytruda). The combination of T-VEC with anti-CTLA-4 (Puzanov et al., 2016) showed no added dose-limiting toxicities (DLTs) of T-VEC and raised the ORR from 26% seen in the Phase-III OPTiM trial to greater than 50%. A greater increase in CD8+ T-cells was detected in patients that were considered to be complete responders (CR) compared to partial responders (PR) or patients with progressive disease. T-VEC combination with pembrolizumab in a Phase-Ib/II trial (Ribas et al., 2017) was found to increase the frequency of complete responses to 62% compared to 24% for T-VEC alone, and a dramatic increase in circulating CD3+/CD8+ T-cells was observed. The field is eagerly awaiting the outcome of the large phase III clinical trial Keynote 034 (NCT02263508) combining T-VEC and Keytruda to hopefully confirm the promising phase Ib/II data. Additional combination studies with T-VEC, checkpoint-inhibitory antibodies and chemo- or radio-therapy are underway. Radiotherapy can increase the levels of tumor antigens and, in combination with T-VEC, may increase the number of tumor-specific targets for activated CTLs.

In addition to T-VEC, other oHSVs have been tested in non-CNS tumor clinical trials. HSV1716 has been evaluated in Phase-I trials for melanoma (MacKie et al., 2001), HNSCC (Mace et al., 2007; Mace et al., 2008) and pediatric non-CNS tumors (Streby et al., 2017). In these trials, doses of $1x10^5$ - $2x10^6$ pfu proved safe and led to virus replication in the tumor, yet failed to provide compelling evidence of efficacy. NV1020 (Figure 1J) contains deletions of the γ34.5 genes, the U_L24 promoter, the complete internal repeat region, and the U_L56 gene that encodes a protein likely involved in Golgi vesicular trafficking of virions (Koshizuka et al., 2005). These deletions resulted in a severely attenuated recombinant virus that can still replicate in tumor cells. *In-vivo* analysis with this vector demonstrated tolerability and efficacy in a number of experimental models of bladder carcinoma (Cozzi et al., 2001), HNSCC (Wong et al., 2001), and non-small cell lung cancer (Ebright et al., 2002). These strong efficacy and safety data in mice led to Phase-I trials conducted by hepatic artery infusion of the virus for colorectal cancer (CRC) metastatic to the liver. The results demonstrated safety of NV1020 with only mild to moderate AEs, such as fever and transient lymphopenia, and no patients displayed evidence of virus shedding (Kemeny et al., 2006; Geevarghese et al., 2010; Sze et al., 2012). Antitumor efficacy was limited in this refractory setting, with 50% of patients showing stable disease. Another oHSV, HF10 (Figure 1K), is a natural HSV-1 variant defective for expression of U_L43, $U_L49.5$, U_L55, U_L56 and LAT (Ushijima et al., 2007). HF10 replicates to high efficiency in tumor cells while demonstrating substantially reduced toxicity compared to wild-type HSV-1. Takara-sponsored Phase-I trials have been conducted in Japan for solid tumors with superficial lesions (Ferris et al., 2014), breast cancer (Nakao et al., 2004; Kimata et al., 2006; Nakao et al., 2007; Sahin et al., 2012), unresectable pancreatic cancer (Nakao et al., 2011; Kasuya et al., 2014; Hirooka et al., 2018), and resectable HNSCC (Fujimoto et al., 2006). In these Phase-I trials, HF10 was well-tolerated without virus-attributable serious adverse events (SAEs). While viral replication was detected in the tumor mass, no evidence of virus shedding from the injection sites was observed. Some evidence was obtained of moderate tumor cell killing along with CD4+ and CD8+ T-cell infiltrates within the TME. These results led to a Phase-II combination trial (NCT02272855) of HF10 with ipilimumab in unresectable stage III-IV malignant melanoma that demonstrated an immune-related response criteria (irRC) best overall response rate (BORR) at 24 weeks of 41% [18% complete disappearance of all lesions (irCR) and 23% decrease in tumor burden by ≥50% (irPR)] (Andtbacka et al., 2018). G47Δ (Figure 1D) has been evaluated in a Phase-I dose-escalation study for the treatment of castration-resistant prostate cancer (UMIN000010463) and in a Phase-I trial (UMIN000011636) for recurrent olfactory neuroblastoma (Fukuhara et al., 2016). Finally, oHSV rRp450, containing a rat cytochrome P450 gene in place of the U_L39 RR gene (Figure 1L) (Chase et al., 1998), is under study in a

Phase-I trial (NCT01071941) for primary liver cancer and liver metastases in 40 patients, comparing single and repeated infusions into the hepatic artery.

4. Summary

To date, clinical trials using oHSVs have demonstrated that all of the viral recombinants (i) were well-tolerated; (ii) had limited risk of shedding; (iii) tended to induce some level of T-cell recruitment; (iv) caused sero-conversion of patients who were HSV-seronegative at the time of oHSV administration; and (v) displayed evidence of intra-tumoral virus replication. When evaluated as monotherapy, clinical efficacy of most oHSVs was limited for indications other than melanoma. Currently, the data evaluating clinical benefit of oHSVs in combination with chemo-, radio- or immunotherapy is limited by the absence of controlled Phase-III studies. Still, Phase-II clinical data evaluating T-VEC and HF10 in combination with ipilimumab or pembrolizumab appear very promising, with roughly a doubling of the response rate achieved by these checkpoint inhibitors alone. More importantly, in the case of T-VEC combined with pembrolizumab, complete responses were achieved with patients presenting with low intratumoral CD8+ T-cell counts and low expression of PD-L1 and IFN-responsive genes, parameters that are indicative of immunologically "cold" tumors that are not expected to respond to immunotherapy (Ribas et al., 2017). These data, on the heels of T-VEC/ Imlygic's approval by the FDA, are fueling the current enthusiasm for oncolytic viruses and the efforts to improve upon the first-generation oHSV vectors.

C. Advances in oHSV design for increased specificity and efficacy

1. Selective HSV attenuation by microRNAs

As described above, the primary method of attenuating oHSV has been to remove or inactivate genes that contribute to innate immune evasion in order to render the virus susceptible to the innate cellular defenses that are active in normal cells but compromised in tumor cells. However, tumors can vary widely in their loss of type I IFN responses (e.g., (Alain et al., 2010)) and IFN can be produced by myeloid cell types within the TME. Therefore, the maintenance of the complete set of viral defense functions is thought to be beneficial to retain full vector replication activity *in vivo*, thus prompting the development of novel approaches for virus attenuation. The maintenance of normal cellular function by microRNAs (miRNAs) is frequently disrupted in tumors, and this dysregulation of miRNA expression represents a distinct signature of the tumor that has been exploited to attenuate oncolytic vector replication in normal cells. The use of differential miRNA expression to limit oHSV replication to tumor cells was first described in a prostate cancer model using an amplicon vector with recognition sequences for miR-143 and miR-145 engineered into the 3'UTR of the ICP4 gene; both miRNAs are expressed in normal tissues but are significantly down-regulated in prostate cancer cells (Lee et al., 2009). The promise of this strategy was confirmed in models of other tumors, including liver tumors where miR-222 expression is highly reduced compared to hepatocytes (Fu et al., 2012), non-small-cell lung cancer (miR-145) (Li et al., 2013), and urothelial bladder cancer (miR-143 + miR-124) (Zhang et al., 2016). In GBM, tumor cells have low or undetectable expression of miR-124, an miRNA that is highly expressed in normal neurons (Sempere et al., 2004; Gaur et al., 2007; Silber et al., 2008). Incorporation of miR-124 target sites into the 3'UTR of the ICP4 gene blocked virus replication in miR-124 expressing cells in culture and virus replication in normal brain, preventing HSV-induced encephalitis (Mazzacurati et al., 2015). The field is only beginning to harness the power of miRNAs to control off-target oHSV replication, which can be combined with additional safety measures, such as the use of tumor-specific promoters to control the expression of virulence genes (Kambara et al., 2005; Nakashima et al., 2018) and, as described below, retargeting of the virus attachment and entry machinery to tumor-related cell surface receptors (Campadelli-Fiume et al., 2016). These combinatorial approaches will be particularly important in the development of vectors for systemic treatment of metastatic cancer.

2. Retargeting HSV infection to tumor-cell receptors.

Vector retargeting is a promising strategy to achieve both oHSV safety and efficacy by restricting virus infection to cells expressing tumor-associated cell surface receptors. By limiting off-target infection, this approach provides opportunities for the use of unattenuated oHSVs in cancer therapy. HSV vectors have been retargeted to a number of cell surface receptors, including the epidermal growth factor receptor (EGFR/EGFRvIII) (Uchida et al., 2013), epithelial cell adhesion molecule (EpCAM) (Shibata et al., 2016), and human epidermal growth factor receptor 2 (HER2) (Menotti et al., 2008) (Figure 2), and several of these retargeted viruses have demonstrated considerable efficacy in *in-vivo* mouse models of solid tumors. Although efficacy following systemic virus administration has yet to be reported, preferential virus homing to tumor sites has been observed (Uchida et al., 2013). Retargeting of HSV is thus emerging as a promising strategy to accomplish unimpaired tumor-specific lytic activity.

a. Virus attachment and entry into host cells

To accomplish vector retargeting, the virus must be modified to eliminate entry through its cognate receptors. In our current understanding, HSV adsorption and entry involves the coordinated activities of glycoproteins gB, gC, gD and the gH/gL heterodimer (see Chapter 2). Virus attachment is initially mediated by binding of gC and gB to cell surface glycosaminoglycans (GAGs), primarily heparan sulfate (HS) (WuDunn and Spear, 1989; Herold et al., 1991; Shieh et al., 1992; Spear et al., 1992; Gruenheid et al., 1993; Herold et al., 1994; Tal-Singer et al., 1995; Trybala et al., 2002). Deletion of gC and the HS binding domain of gB in a single mutant virus impairs virus adsorption and reduces but does not eliminate entry, and wild-type HSV can infect HS-deficient cells (Gruenheid et al., 1993; Laquerre et al., 1998b). Retargeting of gC to a model receptor (erythropoietin receptor) combined with disruption of the HS-binding elements of gC and gB was first reported by Laquerre and colleagues (Laquerre et al., 1998a) and has since been demonstrated by others for additional receptors (Argnani et al., 2004; Grandi et al., 2004; Wang et al., 2005; Kouvatsis et al., 2007; Cao et al., 2010; Grandi et al., 2010). However, while these viruses may preferentially attach to their target cells, their entry specificity is determined by another envelope glycoprotein, gD. In a typical HSV infection, virion attachment to HS is followed by gD-mediated binding to one of several entry receptors: (1) Herpesvirus entry mediator (HVEM/HveA or

TNFRSF14), a member of the TNFα receptor superfamily (Montgomery et al., 1996); (2) nectin-1 (CD111/HveC), a cell adhesion molecule and a member of the immunoglobulin superfamily (Geraghty et al., 1998; Krummenacher et al., 1998); (3) nectin-2 (CD112/HveB), a receptor for HSV-2 and variant HSV-1 strains that carry a specific amino acid change at position 27 of gD (Q27P or Q27R) (Warner et al., 1998); and (4) 3-*O*-sulfated HS (3-*OS* HS), a carbohydrate receptor for HSV-1 (Shukla et al., 1999). Interaction with any of these receptors causes a conformational change in gD that is currently thought to activate the gH/gL heterodimer, which, in turn, activates the fusogenic activity of gB (Atanasiu et al., 2010; Eisenberg et al., 2012). Depending on the cell type, HSV entry takes place either by envelope fusion with the plasma membrane (Fuller and Spear, 1987; Fuller et al., 1989) or by virion endocytosis followed by envelope fusion with the endosomal membrane (Nicola et al., 2003; Nicola and Straus, 2004).

b. HSV retargeting via modification of glycoprotein D

As the receptor-dependent initiator of virus entry, gD is the key determinant of HSV tropism. Retargeting, therefore, requires both disruption of the ability of gD to interact with its cognate receptors and establishment of entry-initiating interactions with a desired target receptor through incorporation of a cognate ligand into gD. Substantial insight into the structure-function relationship of gD has come from extensive mutagenesis studies, mapping of monoclonal antibody binding sites, and the crystal structures of gD alone and complexed with HVEM or nectin-1 (Feenstra et al., 1990; Krummenacher et al., 1998; Whitbeck et al., 1999; Carfi et al., 2001; Connolly et al., 2003; Milne et al., 2003; Yoon et al., 2003; Jogger et al., 2004; Manoj et al., 2004; Krummenacher et al., 2005; Di Giovine et al., 2011). Pertinent to the goal of altering the binding specificity of gD without disrupting its entry-initiating response to receptor binding are the following. First, the membrane-proximal (C-terminal) region of the ectodomain, roughly between residues 260 and 310 and referred to as the profusion domain (Cocchi et al., 2004; Fusco et al., 2005), is required for receptor-dependent signal transduction to gH/gL and gB. Second, the N-terminal region contains multiple residues required for binding to HVEM

(Connolly et al., 2003) and 3-*OS* HS (Yoon et al., 2003), and complete disruption of this binding can be accomplished by mutations, insertions, and small or large deletions. Third, certain substitutions or deletion of residue 38 abolishes nectin-1 binding. Fourth, the central core of the gD ectodomain (residues 61-218) that includes much of the immunoglobulin (Ig) fold of the ectodomain [residues 56-184 (Carfi et al., 2001)], is dispensable for gD activity (Zhou and Roizman, 2007). Accordingly, HSV retargeting by the insertion of a novel ligand into gD has thus far relied almost exclusively on disruption of natural receptor binding by mutations and ligand insertion at the amino terminus, or on replacement of the central Ig domain, which also eliminates natural receptor binding. Recent studies have additionally shown that while short insertions (≤ 20 amino acids) C-terminal to the Ig fold are readily tolerated, longer insertions [62 aa or a single-chain antibody (scFv) of nearly the same size as the gD ectodomain] typically curtail cell surface expression and, consequently, membrane fusion (Fan et al., 2017).

Early attempts at HSV retargeting succeeded in ablating HVEM recognition and facilitating gD engagement of non-HSV receptors through insertion of cognate foreign ligands, including 247- and 256-amino-acid long scFvs. However, elimination of the interaction of gD with nectin-1 proved to be more challenging as the nectin-1-binding site was poorly defined and mutations that selectively eliminated nectin-1 binding were elusive (Zhou et al., 2002; Kamiyama et al., 2006; Menotti et al., 2006). The first fully retargeted HSV recombinant, R5141 (Figure 2), was identified in a stepwise process that included a screen of point mutations in a targeted gD construct for loss of virus entry through nectin-1 (Zhou and Roizman, 2006). The ligand in this construct, IL-13, was genetically fused to gD residue 33 to promote gD interaction with the IL-13 receptor α2 (IL-13Rα2), which is specifically overexpressed in malignant brain tumors. The chimeric IL-13-gD protein lacked residues 1-32 of gD, ablating HVEM binding, and used the IL-13 signal peptide instead of the native signal peptide. The screen identified V34S as a substitution that abolished nectin-1 binding while preserving the ability of the virus to use IL-13Rα2 as an entry receptor. The authors were able to grow this fully retargeted virus on HSV receptor-deficient J1.1-2 cells transduced with IL-13Rα2 (J-13R), thereby eliminating selective pressure for reversion or complementation of the V34S substitution during propagation, although presumably not during replication in nectin-1-positive tumors. Additionally, R5141 could be grown on IL-13Rα2-transduced Vero cells (Vero-13R), albeit to lower titers than parental wild-type HSV-1, but not on standard Vero cells. Interestingly, while a variety of HSV gene products, including other glycoproteins, were abundantly visible on Western blots of R5141-infected Vero-13R cells, the retargeted gD protein was undetectable, suggesting that very low levels of gD were sufficient for infection, but also that the protein was improperly processed, undermining robust production of progeny virus.

A number of HSV-1 recombinants, fully retargeted to HER2 using a HER2-specific scFv, have since been described (Menotti et al., 2008; Menotti et al., 2009) (Figure 2). In recombinant R-LM113, the anti-HER2 scFv replaced gD residues 6-38, eliminating both HVEM and nectin-1 binding, after attempts to do so by ligand insertion between residues 24 and 25, combined with the V34S mutation, failed (Menotti et al., 2008). Replacement of the entire Ig-like core of the gD ectodomain (residues 61-218) with the anti-HER2 scFv, comprising two Ig domains, also yielded a fully retargeted virus, referred to as R-LM249 (Menotti et al., 2009), in agreement with the evidence that the gD core is dispensable for retargeted HSV entry (Zhou and Roizman, 2007). The authors implied that the deletion of the core ensured resistance to reacquisition of natural receptor binding ability, but no specific evidence supporting this claim was presented.

An orthogonal approach to HSV retargeting involved "bridging" gD with non-cognate tumor-specific receptors by using soluble bi-specific adapters (Nakano et al., 2005; Baek et al., 2011). One such adapter, comprising the N-terminal gD-binding Ig domain of nectin-1 and an EGFR-specific scFv that also recognizes the tumor cell-specific EGFR variant III (EGFRvIII), mediated EGFR-dependent HSV-1 infection (Nakano et al., 2005). Since complete detargeting from cellular nectin-1 proved problematic and since extracellular

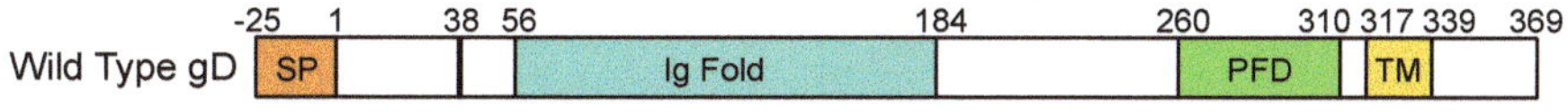

Name	Targeting Ligand	Insertion Site	Signal Peptide	Additional gD Detargeting Mutations	Strain	Reference
R5141[1]	IL13	gD Δ1-32	IL13	gD V34S	F	Zhou and Roizman 2006
R-LM113	scFv HER2	gD Δ6-38	gD	NA	F	Menotti et al, 2008
R-LM249	scFv HER2	gD Δ61-218	gD	NA	F	Menotti et al, 2009
KNE[2]	scFv EGFR/EGFRvIII	gD Δ2-24	gD	gD Y38C	KOS	Uchida et al, 2013
KGNEp[2]	scFv EpCAM	gD Δ2-24	gD	gD Y38C	KOS	Shibata et al, 2016
R-593	scFv PSMA	gD Δ6-38	gD	NA	F	Menotti et al, 2018
R-611	scFv EGFR	gD Δ6-38	gD	NA	F	Menotti et al, 2018
R-613	scFv EGFRvIII	gD Δ6-38	gD	NA	F	Menotti et al, 2018

[1]R5141 is also retargeted via gC with IL13 replacing gC residues 1-140.
[2] The viral backbone contains two point mutations in gB, referred to as gB:NT, that enhance virus entry.

Figure 2. oHSV retargeting via modification of gD-receptor interactions. (top) Schematic representation of gD, illustrating the domains described in the text: signal peptide (SP), Ig fold, profusion domain (PFD), transmembrane domain (TM). The amino acid positions defining each domain are indicated above the schematic; mature gD starts at amino acid 1. The table underneath describes viruses bearing fully retargeted recombinant gD molecules.

adapters cannot redirect lateral virus spread, the scFv was inserted into gD between residues 24 and 25, as in the early gD constructs described above. gD detargeting from HVEM was enforced by deletion of gD residues 7-11, while detargeting from nectin-1 was achieved by changing positions 3 and 38 to cysteines [A3C/Y38C (Connolly et al., 2005)]. Phenotypic complementation assays of a gD-null virus confirmed uniquely EGFR-dependent viral entry that was enhanced by a pair of entry-accelerating mutations in gB (gB:NT) (Uchida et al., 2010). In contrast, replacement of residues 7-39 or 61-218 of gD with the same scFv, similar to the anti-HER2 scFv position in the R-LM113 and R-LM249 viruses, respectively, failed to enable entry through EGFR engagement. Indeed, it was recently reported that replacing the 61-218 region of gD with a different EGFR-specific scFv or with scFvs recognizing EGFRvIII or prostate-specific membrane antigen (PSMA), failed to enable virus entry through the respective cognate receptors although all three scFvs were effective when replacing gD residues 6-38 (R-593, R-611, R-613, Figure 2) (Menotti et al., 2018). The authors of this study proposed that the failure of these new scFvs to function within the 61-218 deletion region might be due to the arrangement of each of these scFvs as V_H-spacer-V_L, as opposed to the V_L-spacer-V_H arrangement of the successful HER2-specific scFv. However, the earlier anti-EGFR/EGFRvIII scFv had the V_L-spacer-V_H arrangement (Uchida et al., 2010), suggesting that the relative positioning of the two antibody V domains alone cannot account for the unique ability of the HER2-specific scFv to function from within the 61-218 deletion. More recently, another effective, fully retargeted virus (KNE, Figure 2) was obtained by substitution of gD residues 2-24 with the anti-EGFR/EGFRvIII scFv, in combination with the Y38C mutation (Uchida et al., 2013). This virus could be grown on standard Vero cells without noticeable reversion of the nectin-1-detargeted phenotype (Uchida et al., 2013). Nevertheless, reversion was observed for other retargeted viruses grown on cognate receptor-transduced nectin-1+ cell lines, suggesting that while entry of the EGFR/EGFRvIII-retargeted virus through endogenous (simian) Vero-cell EGFR may be sufficiently efficient to minimize selective pressure for reversion, avoidance of nectin-1+ cell lines for virus production and rigorous detargeting from nectin-1 are critical. It was recently reported that deletion of gD residue 38 from the reversion-sensitive viruses provides reliable resistance to reversion (Uchida et al., 2018).

c. Retargeting virus infection by modification of other glycoproteins

In recent years, both gH/gL and gB have been found to interact with specific cellular receptors (Parry et al., 2005; Satoh et al., 2008; Arii et al., 2010; Suenaga et al., 2010; Arii et al., 2015), and these interactions may facilitate HSV entry. To clarify the role of gH/gL in HSV-1 entry and to explore whether the receptor specificity of this heterodimer affects HSV tropism, the anti-HER2 scFv was inserted into the amino terminus of gH (gH-HER2) in a wild-type gD virus that could be grown on nectin-1+ cells (Gatta et al., 2015). Remarkably, this virus could both infect and spread on J-HER2 cells that expressed HER2 but lacked functional gD receptors. In addition, a derivative detargeted from HVEM/nectin-1 by deletion of gD residues 6-38 infected a HER2+/nectin-1+ cancer cell line but not HER2-negative cell lines that were permissive for wild-type HSV-1. Similarly, insertion of the anti-HER2 scFv near the amino terminus of gB (gB-HER2) enabled infection of J-HER2 cells and gD detargeting of this virus did not substantially diminish infection of HER2+ cancer cell lines (Petrovic et al., 2017). While it might be concluded from this result that gB-retargeted virus would no longer require gD and, perhaps, gH/gL for entry, neutralizing antibodies to gD or gH were found to block HER2-dependent entry of gD-detargeted gB-HER2 virus. This observation suggested that gD performs a required function in entry that is independent of its binding to a receptor. In support of this, fusion of glycoprotein-transfected cells with J-HER2 cells was equally efficient when gB-HER2 and gH/gL were transfected in the presence of wt gD or detargeted gD but was abolished in the absence of gD (Petrovic et al., 2017). The same results were obtained when gH-HER2 was used with gB and gD (Petrovic et al., 2017). Together, these results not only demonstrated that HSV can be fully retargeted by insertion of a foreign ligand into gD, gH or gB in viruses that cannot utilize the natural gD receptors, but also indicated that none of the four essential entry glycoproteins are dispensable, with the implication that these viruses remain susceptible to neutralization by anti-glycoprotein antibodies in HSV-seropositive cancer patients (Cairns et al., 2015).

d. Production of retargeted viruses

Until recently, retargeted HSV was preferably generated and propagated on gD-receptor-deficient cells transgenic for the target receptor to avoid selective pressure for mutations that could restore gD binding to one of its natural receptors. To obviate the laborious generation of new cell lines for each new target receptor, a standardized system to produce retargeted viruses without dependence on expression of the target receptor on the producer cells was developed (Leoni et al., 2017). This system is analogous to the peptide tag/ anti-tag pseudoreceptor systems previously used to grow fully retargeted adenoviruses and measles viruses (Douglas et al., 1999; Nakamura et al.,

2005). A yeast GCN4 peptide was inserted into the N-terminal region of gH while Vero cells (nectin-1+), which are FDA-approved for clinical-grade virus preparation, were engineered to express a high-affinity scFv to the GCN4 peptide as an artificial GCN4 peptide receptor (GCN4R). This tag/ pseudoreceptor system was shown to enable nectin-1-detargeted virus entry into Vero cells (Leoni et al., 2017). In a subsequent study, the same group reported that gD could simultaneously accommodate the HER2 scFv and the GCN4 peptide for dual virus retargeting by a single recombinant glycoprotein, and described smaller and novel deletions suitable for nectin-1 detargeting and foreign ligand insertion (gD residues 35-39, 214-223 and 219-223) (Leoni et al., 2018a). Moreover, they showed that double retargeting to HER2 and GCN4R could also be accomplished by insertion of the GCN4 peptide at any of four different positions within the N-terminal 100 residues of gB, in combination with either the 6-38 deleted version of HER2-retargeted gD or a novel retargeted gD design combining a deletion of residue 30 with insertion of the anti-HER2 scFv in place of residue 38 (Petrovic et al., 2018). Although differences in replication and spread efficiencies were noted on either or both Vero-GCN4R and HER2+/nectin-1+ cancer cells between different doubly retargeted viruses, all were detargeted from the natural gD receptors and cytotoxic for both cell lines (Leoni et al., 2018a; Petrovic et al., 2018).

e. Retargeted virus efficacy in preclinical models

Ultimately, retargeted viruses are intended to increase tumor cell specificity and decrease off-target cell infection and toxicity, thereby improving vector safety while retaining efficient vector replication in tumor cells *in vivo*. The EGFR- and HER2-retargeted viruses have both been tested in preclinical mouse models of human tumors. The retargeted virus bearing an scFv against EGFR (KNE, Figure 2) recognizes both wild-type EGFR and its uniquely tumor-associated variant EGFRvIII prevalent on human GBM cells. This virus was effective in an orthotopic human-tumor model of GBM, i.e., a model where the tumor is located at its normal place (brain for GBM), with approximately 75% of oHSV-treated mice surviving beyond 90 days following a single intratumoral virus injection. No toxicity was observed upon direct injection of the virus into mouse brain, and systemic injection demonstrated preferential homing of the virus in nude mice to GBM flank tumors over non-tumor tissues (Uchida et al., 2013). The HER2-retargeted virus R-LM113 (Figure 2) was also found to be safe upon direct brain injection. In murine GBM tumors engineered to express human HER2, a single injection of R-LM113 lengthened median survival time of treated animals by 21 days compared to controls (Gambini et al., 2012). Both of these studies illustrated that retargeted oHSVs abolished neuro-invasiveness in the brain. However, to further ensure tumor specificity, the EGFR-retargeted gD was coupled with insertion of miR-124 binding sites in the 3' UTR of the essential ICP4 gene, abolishing virus (KGE-4:T124) replication in miR-124-expressing neurons while permitting virus replication in miR-124 deficient GBM tumor cells (Mazzacurati et al., 2015). KGE-4:T124 was as efficacious in the treatment of orthotopic human tumors as its retargeted parent virus, lacking the miR-124 binding sites (KGE), while providing increased safety (Mazzacurati et al., 2015).

The HER2-retargeted virus R-LM249, bearing a different insertion site for the anti-HER2 scFv than R-LM113 (Figure 2), readily infected a HER2+ human ovarian carcinoma cell line, SK-OV-3. R-LM249 demonstrated a dose-dependent reduction in tumor growth following a single administration of virus directly into subcutaneous SK-OV-3 flank tumors, while repeat administration of the virus (4 doses) led to a significant reduction in tumor burden, with 60% of mice being tumor-free past 200 days (Menotti et al., 2009). Treatment of intraperitoneal (i.p.) SK-OV-3 tumors with weekly i.p. delivery of R-LM249 (5 doses) also reduced tumor burden and increased median survival from 103 days to 440 days (Nanni et al., 2013). HER2 is also expressed in a subset of human breast cancers, and weekly intratumoral administration of R-LM249 was effective against subcutaneous flank tumors formed by HER2+ human breast cancer cells (Nanni et al., 2013). Moreover, weekly i.p. injection of R-LM249 in a multiorgan metastasis model induced by intravenous injection of the same cells significantly reduced ovarian and brain metastatic burden (Nanni et al., 2013). These studies demonstrated that R-LM249 in repeated doses was able to impair the growth of both local and disseminated HER2-expressing tumors.

While the work described above has shown that treatment with retargeted viruses can effectively reduce tumor burden and extend the survival of tumor-bearing immune-deficient mice, the current approach to oncolytic virotherapy seeks to expand immune-system engagement by arming vectors with immune modulatory transgenes. Recently, HER2-targeted R-LM113 armed with IL-12 (R-LM115) was tested in immune-competent mice in a flank tumor model induced by implantation of HER2-transduced murine Lewis lung carcinoma (LLC) cells (Leoni et al., 2018b). The results demonstrated that 4 doses of either R-LM113 or R-LM115 extended animal survival compared to controls, particularly when treatment was given at the early time point of 3 days post tumor cell injection. Of note, while the R-LM115 virus expressing IL-12 was significantly more effective than the unarmed virus at extending animal survival and reducing tumor volume, both viruses inhibited formation of tumors after rechallenge of animals that survived the primary tumor. Analysis of an array of immune cell markers and cytokines illustrated key changes in the tumor microenvironment in response to virus treatment and, in particular, to IL-12 expression, suggesting increased immune-stimulatory factors and tumor infiltration by activated immune cells in responder mice. These data highlight the importance of the immune system in tumor treatment, stimulated, in part, by the virus and its lytic activity, and, in part, by the expression of immune-stimulatory transgenes as discussed in detail below.

D. oHSVs as the next wave of immuno-oncology therapies

1. oHSV and immunogenic tumor cell killing

HSV replication causes tumor cell death resulting in the release of tumor-associated antigens/neoantigens (Bommareddy et al., 2018; Thomas et al., 2019) (Fig 3), viral pathogen-associated molecular patterns (PAMPs), and cell-derived damage-associated molecular patterns (DAMPs). Induction of innate immunity, assessed by the production of type I IFN and other proinflammatory cytokines, is triggered by these DAMPs, PAMPs, and various sensing mechanisms that recognize viral proteins, DNA and RNA [also see section B. 2b; reviewed in (Paludan et al., 2011)]. For example, viral dsDNA is sensed by the cGAS/STING pathway (Ishikawa et al., 2009; Sun et al., 2013; Wu et al., 2013; Zhang et al., 2013) and the PYHIN domain-containing protein IFI16 (Unterholzner et al., 2010; Almine et al., 2017), which induce type I IFN and the expression of proinflammatory cytokines such as CXCL9 and CXCL10. HSV-1 infection also leads to the production of prototypic DAMPs, including the extracellular release of high mobility group protein B1 (HMGB1) and translocation of the ER chaperone calreticulin to the surface of infected cells, which, in conjunction with IFN and cytokines, results in recruitment and activation of NK cells, phagocytic cells and APCs to the site of infection (Obeid et al., 2007; Paolini et al., 2015; Takasu et al., 2016; Muller et al., 2017).

NK cells are innate immune cells recruited to the site of viral infection that are cytotoxic and can produce cytokines without prior antigen stimulation (Orr and Lanier, 2010; Abel et al., 2018). In part, NK cell cytotoxicity is regulated by binding of NK cell activation receptors, such as NKG2D and the natural cytotoxicity receptors (NCRs) NKp30 and NKp46, to various cell surface ligands expressed on target cells (Sauer et al., 2017). Humans express eight NKG2D ligands [U_L16 binding proteins (ULBPs) 1 to 6 and MHC class I-related chains A and B (MICA, MICB)] while mice express 5-6 different ligands among the group of RAE-1(α-ε), H60(a-c), and MULT1, depending on the mouse strain (Vivier et al., 2002; Abel et al., 2018). Healthy cells express MHC class I and typically do not display NK cell-activating ligands on their surface and are shielded from NK cell-mediated lysis. However, cellular stresses associated with transformation, viral infection or other danger signals to the host cause an up-regulation of NKG2D/NCR ligand expression and NK cell activation. The anti-tumor innate immune response mediated by NK cells can be directly inhibited by tumor-expressed metalloproteinases that cleave NKG2D and NKp30 ligands on the cell surface or by tumor-mediated suppression of the expression of NK cell-activating ligands (Raulet et al., 2013; Schlecker et al., 2014; Pogge von Strandmann et al., 2015). However, these effects may be tempered by oHSV infection, which causes upregulation of NCR ligand expression (Chisholm et al., 2007). In addition, NK cells can eliminate HSV-infected tumor cells by HSV-antibody-dependent cell-mediated cytotoxicity (ADCC) involving simultaneous binding of anti-HSV IgG to viral proteins on the

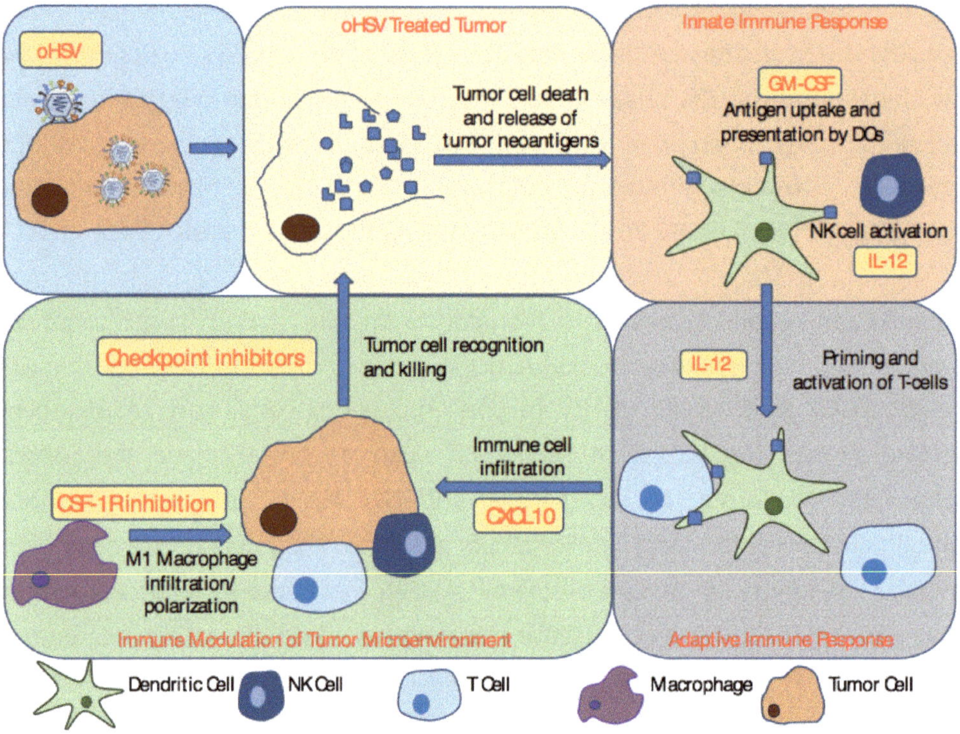

Figure 3. Armed oHSV treatment of GBM. Illustration of the immune response to oHSV infection of GBM tumor cells and examples of arming genes or gene functions (yellow boxes) for oHSV to augment the immune response to tumor antigens. Tumor and immune cell types are depicted at the bottom. Clockwise from the top left, oHSV infection of tumor cells results in virus replication and cell lysis, releasing tumor and viral antigens. Viral infection/cell death elicits an innate immune response, recruiting denditic cells (DCs) and natural killer (NK) cells to the site of infection. The adaptive immune response is stimulated via antigen uptake and presentation by DCs and results in priming and activation of T cells that infiltrate the TME. Tumor infiltration by immune cells results in further tumor cell killing by NK cells, T cells, or macrophages. Arming of oHSV with immune modulatory transgenes can influence different steps of the process: recruitment of DCs (GM-CSF), activation of NK and T cells (IL-12), recruitment of activated immune cells to the tumor (CXCL10), polarization and/or recruitment of pro-inflammatory macrophages (CSF-1R inhibition), and increased tumor cell recognition and killing within the TME (checkpoint inhibitors).

tumor cell surface and the Fc receptor CD16a on NK cells (Dai and Caligiuri, 2018). Furthermore, recent evidence suggests that NK cells can recognize and destroy oHSV-infected GBM in the absence of specific anti-viral antibodies in a process referred to as Fc-bridging cellular cytotoxicity (FcBCC) (Dai et al., 2017a; Dai and Caligiuri, 2018). Although NK cells can lead to premature virus clearance in GBM xenograft models (Alvarez-Breckenridge et al., 2012; Alvarez-Breckenridge et al., 2013), the combination of oHSV and NK cells can be exploited to induce tumor cell killing (Chen et al., 2016; Yoo et al., 2016). The outcome of this complex dual role of NK cells in response to virotherapy may be a function of the ratio of effector to target cells in the tumor microenvironment (Kim et al., 2018).

A key feature of oncolytic viral therapies is the capacity to induce both anti-viral and anti-tumor adaptive immune responses (Figure 3). The initial infection of tumor cells by oHSV causes tumor cell virolysis and an innate immune response to virus infection mediated by NK cells and DCs, ultimately creating a pro-inflammatory, antiviral state. Activated NK cells are a primary source of cytokines (e.g., FLT3L) and chemokines (e.g., CCL5) that can recruit and activate cross-presenting cDC1 cells (Barry et al., 2018; Bottcher et al., 2018), which prime an adaptive immune response by activating antigen-specific CD8+ T cells. cDC1 cells, also referred to as conventional type 1 DCs, are a subset of migratory DCs that are CD103+ (mice)/CD141+ (human), depend on the basic leucine zipper ATF-like transcription factor 3 (BATF3), and have been identified as key players in the cross presentation of both viral and tumor antigens (Hildner et al., 2008; Schlitzer and Ginhoux, 2014; Schlitzer et al., 2015; Roberts et al., 2016; Dai et al., 2017b). This subset has been associated with enhanced therapeutic responses in animal models (Roberts et al., 2016; Salmon et al., 2016; Spranger et al., 2017; Bottcher et al., 2018), and the prevalence of its gene expression signature in human tumors is positively correlated with T cell infiltration and clinical outcome (Broz et al., 2014; Spranger et al., 2016). Together, these innate and adaptive immune responses to viral infection offer a mechanism to explain how oHSVs can promote the generation of a 'hot' tumor environment and a clinical response in combination with immune checkpoint inhibitors, as observed by Ribas and colleagues (Ribas et al., 2017).

2. Immunotherapy and oHSV arming with immune-modulatory genes

Current immunotherapy approaches fall into two broad categories. In the first, specific antigens are targeted while the second is agnostic in relation to tumor antigens (Kamran et al., 2016). Antigen-targeted therapies such as CAR-T therapy, T-cell receptor (TCR)-transduced T-cell therapy, and peptide/RNA based-vaccine immunotherapy generate T cells recognizing known or predicted tumor antigens or antigens that would not be normally expressed in adult diploid tissue (Coulie et al., 2014). While these strategies may be effective, they require advance knowledge of the tumor antigen, autologous immune cells, complex and individualized manufacturing processes, and are often restricted to a specific MHC haplotype. They also depend on functioning active T cells, and T-cell exhaustion within the TME is an important problem so far limiting the clinical efficacy of these agents to hematological malignancies (Kalos et al., 2011; Grupp et al., 2013; Kamran et al., 2016; Bagley et al., 2018). Conversely, therapies such as checkpoint pathway inhibitors, cytokine gene therapy, and oncolytic viral therapies can counter the immunosuppression created in the TME and enhance T-cell recognition of tumor cells without the knowledge of specific antigens.

Checkpoint inhibition therapies have demonstrated remarkable success in recent years. These therapies block T-cell inhibitory receptors that function to turn off inflammation, in an attempt to maintain CD8+ T-cell anti-tumor activity. The interaction between PD-L1 on tumor cells and PD-1 on T cells results in T cell exhaustion (Zhang et al., 2018), and CTLA-4 binding to B7-1 on antigen-presenting cells (APCs) blocks the expansion of T cells and interferes with activation of the T-cell co-stimulatory protein CD28 (Chikuma, 2017). Targeting the PD-1/PD-L1 axis and CTLA-4 has received FDA approval in multiple cancer indications (Hargadon et al., 2018). Several factors have been considered in determining whether a patient will be responsive to checkpoint therapy, including expression of PD-L1 on tumor cells and the presence of intra-tumoral CD8+ T cells (Tumeh et al., 2014; Shields et al., 2017; Zhang et al., 2018). In the TME, CD8+ T cells may be tumor-reactive, but exhausted by checkpoint ligand-receptor interactions. Blocking of these interactions can lead to CTL reactivation and tumor-cell killing. In addition, increased tumor mutational burden has been linked to response to immunotherapy (Rizvi et al., 2015; Colli et al., 2016; Hellmann et al., 2018). In a first-of-its-kind decision, the FDA approved the anti-PD-1 antibody pembrolizumab (Keytruda) for use in any cancer that is high in microsatellite instability or deficient in mismatch repair, has progressed following prior treatment, and has no suitable alternative treatment options (Prasad et al., 2018). These tumors typically have an extremely high mutation load (Le et al., 2015; Le et al., 2017). While checkpoint inhibitors have mostly failed in brain tumor patients, reported cases of response have included patients with mismatch-repair deficiency (Bouffet et al., 2016). The concepts of immunosuppression, mutational burden, and immune evasion can help explain why a tumor may be immunologically "cold" and refractory to immunotherapy or immunologically "hot", and thereby responsiveness to immunotherapy. Cold tumors typically contain large numbers of immunosuppressive monocytes and few tumor-specific cytotoxic T cells whereas hot tumors are seeded with immunologically active macrophages and lymphocytes.

The promise of OVs turning up the heat in cold tumors and enabling a response to immune checkpoint inhibitors is under evaluation in multiple large clinical studies (Chen et al., 2018; Gujar et al., 2018b). The MASTERKEY-265/ KEYNOTE-034 trial (NCT02263508) is sponsored by Merck Sharp and Dohme Corp. and Amgen, Inc. and will report the outcome of the combination of Keytruda with and without T-VEC/Imlygic in 713 patients with unresected melanoma (Dummer et al., 2017; Ribas et al., 2017; Sun et al., 2018). This combination is also being tested in recurrent metastatic head and neck squamous cell carcinoma (MASTERKEY232/KEYNOTE-137 (NCT02626000), in sarcoma (NCT03069378), and in liver cancer (NCT02509507). T-VEC is also under evaluation in smaller studies in combination with anti-PD-1 antibody nivolumab (Opdivo) in lymphoma and pleural effusion and with anti-PD-L1 antibody atezolizumab (Tecentriq) in triple-negative breast cancer (TNBC) and colorectal cancer (CRC) with liver metastasis. T-VEC in combination with ipilimumab (Yervoy) displayed slightly improved efficacy over T-VEC or ipilimumab monotherapies for stage II-IV melanoma (Puzanov et al., 2016). HF10 in combination with nivolumab is being tested in the neoadjuvant setting in resectable melanoma (NCT03259425).

Preclinically, many groups are actively engaged in establishing the potential benefit of adding checkpoint blockade to HSV therapy for GBM because checkpoint inhibition monotherapies have provided convincing evidence of increased survival and immune activation in GBM mouse models (Reardon et al., 2016; Garg et al., 2017; Garzon-Muvdi et al., 2018; Hung et al., 2018). In two different syngeneic models of murine glioma, using G47Δ expressing murine IL-12 (G47Δ-mIL12) in combination with anti-PD-1 and anti-CTLA-4 antibody administration, long term survival was only observed in mice that received the triple combination therapy for CT-2A tumors (50% long-term survival) or glioblastoma stem cell line-derived tumors (89% long-term survival) (Saha et al., 2017b). In this model, CD4+ T cells, CD8+ T cells and macrophages were required for the efficacy of the triple combination therapy. Additionally, triple combination therapy resulted in an influx of pro-inflammatory and anti-tumor M1-polarized macrophages (Saha et al., 2017b). While clinically little therapeutic benefit was afforded by anti-PD-1 monotherapy or the combination of PD-1 and CLTLA4 blockade for GBM (Filley et al., 2017; Omuro et al., 2018), these preclinical data suggest that a greater benefit may be provided by a combination of intratumoral injection of oHSV and systemic treatment with anti-checkpoint monoclonal antibody.

Alternative checkpoint pathways exist that can also be targeted for antibody-mediated inhibition to help maintain CD8+ T-cell anti-tumor activity (Anderson et al., 2016). These include (i) LAG-3 (lymphocyte activation gene 3) on T cells and NK cells that can inhibit T cell expansion (He et al., 2016), (ii) TIGIT (T cell immunoreceptor with Ig and ITIM domains) on T cells and NK cells that can bind CD155 [PVR (poliovirus receptor) or necl-5] and CD112 (nectin-2) to suppress immune activation (Solomon and Garrido-Laguna, 2018), and (iii) TIM-3 (T cell immunoglobulin and mucin domain 3) on multiple immune cell types that can trigger T cell and NK cell exhaustion and immune suppression through different pathways involving DCs, MDSCs and Tregs (He et al., 2018). Tim-3, TIGIT and Lag-3 expression have been associated with glioblastoma in mouse tumor models or in human tumor samples (Li et al., 2017; Harris-Bookman et al., 2018; Hung et al., 2018), and preclinical studies in a GL261 mouse model of GBM demonstrated that combining PD-1 and TIGIT checkpoint blockade resulted in improved survival over each monotherapy (Hung et al., 2018). Future studies will likely engage antibodies that antagonize these interactions in combination with oHSV therapy. While systemic use of anti-checkpoint antibodies could provoke autoimmunity, vector-mediated local expression of antibodies will likely be effective and minimize potential side effects. Attempts are currently underway to locally deliver an anti-PD1 single-chain variable fragment antibody (scFv) by expression from an oHSV (oHSVscFvPD-1). In early reports, oHSVscFvPD-1 was able to express the scFv in multiple glioma models and delayed tumor growth in mice bearing orthotopic gliomas (Passaro et al., 2018). Other efforts in this direction include the exploitation of PTEN to block oHSV infection-induced upregulation of PD-L1 expression on tumor cells to augment anti-tumor immunotherapy in conjunction with oncolysis (Russell et al., 2018). As described above, the effectiveness of checkpoint inhibitors relies on the reactivation of CD8+ T cells that recognize tumor antigens but are exhausted by checkpoint ligand-receptor interactions. In the event that these cells are not available or the number of TAAs is limited, the use of oHSV to stimulate a T-cell response to tumor antigens could potentiate the effect of checkpoint antibodies, whether delivered systemically or as oHSV cargo.

When considering potential targets within the TME for further oHSV vector arming (Figure 3), it is important to recognize that the composition of the TME differs between tumor types such that arming strategies will likely require type-specific optimization. The immune stimulatory cytokine IL-12 plays an important role in activating both adaptive (NK cell) and acquired (T cell) immune responses and, as already referred to, IL-12-expressing oHSV alone or in combination with other therapies has delivered promising results in preclinical studies (Toda et al., 1998; Parker et al., 2000; Chiu et al., 2012; Cody et al., 2012; Markert et al., 2012; Cheema et al., 2013; Roth et al., 2014; Thaci et al., 2014; Bauer et al., 2016; Patel et al., 2016; Ring et al., 2017; Saha et al., 2017a; Friedman et al., 2018; Leoni et al., 2018b), including an increase of pro-inflammatory M1-polarized macrophages. Multiple studies have highlighted the immunosuppressive and tumor-supportive activities of M2-like macrophages as opposed to the anti-tumor activities of M1-like macrophages and their association with poor and improved prognosis, respectively (Lisi et al., 2017; Saha et al., 2017b; van den Bossche et al., 2018). In GBM, tumor associated macrophages, in addition to CNS-resident microglia, are particularly prevalent, representing more than half of the cell content (Darmanis et al., 2017). Macrophages in GBM express transcripts that are associated with both M1 and M2 phenotypes, suggestive of a wound healing and tolerogenic program (Bowman et al., 2016; Takenaka et al., 2019). The chemokine CCL2, through its receptor CCR2, and the growth factor CSF-1 via binding to the macrophage CSF-1 receptor (CSF-1R), are thought to play major roles in macrophage recruitment and maintenance, respectively. Small-molecule inhibitors as well as monoclonal antibodies are under development to block these interactions [reviewed in (Ries et al., 2015)]. In a mouse GBM model, the small molecule CCR2 inhibitor CCX872 reduced myeloid cells recruitment and enhanced the survival benefit of a PD-1 inhibitor antibody (Flores-Toro et al., 2020). CSF-1R inhibition using the receptor tyrosine kinase inhibitor BLZ945 counteracted M2 polarization while increasing phagocytosis, consistent with macrophage/microglia "re-education" (Pyonteck et al., 2013) (Figure 3). Unfortunately, IGF1-driven activation of the PI3K pathway led to resistance to CSF-1R inhibitor and the evaluation of this pathway in the clinic using PLX3397 (NCT01349036) did not yield a significant therapeutic benefit despite reaching pharmacologically active CNS concentration (Butowski et al., 2016; Quail et al., 2016). The investigation of BLZ945 is ongoing in solid tumors in combination with spartalizumab, Novartis' anti-PD-1 antibody (NCT02829723). Expression of indoleamine-2,3-dioxygenase (IDO) in the myeloid cell line

THP-1 promotes adoption of the immunosuppressive M2 phenotype (Wang et al., 2014). oHSV infection can downregulate IDO synthesis in IFNγ-treated glioma cells (Reinhart et al., 2012), raising the possibility that it can promote macrophage M2-to-M1 repolarization. The failure of the IDO1 inhibitor epacadostat to show a benefit in combination with pembrolizumab in the ECHO301/Keynote 252 randomized phase III melanoma trial led to a global demise of the investigation of this axis (Long et al., 2019). Although the CSF-1 and IDO axes may represent viable targets for combining with oHSV to repolarize M2 macrophages to an M1-like tumoricidal phenotype, clearly other therapeutic approaches are warranted in light of the lackluster activity of the drugs targeting these pathways.

Recently, the publication of an armed HSV-1 highlighted the potential for the KLRK1 ligand ULBP3 to modulate tumor associated macrophages and prolong survival in a genetically engineered model of glioblastoma (Wirsching et al., 2019b). oHSV expressing ULBP3, but not the empty oHSV, stimulated abscopal activity in a dual GBM tumor model and its activity was enhanced by the administration of a PD-1 antagonist antibody. Surprisingly, while the cognate receptor for ULBP3 is not expressed in myeloid cells and ULBP3 has no mouse ortholog, oHSV-ULBP3 upregulated pathways linked to antigen processing and presentation in tumor myeloid cells. Although much remains to be understood regarding the mode of action of oHSV-ULBP3, this data suggest a different route for macrophage modulation.

In addition to tumor associated macrophages, the immunosuppressive TME is supported by myeloid-derived suppressor cells (MDSCs) (Ostrand-Rosenberg and Sinha, 2009; Youn and Gabrilovich, 2010), a poorly defined myeloid population in humans, which have been shown to suppress anti-tumor immunity while stimulating tumor growth, angiogenesis and metastasis (Kohanbash and Okada, 2012; Gabrilovich, 2017). In mice, MDSCs are defined as a diverse population of CD11b+/Gr1+ immature myeloid cells that include precursors of granulocytes, macrophages and dendritic cells (Fujita et al., 2011; Zhu et al., 2011). MDSC-mediated immunosuppression involves the production of IL-10 and IL-4 (Kohanbash et al., 2013), which promotes M2 macrophage polarization and inhibits interferon production. MDSC, through high expression of arginase, may lead to the local depletion of arginine, an essential amino acid for T-cell activation and proliferation (Lisi et al., 2017). Additionally, MDSCs can inhibit anti-tumor immunity through production of iNOS, reactive oxygen species, and TGFβ, and can recruit M2 macrophages through the secretion of chemokines that act via CCR2 (Chang et al., 2016; Kamran et al., 2018). Three primary ways can be considered to target human MDSCs: (i) depletion by antibodies targeting MDSC, (ii) maturing the cells to an M1 macrophage phenotype, and (iii) targeting the specific immunosuppressive pathways, such as arginase activity. Antibody-mediated depletion of MDSCs is effective in mice (Fujita et al., 2011), although as yet not fully specific (Kamran et al., 2018). However, so far no antibodies have been described for selective human MDSC depletion as the definition of human MDSC remains unsettled (Elliott et al., 2017). It is unclear which of these strategies may ultimately provide the most therapeutic value but each may potentially be accomplished through the selection of suitable oHSV-arming genes (Kamran et al., 2017; Lowenstein and Castro, 2018).

oHSV infection must recruit APCs to traffic tumor antigens to regional lymph nodes where they may be presented by DCs for T-cell activation (Alvarez-Breckenridge et al., 2015; Roberts et al., 2016). Therefore, arming vectors with cytokines/chemokines such as FLT3L, XCL1 or CXCL10 could enhance APC recruitment (Figure 3) while inhibition of CD47, a "don't-eat-me" protein expressed on various tumor cells, may improve antigen presentation (Chao et al., 2011; Willingham et al., 2012; Wang et al., 2013; Cioffi et al., 2015). Toll-like receptor agonists may be useful as direct tumor immune-therapeutics and can act as adjuvants for T-cell activation (Kaczanowska et al., 2013; Sato-Kaneko et al., 2017). The large payload capacity of some oHSVs (Mazzacurati et al., 2015), approximately 25 kb, allows combinations of arming genes to be introduced into a single vector, but the most effective combinations have yet to be determined. Replimune, Inc., a company funded by the T-VEC developers, is leading the way with the progression in clinical trials of a first oHSV expressing the fusogenic protein GALV (Fu et al., 2003; Price et al., 2010; Thomas et al., 2019), in addition to GM-CSF, and a second oHSV candidate in addition expressing a CTLA-4 antagonist antibody.

3. GBM models for armed oHSV evaluation

The initial studies of oncolytic vectors employed orthotopic implantation of various genetically distinct human tumor cell lines in immune-deficient mice. While appropriate for the evaluation of oncolytic activity, the lack of an adaptive immune response in these mice does not allow for an assessment of the impact of the immune system on both virus replication and tumor cell destruction. Preclinical studies of the ability of armed oHSVs to induce anti-tumor immunity are essential for success in treatment of highly aggressive GBM and necessitate the use of immunocompetent model systems. To achieve this goal, models have been established using murine cells implanted in syngeneic backgrounds [reviewed by (Oh et al., 2014)]. For example, models of GBM have been created by intracranial injection of murine glioma tumor cell lines grown *in vitro* and by using genetic systems where tumors arise locally from neuronal precursor cells within the brain parenchyma.

The simplest model systems rely on the injection of well-characterized murine tumor cell lines. The CT-2A cell line was derived by chemical mutagenesis and has been shown to consistently form heterogenous, invasive tumors that demonstrate multiple characteristics of GBM in the C57BL/6 mouse strain background. The GL261 model, widely used in preclinical testing of GBM therapeutics, is another C57BL/6-based, chemically induced model of GBM that is more immunogenic than most human GBM. GL261 lacks MHC class II molecules while displaying low levels of B7-1 and B7-2, and is sensitive to radiotherapy and to immune checkpoint inhibitor treatment (Szatmari et al., 2006; Wu et al., 2016). More recently, a glioma stem cell (GSC)-based model was developed by Inder Verma (Salk Institute, San Diego) and co-authors from GBM tumors induced by Cre-mediated expression of activated H-Ras and Akt into the stem cell niche of GFAP-Cre Tp53$^{+/-}$ immunocompetent C57BL/6 mice (Marumoto et al., 2009). This model possesses such characteristics of GBM as invasiveness and genetic heterogeneity as well as an immunosuppressive TME and moderate response to G47Δ-IL-12 oHSV (Cheema et al., 2013).

One genetic model system, developed by Eric Holland (Fred Hutchinson Cancer Research Center), uses an HSV-permissive transgenic mouse strain with genotype N/t-va;Ink4A/Arf-/-;PTENfl/fl that forms brain tumors on intracranial administration of a pair of engineered avian retroviruses (RCAS) expressing Cre recombinase and PDGF B-chain, respectively. The mice express the RCAS receptor (TVA) from the nestin promoter, which restricts Cre and PDGFB expression to neuronal precursor cells. Cre expression results in deletion of the loxP-flanked ("floxed", fl/fl) PTEN tumor-suppressor gene, and the resulting tumors resemble the proneural variant of glioblastoma with few, if any, T cells. In mice with a N/t-va;Ink4A/Arf-/-;PTENfl/fl;LSL-EGFRvIII/vIII background, Cre expression additionally induces EGFRvIII expression by removal of a transcription termination element (lox-stop-lox, LSL) between the promoter and coding sequence. EGFRvIII leads to a more classical or mesenchymal phenotype that includes the presence of a few T cells primarily in the perivascular region. Both models give rise to rapidly developing, predominantly mono-focal murine GBM with a TME that closely resembles that in human patients. Tumors arising in these models mimic many other aspects of human GBM, including tumor invasion of the brain parenchyma. This and similar systems with varying genetic mutations have been used to model GBM for oncolytic-vector and other therapeutic treatments (Huse and Holland, 2009, 2010; Pyonteck et al., 2013; Quail et al., 2016).

Another genetic model system that has been utilized to study gliomas, including GBM, is the Sleeping Beauty (SB) glioma model (Wiesner et al., 2009). In this model, neonatal mice receive intracerebroventricular injections of an SB transposase together with SB-compatible vectors containing glioma-relevant oncogenes of the investigator preference, such as EGFRvIII, NRAS(V12), PDGFB, shP53 and many others. In time following the injection, the animals develop tumors and, depending on the gene combination and site of injection, different glioma types can be generated. Additionally, a SB-compatible luciferase vector can be included to allow for non-invasive bioluminescence imaging. The benefits of this model are that it is compatible with any strain or background of mice, and that cell lines can be derived from resulting tumors to create genetically relevant cell-line injection models (Fujita et al., 2010; Fujita et al., 2011; Kohanbash et al., 2013). The pathobiology of the tumors arising in these models is highly reminiscent of human GBM, with a high frequency of Iba-1$^+$ myeloid cells and a low frequency of TILs (Wirsching et al., 2019a).

These models of GBM have all been used to characterize the efficacy of either armed or unarmed oHSV vectors, and comparisons among these models will hopefully shed light on the ability of oHSV to alter the TME in ways to promote an anti-tumor immune environment that can also contribute to the prevention of secondary tumor formation.

E. Conclusions and future directions

HSV has several important attributes that make it a desirable oncolytic vector, including broad tumor cell permissivity, the ability to engineer variants that specifically lyse tumors and a large transgene capacity that allows for *in situ* delivery of multiple payload products by a single vector. With the approval of Imlygic® (T-VEC) in the US for the treatment of malignant melanoma, oHSVs are gaining prominence as an addition to standard-of-care, i.e., current clinical protocols including immune therapy and radiation therapy of cancer (Harrington et al., 2010; Andtbacka et al., 2015; Pol et al., 2016; Blake et al., 2018). Combined with checkpoint inhibitors, further improvement of tumor treatment outcome may be achieved, a result pending the completion of phase 3 randomized clinical trails such as Keynote034/Masterkey265. A current focus in the development of oHSVs for the treatment of solid tumors is to demonstrate that arming genes expressed from a lytic vector can overcome immunosuppression and enhance innate and adaptive anti-tumor immunity. Ultimately, the aim is to identify the most effective genes and combinations that create a potent immunostimulatory OV vector. By combining virus-mediated tumor destruction with the activities of specific arming-gene products, oHSVs have considerable potential to create tumor vaccines that rely on the ability of the immune system to reject tumors. Ideally, payloads that are tailored to the TME of specific tumors should be developed. At present, several clinical trials are investigating the TME in biopsy material by RNA or proteomics profiling prior to treatment. This information could be very important for selection of arming genes.

Increasing preclinical evidence suggests that armed oHSVs can not only promote the conversion of a tumor-supportive environment in the injected primary tumor to a tumor-hostile environment, but can also bring about rejection of untreated secondary distal or visceral tumors in what has been termed an abscopal effect, the essence of an effective tumor vaccine. While much remains to be learned about the timing, reach and potency of these effects, they raise hope for the treatment of metastatic disease. As an additional strategy to attack metastatic tumor foci, there is also increasing interest in the development of oncolytic vectors for systemic application. These vectors would be (i) engineered for selective infection and replication in tumor cells and (ii) antigenically "stealthed" and thus resistant to neutralization by pre-existing or acquired virus-neutralizing antibodies. As described above, oHSV vectors have been developed that are uniquely retargeted to tumor-associated cell surface markers for attachment and entry in combination with miRNA-dependent attenuation to block virus replication in off-target tissue. While these "retargeted" oHSVs can home in on tumors bearing the target receptor in immune-compromised mice (Uchida et al., 2013), a large fraction of the human population is HSV-seropositive, and systemic treatment of metastatic cancer with even tumor-targeted oHSV will likely be impaired in these patients. Thus, another important future goal is to create retargeted oHSVs that are concealed from pre-existing immunity. Successful development of this novel class of oHSVs would represent a next-generation OV applicable to the systemic treatment of metastatic cancer.

F. Acknowledgements

Studies in laboratories 1-6 were supported in part by NIH grant P01 CA163205 (MAC/EAC, co-PIs). Additional support for the JCG lab and collaborators (GK, NA, ECH) was provided by NIH R01 grants CA222804 and CA175052 to JCG. The JCG lab is also funded by a grant from the Alliance for Cancer Gene Therapy (ACGT).

G. Web resources

1. http://www.abta.org/about-us/news/brain-tumor-statistics/
2. https://clinicaltrials.gov/

H. References

Abel, A.M., Yang, C., Thakar, M.S., and Malarkannan, S. (2018). Natural Killer Cells: Development, Maturation, and Clinical Utilization. Front Immunol 9, 1869. http://dx.doi.org/10.3389/fimmu.2018.01869.

Advani, S.J., Sibley, G.S., Song, P.Y., Hallahan, D.E., Kataoka, Y., Roizman, B., and Weichselbaum, R.R. (1998). Enhancement of replication of genetically engineered herpes simplex viruses by ionizing radiation: a new paradigm for destruction of therapeutically intractable tumors. Gene Ther 5, 160-165. https://www.ncbi.nlm.nih.gov/pubmed/9578834

Aghi, M., Visted, T., Depinho, R.A., and Chiocca, E.A. (2008). Oncolytic herpes virus with defective ICP6 specifically replicates in quiescent cells with homozygous genetic mutations in p16. Oncogene 27, 4249-4254. http://dx.doi.org/10.1038/onc.2008.53.

Alain, T., Lun, X., Martineau, Y., Sean, P., Pulendran, B., Petroulakis, E., Zemp, F.J., Lemay, C.G., Roy, D., Bell, J.C., *et al.* (2010). Vesicular stomatitis virus oncolysis is potentiated by impairing mTORC1-dependent type I IFN production. Proc Natl Acad Sci U S A 107, 1576-1581. https://www.ncbi.nlm.nih.gov/pubmed/20080710

Almine, J.F., O'Hare, C.A., Dunphy, G., Haga, I.R., Naik, R.J., Atrih, A., Connolly, D.J., Taylor, J., Kelsall, I.R., Bowie, A.G., *et al.* (2017). IFI16 and cGAS cooperate in the activation of STING during DNA sensing in human keratinocytes. Nat Commun 8, 14392. http://dx.doi.org/10.1038/ncomms14392.

Alvarez-Breckenridge, C.A., Choi, B.D., Suryadevara, C.M., and Chiocca, E.A. (2015). Potentiating oncolytic viral therapy through an understanding of the initial immune responses to oncolytic viral infection. Curr Opin Virol 13, 25-32. https://www.ncbi.nlm.nih.gov/pubmed/25846988

Alvarez-Breckenridge, C.A., Yu, J., Caligiuri, M.A., and Chiocca, E.A. (2013). Uncovering a novel mechanism whereby NK cells interfere with glioblastoma virotherapy. OncoImmunology 2, e23658. https://doi.org/10.4161/onci.23658

Alvarez-Breckenridge, C.A., Yu, J., Price, R., Wojton, J., Pradarelli, J., Mao, H., Wei, M., Wang, Y., He, S., Hardcastle, J., *et al.* (2012). NK cells impede glioblastoma virotherapy through NKp30 and NKp46 natural cytotoxicity receptors. Nat Med 18, 1827-1834. https://www.ncbi.nlm.nih.gov/pubmed/23178246

Anderson, A.C., Joller, N., and Kuchroo, V.K. (2016). Lag-3, Tim-3, and TIGIT: Co-inhibitory Receptors with Specialized Functions in Immune Regulation. Immunity 44, 989-1004. http://dx.doi.org/10.1016/j.immuni.2016.05.001.

Andreansky, S., He, B., van Cott, J., McGhee, J., Markert, J.M., Gillespie, G.Y., Roizman, B., and Whitley, R.J. (1998). Treatment of intracranial gliomas in immunocompetent mice using herpes simplex viruses that express murine interleukins. Gene Ther 5, 121-130. https://www.ncbi.nlm.nih.gov/pubmed/9536273

Andtbacka, R.H., Kaufman, H.L., Collichio, F., Amatruda, T., Senzer, N., Chesney, J., Delman, K.A., Spitler, L.E., Puzanov, I., Agarwala, S.S., *et al.* (2015). Talimogene Laherparepvec Improves Durable Response Rate in Patients With Advanced Melanoma. J Clin Oncol 33, 2780-2788. https://www.ncbi.nlm.nih.gov/pubmed/26014293

Andtbacka, R.H.I., Collichio, F., Harrington, K.J., Middleton, M.R., Downey, G., hrling, K., and Kaufman, H.L. (2019). Final analyses of OPTiM: a randomized phase III trial of talimogene laherparepvec versus granulocyte-macrophage colony-stimulating factor in unresectable stage III-IV melanoma. J Immunother Cancer 7, 145. https://www.ncbi.nlm.nih.gov/pubmed/31171039

Andtbacka, R.H.I., Ross, M.I., Agarwala, S.S., Taylor, M.H., Vetto, J.T., Neves, R.I., Daud, A., Khong, H.T., Ungerleider, R.S., and Tanaka, M. (2018). Efficacy and genetic analysis for a phase II multicenter trial of HF10, a replication-competent HSV-1 oncolytic immunotherapy, and ipilimumab combination treatment in patients with stage IIIb-IV unresectable or metastatic melanoma. J Clin Oncol 36, 9541-9541. http://ascopubs.org/doi/abs/10.1200/JCO.2018.36.15_suppl.9541

Argnani, R., Boccafogli, L., Marconi, P.C., and Manservigi, R. (2004). Specific targeted binding of herpes simplex virus type 1 to hepatocytes via the human hepatitis B virus preS1 peptide. Gene Ther 11, 1087-1098. https://www.ncbi.nlm.nih.gov/pubmed/15057264

Arii, J., Goto, H., Suenaga, T., Oyama, M., Kozuka-Hata, H., Imai, T., Minowa, A., Akashi, H., Arase, H., Kawaoka, Y., *et al.* (2010). Non-muscle myosin IIA is a functional entry receptor for herpes simplex virus-1. Nature 467, 859-862. https://www.ncbi.nlm.nih.gov/pubmed/20944748

Arii, J., Hirohata, Y., Kato, A., and Kawaguchi, Y. (2015). Nonmuscle myosin heavy chain IIb mediates herpes simplex virus 1 entry. J Virol 89, 1879-1888. https://www.ncbi.nlm.nih.gov/pubmed/25428876

Atanasiu, D., Saw, W.T., Cohen, G.H., and Eisenberg, R.J. (2010). Cascade of events governing cell-cell fusion induced by herpes simplex virus glycoproteins gD, gH/gL, and gB. J Virol 84, 12292-12299. https://www.ncbi.nlm.nih.gov/pubmed/20861251

Baek, H., Uchida, H., Jun, K., Kim, J.H., Kuroki, M., Cohen, J.B., Glorioso, J.C., and Kwon, H. (2011). Bispecific adapter-mediated retargeting of a receptor-restricted HSV-1 vector to CEA-bearing tumor cells. Mol Ther 19, 507-514. https://www.ncbi.nlm.nih.gov/pubmed/20924362

Bagley, S.J., Desai, A.S., Linette, G.P., June, C.H., and O'Rourke, D.M. (2018). CAR T-cell therapy for glioblastoma: recent clinical advances and future challenges. Neuro Oncol 20, 1429-1438. http://dx.doi.org/10.1093/neuonc/noy032.

Barry, K.C., Hsu, J., Broz, M.L., Cueto, F.J., Binnewies, M., Combes, A.J., Nelson, A.E., Loo, K., Kumar, R., Rosenblum, M.D., *et al.* (2018). A natural killer-dendritic cell axis defines checkpoint therapy-responsive tumor microenvironments. Nat Med 24, 1178-1191. https://www.ncbi.nlm.nih.gov/pubmed/29942093

Bauer, D.F., Pereboeva, L., Gillespie, G.Y., Cloud, G.A., Elzafarany, O., Langford, C., Markert, J.M., and Lamb, L.S., Jr. (2016). Effect of HSV-IL12 Loaded Tumor Cell-Based Vaccination in a Mouse Model of High-Grade Neuroblastoma. J Immunol Res 2016, 2568125. https://www.ncbi.nlm.nih.gov/pubmed/27610392

Blake, Z., Marks, D.K., Gartrell, R.D., Hart, T., Horton, P., Cheng, S.K., Taback, B., Horst, B.A., and Saenger, Y.M. (2018). Complete intracranial response to talimogene laherparepvec (T-Vec), pembrolizumab and whole brain radiotherapy in a patient with melanoma brain metastases refractory to dual checkpoint-inhibition. J Immunother Cancer 6, 25. https://www.ncbi.nlm.nih.gov/pubmed/29622046

Blank, S.V., Rubin, S.C., Coukos, G., Amin, K.M., Albelda, S.M., and Molnar-Kimber, K.L. (2002). Replication-selective herpes simplex virus type 1 mutant therapy of cervical cancer is enhanced by low-dose radiation. Hum Gene Ther 13, 627-639. https://www.ncbi.nlm.nih.gov/pubmed/11916486

Blumenthal, D.T., Gorlia, T., Gilbert, M.R., Kim, M.M., Burt Nabors, L., Mason, W.P., Hegi, M.E., Zhang, P., Golfinopoulos, V., Perry, J.R., *et al.* (2017). Is more better? The impact of extended adjuvant temozolomide in newly diagnosed glioblastoma: a secondary analysis of EORTC and NRG Oncology/RTOG. Neuro Oncol 19, 1119-1126. https://www.ncbi.nlm.nih.gov/pubmed/28371907

Bommareddy, P.K., Shettigar, M., and Kaufman, H.L. (2018). Integrating oncolytic viruses in combination cancer immunotherapy. Nat Rev Immunol 18, 498-513. https://www.ncbi.nlm.nih.gov/pubmed/29743717

Bottcher, J.P., Bonavita, E., Chakravarty, P., Blees, H., Cabeza-Cabrerizo, M., Sammicheli, S., Rogers, N.C., Sahai, E., Zelenay, S., and Reis e Sousa, C. (2018). NK Cells Stimulate Recruitment of cDC1 into the Tumor Microenvironment Promoting Cancer Immune Control. Cell 172, 1022-1037 e1014. https://www.ncbi.nlm.nih.gov/pubmed/29429633

Bouffet, E., Larouche, V., Campbell, B.B., Merico, D., de Borja, R., Aronson, M., Durno, C., Krueger, J., Cabric, V., Ramaswamy, V., *et al.* (2016). Immune Checkpoint Inhibition for Hypermutant Glioblastoma Multiforme Resulting

From Germline Biallelic Mismatch Repair Deficiency. J Clin Oncol 34, 2206-2211. https://www.ncbi.nlm.nih.gov/pubmed/27001570

Boviatsis, E.J., Park, J.S., Sena-Esteves, M., Kramm, C.M., Chase, M., Efird, J.T., Wei, M.X., Breakefield, X.O., and Chiocca, E.A. (1994). Long-term survival of rats harboring brain neoplasms treated with ganciclovir and a herpes simplex virus vector that retains an intact thymidine kinase gene. Cancer Res 54, 5745-5751. https://www.ncbi.nlm.nih.gov/pubmed/7954393

Bowman, R.L., Klemm, F., Akkari, L., Pyonteck, S.M., Sevenich, L., Quail, D.F., Dhara, S., Simpson, K., Gardner, E.E., Iacobuzio-Donahue, C.A., *et al.* (2016). Macrophage Ontogeny Underlies Differences in Tumor-Specific Education in Brain Malignancies. Cell Rep 17, 2445-2459. https://www.ncbi.nlm.nih.gov/pubmed/27840052

Bradley, J.D., Kataoka, Y., Advani, S., Chung, S.M., Arani, R.B., Gillespie, G.Y., Whitley, R.J., Markert, J.M., Roizman, B., and Weichselbaum, R.R. (1999). Ionizing radiation improves survival in mice bearing intracranial high-grade gliomas injected with genetically modified herpes simplex virus. Clin Cancer Res 5, 1517-1522. https://www.ncbi.nlm.nih.gov/pubmed/10389941

Broz, M.L., Binnewies, M., Boldajipour, B., Nelson, A.E., Pollack, J.L., Erle, D.J., Barczak, A., Rosenblum, M.D., Daud, A., Barber, D.L., *et al.* (2014). Dissecting the Tumor Myeloid Compartment Reveals Rare Activating Antigen-Presenting Cells Critical for T Cell Immunity. Cancer Cell 26, 938. http://dx.doi.org/10.1016/j.ccell.2014.11.010.

Butowski, N., Colman, H., De Groot, J.F., Omuro, A.M., Nayak, L., Wen, P.Y., Cloughesy, T.F., Marimuthu, A., Haidar, S., Perry, A., *et al.* (2016). Orally administered colony stimulating factor 1 receptor inhibitor PLX3397 in recurrent glioblastoma: an Ivy Foundation Early Phase Clinical Trials Consortium phase II study. Neuro Oncol 18, 557-564. https://www.ncbi.nlm.nih.gov/pubmed/26449250

Cairns, T.M., Huang, Z.Y., Gallagher, J.R., Lin, Y., Lou, H., Whitbeck, J.C., Wald, A., Cohen, G.H., and Eisenberg, R.J. (2015). Patient-Specific Neutralizing Antibody Responses to Herpes Simplex Virus Are Attributed to Epitopes on gD, gB, or Both and Can Be Type Specific. J Virol 89, 9213-9231. http://dx.doi.org/10.1128/jvi.01213-15.

Campadelli-Fiume, G., Petrovic, B., Leoni, V., Gianni, T., Avitabile, E., Casiraghi, C., and Gatta, V. (2016). Retargeting Strategies for Oncolytic Herpes Simplex Viruses. Viruses 8, 63. https://www.ncbi.nlm.nih.gov/pubmed/26927159

Cao, H., Zhang, G.R., and Geller, A.I. (2010). Antibody-mediated targeted gene transfer to NMDA NR1-containing neurons in rat neocortex by helper virus-free HSV-1 vector particles containing a chimeric HSV-1 glycoprotein C-staphylococcus A protein. Brain Res 1351, 1-12. https://www.ncbi.nlm.nih.gov/pubmed/20599821

Carfi, A., Willis, S.H., Whitbeck, J.C., Krummenacher, C., Cohen, G.H., Eisenberg, R.J., and Wiley, D.C. (2001). Herpes simplex virus glycoprotein D bound to the human receptor HveA. Mol Cell 8, 169-179. https://www.ncbi.nlm.nih.gov/pubmed/11511370

Cassady, K.A., Bauer, D.F., Roth, J., Chambers, M.R., Shoeb, T., Coleman, J., Prichard, M., Gillespie, G.Y., and Markert, J.M. (2017). Pre-clinical Assessment of C134, a Chimeric Oncolytic Herpes Simplex Virus, in Mice and Non-human Primates. Mol Ther Oncolytics 5, 1-10. https://www.ncbi.nlm.nih.gov/pubmed/28345027

Cassady, K.A., Gross, M., and Roizman, B. (1998). The herpes simplex virus US11 protein effectively compensates for the gamma1(34.5) gene if present before activation of protein kinase R by precluding its phosphorylation and that of the alpha subunit of eukaryotic translation initiation factor 2. J Virol 72, 8620-8626. https://www.ncbi.nlm.nih.gov/pubmed/9765401

Chahlavi, A., Todo, T., Martuza, R.L., and Rabkin, S.D. (1999). Replication-competent herpes simplex virus vector G207 and cisplatin combination therapy for head and neck squamous cell carcinoma. Neoplasia 1, 162-169. https://www.ncbi.nlm.nih.gov/pubmed/10933051

Chaichana, K.L., Cabrera-Aldana, E.E., Jusue-Torres, I., Wijesekera, O., Olivi, A., Rahman, M., and Quinones-Hinojosa, A. (2014). When gross total resection of a glioblastoma is possible, how much resection should be achieved? World Neurosurg 82, e257-265. http://dx.doi.org/10.1016/j.wneu.2014.01.019.

Chang, A.L., Miska, J., Wainwright, D.A., Dey, M., Rivetta, C.V., Yu, D., Kanojia, D., Pituch, K.C., Qiao, J., Pytel, P., *et al.* (2016). CCL2 Produced by the Glioma Microenvironment Is Essential for the Recruitment of Regulatory T Cells and Myeloid-Derived Suppressor Cells. Cancer Res 76, 5671-5682. https://www.ncbi.nlm.nih.gov/pubmed/27530322

Chao, M.P., Alizadeh, A.A., Tang, C., Jan, M., Weissman-Tsukamoto, R., Zhao, F., Park, C.Y., Weissman, I.L., and Majeti, R. (2011). Therapeutic antibody targeting of CD47 eliminates human acute lymphoblastic leukemia. Cancer Res 71, 1374-1384. https://www.ncbi.nlm.nih.gov/pubmed/21177380

Chase, M., Chung, R.Y., and Chiocca, E.A. (1998). An oncolytic viral mutant that delivers the CYP2B1 transgene and augments cyclophosphamide chemotherapy. Nat Biotechnol 16, 444-448. https://www.ncbi.nlm.nih.gov/pubmed/9592392

Cheema, T.A., Wakimoto, H., Fecci, P.E., Ning, J., Kuroda, T., Jeyaretna, D.S., Martuza, R.L., and Rabkin, S.D. (2013). Multifaceted oncolytic virus therapy for glioblastoma in an immunocompetent cancer stem cell model. Proc Natl Acad Sci U S A 110, 12006-12011. https://www.ncbi.nlm.nih.gov/pubmed/23754388

Chen, C.Y., Hutzen, B., Wedekind, M.F., and Cripe, T.P. (2018). Oncolytic virus and PD-1/PD-L1 blockade combination therapy. Oncolytic Virother 7, 65-77. https://www.ncbi.nlm.nih.gov/pubmed/30105219

Chen, X., Han, J., Chu, J., Zhang, L., Zhang, J., Chen, C., Chen, L., Wang, Y., Wang, H., Yi, L., *et al.* (2016). A combinational therapy of EGFR-CAR NK cells and oncolytic herpes simplex virus 1 for breast cancer brain metastases. Oncotarget 7, 27764-27777. https://www.ncbi.nlm.nih.gov/pubmed/27050072

Cheng, G., Yang, K., and He, B. (2003). Dephosphorylation of eIF-2alpha mediated by the gamma(1)34.5 protein of herpes simplex virus type 1 is required for viral response to interferon but is not sufficient for efficient viral replication. J Virol 77, 10154-10161. https://www.ncbi.nlm.nih.gov/pubmed/12941928

Chikuma, S. (2017). CTLA-4, an Essential Immune-Checkpoint for T-Cell Activation. In Emerging Concepts Targeting Immune Checkpoints in Cancer and Autoimmunity, A. Yoshimura, ed. (Cham: Springer International Publishing), pp. 99-126.

Chiocca, E.A., Yu, J.S., Lukas, R.V., Solomon, I.H., Ligon, K.L., Nakashima, H., Triggs, D.A., Reardon, D.A., Wen, P., Stopa, B.M., *et al.* (2019). Regulatable interleukin-12 gene therapy in patients with recurrent high-grade glioma: Results of a phase 1 trial. Sci Transl Med 11. https://www.ncbi.nlm.nih.gov/pubmed/31413142

Chisholm, S.E., Howard, K., Gomez, M.V., and Reyburn, H.T. (2007). Expression of ICP0 is sufficient to trigger natural killer cell recognition of herpes simplex virus-infected cells by natural cytotoxicity receptors. J Infect Dis 195, 1160-1168. http://dx.doi.org/10.1086/512862.

Chiu, T.L., Wang, M.J., and Su, C.C. (2012). The treatment of glioblastoma multiforme through activation of microglia and TRAIL induced by rAAV2-mediated IL-12 in a syngeneic rat model. J Biomed Sci 19, 45.

Cioffi, M., Trabulo, S., Hidalgo, M., Costello, E., Greenhalf, W., Erkan, M., Kleeff, J., Sainz, B., Jr., and Heeschen, C. (2015). Inhibition of CD47 Effectively Targets Pancreatic Cancer Stem Cells via Dual Mechanisms. Clin Cancer Res 21, 2325-2337. https://www.ncbi.nlm.nih.gov/pubmed/25717063

Cocchi, F., Fusco, D., Menotti, L., Gianni, T., Eisenberg, R.J., Cohen, G.H., and Campadelli-Fiume, G. (2004). The soluble ectodomain of herpes simplex virus gD contains a membrane-proximal pro-fusion domain and suffices to mediate virus entry. Proc Natl Acad Sci U S A 101, 7445-7450. https://www.ncbi.nlm.nih.gov/pubmed/15123804

Cody, J.J., Scaturro, P., Cantor, A.B., Yancey Gillespie, G., Parker, J.N., and Markert, J.M. (2012). Preclinical evaluation of oncolytic deltagamma(1)34.5 herpes simplex virus expressing interleukin-12 for therapy of breast cancer brain metastases. Int J Breast Cancer 2012, 628697. https://www.ncbi.nlm.nih.gov/pubmed/23346408

Colli, L.M., Machiela, M.J., Myers, T.A., Jessop, L., Yu, K., and Chanock, S.J. (2016). Burden of Nonsynonymous Mutations among TCGA Cancers and Candidate Immune Checkpoint Inhibitor Responses. Cancer Res 76, 3767-3772. http://dx.doi.org/10.1158/0008-5472.Can-16-0170.

Connolly, S.A., Landsburg, D.J., Carfi, A., Whitbeck, J.C., Zuo, Y., Wiley, D.C., Cohen, G.H., and Eisenberg, R.J. (2005). Potential nectin-1 binding site on herpes simplex virus glycoprotein D. J Virol 79, 1282-1295. https://www.ncbi.nlm.nih.gov/pubmed/15613355

Connolly, S.A., Landsburg, D.J., Carfi, A., Wiley, D.C., Cohen, G.H., and Eisenberg, R.J. (2003). Structure-based mutagenesis of herpes simplex virus glycoprotein D defines three critical regions at the gD-HveA/HVEM binding interface. J Virol 77, 8127-8140. https://www.ncbi.nlm.nih.gov/pubmed/12829851

Coulie, P.G., Van den Eynde, B.J., van der Bruggen, P., and Boon, T. (2014). Tumour antigens recognized by T lymphocytes: at the core of cancer immunotherapy. Nat Rev Cancer 14, 135-146. https://www.ncbi.nlm.nih.gov/pubmed/24457417

Cozzi, P.J., Malhotra, S., McAuliffe, P., Kooby, D.A., Federoff, H.J., Huryk, B., Johnson, P., Scardino, P.T., Heston, W.D., and Fong, Y. (2001). Intravesical oncolytic viral therapy using attenuated, replication-competent herpes simplex viruses G207 and Nv1020 is effective in the treatment of bladder cancer in an orthotopic syngeneic model. FASEB J 15, 1306-1308. https://www.ncbi.nlm.nih.gov/pubmed/11344122

Dai, H.S., and Caligiuri, M.A. (2018). Molecular Basis for the Recognition of Herpes Simplex Virus Type 1 Infection by Human Natural Killer Cells. Front Immunol 9, 183. https://www.ncbi.nlm.nih.gov/pubmed/29483911

Dai, H.S., Griffin, N., Bolyard, C., Mao, H.C., Zhang, J., Cripe, T.P., Suenaga, T., Arase, H., Nakano, I., Chiocca, E.A., *et al.* (2017a). The Fc Domain of Immunoglobulin Is Sufficient to Bridge NK Cells with Virally Infected Cells. Immunity 47, 159-170 e110. https://www.ncbi.nlm.nih.gov/pubmed/28723548

Dai, P., Wang, W., Yang, N., Serna-Tamayo, C., Ricca, J.M., Zamarin, D., Shuman, S., Merghoub, T., Wolchok, J.D., and Deng, L. (2017b). Intratumoral delivery of inactivated modified vaccinia virus Ankara (iMVA) induces systemic antitumor immunity via STING and Batf3-dependent dendritic cells. Sci Immunol 2, eaal1713. http://dx.doi.org/10.1126/sciimmunol.aal1713.

Darmanis, S., Sloan, S.A., Croote, D., Mignardi, M., Chernikova, S., Samghababi, P., Zhang, Y., Neff, N., Kowarsky, M., Caneda, C., *et al.* (2017).

Single-Cell RNA-Seq Analysis of Infiltrating Neoplastic Cells at the Migrating Front of Human Glioblastoma. Cell Rep 21, 1399-1410. https://www.ncbi.nlm.nih.gov/pubmed/29091775

DeGregori, J., Kowalik, T., and Nevins, J.R. (1995). Cellular targets for activation by the E2F1 transcription factor include DNA synthesis- and G1/S-regulatory genes. Mol Cell Biol 15, 4215-4224. https://www.ncbi.nlm.nih.gov/pubmed/7623816

Delgado-Lopez, P.D., and Corrales-Garcia, E.M. (2016). Survival in glioblastoma: a review on the impact of treatment modalities. Clin Transl Oncol 18, 1062-1071. https://www.ncbi.nlm.nih.gov/pubmed/26960561

Desjardins, A., Reardon, D.A., and Vredenburgh, J.J. (2009). Current available therapies and future directions in the treatment of malignant gliomas. Biologics 3, 15-25. https://www.ncbi.nlm.nih.gov/pubmed/19707392

Di Giovine, P., Settembre, E.C., Bhargava, A.K., Luftig, M.A., Lou, H., Cohen, G.H., Eisenberg, R.J., Krummenacher, C., and Carfi, A. (2011). Structure of herpes simplex virus glycoprotein D bound to the human receptor nectin-1. PLoS Pathog 7, e1002277. https://www.ncbi.nlm.nih.gov/pubmed/21980294

Dmitrieva, N., Yu, L., Viapiano, M., Cripe, T.P., Chiocca, E.A., Glorioso, J.C., and Kaur, B. (2011). Chondroitinase ABC I-mediated enhancement of oncolytic virus spread and antitumor efficacy. Clin Cancer Res 17, 1362-1372. https://www.ncbi.nlm.nih.gov/pubmed/21177410

Douglas, J.T., Miller, C.R., Kim, M., Dmitriev, I., Mikheeva, G., Krasnykh, V., and Curiel, D.T. (1999). A system for the propagation of adenoviral vectors with genetically modified receptor specificities. Nat Biotechnol 17, 470-475. https://www.ncbi.nlm.nih.gov/pubmed/10331807

Dummer, R., Hoeller, C., Gruter, I.P., and Michielin, O. (2017). Combining talimogene laherparepvec with immunotherapies in melanoma and other solid tumors. Cancer Immunol Immunother 66, 683-695. https://www.ncbi.nlm.nih.gov/pubmed/28238174

Ebright, M.I., Zager, J.S., Malhotra, S., Delman, K.A., Weigel, T.L., Rusch, V.W., and Fong, Y. (2002). Replication-competent herpes virus NV1020 as direct treatment of pleural cancer in a rat model. J Thorac Cardiovasc Surg 124, 123-129. https://www.ncbi.nlm.nih.gov/pubmed/12091817

Eisenberg, R.J., Atanasiu, D., Cairns, T.M., Gallagher, J.R., Krummenacher, C., and Cohen, G.H. (2012). Herpes virus fusion and entry: a story with many characters. Viruses 4, 800-832. https://www.ncbi.nlm.nih.gov/pubmed/22754650

Elliott, L.A., Doherty, G.A., Sheahan, K., and Ryan, E.J. (2017). Human Tumor-Infiltrating Myeloid Cells: Phenotypic and Functional Diversity. Front Immunol 8, 86. http://dx.doi.org/10.3389/fimmu.2017.00086.

Fan, Q., Kopp, S., Connolly, S.A., Muller, W.J., and Longnecker, R. (2017). Mapping sites of herpes simplex virus type 1 glycoprotein D that permit insertions and impact gD and gB receptors usage. Sci Rep 7, 43712. https://www.ncbi.nlm.nih.gov/pubmed/28255168

Feenstra, V., Hodaie, M., and Johnson, D.C. (1990). Deletions in herpes simplex virus glycoprotein D define nonessential and essential domains. J Virol 64, 2096-2102. https://www.ncbi.nlm.nih.gov/pubmed/2157872

Feng, E., Sui, C., Wang, T., and Sun, G. (2017). Temozolomide with or without Radiotherapy in Patients with Newly Diagnosed Glioblastoma Multiforme: A Meta-Analysis. Eur Neurol 77, 201-210. https://www.ncbi.nlm.nih.gov/pubmed/28192785

Ferris, R.L., Gross, N.D., Nemunaitis, J.J., Andtbacka, R.H., Argiris, A., Ohr, J., Vetto, J.T., Senzer, N., Bedell, C., Ungerleider, R.S., *et al.* (2014). PhaseI trial of intratumoral therapy using HF10, an oncolytic HSV-1, demonstrates safety in HSV+/HSV- patients with refractory and superficial cancers. J Clin Oncol 32, 6082. http://dx.doi.org/10.1200/jco.2014.32.15_suppl.6082.

Filley, A.C., Henriquez, M., and Dey, M. (2017). Recurrent glioma clinical trial, CheckMate-143: the game is not over yet. Oncotarget 8, 91779-91794. https://www.ncbi.nlm.nih.gov/pubmed/29207684

Flores-Toro, J.A., Luo, D., Gopinath, A., Sarkisian, M.R., Campbell, J.J., Charo, I.F., Singh, R., Schall, T.J., Datta, M., Jain, R.K., *et al.* (2020). CCR2 inhibition reduces tumor myeloid cells and unmasks a checkpoint inhibitor effect to slow progression of resistant murine gliomas. Proc Natl Acad Sci U S A 117, 1129-1138. https://www.ncbi.nlm.nih.gov/pubmed/31879345

Friedman, G.K., Bernstock, J.D., Chen, D., Nan, L., Moore, B.P., Kelly, V.M., Youngblood, S.L., Langford, C.P., Han, X., Ring, E.K., *et al.* (2018).

Enhanced Sensitivity of Patient-Derived Pediatric High-Grade Brain Tumor Xenografts to Oncolytic HSV-1 Virotherapy Correlates with Nectin-1 Expression. Sci Rep 8, 13930. https://www.ncbi.nlm.nih.gov/pubmed/30224769

Fruh, K., Ahn, K., Djaballah, H., Sempe, P., van Endert, P.M., Tampe, R., Peterson, P.A., and Yang, Y. (1995). A viral inhibitor of peptide transporters for antigen presentation. Nature 375, 415-418. https://www.ncbi.nlm.nih.gov/pubmed/7760936

Fu, X., Rivera, A., Tao, L., De Geest, B., and Zhang, X. (2012). Construction of an oncolytic herpes simplex virus that precisely targets hepatocellular carcinoma cells. Mol Ther 20, 339-346. https://www.ncbi.nlm.nih.gov/pubmed/22146341

Fu, X., Tao, L., Jin, A., Vile, R., Brenner, M.K., and Zhang, X. (2003). Expression of a fusogenic membrane glycoprotein by an oncolytic herpes simplex virus potentiates the viral antitumor effect. Mol Ther 7, 748-754. https://www.ncbi.nlm.nih.gov/pubmed/12788648

Fujii, K., Kurozumi, K., Ichikawa, T., Onishi, M., Shimazu, Y., Ishida, J., Chiocca, E.A., Kaur, B., and Date, I. (2013). The integrin inhibitor cilengitide enhances the anti-glioma efficacy of vasculostatin-expressing oncolytic virus. Cancer Gene Ther 20, 437-444. https://www.ncbi.nlm.nih.gov/pubmed/23827879

Fujimoto, Y., Mizuno, T., Sugiura, S., Goshima, F., Kohno, S., Nakashima, T., and Nishiyama, Y. (2006). Intratumoral injection of herpes simplex virus HF10 in recurrent head and neck squamous cell carcinoma. Acta Otolaryngol 126, 1115-1117. https://www.ncbi.nlm.nih.gov/pubmed/16923721

Fujita, M., Kohanbash, G., Fellows-Mayle, W., Hamilton, R.L., Komohara, Y., Decker, S.A., Ohlfest, J.R., and Okada, H. (2011). COX-2 blockade suppresses gliomagenesis by inhibiting myeloid-derived suppressor cells. Cancer Res 71, 2664-2674. https://www.ncbi.nlm.nih.gov/pubmed/21324923

Fujita, M., Scheurer, M.E., Decker, S.A., McDonald, H.A., Kohanbash, G., Kastenhuber, E.R., Kato, H., Bondy, M.L., Ohlfest, J.R., and Okada, H. (2010). Role of type 1 IFNs in antiglioma immunosurveillance--using mouse studies to guide examination of novel prognostic markers in humans. Clin Cancer Res 16, 3409-3419. https://www.ncbi.nlm.nih.gov/pubmed/20472682

Fukuhara, H., Ino, Y., and Todo, T. (2016). Oncolytic virus therapy: A new era of cancer treatment at dawn. Cancer Sci 107, 1373-1379. https://www.ncbi.nlm.nih.gov/pubmed/27486853

Fulci, G., and Chiocca, E.A. (2007). The status of gene therapy for brain tumors. Expert Opin Biol Ther 7, 197-208. https://www.ncbi.nlm.nih.gov/pubmed/17250458

Fulci, G., Dmitrieva, N., Gianni, D., Fontana, E.J., Pan, X., Lu, Y., Kaufman, C.S., Kaur, B., Lawler, S.E., Lee, R.J., *et al.* (2007). Depletion of peripheral macrophages and brain microglia increases brain tumor titers of oncolytic viruses. Cancer Res 67, 9398-9406. https://www.ncbi.nlm.nih.gov/pubmed/17909049

Fuller, A.O., Santos, R.E., and Spear, P.G. (1989). Neutralizing antibodies specific for glycoprotein H of herpes simplex virus permit viral attachment to cells but prevent penetration. J Virol 63, 3435-3443. https://www.ncbi.nlm.nih.gov/pubmed/2545914

Fuller, A.O., and Spear, P.G. (1987). Anti-glycoprotein D antibodies that permit adsorption but block infection by herpes simplex virus 1 prevent virion-cell fusion at the cell surface. Proc Natl Acad Sci U S A 84, 5454-5458. https://www.ncbi.nlm.nih.gov/pubmed/3037552

Fusco, D., Forghieri, C., and Campadelli-Fiume, G. (2005). The pro-fusion domain of herpes simplex virus glycoprotein D (gD) interacts with the gD N terminus and is displaced by soluble forms of viral receptors. Proc Natl Acad Sci U S A 102, 9323-9328. https://www.ncbi.nlm.nih.gov/pubmed/15972328

Gabrilovich, D.I. (2017). Myeloid-Derived Suppressor Cells. Cancer Immunol Res 5, 3-8. http://dx.doi.org/10.1158/2326-6066.Cir-16-0297.

Gambini, E., Reisoli, E., Appolloni, I., Gatta, V., Campadelli-Fiume, G., Menotti, L., and Malatesta, P. (2012). Replication-competent herpes simplex virus retargeted to HER2 as therapy for high-grade glioma. Mol Ther 20, 994-1001. https://www.ncbi.nlm.nih.gov/pubmed/22354378

Garg, A.D., Vandenberk, L., Van Woensel, M., Belmans, J., Schaaf, M., Boon, L., De Vleeschouwer, S., and Agostinis, P. (2017). Preclinical efficacy of immune-checkpoint monotherapy does not recapitulate corresponding biomarkers-based clinical predictions in glioblastoma. Oncoimmunology 6, e1295903. https://www.ncbi.nlm.nih.gov/pubmed/28507806

Garzon-Muvdi, T., Theodros, D., Luksik, A.S., Maxwell, R., Kim, E., Jackson, C.M., Belcaid, Z., Ganguly, S., Tyler, B., Brem, H., *et al.* (2018). Dendritic cell activation enhances anti-PD-1 mediated immunotherapy against glioblastoma. Oncotarget 9, 20681-20697. https://www.ncbi.nlm.nih.gov/pubmed/29755681

Gatta, V., Petrovic, B., and Campadelli-Fiume, G. (2015). The Engineering of a Novel Ligand in gH Confers to HSV an Expanded Tropism Independent of gD Activation by Its Receptors. PLoS Pathog 11, e1004907. https://www.ncbi.nlm.nih.gov/pubmed/25996983

Gaur, A., Jewell, D.A., Liang, Y., Ridzon, D., Moore, J.H., Chen, C., Ambros, V.R., and Israel, M.A. (2007). Characterization of microRNA expression levels and their biological correlates in human cancer cell lines. Cancer Res 67, 2456-2468. https://www.ncbi.nlm.nih.gov/pubmed/17363563

Geevarghese, S.K., Geller, D.A., de Haan, H.A., Horer, M., Knoll, A.E., Mescheder, A., Nemunaitis, J., Reid, T.R., Sze, D.Y., Tanabe, K.K., *et al.* (2010). Phase I/II study of oncolytic herpes simplex virus NV1020 in patients with extensively pretreated refractory colorectal cancer metastatic to the liver. Hum Gene Ther 21, 1119-1128. https://www.ncbi.nlm.nih.gov/pubmed/20486770

Geraghty, R.J., Krummenacher, C., Cohen, G.H., Eisenberg, R.J., and Spear, P.G. (1998). Entry of alphaherpesviruses mediated by poliovirus receptor-related protein 1 and poliovirus receptor. Science 280, 1618-1620. https://www.ncbi.nlm.nih.gov/pubmed/9616127

Gilbert, M.R., Wang, M., Aldape, K.D., Stupp, R., Hegi, M.E., Jaeckle, K.A., Armstrong, T.S., Wefel, J.S., Won, M., Blumenthal, D.T., *et al.* (2013). Dose-dense temozolomide for newly diagnosed glioblastoma: a randomized phase III clinical trial. J Clin Oncol 31, 4085-4091. https://www.ncbi.nlm.nih.gov/pubmed/24101040

Goldsmith, K., Chen, W., Johnson, D.C., and Hendricks, R.L. (1998). Infected cell protein (ICP)47 enhances herpes simplex virus neurovirulence by blocking the CD8+ T cell response. J Exp Med 187, 341-348. https://www.ncbi.nlm.nih.gov/pubmed/9449714

Goodwin, J.M., Schmitt, A.D., McGinn, C.M., Fuchs, B.C., Kuruppu, D., Tanabe, K.K., and Lanuti, M. (2012). Angiogenesis inhibition using an oncolytic herpes simplex virus expressing endostatin in a murine lung cancer model. Cancer Invest 30, 243-250. https://www.ncbi.nlm.nih.gov/pubmed/22360364

Grandi, P., Fernandez, J., Szentirmai, O., Carter, R., Gianni, D., Sena-Esteves, M., and Breakefield, X.O. (2010). Targeting HSV-1 virions for specific binding to epidermal growth factor receptor-vIII-bearing tumor cells. Cancer Gene Ther 17, 655-663. https://www.ncbi.nlm.nih.gov/pubmed/20508670

Grandi, P., Wang, S., Schuback, D., Krasnykh, V., Spear, M., Curiel, D.T., Manservigi, R., and Breakefield, X.O. (2004). HSV-1 virions engineered for specific binding to cell surface receptors. Mol Ther 9, 419-427. https://www.ncbi.nlm.nih.gov/pubmed/15006609

Grossman, S.A., Ye, X., Piantadosi, S., Desideri, S., Nabors, L.B., Rosenfeld, M., Fisher, J., and Consortium, N.C. (2010). Survival of patients with newly diagnosed glioblastoma treated with radiation and temozolomide in research studies in the United States. Clin Cancer Res 16, 2443-2449. https://www.ncbi.nlm.nih.gov/pubmed/20371685

Gruenheid, S., Gatzke, L., Meadows, H., and Tufaro, F. (1993). Herpes simplex virus infection and propagation in a mouse L cell mutant lacking heparan sulfate proteoglycans. J Virol 67, 93-100. https://www.ncbi.nlm.nih.gov/pubmed/8380101

Grupp, S.A., Kalos, M., Barrett, D., Aplenc, R., Porter, D.L., Rheingold, S.R., Teachey, D.T., Chew, A., Hauck, B., Wright, J.F., *et al.* (2013). Chimeric antigen receptor-modified T cells for acute lymphoid leukemia. N Engl J Med 368, 1509-1518. http://dx.doi.org/10.1056/NEJMoa1215134.

Gujar, S., Pol, J.G., Kim, Y., Lee, P.W., and Kroemer, G. (2018a). Antitumor Benefits of Antiviral Immunity: An Underappreciated Aspect of Oncolytic Virotherapies. Trends Immunol 39, 209-221. https://www.ncbi.nlm.nih.gov/pubmed/29275092

Gujar, S., Pol, J.G., and Kroemer, G. (2018b). Heating it up: Oncolytic viruses make tumors 'hot' and suitable for checkpoint blockade immunotherapies. Oncoimmunology 7, e1442169. https://www.ncbi.nlm.nih.gov/pubmed/30221036

Han, J., Chu, J., Keung Chan, W., Zhang, J., Wang, Y., Cohen, J.B., Victor, A., Meisen, W.H., Kim, S.H., Grandi, P., *et al.* (2015). CAR-Engineered NK Cells Targeting Wild-Type EGFR and EGFRvIII Enhance Killing of Glioblastoma and Patient-Derived Glioblastoma Stem Cells. Sci Rep 5, 11483. https://www.ncbi.nlm.nih.gov/pubmed/26155832

Hanson, N., Henderson, G., and Jones, C. (1994). The herpes simplex virus type 2 gene which encodes the large subunit of ribonucleotide reductase has unusual regulatory properties. Virus Res 34, 265-280. https://www.ncbi.nlm.nih.gov/pubmed/7856314

Hargadon, K.M., Johnson, C.E., and Williams, C.J. (2018). Immune checkpoint blockade therapy for cancer: An overview of FDA-approved immune checkpoint inhibitors. Int Immunopharmacol 62, 29-39. https://www.ncbi.nlm.nih.gov/pubmed/29990692

Harrington, K.J., Andtbacka, R.H., Collichio, F., Downey, G., Chen, L., Szabo, Z., and Kaufman, H.L. (2016). Efficacy and safety of talimogene laherparepvec versus granulocyte-macrophage colony-stimulating factor in patients with stage IIIB/C and IVM1a melanoma: subanalysis of the Phase III OPTiM trial. Onco Targets Ther 9, 7081-7093. https://www.ncbi.nlm.nih.gov/pubmed/27895500

Harrington, K.J., Hingorani, M., Tanay, M.A., Hickey, J., Bhide, S.A., Clarke, P.M., Renouf, L.C., Thway, K., Sibtain, A., McNeish, I.A., *et al.* (2010). Phase I/II study of oncolytic HSV GM-CSF in combination with radiotherapy and cisplatin in untreated stage III/IV squamous cell cancer of the head and neck. Clin Cancer Res 16, 4005-4015. https://www.ncbi.nlm.nih.gov/pubmed/20670951

Harris-Bookman, S., Mathios, D., Martin, A.M., Xia, Y., Kim, E., Xu, H., Belcaid, Z., Polanczyk, M., Barberi, T., Theodros, D., *et al.* (2018). Expression of LAG-3 and efficacy of combination treatment with anti-LAG-3 and anti-PD-1 monoclonal antibodies in glioblastoma. Int J Cancer 143, 3201-3208. http://dx.doi.org/10.1002/ijc.31661.

Harrow, S., Papanastassiou, V., Harland, J., Mabbs, R., Petty, R., Fraser, M., Hadley, D., Patterson, J., Brown, S.M., and Rampling, R. (2004). HSV1716 injection into the brain adjacent to tumour following surgical resection of high-grade glioma: safety data and long-term survival. Gene Ther 11, 1648-1658. https://www.ncbi.nlm.nih.gov/pubmed/15334111

He, Y., Cao, J., Zhao, C., Li, X., Zhou, C., and Hirsch, F.R. (2018). TIM-3, a promising target for cancer immunotherapy. Onco Targets Ther 11, 7005-7009. http://dx.doi.org/10.2147/ott.S170385.

He, Y., Rivard, C.J., Rozeboom, L., Yu, H., Ellison, K., Kowalewski, A., Zhou, C., and Hirsch, F.R. (2016). Lymphocyte-activation gene-3, an important immune checkpoint in cancer. Cancer Sci 107, 1193-1197. http://dx.doi.org/10.1111/cas.12986.

Hellmann, M.D., Callahan, M.K., Awad, M.M., Calvo, E., Ascierto, P.A., Atmaca, A., Rizvi, N.A., Hirsch, F.R., Selvaggi, G., Szustakowski, J.D., *et al.* (2018). Tumor Mutational Burden and Efficacy of Nivolumab Monotherapy and in Combination with Ipilimumab in Small-Cell Lung Cancer. Cancer Cell 33, 853-861. http://dx.doi.org/10.1016/j.ccell.2018.04.001.

Herold, B.C., Visalli, R.J., Susmarski, N., Brandt, C.R., and Spear, P.G. (1994). Glycoprotein C-independent binding of herpes simplex virus to cells requires cell surface heparan sulphate and glycoprotein B. J Gen Virol 75 1211-1222. https://www.ncbi.nlm.nih.gov/pubmed/8207388

Herold, B.C., WuDunn, D., Soltys, N., and Spear, P.G. (1991). Glycoprotein C of herpes simplex virus type 1 plays a principal role in the adsorption of virus to cells and in infectivity. J Virol 65, 1090-1098. http://dx.doi.org/10.1099/0022-1317-75-6-1211.

Hildner, K., Edelson, B.T., Purtha, W.E., Diamond, M., Matsushita, H., Kohyama, M., Calderon, B., Schraml, B.U., Unanue, E.R., Diamond, M.S., *et al.* (2008). Batf3 deficiency reveals a critical role for CD8alpha+ dendritic cells in cytotoxic T cell immunity. Science 322, 1097-1100. http://dx.doi.org/10.1126/science.1164206.

Hirooka, Y., Kasuya, H., Ishikawa, T., Kawashima, H., Ohno, E., Villalobos, I.B., Naoe, Y., Ichinose, T., Koyama, N., Tanaka, M., *et al.* (2018). A Phase I clinical trial of EUS-guided intratumoral injection of the oncolytic virus, HF10 for unresectable locally advanced pancreatic cancer. BMC Cancer 18, 596. https://www.ncbi.nlm.nih.gov/pubmed/29801474

Hodges, T.R., Ott, M., Xiu, J., Gatalica, Z., Swensen, J., Zhou, S., Huse, J.T., de Groot, J., Li, S., Overwijk, W.W., *et al.* (2017). Mutational burden, immune checkpoint expression, and mismatch repair in glioma: implications for immune checkpoint immunotherapy. Neuro Oncol 19, 1047-1057. https://www.ncbi.nlm.nih.gov/pubmed/28371827

Hu, J.C., Coffin, R.S., Davis, C.J., Graham, N.J., Groves, N., Guest, P.J., Harrington, K.J., James, N.D., Love, C.A., McNeish, I., *et al.* (2006). A phase I study of OncoVEXGM-CSF, a second-generation oncolytic herpes simplex virus expressing granulocyte macrophage colony-stimulating factor. Clin Cancer Res 12, 6737-6747. https://www.ncbi.nlm.nih.gov/pubmed/17121894

Hung, A.L., Maxwell, R., Theodros, D., Belcaid, Z., Mathios, D., Luksik, A.S., Kim, E., Wu, A., Xia, Y., Garzon-Muvdi, T., *et al.* (2018). TIGIT and PD-1 dual checkpoint blockade enhances antitumor immunity and survival in GBM. Oncoimmunology 7, e1466769. https://www.ncbi.nlm.nih.gov/pubmed/30221069

Hunter, W.D., Martuza, R.L., Feigenbaum, F., Todo, T., Mineta, T., Yazaki, T., Toda, M., Newsome, J.T., Platenberg, R.C., Manz, H.J., *et al.* (1999). Attenuated, replication-competent herpes simplex virus type 1 mutant G207: safety evaluation of intracerebral injection in nonhuman primates. J Virol 73, 6319-6326. https://www.ncbi.nlm.nih.gov/pubmed/10400723

Huse, J.T., and Holland, E.C. (2009). Genetically engineered mouse models of brain cancer and the promise of preclinical testing. Brain Pathol 19, 132-143. https://www.ncbi.nlm.nih.gov/pubmed/19076778

Huse, J.T., and Holland, E.C. (2010). Targeting brain cancer: advances in the molecular pathology of malignant glioma and medulloblastoma. Nat Rev Cancer 10, 319-331. https://www.ncbi.nlm.nih.gov/pubmed/20414201

Huszthy, P.C., Goplen, D., Thorsen, F., Immervoll, H., Wang, J., Gutermann, A., Miletic, H., and Bjerkvig, R. (2008). Oncolytic herpes simplex virus type-1 therapy in a highly infiltrative animal model of human glioblastoma. Clin Cancer Res 14, 1571-1580. https://www.ncbi.nlm.nih.gov/pubmed/18316582

Ishikawa, H., Ma, Z., and Barber, G.N. (2009). STING regulates intracellular DNA-mediated, type I interferon-dependent innate immunity. Nature 461, 788-792. https://www.ncbi.nlm.nih.gov/pubmed/19776740

Jaime-Ramirez, A.C., Dmitrieva, N., Yoo, J.Y., Banasavadi-Siddegowda, Y., Zhang, J., Relation, T., Bolyard, C., Wojton, J., and Kaur, B. (2017). Humanized chondroitinase ABC sensitizes glioblastoma cells to temozolomide. J Gene Med 19. https://www.ncbi.nlm.nih.gov/pubmed/28087981

Jogger, C.R., Montgomery, R.I., and Spear, P.G. (2004). Effects of linker-insertion mutations in herpes simplex virus 1 gD on glycoprotein-induced fusion with cells expressing HVEM or nectin-1. Virology 318, 318-326. https://www.ncbi.nlm.nih.gov/pubmed/14972557

Jorgensen, T.J., Katz, S., Wittmack, E.K., Varghese, S., Todo, T., Rabkin, S.D., and Martuza, R.L. (2001). Ionizing radiation does not alter the antitumor activity of herpes simplex virus vector G207 in subcutaneous tumor models of human and murine prostate cancer. Neoplasia 3, 451-456. https://www.ncbi.nlm.nih.gov/pubmed/11687957

Kaczanowska, S., Joseph Am Fau - Davila, E., and Davila, E. (2013). TLR agonists: our best frenemy in cancer immunotherapy. J Leukoc Biol 93, 847-863. http://dx.doi.org/10.1189/jlb.1012501.

Kalos, M., Levine, B.L., Porter, D.L., Katz, S., Grupp, S.A., Bagg, A., and June, C.H. (2011). T cells with chimeric antigen receptors have potent antitumor effects and can establish memory in patients with advanced leukemia. Sci Transl Med 3, 95ra73. http://dx.doi.org/10.1126/scitranslmed.3002842.

Kambara, H., Okano, H., Chiocca, E.A., and Saeki, Y. (2005). An oncolytic HSV-1 mutant expressing ICP34.5 under control of a nestin promoter increases survival of animals even when symptomatic from a brain tumor. Cancer Res 65, 2832-2839. https://www.ncbi.nlm.nih.gov/pubmed/15805284

Kamiyama, H., Zhou, G., and Roizman, B. (2006). Herpes simplex virus 1 recombinant virions exhibiting the amino terminal fragment of urokinase-type plasminogen activator can enter cells via the cognate receptor. Gene Ther 13, 621-629. https://www.ncbi.nlm.nih.gov/pubmed/16292350

Kamran, N., Calinescu, A., Candolfi, M., Chandran, M., Mineharu, Y., Asad, A.S., Koschmann, C., Nunez, F.J., Lowenstein, P.R., and Castro, M.G. (2016). Recent advances and future of immunotherapy for glioblastoma. Expert Opin Biol Ther 16, 1245-1264. https://www.ncbi.nlm.nih.gov/pubmed/27411023

Kamran, N., Chandran, M., Lowenstein, P.R., and Castro, M.G. (2018). Immature myeloid cells in the tumor microenvironment: Implications for immunotherapy. Clin Immunol 189, 34-42. https://www.ncbi.nlm.nih.gov/pubmed/27777083

Kamran, N., Kadiyala, P., Saxena, M., Candolfi, M., Li, Y., Moreno-Ayala, M.A., Raja, N., Shah, D., Lowenstein, P.R., and Castro, M.G. (2017). Immunosuppressive Myeloid Cells' Blockade in the Glioma Microenvironment Enhances the Efficacy of Immune-Stimulatory Gene Therapy. Mol Ther 25, 232-248. https://www.ncbi.nlm.nih.gov/pubmed/28129117

Kasuya, H., Kodera, Y., Nakao, A., Yamamura, K., Gewen, T., Zhiwen, W., Hotta, Y., Yamada, S., Fujii, T., Fukuda, S., *et al.* (2014). Phase I Dose-escalation Clinical Trial of HF10 Oncolytic Herpes Virus in 17 Japanese Patients with Advanced Cancer. Hepatogastroenterology 61, 599-605. https://www.ncbi.nlm.nih.gov/pubmed/26176043

Kaufman, H.E., Azcuy, A.M., Varnell, E.D., Sloop, G.D., Thompson, H.W., and Hill, J.M. (2005). HSV-1 DNA in tears and saliva of normal adults. Invest Ophthalmol Vis Sci 46, 241-247. https://www.ncbi.nlm.nih.gov/pubmed/15623779

Kaufman, H.L., Kim, D.W., DeRaffele, G., Mitcham, J., Coffin, R.S., and Kim-Schulze, S. (2010). Local and distant immunity induced by intralesional vaccination with an oncolytic herpes virus encoding GM-CSF in patients with stage IIIc and IV melanoma. Ann Surg Oncol 17, 718-730. https://www.ncbi.nlm.nih.gov/pubmed/19915919

Kemeny, N., Brown, K., Covey, A., Kim, T., Bhargava, A., Brody, L., Guilfoyle, B., Haag, N.P., Karrasch, M., Glasschroeder, B., *et al.* (2006). Phase I, open-label, dose-escalating study of a genetically engineered herpes simplex virus,

NV1020, in subjects with metastatic colorectal carcinoma to the liver. Hum Gene Ther 17, 1214-1224. https://www.ncbi.nlm.nih.gov/pubmed/17107303

Kesari, S., Lasner, T.M., Balsara, K.R., Randazzo, B.P., Lee, V.M., Trojanowski, J.Q., and Fraser, N.W. (1998). A neuroattenuated ICP34.5-deficient herpes simplex virus type 1 replicates in ependymal cells of the murine central nervous system. J Gen Virol 79 (Pt 3), 525-536. https://www.ncbi.nlm.nih.gov/pubmed/9519831

Kesari, S., Randazzo, B.P., Valyi-Nagy, T., Huang, Q.S., Brown, S.M., MacLean, A.R., Lee, V.M., Trojanowski, J.Q., and Fraser, N.W. (1995). Therapy of experimental human brain tumors using a neuroattenuated herpes simplex virus mutant. Lab Invest 73, 636-648. https://www.ncbi.nlm.nih.gov/pubmed/7474937

Kim, Y., Lee, H.G., Dmitrieva, N., Kim, J., Kaur, B., and Friedman, A. (2014). Choindroitinase ABC I-mediated enhancement of oncolytic virus spread and anti tumor efficacy: a mathematical model. PLoS One 9, e102499. https://www.ncbi.nlm.nih.gov/pubmed/25047810

Kim, Y., Yoo, J.Y., Lee, T.J., Liu, J., Yu, J., Caligiuri, M.A., Kaur, B., and Friedman, A. (2018). Complex role of NK cells in regulation of oncolytic virus-bortezomib therapy. Proc Natl Acad Sci U S A 115, 4927-4932. https://www.ncbi.nlm.nih.gov/pubmed/29686060

Kimata, H., Imai, T., Kikumori, T., Teshigahara, O., Nagasaka, T., Goshima, F., Nishiyama, Y., and Nakao, A. (2006). Pilot study of oncolytic viral therapy using mutant herpes simplex virus (HF10) against recurrent metastatic breast cancer. Ann Surg Oncol 13, 1078-1084. https://www.ncbi.nlm.nih.gov/pubmed/16865590

Kohanbash, G., McKaveney, K., Sakaki, M., Ueda, R., Mintz, A.H., Amankulor, N., Fujita, M., Ohlfest, J.R., and Okada, H. (2013). GM-CSF promotes the immunosuppressive activity of glioma-infiltrating myeloid cells through interleukin-4 receptor-alpha. Cancer Res 73, 6413-6423. https://www.ncbi.nlm.nih.gov/pubmed/24030977

Kohanbash, G., and Okada, H. (2012). Myeloid-derived suppressor cells (MDSCs) in gliomas and glioma-development. Immunol Invest 41, 658-679. https://www.ncbi.nlm.nih.gov/pubmed/23017140

Koshizuka, T., Kawaguchi, Y., and Nishiyama, Y. (2005). Herpes simplex virus type 2 membrane protein UL56 associates with the kinesin motor protein KIF1A. J Gen Virol 86, 527-533. https://www.ncbi.nlm.nih.gov/pubmed/15722511

Kouvatsis, V., Argnani, R., Tsitoura, E., Arsenakis, M., Georgopoulou, U., Mavromara, P., and Manservigi, R. (2007). Characterization of herpes simplex virus type 1 recombinants that express and incorporate high levels of HCV E2-gC chimeric proteins. Virus Res 123, 40-49. https://www.ncbi.nlm.nih.gov/pubmed/16989918

Kramm, C.M., Chase, M., Herrlinger, U., Jacobs, A., Pechan, P.A., Rainov, N.G., Sena-Esteves, M., Aghi, M., Barnett, F.H., Chiocca, E.A., *et al.* (1997). Therapeutic efficiency and safety of a second-generation replication-conditional HSV1 vector for brain tumor gene therapy. Hum Gene Ther 8, 2057-2068. https://www.ncbi.nlm.nih.gov/pubmed/9414254

Krummenacher, C., Nicola, A.V., Whitbeck, J.C., Lou, H., Hou, W., Lambris, J.D., Geraghty, R.J., Spear, P.G., Cohen, G.H., and Eisenberg, R.J. (1998). Herpes simplex virus glycoprotein D can bind to poliovirus receptor-related protein 1 or herpesvirus entry mediator, two structurally unrelated mediators of virus entry. J Virol 72, 7064-7074. https://www.ncbi.nlm.nih.gov/pubmed/9696799

Krummenacher, C., Supekar, V.M., Whitbeck, J.C., Lazear, E., Connolly, S.A., Eisenberg, R.J., Cohen, G.H., Wiley, D.C., and Carfi, A. (2005). Structure of unliganded HSV gD reveals a mechanism for receptor-mediated activation of virus entry. EMBO J 24, 4144-4153. https://www.ncbi.nlm.nih.gov/pubmed/16292345

Laquerre, S., Anderson, D.B., Stolz, D.B., and Glorioso, J.C. (1998a). Recombinant herpes simplex virus type 1 engineered for targeted binding to erythropoietin receptor-bearing cells. J Virol 72, 9683-9697. https://www.ncbi.nlm.nih.gov/pubmed/9811702

Laquerre, S., Argnani, R., Anderson, D.B., Zucchini, S., Manservigi, R., and Glorioso, J.C. (1998b). Heparan sulfate proteoglycan binding by herpes simplex virus type 1 glycoproteins B and C, which differ in their contributions to virus attachment, penetration, and cell-to-cell spread. J Virol 72, 6119-6130. https://www.ncbi.nlm.nih.gov/pubmed/9621076

Le, D.T., Durham, J.N., Smith, K.N., Wang, H., Bartlett, B.R., Aulakh, L.K., Lu, S., Kemberling, H., Wilt, C., Luber, B.S., *et al.* (2017). Mismatch repair deficiency predicts response of solid tumors to PD-1 blockade. Science 357, 409-413. http://dx.doi.org/10.1126/science.aan6733.

Le, D.T., Uram, J.N., Wang, H., Bartlett, B.R., Kemberling, H., Eyring, A.D., Skora, A.D., Luber, B.S., Azad, N.S., Laheru, D., *et al.* (2015). PD-1 Blockade in Tumors with Mismatch-Repair Deficiency. N Engl J Med 372, 2509-2520. https://www.ncbi.nlm.nih.gov/pubmed/26028255

Lee, C.Y., Rennie, P.S., and Jia, W.W. (2009). MicroRNA regulation of oncolytic herpes simplex virus-1 for selective killing of prostate cancer cells. Clin Cancer Res 15, 5126-5135. https://www.ncbi.nlm.nih.gov/pubmed/19671871

Leib, D.A., Alexander, D.E., Cox, D., Yin, J., and Ferguson, T.A. (2009). Interaction of ICP34.5 with Beclin 1 modulates herpes simplex virus type 1 pathogenesis through control of CD4+ T-cell responses. J Virol 83, 12164-12171. https://www.ncbi.nlm.nih.gov/pubmed/19759141

Leoni, V., Gatta, V., Casiraghi, C., Nicosia, A., Petrovic, B., and Campadelli-Fiume, G. (2017). A Strategy for Cultivation of Retargeted Oncolytic Herpes Simplex Viruses in Non-cancer Cells. J Virol 91, e00067-00017. https://www.ncbi.nlm.nih.gov/pubmed/28250120

Leoni, V., Petrovic, B., Gianni, T., Gatta, V., and Campadelli-Fiume, G. (2018a). Simultaneous Insertion of Two Ligands in gD for Cultivation of Oncolytic Herpes Simplex Viruses in Noncancer Cells and Retargeting to Cancer Receptors. J Virol 92, e02132-02117. https://www.ncbi.nlm.nih.gov/pubmed/29263255

Leoni, V., Vannini, A., Gatta, V., Rambaldi, J., Sanapo, M., Barboni, C., Zaghini, A., Nanni, P., Lollini, P.L., Casiraghi, C., *et al.* (2018b). A fully-virulent retargeted oncolytic HSV armed with IL-12 elicits local immunity and vaccine therapy towards distant tumors. PLoS Pathog 14, e1007209. https://www.ncbi.nlm.nih.gov/pubmed/30080893

Li, G., Wang, Z., Zhang, C., Liu, X., Cai, J., Wang, Z., Hu, H., Wu, F., Bao, Z., Liu, Y., *et al.* (2017). Molecular and clinical characterization of TIM-3 in glioma through 1,024 samples. Oncoimmunology 6, e1328339. http://dx.doi.org/10.1080/2162402x.2017.1328339.

Li, J.M., Kao, K.C., Li, L.F., Yang, T.M., Wu, C.P., Horng, Y.M., Jia, W.W., and Yang, C.T. (2013). MicroRNA-145 regulates oncolytic herpes simplex virus-1 for selective killing of human non-small cell lung cancer cells. Virol J 10, 241. https://www.ncbi.nlm.nih.gov/pubmed/23876001

Li, Y., Zhang, C., Chen, X., Yu, J., Wang, Y., Yang, Y., Du, M., Jin, H., Ma, Y., He, B., *et al.* (2011). ICP34.5 protein of herpes simplex virus facilitates the initiation of protein translation by bridging eukaryotic initiation factor 2alpha (eIF2alpha) and protein phosphatase 1. J Biol Chem 286, 24785-24792. https://www.ncbi.nlm.nih.gov/pubmed/21622569

Lisi, L., Ciotti, G.M., Braun, D., Kalinin, S., Curro, D., Dello Russo, C., Coli, A., Mangiola, A., Anile, C., Feinstein, D.L., *et al.* (2017). Expression of iNOS, CD163 and ARG-1 taken as M1 and M2 markers of microglial polarization in human glioblastoma and the surrounding normal parenchyma. Neurosci Lett 645, 106-112. https://www.ncbi.nlm.nih.gov/pubmed/28259657

Liu, B.L., Robinson, M., Han, Z.Q., Branston, R.H., English, C., Reay, P., McGrath, Y., Thomas, S.K., Thornton, M., Bullock, P., *et al.* (2003). ICP34.5 deleted herpes simplex virus with enhanced oncolytic, immune stimulating, and anti-tumour properties. Gene Ther 10, 292-303. https://www.ncbi.nlm.nih.gov/pubmed/12595888

Long, G.V., Dummer, R., Hamid, O., Gajewski, T.F., Caglevic, C., Dalle, S., Arance, A., Carlino, M.S., Grob, J.J., Kim, T.M., *et al.* (2019). Epacadostat plus pembrolizumab versus placebo plus pembrolizumab in patients with unresectable or metastatic melanoma (ECHO-301/KEYNOTE-252): a phase 3, randomised, double-blind study. Lancet Oncol 20, 1083-1097. https://www.ncbi.nlm.nih.gov/pubmed/31221619

Lowenstein, P.R., and Castro, M.G. (2018). Evolutionary basis of a new gene- and immune-therapeutic approach for the treatment of malignant brain tumors: from mice to clinical trials for glioma patients. Clin Immunol 189, 43-51. https://www.ncbi.nlm.nih.gov/pubmed/28720549

Luo, C., Mori, I., Goshima, F., Ushijima, Y., Nawa, A., Kimura, H., and Nishiyama, Y. (2007). Replication-competent, oncolytic herpes simplex virus type 1 mutants induce a bystander effect following ganciclovir treatment. J Gene Med 9, 875-883. https://www.ncbi.nlm.nih.gov/pubmed/17685493

Mace, A.T., Ganly, I., Soutar, D.S., and Brown, S.M. (2008). Potential for efficacy of the oncolytic Herpes simplex virus 1716 in patients with oral squamous cell carcinoma. Head Neck 30, 1045-1051. https://www.ncbi.nlm.nih.gov/pubmed/18615711

Mace, A.T., Harrow, S.J., Ganly, I., and Brown, S.M. (2007). Cytotoxic effects of the oncolytic herpes simplex virus HSV1716 alone and in combination with cisplatin in head and neck squamous cell carcinoma. Acta Otolaryngol 127, 880-887. https://www.ncbi.nlm.nih.gov/pubmed/17763002

MacKie, R.M., Stewart, B., and Brown, S.M. (2001). Intralesional injection of herpes simplex virus 1716 in metastatic melanoma. Lancet 357, 525-526. https://www.ncbi.nlm.nih.gov/pubmed/11229673

MacLean, A.R., ul-Fareed, M., Robertson, L., Harland, J., and Brown, S.M. (1991). Herpes simplex virus type 1 deletion variants 1714 and 1716 pinpoint neurovirulence-related sequences in Glasgow strain 17+ between immediate early gene 1 and the 'a' sequence. J Gen Virol 72 (Pt 3), 631-639. https://www.ncbi.nlm.nih.gov/pubmed/1848598

Manoj, S., Jogger, C.R., Myscofski, D., Yoon, M., and Spear, P.G. (2004). Mutations in herpes simplex virus glycoprotein D that prevent cell entry via nectins and alter cell tropism. Proc Natl Acad Sci U S A 101, 12414-12421. https://www.ncbi.nlm.nih.gov/pubmed/15273289

Markert, J.M., Cody, J.J., Parker, J.N., Coleman, J.M., Price, K.H., Kern, E.R., Quenelle, D.C., Lakeman, A.D., Schoeb, T.R., Palmer, C.A., *et al.* (2012). Preclinical evaluation of a genetically engineered herpes simplex virus expressing interleukin-12. J Virol 86, 5304-5313. https://www.ncbi.nlm.nih.gov/pubmed/22379082

Markert, J.M., Liechty, P.G., Wang, W., Gaston, S., Braz, E., Karrasch, M., Nabors, L.B., Markiewicz, M., Lakeman, A.D., Palmer, C.A., *et al.* (2009). Phase Ib trial of mutant herpes simplex virus G207 inoculated pre-and post-tumor resection for recurrent GBM. Mol Ther 17, 199-207. http://www.ncbi.nlm.nih.gov/entrez/query.fcgi?cmd=Retrieve&db=PubMed&dopt=Citation&list_uids=18957964

Markert, J.M., Medlock, M.D., Rabkin, S.D., Gillespie, G.Y., Todo, T., Hunter, W.D., Palmer, C.A., Feigenbaum, F., Tornatore, C., Tufaro, F., *et al.* (2000). Conditionally replicating herpes simplex virus mutant, G207 for the treatment of malignant glioma: results of a phase I trial. Gene Ther 7, 867-874. http://www.ncbi.nlm.nih.gov/entrez/query.fcgi?cmd=Retrieve&db=PubMed&dopt=Citation&list_uids=10845725

Martuza, R.L., Malick, A., Markert, J.M., Ruffner, K.L., and Coen, D.M. (1991). Experimental therapy of human glioma by means of a genetically engineered virus mutant. Science 252, 854-856. http://www.ncbi.nlm.nih.gov/entrez/query.fcgi?cmd=Retrieve&db=PubMed&dopt=Citation&list_uids=1851332

Marumoto, T., Tashiro, A., Friedmann-Morvinski, D., Scadeng, M., Soda, Y., Gage, F.H., and Verma, I.M. (2009). Development of a novel mouse glioma model using lentiviral vectors. Nat Med 15, 110-116. https://www.ncbi.nlm.nih.gov/pubmed/19122659

Mazzacurati, L., Marzulli, M., Reinhart, B., Miyagawa, Y., Uchida, H., Goins, W.F., Li, A., Kaur, B., Caligiuri, M., Cripe, T., *et al.* (2015). Use of miRNA response sequences to block off-target replication and increase the safety of an unattenuated, glioblastoma-targeted oncolytic HSV. Mol Ther 23, 99-107. https://www.ncbi.nlm.nih.gov/pubmed/25200130

Meignier, B., Longnecker, R., and Roizman, B. (1988). In vivo behavior of genetically engineered herpes simplex viruses R7017 and R7020: construction and evaluation in rodents. J Infect Dis 158, 602-614. http://dx.doi.org/10.1093/infdis/158.3.602.

Meisen, W.H., Wohleb, E.S., Jaime-Ramirez, A.C., Bolyard, C., Yoo, J.Y., Russell, L., Hardcastle, J., Dubin, S., Muili, K., Yu, J., *et al.* (2015). The Impact of Macrophage- and Microglia-Secreted TNFalpha on Oncolytic HSV-1 Therapy in the Glioblastoma Tumor Microenvironment. Clin Cancer Res 21, 3274-3285. https://www.ncbi.nlm.nih.gov/pubmed/25829396

Menotti, L., Avitabile, E., Gatta, V., Malatesta, P., Petrovic, B., and Campadelli-Fiume, G. (2018). HSV as A Platform for the Generation of Retargeted, Armed, and Reporter-Expressing Oncolytic Viruses. Viruses 10, 352. https://www.ncbi.nlm.nih.gov/pubmed/29966356

Menotti, L., Cerretani, A., and Campadelli-Fiume, G. (2006). A herpes simplex virus recombinant that exhibits a single-chain antibody to HER2/neu enters cells through the mammary tumor receptor, independently of the gD receptors. J Virol 80, 5531-5539. https://www.ncbi.nlm.nih.gov/pubmed/16699034

Menotti, L., Cerretani, A., Hengel, H., and Campadelli-Fiume, G. (2008). Construction of a fully retargeted herpes simplex virus 1 recombinant capable of entering cells solely via human epidermal growth factor receptor 2. J Virol 82, 10153-10161. https://www.ncbi.nlm.nih.gov/pubmed/18684832

Menotti, L., Nicoletti, G., Gatta, V., Croci, S., Landuzzi, L., De Giovanni, C., Nanni, P., Lollini, P.L., and Campadelli-Fiume, G. (2009). Inhibition of human tumor growth in mice by an oncolytic herpes simplex virus designed to target solely HER-2-positive cells. Proc Natl Acad Sci U S A 106, 9039-9044. https://www.ncbi.nlm.nih.gov/pubmed/19458262

Milne, R.S., Hanna, S.L., Rux, A.H., Willis, S.H., Cohen, G.H., and Eisenberg, R.J. (2003). Function of herpes simplex virus type 1 gD mutants with different receptor-binding affinities in virus entry and fusion. J Virol 77, 8962-8972. https://www.ncbi.nlm.nih.gov/pubmed/12885913

Mohr, I., Sternberg, D., Ward, S., Leib, D., Mulvey, M., and Gluzman, Y. (2001). A herpes simplex virus type 1 gamma34.5 second-site suppressor mutant that exhibits enhanced growth in cultured glioblastoma cells is severely attenuated in animals. J Virol 75, 5189-5196. https://www.ncbi.nlm.nih.gov/pubmed/11333900

Montgomery, R.I., Warner, M.S., Lum, B.J., and Spear, P.G. (1996). Herpes simplex virus-1 entry into cells mediated by a novel member of the TNF/NGF receptor family. Cell 87, 427-436. https://www.ncbi.nlm.nih.gov/pubmed/8898196

Mossman, K.L., and Smiley, J.R. (2002). Herpes simplex virus ICP0 and ICP34.5 counteract distinct interferon-induced barriers to virus replication. J Virol 76, 1995-1998. https://www.ncbi.nlm.nih.gov/pubmed/11799195

Muller, L., Aigner, P., and Stoiber, D. (2017). Type I Interferons and Natural Killer Cell Regulation in Cancer. Front Immunol 8, 304. http://dx.doi.org/10.3389/fimmu.2017.00304.

Nakamura, T., Peng, K.W., Harvey, M., Greiner, S., Lorimer, I.A., James, C.D., and Russell, S.J. (2005). Rescue and propagation of fully retargeted oncolytic measles viruses. Nat Biotechnol 23, 209-214. https://www.ncbi.nlm.nih.gov/pubmed/15685166

Nakano, K., Asano, R., Tsumoto, K., Kwon, H., Goins, W.F., Kumagai, I., Cohen, J.B., and Glorioso, J.C. (2005). Herpes simplex virus targeting to the EGF receptor by a gD-specific soluble bridging molecule. Mol Ther 11, 617-626. https://www.ncbi.nlm.nih.gov/pubmed/15771964

Nakao, A., Kasuya, H., Sahin, T.T., Nomura, N., Kanzaki, A., Misawa, M., Shirota, T., Yamada, S., Fujii, T., Sugimoto, H., *et al.* (2011). A phase I dose-escalation clinical trial of intraoperative direct intratumoral injection of HF10 oncolytic virus in non-resectable patients with advanced pancreatic cancer. Cancer Gene Ther 18, 167-175. https://www.ncbi.nlm.nih.gov/pubmed/21102422

Nakao, A., Kimata, H., Imai, T., Kikumori, T., Teshigahara, O., Nagasaka, T., Goshima, F., and Nishiyama, Y. (2004). Intratumoral injection of herpes simplex virus HF10 in recurrent breast cancer. Ann Oncol 15, 988-989. https://www.ncbi.nlm.nih.gov/pubmed/15151960

Nakao, A., Takeda, S., Shimoyama, S., Kasuya, H., Kimata, H., Teshigahara, O., Sawaki, M., Kikumori, T., Kodera, Y., Nagasaka, T., *et al.* (2007). Clinical experiment of mutant herpes simplex virus HF10 therapy for cancer. Curr Cancer Drug Targets 7, 169-174. https://www.ncbi.nlm.nih.gov/pubmed/17346108

Nakashima, H., Nguyen, T., Kasai, K., Passaro, C., Ito, H., Goins, W.F., Shaikh, I., Erdelyi, R., Nishihara, R., Nakano, I., *et al.* (2018). Toxicity and Efficacy of a Novel GADD34-expressing Oncolytic HSV-1 for the Treatment of Experimental Glioblastoma. Clin Cancer Res 24, 2574-2584. https://www.ncbi.nlm.nih.gov/pubmed/29511029

Nanni, P., Gatta, V., Menotti, L., De Giovanni, C., Ianzano, M., Palladini, A., Grosso, V., Dall'ora, M., Croci, S., Nicoletti, G., *et al.* (2013). Preclinical therapy of disseminated HER-2(+) ovarian and breast carcinomas with a HER-2-retargeted oncolytic herpesvirus. PLoS Pathog 9, e1003155. https://www.ncbi.nlm.nih.gov/pubmed/23382683

Nicola, A.V., McEvoy, A.M., and Straus, S.E. (2003). Roles for endocytosis and low pH in herpes simplex virus entry into HeLa and Chinese hamster ovary cells. J Virol 77, 5324-5332. https://www.ncbi.nlm.nih.gov/pubmed/12692234

Nicola, A.V., and Straus, S.E. (2004). Cellular and viral requirements for rapid endocytic entry of herpes simplex virus. J Virol 78, 7508-7517. https://www.ncbi.nlm.nih.gov/pubmed/15220424

Norden, A.D., Lesser, G.J., Drappatz, J., Ligon, K.L., Hammond, S.N., Lee, E.Q., Reardon, D.R., Fadul, C.E., Plotkin, S.R., Batchelor, T.T., *et al.* (2013). Phase 2 study of dose-intense temozolomide in recurrent glioblastoma. Neuro Oncol 15, 930-935. https://www.ncbi.nlm.nih.gov/pubmed/23553268

Obeid, M., Tesniere, A., Ghiringhelli, F., Fimia, G.M., Apetoh, L., Perfettini, J.L., Castedo, M., Mignot, G., Panaretakis, T., Casares, N., *et al.* (2007). Calreticulin exposure dictates the immunogenicity of cancer cell death. Nat Med 13, 54-61. http://dx.doi.org/10.1038/nm1523.

Oh, T., Fakurnejad, S., Sayegh, E.T., Clark, A.J., Ivan, M.E., Sun, M.Z., Safaee, M., Bloch, O., James, C.D., and Parsa, A.T. (2014). Immunocompetent murine models for the study of glioblastoma immunotherapy. J Transl Med 12, 107. https://www.ncbi.nlm.nih.gov/pubmed/24779345

Omuro, A., Vlahovic, G., Lim, M., Sahebjam, S., Baehring, J., Cloughesy, T., Voloschin, A., Ramkissoon, S.H., Ligon, K.L., Latek, R., *et al.* (2018). Nivolumab with or without ipilimumab in patients with recurrent glioblastoma: results from exploratory phase I cohorts of CheckMate 143. Neuro Oncol 20, 674-686. https://www.ncbi.nlm.nih.gov/pubmed/29106665

Orr, M.T., and Lanier, L.L. (2010). Natural killer cell education and tolerance. Cell 142, 847-856. https://www.ncbi.nlm.nih.gov/pubmed/20850008

Orvedahl, A., Alexander, D., Talloczy, Z., Sun, Q., Wei, Y., Zhang, W., Burns, D., Leib, D.A., and Levine, B. (2007). HSV-1 ICP34.5 confers neurovirulence by targeting the Beclin 1 autophagy protein. Cell Host Microbe 1, 23-35. https://www.ncbi.nlm.nih.gov/pubmed/18005679

Ostrand-Rosenberg, S., and Sinha, P. (2009). Myeloid-derived suppressor cells: linking inflammation and cancer. J Immunol 182, 4499-4506. https://www.ncbi.nlm.nih.gov/pubmed/19342621

Paludan, S.R., Bowie, A.G., Horan, K.A., and Fitzgerald, K.A. (2011). Recognition of herpesviruses by the innate immune system. Nat Rev Immunol 11, 143-154. https://www.ncbi.nlm.nih.gov/pubmed/21267015

Pan, S., Liu, X., Ma, Y., Cao, Y., and He, B. (2018). Herpes Simplex Virus 1 gamma134.5 Protein Inhibits STING Activation That Restricts Viral Replication. J Virol 92, e01015-01018. https://www.ncbi.nlm.nih.gov/pubmed/30045990

Paolini, R., Bernardini, G., Molfetta, R., and Santoni, A. (2015). NK cells and interferons. Cytokine Growth Factor Rev 26, 113-120. http://dx.doi.org/10.1016/j.cytogfr.2014.11.003.

Papanastassiou, V., Rampling, R., Fraser, M., Petty, R., Hadley, D., Nicoll, J., Harland, J., Mabbs, R., and Brown, M. (2002). The potential for efficacy of the modified (ICP 34.5(-)) herpes simplex virus HSV1716 following intratumoural injection into human malignant glioma: a proof of principle study. Gene Ther 9, 398-406. https://www.ncbi.nlm.nih.gov/pubmed/11960316

Parker, J.N., Gillespie, G.Y., Love, C.E., Randall, S., Whitley, R.J., and Markert, J.M. (2000). Engineered herpes simplex virus expressing IL-12 in the treatment of experimental murine brain tumors. Proc Natl Acad Sci U S A 97, 2208-2213. https://www.ncbi.nlm.nih.gov/pubmed/10681459

Parry, C., Bell, S., Minson, T., and Browne, H. (2005). Herpes simplex virus type 1 glycoprotein H binds to alphavbeta3 integrins. J Gen Virol 86, 7-10. https://www.ncbi.nlm.nih.gov/pubmed/15604426

Passaro, C., Alayo, Q., DeLaura, I., McNulty, J.J., Grauwet, K., Ito, H., Bhaskaran, V., Mineo, M., Lawler, S.E., Shah, K., *et al.* (2018). Arming an oncolytic herpes simplex virus Type 1 with a single chain fragment variable antibody against PD-1 for experimental glioblastoma therapy. Clin Cancer Res. https://www.ncbi.nlm.nih.gov/pubmed/30279232

Patel, D.M., Foreman, P.M., Nabors, L.B., Riley, K.O., Gillespie, G.Y., and Markert, J.M. (2016). Design of a Phase I Clinical Trial to Evaluate M032, a Genetically Engineered HSV-1 Expressing IL-12, in Patients with Recurrent/Progressive Glioblastoma Multiforme, Anaplastic Astrocytoma, or Gliosarcoma.

Hum Gene Ther Clin Dev 27, 69-78. https://www.ncbi.nlm.nih.gov/pubmed/27314913

Petrovic, B., Gianni, T., Gatta, V., and Campadelli-Fiume, G. (2017). Insertion of a ligand to HER2 in gB retargets HSV tropism and obviates the need for activation of the other entry glycoproteins. PLoS Pathog 13, e1006352. https://www.ncbi.nlm.nih.gov/pubmed/28423057

Petrovic, B., Leoni, V., Gatta, V., Zaghini, A., Vannini, A., and Campadelli-Fiume, G. (2018). Dual Ligand Insertion in gB and gD of Oncolytic Herpes Simplex Viruses for Retargeting to a Producer Vero Cell Line and to Cancer Cells. J Virol 92, e02122-02117. https://www.ncbi.nlm.nih.gov/pubmed/29263257

Poffenberger, K.L., Tabares, E., and Roizman, B. (1983). Characterization of a viable, noninverting herpes simplex virus 1 genome derived by insertion and deletion of sequences at the junction of components L and S. Proc Natl Acad Sci U S A 80, 2690-2694. http://dx.doi.org/10.1073/pnas.80.9.2690.

Pogge von Strandmann, E., Shatnyeva, O., and Hansen, H.P. (2015). NKp30 and its ligands: emerging players in tumor immune evasion from natural killer cells. Ann Transl Med 3, 314. http://dx.doi.org/10.3978/j.issn.2305-5839.2015.09.08.

Pol, J., Kroemer, G., and Galluzzi, L. (2016). First oncolytic virus approved for melanoma immunotherapy. Oncoimmunology 5, e1115641. https://www.ncbi.nlm.nih.gov/pubmed/26942095

Poppers, J., Mulvey, M., Khoo, D., and Mohr, I. (2000). Inhibition of PKR activation by the proline-rich RNA binding domain of the herpes simplex virus type 1 Us11 protein. J Virol 74, 11215-11221. https://www.ncbi.nlm.nih.gov/pubmed/11070019

Prasad, V., Kaestner, V., and Mailankody, S. (2018). Cancer Drugs Approved Based on Biomarkers and Not Tumor Type-FDA Approval of Pembrolizumab for Mismatch Repair-Deficient Solid Cancers. JAMA Oncol 4, 157-158. http://dx.doi.org/10.1001/jamaoncol.2017.4182.

Preusser, M., Lim, M., Hafler, D.A., Reardon, D.A., and Sampson, J.H. (2015). Prospects of immune checkpoint modulators in the treatment of glioblastoma. Nat Rev Neurol 11, 504-514. https://www.ncbi.nlm.nih.gov/pubmed/26260659

Price, D.L., Lin, S.F., Han, Z., Simpson, G., Coffin, R.S., Wong, J., Li, S., Fong, Y., and Wong, R.J. (2010). Oncolysis using herpes simplex virus type 1 engineered to express cytosine deaminase and a fusogenic glycoprotein for head and neck squamous cell carcinoma. Arch Otolaryngol Head Neck Surg 136, 151-158. https://www.ncbi.nlm.nih.gov/pubmed/20157061

Puzanov, I., Milhem, M.M., Minor, D., Hamid, O., Li, A., Chen, L., Chastain, M., Gorski, K.S., Anderson, A., Chou, J., *et al.* (2016). Talimogene Laherparepvec in Combination With Ipilimumab in Previously Untreated, Unresectable Stage IIIB-IV Melanoma. J Clin Oncol 34, 2619-2626. https://www.ncbi.nlm.nih.gov/pubmed/27298410

Pyonteck, S.M., Akkari, L., Schuhmacher, A.J., Bowman, R.L., Sevenich, L., Quail, D.F., Olson, O.C., Quick, M.L., Huse, J.T., Teijeiro, V., *et al.* (2013). CSF-1R inhibition alters macrophage polarization and blocks glioma progression. Nat Med 19, 1264-1272. https://www.ncbi.nlm.nih.gov/pubmed/24056773

Quail, D.F., Bowman, R.L., Akkari, L., Quick, M.L., Schuhmacher, A.J., Huse, J.T., Holland, E.C., Sutton, J.C., and Joyce, J.A. (2016). The tumor microenvironment underlies acquired resistance to CSF-1R inhibition in gliomas. Science 352, aad3018. https://www.ncbi.nlm.nih.gov/pubmed/27199435

Rampling, R., Cruickshank, G., Papanastassiou, V., Nicoll, J., Hadley, D., Brennan, D., Petty, R., MacLean, A., Harland, J., McKie, E., *et al.* (2000). Toxicity evaluation of replication-competent herpes simplex virus (ICP 34.5 null mutant 1716) in patients with recurrent malignant glioma. Gene Ther 7, 859-866. https://www.ncbi.nlm.nih.gov/pubmed/10845724

Randazzo, B.P., Kesari, S., Gesser, R.M., Alsop, D., Ford, J.C., Brown, S.M., Maclean, A., and Fraser, N.W. (1995). Treatment of experimental intracranial murine melanoma with a neuroattenuated herpes simplex virus 1 mutant. Virology 211, 94-101. https://www.ncbi.nlm.nih.gov/pubmed/7645240

Raulet, D.H., Gasser, S., Gowen, B.G., Deng, W., and Jung, H. (2013). Regulation of ligands for the NKG2D activating receptor. Annu Rev Immunol 31, 413-441. http://dx.doi.org/10.1146/annurev-immunol-032712-095951.

Reardon, D.A., Gokhale, P.C., Klein, S.R., Ligon, K.L., Rodig, S.J., Ramkissoon, S.H., Jones, K.L., Conway, A.S., Liao, X., Zhou, J., *et al.* (2016). Glioblastoma Eradication Following Immune Checkpoint Blockade in an Orthotopic, Immunocompetent Model. Cancer Immunol Res 4, 124-135. https://www.ncbi.nlm.nih.gov/pubmed/26546453

Reinhart, B., Mazzacurati, L., Forero, A., Hong, C.S., Eguchi, J., Okada, H., Fellows, W., Niranjan, A., Cohen, J.B., Glorioso, J.C., *et al.* (2012). Inhibition of Indoleamine-2,3-dioxygenase (IDO) in Glioblastoma Cells by Oncolytic Herpes Simplex Virus. Adv Virol 2012, 815465. https://www.ncbi.nlm.nih.gov/pubmed/22924042

Ribas, A., Dummer, R., Puzanov, I., VanderWalde, A., Andtbacka, R.H.I., Michielin, O., Olszanski, A.J., Malvehy, J., Cebon, J., Fernandez, E., *et al.* (2017). Oncolytic Virotherapy Promotes Intratumoral T Cell Infiltration and Improves Anti-PD-1 Immunotherapy. Cell 170, 1109-1119 e1110. https://www.ncbi.nlm.nih.gov/pubmed/28886381

Richards, A.L., Sollars, P.J., Pitts, J.D., Stults, A.M., Heldwein, E.E., Pickard, G.E., and Smith, G.A. (2017). The pUL37 tegument protein guides alpha-herpesvirus retrograde axonal transport to promote neuroinvasion. PLoS Pathog 13, e1006741. https://www.ncbi.nlm.nih.gov/pubmed/29216315

Ries, C.H., Hoves, S., Cannarile, M.A., and Ruttinger, D. (2015). CSF-1/CSF-1R targeting agents in clinical development for cancer therapy. Curr Opin Pharmacol 23, 45-51. https://www.ncbi.nlm.nih.gov/pubmed/26051995

Ring, E.K., Li, R., Moore, B.P., Nan, L., Kelly, V.M., Han, X., Beierle, E.A., Markert, J.M., Leavenworth, J.W., Gillespie, G.Y., *et al.* (2017). Newly Characterized Murine Undifferentiated Sarcoma Models Sensitive to Virotherapy with Oncolytic HSV-1 M002. Mol Ther Oncolytics 7, 27-36. https://www.ncbi.nlm.nih.gov/pubmed/29034313

Rizvi, N.A., Hellmann, M.D., Snyder, A., Kvistborg, P., Makarov, V., Havel, J.J., Lee, W., Yuan, J., Wong, P., Ho, T.S., *et al.* (2015). Cancer immunology. Mutational landscape determines sensitivity to PD-1 blockade in non-small cell lung cancer. Science 348, 124-128. http://dx.doi.org/10.1126/science.aaa1348.

Roberts, E.W., Broz, M.L., Binnewies, M., Headley, M.B., Nelson, A.E., Wolf, D.M., Kaisho, T., Bogunovic, D., Bhardwaj, N., and Krummel, M.F. (2016). Critical Role for CD103(+)/CD141(+) Dendritic Cells Bearing CCR7 for Tumor Antigen Trafficking and Priming of T Cell Immunity in Melanoma. Cancer Cell 30, 324-336. https://www.ncbi.nlm.nih.gov/pubmed/27424807

Roizman, B., and Zhou, G. (2015). The 3 facets of regulation of herpes simplex virus gene expression: A critical inquiry. Virology 479-480, 562-567. https://www.ncbi.nlm.nih.gov/pubmed/25771487

Roizman, B., and Knipe, D.M. (2001). Herpes simplex viruses and their replication. In Fields Virology, Knipe, D.M. and Howley, P.M., eds. (Philadelphia: Lippincott Williams and Wilkins), pp. 2399-2459.

Roth, J.C., Cassady, K.A., Cody, J.J., Parker, J.N., Price, K.H., Coleman, J.M., Peggins, J.O., Noker, P.E., Powers, N.W., Grimes, S.D., *et al.* (2014). Evaluation of the safety and biodistribution of M032, an attenuated herpes simplex virus type 1 expressing hIL-12, after intracerebral administration to aotus nonhuman primates. Hum Gene Ther Clin Dev 25, 16-27. https://www.ncbi.nlm.nih.gov/pubmed/24649838

Saha, D., Martuza, R.L., and Rabkin, S.D. (2017a). Curing glioblastoma: oncolytic HSV-IL12 and checkpoint blockade. Oncoscience 4, 67-69. https://www.ncbi.nlm.nih.gov/pubmed/28966936

Saha, D., Martuza, R.L., and Rabkin, S.D. (2017b). Macrophage Polarization Contributes to Glioblastoma Eradication by Combination Immunovirotherapy and Immune Checkpoint Blockade. Cancer Cell 32, 253-267 e255. https://www.ncbi.nlm.nih.gov/pubmed/28810147

Sahin, T.T., Kasuya, H., Nomura, N., Shikano, T., Yamamura, K., Gewen, T., Kanzaki, A., Fujii, T., Sugae, T., Imai, T., *et al.* (2012). Impact of novel oncolytic virus HF10 on cellular components of the tumor microenviroment in patients with recurrent breast cancer. Cancer Gene Ther 19, 229-237. https://www.ncbi.nlm.nih.gov/pubmed/22193629

Salmon, H., Idoyaga, J., Rahman, A., Leboeuf, M., Remark, R., Jordan, S., Casanova-Acebes, M., Khudoynazarova, M., Agudo, J., Tung, N., *et al.* (2016). Expansion and Activation of CD103(+) Dendritic Cell Progenitors at the Tumor Site Enhances Tumor Responses to Therapeutic PD-L1 and BRAF

Inhibition. Immunity 44, 924-938. https://www.ncbi.nlm.nih.gov/pubmed/27096321

Sato-Kaneko, F., Yao, S., Ahmadi, A., Zhang, S.S., Hosoya, T., Kaneda, M.M., Varner, J.A., Pu, M., Messer, K.S., Guiducci, C., *et al.* (2017). Combination immunotherapy with TLR agonists and checkpoint inhibitors suppresses head and neck cancer. JCI Insight 2, e93397. https://www.ncbi.nlm.nih.gov/pubmed/28931759

Satoh, T., Arii, J., Suenaga, T., Wang, J., Kogure, A., Uehori, J., Arase, N., Shiratori, I., Tanaka, S., Kawaguchi, Y., *et al.* (2008). PILRalpha is a herpes simplex virus-1 entry coreceptor that associates with glycoprotein B. Cell 132, 935-944. https://www.ncbi.nlm.nih.gov/pubmed/18358807

Sauer, M., Schuldner, M., Hoffmann, N., Cetintas, A., Reiners, K.S., Shatnyeva, O., Hallek, M., Hansen, H.P., Gasser, S., and von Strandmann, E.P. (2017). CBP/p300 acetyltransferases regulate the expression of NKG2D ligands on tumor cells. Oncogene 36, 933-941. https://www.ncbi.nlm.nih.gov/pubmed/27477692

Schlecker, E., Fiegler, N., Arnold, A., Altevogt, P., Rose-John, S., Moldenhauer, G., Sucker, A., Paschen, A., von Strandmann, E.P., Textor, S., *et al.* (2014). Metalloprotease-mediated tumor cell shedding of B7-H6, the ligand of the natural killer cell-activating receptor NKp30. Cancer Res 74, 3429-3440. http://dx.doi.org/10.1158/0008-5472.Can-13-3017.

Schlitzer, A., and Ginhoux, F. (2014). Organization of the mouse and human DC network. Curr Opin Immunol 26, 90-99. http://dx.doi.org/10.1016/j.coi.2013.11.002.

Schlitzer, A., McGovern, N., and Ginhoux, F. (2015). Dendritic cells and monocyte-derived cells: Two complementary and integrated functional systems. Semin Cell Dev Biol 41, 9-22. http://dx.doi.org/10.1016/j.semcdb.2015.03.011.

Sempere, L.F., Freemantle, S., Pitha-Rowe, I., Moss, E., Dmitrovsky, E., and Ambros, V. (2004). Expression profiling of mammalian microRNAs uncovers a subset of brain-expressed microRNAs with possible roles in murine and human neuronal differentiation. Genome Biol 5, R13. https://www.ncbi.nlm.nih.gov/pubmed/15003116

Senzer, N.N., Kaufman, H.L., Amatruda, T., Nemunaitis, M., Reid, T., Daniels, G., Gonzalez, R., Glaspy, J., Whitman, E., Harrington, K., *et al.* (2009). Phase II clinical trial of a granulocyte-macrophage colony-stimulating factor-encoding, second-generation oncolytic herpesvirus in patients with unresectable metastatic melanoma. J Clin Oncol 27, 5763-5771. https://www.ncbi.nlm.nih.gov/pubmed/19884534

Sette, P., Amankulor, N., Li, A., Marzulli, M., Leronni, D., Zhang, M., Goins, W.F., Kaur, B., Bolyard, C., Cripe, T.P., *et al.* (2019). GBM-Targeted oHSV Armed with Matrix Metalloproteinase 9 Enhances Anti-tumor Activity and Animal Survival. Mol Ther Oncolytics 15, 214-222. https://www.ncbi.nlm.nih.gov/pubmed/31890868

Shibata, T., Uchida, H., Shiroyama, T., Okubo, Y., Suzuki, T., Ikeda, H., Yamaguchi, M., Miyagawa, Y., Fukuhara, T., Cohen, J.B., *et al.* (2016). Development of an oncolytic HSV vector fully retargeted specifically to cellular EpCAM for virus entry and cell-to-cell spread. Gene Ther 23, 479-488. https://www.ncbi.nlm.nih.gov/pubmed/26905369

Shieh, M.T., WuDunn, D., Montgomery, R.I., Esko, J.D., and Spear, P.G. (1992). Cell surface receptors for herpes simplex virus are heparan sulfate proteoglycans. J Cell Biol 116, 1273-1281. https://www.ncbi.nlm.nih.gov/pubmed/1310996

Shields, B.D., Mahmoud, F., Taylor, E.M., Byrum, S.D., Sengupta, D., Koss, B., Baldini, G., Ransom, S., Cline, K., Mackintosh, S.G., *et al.* (2017). Indicators of responsiveness to immune checkpoint inhibitors. Sci Rep 7, 807. http://dx.doi.org/10.1038/s41598-017-01000-2.

Shukla, D., Liu, J., Blaiklock, P., Shworak, N.W., Bai, X., Esko, J.D., Cohen, G.H., Eisenberg, R.J., Rosenberg, R.D., and Spear, P.G. (1999). A novel role for 3-O-sulfated heparan sulfate in herpes simplex virus 1 entry. Cell 99, 13-22. https://www.ncbi.nlm.nih.gov/pubmed/10520990

Silber, J., Lim, D.A., Petritsch, C., Persson, A.I., Maunakea, A.K., Yu, M., Vandenberg, S.R., Ginzinger, D.G., James, C.D., Costello, J.F., *et al.* (2008). miR-124 and miR-137 inhibit proliferation of glioblastoma multiforme cells and induce differentiation of brain tumor stem cells. BMC Med 6, 14. https://www.ncbi.nlm.nih.gov/pubmed/18577219

Solomon, B.L., and Garrido-Laguna, I. (2018). TIGIT: a novel immunotherapy target moving from bench to bedside. Cancer Immunol Immunother 67, 1659-1667. http://dx.doi.org/10.1007/s00262-018-2246-5.

Spear, P.G., Shieh, M.T., Herold, B.C., WuDunn, D., and Koshy, T.I. (1992). Heparan sulfate glycosaminoglycans as primary cell surface receptors for herpes simplex virus. Adv Exp Med Biol 313, 341-353. https://www.ncbi.nlm.nih.gov/pubmed/1332443

Spranger, S., Dai, D., Horton, B., and Gajewski, T.F. (2017). Tumor-Residing Batf3 Dendritic Cells Are Required for Effector T Cell Trafficking and Adoptive T Cell Therapy. Cancer Cell 31, 711-723 e714. https://www.ncbi.nlm.nih.gov/pubmed/28486109

Spranger, S., Luke, J.J., Bao, R., Zha, Y., Hernandez, K.M., Li, Y., Gajewski, A.P., Andrade, J., and Gajewski, T.F. (2016). Density of immunogenic antigens does not explain the presence or absence of the T-cell-inflamed tumor microenvironment in melanoma. Proc Natl Acad Sci U S A 113, E7759-E7768. https://www.ncbi.nlm.nih.gov/pubmed/27837020

St Leger, A.J., and Hendricks, R.L. (2011). CD8+ T cells patrol HSV-1-infected trigeminal ganglia and prevent viral reactivation. J Neurovirol 17, 528-534. https://www.ncbi.nlm.nih.gov/pubmed/22161682

St Leger, A.J., Jeon, S., and Hendricks, R.L. (2013). Broadening the repertoire of functional herpes simplex virus type 1-specific CD8+ T cells reduces viral reactivation from latency in sensory ganglia. J Immunol 191, 2258-2265. https://www.ncbi.nlm.nih.gov/pubmed/23878317

Stanziale, S.F., Petrowsky, H., Joe, J.K., Roberts, G.D., Zager, J.S., Gusani, N.J., Ben-Porat, L., Gonen, M., and Fong, Y. (2002). Ionizing radiation potentiates the antitumor efficacy of oncolytic herpes simplex virus G207 by upregulating ribonucleotide reductase. Surgery 132, 353-359. https://www.ncbi.nlm.nih.gov/pubmed/12219034

Streby, K.A., Geller, J.I., Currier, M.A., Warren, P.S., Racadio, J.M., Towbin, A.J., Vaughan, M.R., Triplet, M., Ott-Napier, K., Dishman, D.J., *et al.* (2017). Intratumoral Injection of HSV1716, an Oncolytic Herpes Virus, Is Safe and Shows Evidence of Immune Response and Viral Replication in Young Cancer

Patients. Clin Cancer Res 23, 3566-3574. https://www.ncbi.nlm.nih.gov/pubmed/28495911

Stupp, R., Hegi, M.E., Mason, W.P., van den Bent, M.J., Taphoorn, M.J., Janzer, R.C., Ludwin, S.K., Allgeier, A., Fisher, B., Belanger, K., *et al.* (2009). Effects of radiotherapy with concomitant and adjuvant temozolomide versus radiotherapy alone on survival in glioblastoma in a randomised phase III study: 5-year analysis of the EORTC-NCIC trial. Lancet Oncol 10, 459-466. https://www.ncbi.nlm.nih.gov/pubmed/19269895

Stupp, R., Mason, W.P., van den Bent, M.J., Weller, M., Fisher, B., Taphoorn, M.J., Belanger, K., Brandes, A.A., Marosi, C., Bogdahn, U., *et al.* (2005). Radiotherapy plus concomitant and adjuvant temozolomide for glioblastoma. N Engl J Med 352, 987-996. https://www.ncbi.nlm.nih.gov/pubmed/15758009

Suenaga, T., Satoh, T., Somboonthum, P., Kawaguchi, Y., Mori, Y., and Arase, H. (2010). Myelin-associated glycoprotein mediates membrane fusion and entry of neurotropic herpesviruses. Proc Natl Acad Sci U S A 107, 866-871. https://www.ncbi.nlm.nih.gov/pubmed/20080767

Sun, L., Funchain, P., Song, J.M., Rayman, P., Tannenbaum, C., Ko, J., McNamara, M., Marcela Diaz-Montero, C., and Gastman, B. (2018). Talimogene Laherparepvec combined with anti-PD-1 based immunotherapy for unresectable stage III-IV melanoma: a case series. J Immunother Cancer 6, 36. https://www.ncbi.nlm.nih.gov/pubmed/29764498

Sun, L., Wu, J., Du, F., Chen, X., and Chen, Z.J. (2013). Cyclic GMP-AMP synthase is a cytosolic DNA sensor that activates the type I interferon pathway. Science 339, 786-791. https://www.ncbi.nlm.nih.gov/pubmed/23258413

Szatmari, T., Lumniczky, K., Desaknai, S., Trajcevski, S., Hidvegi, E.J., Hamada, H., and Safrany, G. (2006). Detailed characterization of the mouse glioma 261 tumor model for experimental glioblastoma therapy. Cancer Sci 97, 546-553. https://www.ncbi.nlm.nih.gov/pubmed/16734735

Sze, D.Y., Iagaru, A.H., Gambhir, S.S., De Haan, H.A., and Reid, T.R. (2012). Response to intra-arterial oncolytic virotherapy with the herpes virus NV1020 evaluated by [18F]fluorodeoxyglucose positron emission tomography and computed tomography. Hum Gene Ther 23, 91-97. https://www.ncbi.nlm.nih.gov/pubmed/21895536

Takasu, A., Masui, A., Hamada, M., Imai, T., Iwai, S., and Yura, Y. (2016). Immunogenic cell death by oncolytic herpes simplex virus type 1 in squamous cell carcinoma cells. Cancer Gene Ther 23, 107-113. http://dx.doi.org/10.1038/cgt.2016.8.

Takenaka, M.C., Gabriely, G., Rothhammer, V., Mascanfroni, I.D., Wheeler, M.A., Chao, C.C., Gutierrez-Vazquez, C., Kenison, J., Tjon, E.C., Barroso, A., *et al.* (2019). Control of tumor-associated macrophages and T cells in glioblastoma via AHR and CD39. Nat Neurosci 22, 729-740. https://www.ncbi.nlm.nih.gov/pubmed/30962630

Tal-Singer, R., Peng, C., Ponce De Leon, M., Abrams, W.R., Banfield, B.W., Tufaro, F., Cohen, G.H., and Eisenberg, R.J. (1995). Interaction of herpes simplex virus glycoprotein gC with mammalian cell surface molecules. J Virol 69, 4471-4483. https://www.ncbi.nlm.nih.gov/pubmed/7769707

Tamimi, A.F., and Juweid, M. (2017). Epidemiology and Outcome of Glioblastoma. In Glioblastoma, S. De Vleeschouwer, ed. (Brisbane (AU): Codon Publications).

Thaci, B., Ahmed, A.U., Ulasov, I.V., Wainwright, D.A., Nigam, P., Auffinger, B., Tobias, A.L., Han, Y., Zhang, L., Moon, K.S., *et al.* (2014). Depletion of myeloid-derived suppressor cells during interleukin-12 immunogene therapy does not confer a survival advantage in experimental malignant glioma. Cancer Gene Ther 21, 38-44. https://www.ncbi.nlm.nih.gov/pubmed/24434573

Thaker, N.G., and Pollack, I.F. (2009). Molecularly targeted therapies for malignant glioma: rationale for combinatorial strategies. Expert Rev Neurother 9, 1815-1836. https://www.ncbi.nlm.nih.gov/pubmed/19951140

Thomas, S., Kuncheria, L., Roulstone, V., Kyula, J.N., Mansfield, D., Bommareddy, P.K., Smith, H., Kaufman, H.L., Harrington, K.J., and Coffin, R.S. (2019). Development of a new fusion-enhanced oncolytic immunotherapy platform based on herpes simplex virus type 1. J Immunother Cancer 7, 214. https://www.ncbi.nlm.nih.gov/pubmed/31399043

Toda, M., Martuza, R.L., Kojima, H., and Rabkin, S.D. (1998). In situ cancer vaccination: an IL-12 defective vector/replication-competent herpes simplex virus combination induces local and systemic antitumor activity. J Immunol 160, 4457-4464. https://www.ncbi.nlm.nih.gov/pubmed/9574551

Todo, T. (2019). ATIM-14. Results of Phase Ii Clinical Trial of Oncolytic Herpes Virus G47Δ in Patients with Glioblastoma. Neuro-Oncology 21, vi4-vi4. https://doi.org/10.1093/neuonc/noz175.014

Todo, T., Feigenbaum, F., Rabkin, S.D., Lakeman, F., Newsome, J.T., Johnson, P.A., Mitchell, E., Belliveau, D., Ostrove, J.M., and Martuza, R.L. (2000). Viral shedding and biodistribution of G207, a multimutated, conditionally replicating herpes simplex virus type 1, after intracerebral inoculation in aotus. Mol Ther 2, 588-595. https://www.ncbi.nlm.nih.gov/pubmed/11124059

Todo, T., Martuza, R.L., Rabkin, S.D., and Johnson, P.A. (2001). Oncolytic herpes simplex virus vector with enhanced MHC class I presentation and tumor cell killing. Proc Natl Acad Sci USA 98, 6396-6401. https://www.ncbi.nlm.nih.gov/pubmed/11353831

Trybala, E., Roth, A., Johansson, M., Liljeqvist, J.A., Rekabdar, E., Larm, O., and Bergstrom, T. (2002). Glycosaminoglycan-binding ability is a feature of wild-type strains of herpes simplex virus type 1. Virology 302, 413-419. https://www.ncbi.nlm.nih.gov/pubmed/12441085

Tumeh, P.C., Harview, C.L., Yearley, J.H., Shintaku, I.P., Taylor, E.J., Robert, L., Chmielowski, B., Spasic, M., Henry, G., Ciobanu, V., *et al.* (2014). PD-1 blockade induces responses by inhibiting adaptive immune resistance. Nature 515, 568-571. http://dx.doi.org/10.1038/nature13954.

Uchida, H., Chan, J., Goins, W.F., Grandi, P., Kumagai, I., Cohen, J.B., and Glorioso, J.C. (2010). A double mutation in glycoprotein gB compensates for ineffective gD-dependent initiation of herpes simplex virus type 1 infection. J Virol 84, 12200-12209. https://www.ncbi.nlm.nih.gov/pubmed/20861246

Uchida, H., Hamada, H., Nakano, K., Kwon, H., Tahara, H., Cohen, J.B., and Glorioso, J.C. (2018). Oncolytic Herpes Simplex Virus Vectors Fully Retargeted to Tumor- Associated Antigens. Curr Cancer Drug Targets 18, 162-170. https://www.ncbi.nlm.nih.gov/pubmed/28176649

Uchida, H., Marzulli, M., Nakano, K., Goins, W.F., Chan, J., Hong, C.S., Mazzacurati, L., Yoo, J.Y., Haseley, A., Nakashima, H., *et al.* (2013). Effective treatment of an orthotopic xenograft model of human glioblastoma using an EGFR-retargeted oncolytic herpes simplex virus. Mol Ther 21, 561-569. https://www.ncbi.nlm.nih.gov/pubmed/23070115

Unterholzner, L., Keating, S.E., Baran, M., Horan, K.A., Jensen, S.B., Sharma, S., Sirois, C.M., Jin, T., Latz, E., Xiao, T.S., *et al.* (2010). IFI16 is an innate immune sensor for intracellular DNA. Nat Immunol 11, 997-1004. https://www.ncbi.nlm.nih.gov/pubmed/20890285

Ushijima, Y., Luo, C., Goshima, F., Yamauchi, Y., Kimura, H., and Nishiyama, Y. (2007). Determination and analysis of the DNA sequence of highly attenuated herpes simplex virus type 1 mutant HF10, a potential oncolytic virus. Microbes Infect 9, 142-149. https://www.ncbi.nlm.nih.gov/pubmed/17218138

Valyi-Nagy, T., Fareed, M.U., O'Keefe, J.S., Gesser, R.M., MacLean, A.R., Brown, S.M., Spivack, J.G., and Fraser, N.W. (1994). The herpes simplex virus type 1 strain 17+ gamma 34.5 deletion mutant 1716 is avirulent in SCID mice. J Gen Virol 75 2059-2063. https://www.ncbi.nlm.nih.gov/pubmed/8046409

van den Bossche, W.B.L., Kleijn, A., Teunissen, C.E., Voerman, J.S.A., Teodosio, C., Noske, D.P., van Dongen, J.J.M., Dirven, C.M.F., and Lamfers, M.L.M. (2018). Oncolytic virotherapy in glioblastoma patients induces a tumor macrophage phenotypic shift leading to an altered glioblastoma microenvironment. Neuro Oncol 20, 1494-1504. https://www.ncbi.nlm.nih.gov/pubmed/29796615

Varghese, S., Newsome, J.T., Rabkin, S.D., McGeagh, K., Mahoney, D., Nielsen, P., Todo, T., and Martuza, R.L. (2001). Preclinical safety evaluation of G207, a replication-competent herpes simplex virus type 1, inoculated intraprostatically in mice and nonhuman primates. Hum Gene Ther 12, 999-1010. https://www.ncbi.nlm.nih.gov/pubmed/11387063

Vivier, E., Tomasello, E., and Paul, P. (2002). Lymphocyte activation via NKG2D: towards a new paradigm in immune recognition? Curr Opin Immunol 14, 306-311. https://www.ncbi.nlm.nih.gov/pubmed/11973127

Wang, R.F., and Wang, H.Y. (2017). Immune targets and neoantigens for cancer immunotherapy and precision medicine. Cell Res 27, 11-37. https://www.ncbi.nlm.nih.gov/pubmed/28025978

Wang, X., Kong, L., Zhang, G.R., Sun, M., and Geller, A.I. (2005). Targeted gene transfer to nigrostriatal neurons in the rat brain by helper virus-free HSV-1 vector particles that contain either a chimeric HSV-1 glycoprotein C-GDNF or a gC-BDNF protein. Brain Res Mol Brain Res 139, 88-102. https://www.ncbi.nlm.nih.gov/pubmed/15993510

Wang, X.F., Wang, H.S., Wang, H., Zhang, F., Wang, K.F., Guo, Q., Zhang, G., Cai, S.H., and Du, J. (2014). The role of indoleamine 2,3-dioxygenase (IDO) in immune tolerance: focus on macrophage polarization of THP-1 cells. Cell Immunol 289, 42-48. https://www.ncbi.nlm.nih.gov/pubmed/24721110

Wang, Y., Xu, Z., Guo, S., Zhang, L., Sharma, A., Robertson, G.P., and Huang, L. (2013). Intravenous delivery of siRNA targeting CD47 effectively inhibits melanoma tumor growth and lung metastasis. Mol Ther 21, 1919-1929. https://www.ncbi.nlm.nih.gov/pubmed/23774794

Warner, M.S., Geraghty, R.J., Martinez, W.M., Montgomery, R.I., Whitbeck, J.C., Xu, R., Eisenberg, R.J., Cohen, G.H., and Spear, P.G. (1998). A cell surface protein with herpesvirus entry activity (HveB) confers susceptibility to infection by mutants of herpes simplex virus type 1, herpes simplex virus type 2, and pseudorabies virus. Virology 246, 179-189. https://www.ncbi.nlm.nih.gov/pubmed/9657005

Waters, A.M., Johnston, J.M., Reddy, A.T., Fiveash, J., Madan-Swain, A., Kachurak, K., Bag, A.K., Gillespie, G.Y., Markert, J.M., and Friedman, G.K. (2017). Rationale and Design of a Phase 1 Clinical Trial to Evaluate HSV G207 Alone or with a Single Radiation Dose in Children with Progressive or Recurrent Malignant Supratentorial Brain Tumors. Hum Gene Ther Clin Dev 28, 7-16. https://www.ncbi.nlm.nih.gov/pubmed/28319448

Wen, P.Y., and Kesari, S. (2008). Malignant gliomas in adults. N Engl J Med 359, 492-507. https://www.ncbi.nlm.nih.gov/pubmed/18669428

Whitbeck, J.C., Muggeridge, M.I., Rux, A.H., Hou, W., Krummenacher, C., Lou, H., van Geelen, A., Eisenberg, R.J., and Cohen, G.H. (1999). The major neutralizing antigenic site on herpes simplex virus glycoprotein D overlaps a receptor-binding domain. J Virol 73, 9879-9890. https://www.ncbi.nlm.nih.gov/pubmed/10559300

Wiesner, S.M., Decker, S.A., Larson, J.D., Ericson, K., Forster, C., Gallardo, J.L., Long, C., Demorest, Z.L., Zamora, E.A., Low, W.C., *et al.* (2009). De novo induction of genetically engineered brain tumors in mice using plasmid DNA. Cancer Res 69, 431-439. https://www.ncbi.nlm.nih.gov/pubmed/19147555

Willingham, S.B., Volkmer, J.P., Gentles, A.J., Sahoo, D., Dalerba, P., Mitra, S.S., Wang, J., Contreras-Trujillo, H., Martin, R., Cohen, J.D., *et al.* (2012). The CD47-signal regulatory protein alpha (SIRPa) interaction is a therapeutic target for human solid tumors. Proc Natl Acad Sci U S A 109, 6662-6667. https://www.ncbi.nlm.nih.gov/pubmed/22451913

Wirsching, H.G., Arora, S., Zhang, H., Szulzewsky, F., Cimino, P.J., Queva, C., Houghton, A.M., Glorioso, J.C., Weller, M., and Holland, E.C. (2019a). Cooperation of oncolytic virotherapy with VEGF-neutralizing antibody treatment in IDH wildtype glioblastoma depends on MMP9. Neuro Oncol 21, 1607-1609. https://www.ncbi.nlm.nih.gov/pubmed/31412117

Wirsching, H.G., Zhang, H., Szulzewsky, F., Arora, S., Grandi, P., Cimino, P.J., Amankulor, N., Campbell, J.S., McFerrin, L., Pattwell, S.S., *et al.* (2019b). Arming oHSV with ULBP3 drives abscopal immunity in lymphocyte-depleted glioblastoma. JCI Insight 4. https://www.ncbi.nlm.nih.gov/pubmed/31292299

Wong, R.J., Kim, S.H., Joe, J.K., Shah, J.P., Johnson, P.A., and Fong, Y. (2001). Effective treatment of head and neck squamous cell carcinoma by an oncolytic herpes simplex virus. J Am Coll Surg 193, 12-21. https://www.ncbi.nlm.nih.gov/pubmed/11442249

Wu, J., Jordan, M., and Waxman, D.J. (2016). Metronomic cyclophosphamide activation of anti-tumor immunity: tumor model, mouse host, and drug schedule dependence of gene responses and their upstream regulators. BMC Cancer 16, 623. https://www.ncbi.nlm.nih.gov/pubmed/27515027

Wu, J., Sun, L., Chen, X., Du, F., Shi, H., Chen, C., and Chen, Z.J. (2013). Cyclic GMP-AMP is an endogenous second messenger in innate immune signaling by cytosolic DNA. Science 339, 826-830. https://www.ncbi.nlm.nih.gov/pubmed/23258412

WuDunn, D., and Spear, P.G. (1989). Initial interaction of herpes simplex virus with cells is binding to heparan sulfate. J Virol 63, 52-58. https://www.ncbi.nlm.nih.gov/pubmed/2535752

Xing, J., Wang, S., Lin, R., Mossman, K.L., and Zheng, C. (2012). Herpes simplex virus 1 tegument protein US11 downmodulates the RLR signaling pathway via direct interaction with RIG-I and MDA-5. J Virol 86, 3528-3540. https://www.ncbi.nlm.nih.gov/pubmed/22301138

Yarchoan, M., Johnson, B.A., 3rd, Lutz, E.R., Laheru, D.A., and Jaffee, E.M. (2017). Targeting neoantigens to augment antitumour immunity. Nat Rev Cancer 17, 569. https://www.ncbi.nlm.nih.gov/pubmed/28835723

Yoo, J.Y., Haseley, A., Bratasz, A., Chiocca, E.A., Zhang, J., Powell, K., and Kaur, B. (2012). Antitumor efficacy of 34.5ENVE: a transcriptionally retargeted and "Vstat120"-expressing oncolytic virus. Mol Ther 20, 287-297. https://www.ncbi.nlm.nih.gov/pubmed/22031239

Yoo, J.Y., Jaime-Ramirez, A.C., Bolyard, C., Dai, H., Nallanagulagari, T., Wojton, J., Hurwitz, B.S., Relation, T., Lee, T.J., Lotze, M.T., *et al.* (2016). Bortezomib Treatment Sensitizes Oncolytic HSV-1-Treated Tumors to NK Cell Immunotherapy. Clin Cancer Res 22, 5265-5276. https://www.ncbi.nlm.nih.gov/pubmed/27390350

Yoon, M., Zago, A., Shukla, D., and Spear, P.G. (2003). Mutations in the N termini of herpes simplex virus type 1 and 2 gDs alter functional interactions with the entry/fusion receptors HVEM, nectin-2, and 3-O-sulfated heparan sulfate but not with nectin-1. J Virol 77, 9221-9231. https://www.ncbi.nlm.nih.gov/pubmed/12915538

York, I.A., Roop, C., Andrews, D.W., Riddell, S.R., Graham, F.L., and Johnson, D.C. (1994). A cytosolic herpes simplex virus protein inhibits antigen presentation to CD8+ T lymphocytes. Cell 77, 525-535. https://www.ncbi.nlm.nih.gov/pubmed/8187174

Youn, J.I., and Gabrilovich, D.I. (2010). The biology of myeloid-derived suppressor cells: the blessing and the curse of morphological and functional heterogeneity. Eur J Immunol 40, 2969-2975. https://www.ncbi.nlm.nih.gov/pubmed/21061430

Zeng, W., Hu, P., Wu, J., Wang, J., Li, J., Lei, L., and Liu, R. (2013). The oncolytic herpes simplex virus vector G47 effectively targets breast cancer stem cells. Oncol Rep 29, 1108-1114. https://www.ncbi.nlm.nih.gov/pubmed/23292314

Zhang, J., Dang, F., Ren, J., and Wei, W. (2018). Biochemical Aspects of PD-L1 Regulation in Cancer Immunotherapy. Trends Biochem Sci. 43, 1014-1032. http://dx.doi.org/10.1016/j.tibs.2018.09.004.

Zhang, K.X., Matsui, Y., Lee, C., Osamu, O., Skinner, L., Wang, J., So, A., Rennie, P.S., and Jia, W.W. (2016). Intravesical treatment of advanced urothelial bladder cancers with oncolytic HSV-1 co-regulated by differentially expressed microRNAs. Gene Ther 23, 460-468. https://www.ncbi.nlm.nih.gov/pubmed/26905370

Zhang, X., Shi, H., Wu, J., Zhang, X., Sun, L., Chen, C., and Chen, Z.J. (2013). Cyclic GMP-AMP containing mixed phosphodiester linkages is an endogenous high-affinity ligand for STING. Mol Cell 51, 226-235. https://www.ncbi.nlm.nih.gov/pubmed/23747010

Zhou, G., and Roizman, B. (2006). Construction and properties of a herpes simplex virus 1 designed to enter cells solely via the IL-13alpha2 receptor. Proc Natl Acad Sci U S A 103, 5508-5513. https://www.ncbi.nlm.nih.gov/pubmed/16554374

Zhou, G., and Roizman, B. (2007). Separation of receptor-binding and profusogenic domains of glycoprotein D of herpes simplex virus 1 into distinct interacting proteins. Proc Natl Acad Sci U S A 104, 4142-4146. https://www.ncbi.nlm.nih.gov/pubmed/17360490

Zhou, G., Ye, G.J., Debinski, W., and Roizman, B. (2002). Engineered herpes simplex virus 1 is dependent on IL13Ralpha 2 receptor for cell entry and independent of glycoprotein D receptor interaction. Proc Natl Acad Sci U S A 99, 15124-15129. https://www.ncbi.nlm.nih.gov/pubmed/12417744

Zhu, X., Fujita, M., Snyder, L.A., and Okada, H. (2011). Systemic delivery of neutralizing antibody targeting CCL2 for glioma therapy. J Neurooncol 104, 83-92. https://www.ncbi.nlm.nih.gov/pubmed/21116835

www.ingramcontent.com/pod-product-compliance
Lightning Source LLC
LaVergne TN
LVHW070844210926
848482LV00016B/864
9781913652555